Computational Magnetics

Computational Magnetics

Editor

Jan K. Sykulski

CHAPMAN & HALL

London · Glasgow · Weinheim · New York · Tokyo · Melbourne · Madras

Co-published by James & James (Science Publishers) Ltd, 5 Castle Road, London NW1 8PR, UK and Chapman & Hall, 2-6 Boundary Row, London SE1 8HN, UK

Chapman & Hall, 2-6 Boundary Row, London SE1 8HN, UK

Blackie Academic & Professional, Wester Cleddens Road, Bishopbriggs, Glasgow G64 2NZ, UK

Chapman & Hall GmbH, Pappelallee 3, 69469 Weinheim, Germany

Chapman & Hall USA, One Penn Plaza, 41st Floor, New York, NY10119, USA

Chapman & Hall Japan, ITP-Japan, Kyowa Building, 3F, 2-2-1 Hirakawacho, Chiyoda-ku, Tokyo 102, Japan

Chapman & Hall Australia, Thomas Nelson Australia, 102 Dodds Street, South Melbourne, Victoria 3205, Australia

Chapman & Hall India, R. Seshadri, 32 Second Main Road, CIT East, Madras 600 035, India

First edition 1995

Typeset in 10/12pt Times Roman by Colset Pte Ltd, Singapore
Printed in Great Britain by St Edmundsbury Press, Bury St Edmunds, Suffolk

ISBN 0 412 58570 7

A catalogue record for this book is available from the British Library

∞ Printed on permanent acid-free text paper, manufactured in accordance with ANSI/NISO Z39.48-1992 and ANSI/NISO Z39.48-1984 (Permanence of Paper).

Contents

Preface

Computational Magnetics comprises a spectrum of subjects covering field and potential theory, interpolation and approximation techniques, numerical methods and programming routines, as well as relevant experimental procedures and analogue systems. Its objective is to provide a working environment and a set of tools for a designer of magnetic apparatus, or a system analyst, to facilitate development of appropriate models of devices and processes for the purpose of simulation and to establish equivalent circuits, or other simplified representations, and their parameters. It is a practical tool that aims at helping an engineer in the difficult task of designing magnetic devices and optimizing their performance. Moreover, it aids an understanding of the behaviour of such devices.

Advances in computational magnetics stem from progress in various areas of mathematics, physics and computer science, and, in turn, feed back and contribute to developments in field and potential theory, calculus of variations, optimization methods, numerical analysis, and computational geometry. Scientists, engineers and software developers all over the world actively pursue research in many diverse topics relevant to computational magnetics, ranging from gauge theory, scalar and vector potential formulations, and methods of synthesis to benchmarking techniques, parallel processing, computer graphics and data structures. Error estimation and adaptive solver technology, treatment of exterior circuits and coupling of magnetic to other field systems are also very important.

The notion of computer-aided design (CAD) is central to contemporary systems for magnetic analysis and provides the framework for implementation of key concepts of computational magnetics. Numerous commercial CAD systems are now available for magnetics work and the software has grown to play a fundamental role in the industrial design environment. As a result the designer's approach to problems is inevitably changing too, as conceptual models used in design are now influenced by what is offered by these new powerful tools. New questions arise however on inherent simplification of models and methods used, relevance of computed quantities to the needs of the designer and accuracy of computation. Another important factor is the convenience with which the solution is found, that is the user-friendliness of the program, and how the software communicates with other components of the design system, i.e. versatility and compatibility of data structures. And last but not least the user of any CAD software will be looking very carefully at the total time required to set up the problem, analyze it and interrogate the solution. Modern computers are immensely powerful and fast, but complexity of some real engineering problems, especially if large three-dimensional structures or optimization are involved, may be such that resulting computing times are impractical for design purposes. Approximations and simplifications are then sought so that a compromise may be reached between accuracy and speed of computation. Inter-active design becomes particularly helpful where the designer is in constant 'conversation' with an appropriate software system and thus his knowledge and expertise are directly and immediately used when required.

There is an almost bewildering range of computational methods in magnetics, each with a vast literature, well developed theory and dedicated supporters. The reader is no doubt familiar with some of the methods, such as separation of variables, Laplace transforms, conformal transformations, analogue methods (e.g. transmission-line modelling) or images. Numerical methods have been found particularly helpful for CAD applications and they have indeed dominated the scene in recent years. The most important techniques, all discussed in this book in some depth, are finite differences, finite elements, boundary elements, reluctance networks (based on circuit analogy) and tubes and slices (a graphical computational scheme based on dual energy approach). Although all these methods aim at describing the same physical processes, as defined by appropriate field equations, they differ in both the formulation and practical implementation. A typical CAD system will be firmly rooted in one of such methods and will have an appropriate graphical user interface and possibly a design or optimization shell, as well as some other specialist toolboxes.

Methods of field synthesis are being introduced gradually into computational magnetics, as many of practical engineering designs aim at achieving specific characteristics or particular field levels or distributions. Although ultimately a finite-element technique or other method of analysis is usually employed, the mathematics of field synthesis is somewhat different and requires careful examination. Solutions of such inverse problems are increasingly of interest to designers.

This book has been written by a group of authors who are all actively engaged in research and pursuing developments of various aspects of computational magnetics. The coverage is fairly comprehensive and addresses a number of fundamental issues, some of which have been mentioned above. We do not concentrate on any particular method, nor do we dwell on theory alone. The flavour of the book is the link between underlying mathematical and physical concepts and computational practice, with particular emphasis on engineering applications. We have thus tried to bridge the gap between theory, available computational techniques and engineering and design requirements. It is up to the reader to decide whether we have achieved our objective.

The reader is assumed to be familiar with fundamental concepts of electromagnetics as covered by a typical second year undergraduate course. A recently published textbook and an accompanying computer disk may be helpful in revising the subject (P. Hammond and J.K. Sykulski, Engineering electromagnetism: physical processes and computation, Oxford University Press, New York, 1994, ISBN 0 19 856289 6 or ISBN 0 19 856288 8). Conversely, a reader who wishes to further his studies of a particular method or technique, beyond the coverage offered in one of the chapters, or to learn about other methods not discussed in this book, will find a very extensive list of references of both specialist books and monographs, as well as scientific papers published in learned societies journals and conference proceedings. There are some conferences (e.g. COMPUMAG, ISEF) which regularly report on advances in computational magnetics and are therefore particularly useful to those who are interested in the subject.

Our book presents the state of the art in computational magnetics.. The subject is vast and it would be impossible to treat equally thoroughly every relevant topic. We have thus selected and presented in greater depth some of the issues, but only briefly mentioned or referred the reader to relevant literature in case of others. We think that the book will be particularly useful to practising engineers and designers in various sections of industry, in particular electromechanical, dealing with magnetic devices at low frequency, large and small, and to researchers and postgraduate students involved in magnetic analysis. It will also be helpful in running postgraduate and continuing education courses and to final year undergraduate students taking specialist options and for project work.

The book starts in the first chapter with a brief review of magnetic fields, covering Maxwell's equations, scalar and vector potential formulations, classification of fields, boundary conditions and magnetics materials. It then goes into the discussion of energy storage and forces and gives a brief account of the eddy-current skin effect. Finally, it introduces the geometrical structure of fields and the notion of differential forms, which give rise to a geometrical computational technique known as a method of tubes and slices.

The next three chapters discuss theory, computational aspects and applications of the most popular numerical methods. Chapter 2 deals with the oldest and at one time hugely successful method of finite differences. In its simplest form the method is easily understood and almost intuitive; it is still very useful for special purposes and when the geometry is simple. The discussion concentrates firstly on general matters of stability and convergence of the numerical scheme, and then on more specific questions of nonlinearity, hysteresis, multi-region problems and transient solutions. Many practical solutions are demonstrated. Chapter 3 describes the two techniques which have completely dominated the scene and are used in almost all commercial software: the finite-element and boundary-element methods. The algorithms are much more complicated than the finite-difference formulation. A comprehensive derivation of basic theory is followed by some of the more advanced features, such as nonlinearity, open-boundaries and hierarchical elements. Some examples of practical solutions are given. Finally, Chapter 4 presents a rather unique implementation of an old approach based on equivalent reluctance networks. With the help of a dedicated computer program the method offers great economy of computation, providing approximate answers in a matter of seconds, even for complicated three-dimensional structures. The method is partly empirical and requires experimental verification, but numerous successful solutions have been obtained and are reported, together with a more general discussion of theory and applications.

Chapter 5 deals with field synthesis. After defining the term it then provides a survey of integral equations and discusses types of typical synthesis tasks: inverse problems to Laplace's or Poisson's equations, variable boundary problems and synthesis of environmental parameters. Possible methods of solution are reviewed and discussed in some depth, including one of the most promising techniques of regularization. Some practical examples are also described. Coupling between magnetic and thermal, mechanical and other field systems is the subject of Chapter 6. This is an area of growing interest to engineers as most design problems involve interaction between different types of fields. Relevant theory is explained and various mathematical models are put forward. The approach is very practical and numerous examples are included. Consequences of strong and weak coupling are carefully studied and possibilities of approximations and simplifications of models and equations are explored.

Chapter 7 addresses the fundamental issues of computer aided design in magnetics and attempts to summarize the state of the art in both theory and implementations. The CAD environment is introduced together with its main components: preprocessor, solver and post-processor. These are discussed in some detail and many recent developments are highlighted. The discussion focuses on such topics as error analysis and adaptive meshing, sparse matrices and compact storage schemes, optimization, field visualization and parallel computing. A short survey of currently available commercial CAD systems for magnetics is provided for information. Finally, Chapter 8 looks at experimental methods. Fundamentals of physical modelling and scaling criteria are explained, followed by a discussion of various experimental techniques, and concluded with classification of experimental errors. Measurement is a very important component of a modelling process, both as a verification of theoretically derived results and as an investigating tool in its own right.

As the editor and co-author of this book I have been privileged to cooperate with my distinguished colleagues and I would like to express my gratitude for their innovative approach to the subject and all the hard work they put in to complete the manuscript and meet the deadlines. The book is a collaborative effort of the following group of authors:

- Krystyn Pawluk – Professor of Electrical Engineering, Institute of Electrotechnics, Warsaw, Poland
- Ryszard Sikora – Professor and Head of Department of Electrical Engineering, Technical University of Szczecin, Poland

- Richard L. Stoll - Reader in Electrical Machines, Department of Electrical Engineering, University of Southampton, UK
- Jan K. Sykulski - Lecturer in Electrical Machines, Electromagnetism and CAD at the Department of Electrical Engineering, University of Southampton, UK, and Visiting Professor at the Technical University of Lodz, Poland
- Janusz Turowski - Professor and Vice-Rektor of the Technical University of Lodz, Poland
- Kazimierz Zakrzewski - Professor and Head of the Institute of Electrical Machines and Transformers, Technical University of Lodz, Poland

The collaboration between the Department of Electrical Engineering at Southampton, UK, and the Universities of Lodz, Szczecin and Institute of Electrotechnics in Warsaw, Poland, has a long history and the research links have been active for over fifteen years. The International Symposium on Electromagnetic Fields in Electrical Engineering (ISEF), a biannual event held alternately in Poland and another European venue, was hosted by the University of Southampton in 1991 and was a very successful meeting. The ISEF Symposium is one of the venues where advances in computational magnetics are regularly reported. The present book reflects the strength of the cooperation between researchers and institutions in the two countries.

I hope that the reader will find the book enjoyable and helpful. Suggestions for future changes and improvements will be most welcome.

Jan K. Sykulski
Southampton
May 1994

1
A Brief Review of Magnetic Fields

Jan K. Sykulski and Richard L. Stoll

1.1 MAXWELL'S EQUATIONS

Most of this book addresses the problems of how to find, estimate or compute the distribution of magnetic fields for different geometries and under various particular conditions. Such fields are described, in a most compact form, by Maxwell's equations, and thus for most of the time we shall be seeking solutions to these equations. Readers familiar with the theory may wish to proceed directly to a chapter dealing with a particular method or technique of interest, but we shall nevertheless start by introducing briefly the most fundamental aspects of the underlying theory and basic formulations.

Maxwell was the first investigator to use partial differential equations to describe electromagnetic phenomena. For general time-varying fields, using today's notation, *Maxwell's equations* may be written in differential form as

$$curl\,\mathbf{H} = \mathbf{J} + \frac{\partial \mathbf{D}}{\partial t} \tag{1.1}$$

$$curl\,\mathbf{E} = -\frac{\partial \mathbf{B}}{\partial t} \tag{1.2}$$

$$div\,\mathbf{D} = \rho \tag{1.3}$$

$$div\,\mathbf{B} = 0 \tag{1.4}$$

where:

- **H** magnetic field intensity ($\mathrm{A\,m^{-1}}$);
- **E** electric flux intensity ($\mathrm{V\,m^{-1}}$);
- **B** magnetic flux density ($\mathrm{Wb\,m^{-2}}$, or T);
- **D** electric flux density ($\mathrm{C\,m^{-2}}$);
- **J** electric current density ($\mathrm{A\,m^{-2}}$);
- ρ electric charge density ($\mathrm{C\,m^{-3}}$).

The current density **J** and the charge density ρ are the sources of the field and are related through the *equation of continuity*, which can be written as

$$div\,\mathbf{J} = -\frac{\partial \rho}{\partial t} \tag{1.5}$$

and thus specifies the conservation of charge.

The above equations are supplemented by the *constitutive equations* describing macroscopic properties of the medium

$$\mathbf{D} = \varepsilon \mathbf{E} \tag{1.6}$$

$$\mathbf{B} = \mu \mathbf{H} \tag{1.7}$$

$$\mathbf{J} = \sigma \mathbf{E} \tag{1.8}$$

where the constitutive parameters ε, μ and σ denote, respectively, the *permittivity* ($\mathrm{F\,m^{-1}}$), *permeability* ($\mathrm{H\,m^{-1}}$), and *conductivity* ($\mathrm{S\,m^{-1}}$) of the medium. For isotropic media these parameters are scalars, but for anisotropic materials they become tensors. Moreover, for non-homogeneous materials, they are functions of position.

For *static fields*, where the field quantities do not vary with time, eqns (1.1), (1.2) and (1.5) reduce to

$$\mathit{curl}\,\mathbf{H} = \mathbf{J} \tag{1.9}$$

$$\mathit{curl}\,\mathbf{E} = 0 \tag{1.10}$$

$$\mathit{div}\,\mathbf{J} = 0 \tag{1.11}$$

which shows that there is no interaction between electric and magnetic fields, and thus we can have separately an *electrostatic* case or a *magnetostatic* case.

An important special condition arises also when all field quantities vary sinusoidally with time as $e^{j\omega t}$, where ω is an angular frequency. Using *phasor* notation, the Maxwell's equations involving time derivatives can now be written as

$$\mathit{curl}\,\underline{\mathbf{H}} = \underline{\mathbf{J}} + j\omega\underline{\mathbf{D}} \tag{1.12}$$

$$\mathit{curl}\,\underline{\mathbf{E}} = -j\omega\underline{\mathbf{B}} \tag{1.13}$$

$$\mathit{div}\,\underline{\mathbf{J}} = -j\omega\rho \tag{1.14}$$

The three vector operators *grad*, *div* and *curl* are very helpful in all of the above compact notations. The definitions of these operators are as follows:

$$\mathit{grad}\,\Phi = \frac{\partial \Phi}{\partial n}\hat{\mathbf{n}} \tag{1.15}$$

and is thus the total or maximum slope of the scalar Φ, where the unit vector points in the 'uphill' direction normal to the equipotential surface. Secondly,

$$\mathit{div}\,\mathbf{F} = \lim_{v \to 0} \frac{1}{v} \oint \mathbf{F} \cdot \mathrm{d}\mathbf{s} \tag{1.16}$$

describes the net outflow of vector **F** per unit volume, for a small volume, and finally

$$curl\,\mathbf{F} = \lim_{s\to 0} \frac{1}{S} \oint \mathbf{F} \cdot \mathrm{d}\mathbf{l} \tag{1.17}$$

is the circulation per unit area, for a small area. In a *rectangular* coordinate system (x, y, z) it is convenient to introduce a *nabla* (sometimes called a *del*) operator, defined as

$$\nabla = \hat{\mathbf{i}}\frac{\partial}{\partial x} + \hat{\mathbf{j}}\frac{\partial}{\partial y} + \hat{\mathbf{k}}\frac{\partial}{\partial z} \tag{1.18}$$

and the three operations are reduced to

$$grad\,\Phi = \nabla\Phi \tag{1.19}$$

$$div\,\mathbf{F} = \nabla \cdot \mathbf{F} \tag{1.20}$$

$$curl\,\mathbf{F} = \nabla \times \mathbf{F} \tag{1.21}$$

which also give alternative notation. In other coordinate systems, definitions of eqns (1.15), (1.16) and (1.17) have to be obtained directly, but the nabla notation is still often used, although it will no longer imply operation of eqn (1.18).

Another form of the equations may be found by application of some fundamental integral relationships of vectors. The two most important are Gauss's or divergence theorem

$$\iiint \nabla \cdot \mathbf{F}\,\mathrm{d}v = \oint \mathbf{F} \cdot \mathrm{d}\mathbf{s} \tag{1.22}$$

and Stoke's or circulation theorem

$$\iint \nabla \times \mathbf{F} \cdot \mathrm{d}\mathbf{s} = \oint \mathbf{F} \cdot \mathrm{d}\mathbf{l} \tag{1.23}$$

Hence, for example, for a steady current flow the integral equation equivalent to eqn (1.11) will read

$$\oint_S \mathbf{J} \cdot \mathrm{d}\mathbf{s} = 0 \tag{1.24}$$

Equations (1.11) and (1.24) contain the same information and are vector equivalents of the first Kirchhoff's current law.

In electrostatics we can write

$$\oint_S \mathbf{D} \cdot \mathrm{d}\mathbf{s} = Q \tag{1.25}$$

Thus, Gauss's theorem states that the electric flux over any arbitrary closed surface is always equal to the strength of the enclosed sources. The last equation is equivalent to the statement of eqn (1.3).

In magnetostatics there may be no net polarity as all magnets consist of dipoles, so that

$$\oint_S \mathbf{B} \cdot \mathrm{d}\mathbf{s} = 0 \tag{1.26}$$

which in differential form is expressed by eqn (1.4). Moreover, application of Stoke's theorem to eqn (1.9) yields

$$\oint \mathbf{H} \cdot \mathrm{d}\mathbf{l} = \iint_S \mathbf{J} \cdot \mathrm{d}\mathbf{s} = I \tag{1.27}$$

which is known as Ampère's equation and shows that the magnetic field of a current is non-conservative.

Faraday's law, eqn (1.2), can also be written as

$$\oint \mathbf{E} \cdot \mathrm{d}\mathbf{l} = -\frac{\mathrm{d}\Phi}{\mathrm{d}t} \tag{1.28}$$

where the magnetic flux is given by

$$\Phi = \iint \mathbf{B} \cdot \mathrm{d}\mathbf{s} \tag{1.29}$$

It is also important to notice that the differential formulation of Faraday's law is independent of the motion of the material through the field. This enables us to transfer from a moving reference frame to a stationary one and vice versa.

Finally, we recall the two relationships known as Green's theorem

$$\iiint (\nabla U \cdot \nabla V + U\nabla^2 V)\, \mathrm{d}v = \oint U\nabla V \cdot \mathrm{d}\mathbf{s} \tag{1.30}$$

$$\iiint (U\nabla^2 V - V\nabla^2 U)\, \mathrm{d}v = \oint (U\nabla V - V\nabla U) \cdot \mathrm{d}\mathbf{s} \tag{1.31}$$

which are particularly helpful in the formulation of the finite-element and boundary-element methods.

1.2 SCALAR AND VECTOR POTENTIALS

One way of finding field distributions is to solve Maxwell's equations directly as a system of first-order differential equations in terms of the appropriate field quantities. In most, although not all, important cases, however, it will be more economical to convert the first-order equations into second-order differential equations involving scalar or vector potentials. Any vector field **F** is uniquely defined if its sources are given as

$$\mathit{div}\,\mathbf{F} = w \tag{1.32}$$

and

$$\mathit{curl}\,\mathbf{F} = \mathbf{d} \tag{1.33}$$

where the source distribution w is a scalar field and the source distribution **d** is a vector field. The total field **F** in the presence of both types of sources can be obtained by superposition

$$\mathbf{F} = -grad\,\phi + curl\,\mathbf{A} \tag{1.34}$$

where the two components are independent of each other since

$$div\,curl\,\mathbf{A} = 0 \tag{1.35}$$

and

$$curl\,grad\,\phi = 0 \tag{1.36}$$

so that eqns (1.32) and (1.33) are satisfied through the following second-order differential equations

$$div\,grad\,\phi = -w \tag{1.37}$$

and

$$curl\,curl\,\mathbf{A} = \mathbf{d} \tag{1.38}$$

It is, therefore, convenient to divide the field **F** into a gradient field and a solenoidal field corresponding to the two types of sources, especially because such sources are often easily distinguishable in physical problems.

In *electrostatics* we introduce the *electric scalar potential* in terms of the electric field intensity as

$$\mathbf{E} = -grad\,\phi \tag{1.39}$$

or using the nabla notation

$$\mathbf{E} = -\nabla\phi \tag{1.40}$$

Substituting eqn (1.40) into eqn (1.3) with the aid of eqn (1.6) yields

$$\nabla \cdot (\varepsilon\nabla\phi) = -\rho \tag{1.41}$$

and if ε is constant within a region

$$\nabla^2\phi = -\frac{\rho}{\varepsilon} \tag{1.42}$$

which is a second-order differential equation, known as the *Poisson's equation*.

In *magnetostatics* we often use the *vector potential* **A** defined in terms of the magnetic flux density **B** as

$$\mathbf{B} = \nabla \times \mathbf{A} \tag{1.43}$$

With the help of eqns (1.9) and (1.7) we find

$$\nabla \times \left(\frac{1}{\mu}\nabla \times \mathbf{A}\right) = \mathbf{J} \tag{1.44}$$

In conductors it is likely that $\mu = \mu_0$, so that

$$\nabla \times \nabla \times \mathbf{A} = \nabla\nabla \cdot \mathbf{A} - \nabla^2\mathbf{A} = \mu_0\mathbf{J} \tag{1.45}$$

For convenience we are allowed to put

$$\nabla \cdot \mathbf{A} = 0 \tag{1.46}$$

because there is a free choice in selecting the divergence sources of **A**, since only the curl sources have been defined in eqn (1.43). The condition imposed on the divergence of the vector potential is called a *gauge condition*, and the particular choice of eqn (1.46) is known as the *Coulomb gauge*. Hence if we choose zero divergence

$$\nabla^2 \mathbf{A} = -\mu_0 \mathbf{J} \tag{1.47}$$

In a rectangular coordinate system eqn (1.47) separates into three equations

$$\nabla^2 A_x = -\mu_0 J_x \tag{1.48}$$

$$\nabla^2 A_y = -\mu_0 J_y \tag{1.49}$$

$$\nabla^2 A_z = -\mu_0 J_z \tag{1.50}$$

On the other hand, if the magnetostatic field is not required within the regions of conduction current, eqn (1.9) simplifies to

$$\nabla \times \mathbf{H} = 0 \tag{1.51}$$

and we have an alternative formulation available in terms of **H**. Thus we can write

$$\mathbf{H} = -\nabla \phi_m \tag{1.52}$$

where ϕ_m is a *magnetic scalar potential*. This leads to the following equation

$$\nabla \cdot (\mu \nabla \phi_m) = 0 \tag{1.53}$$

which when expanded yields

$$\mu \nabla^2 \phi_m + \nabla \mu \cdot \nabla \phi_m = 0 \tag{1.54}$$

The formulation in terms of the magnetic scalar potential has some advantages as it reduces the number of unknown functions which have to be solved from three to one. In two-dimensional cases this does not matter as the vector potential typically has only one component (in the direction of current flow), but in three-dimensional fields savings in computing effort may be considerable. Either formulation is complicated by the fact that permeability is not only a function of position but depends also on field strength. Moreover, many practical magnetic materials are anisotropic, and so the permeability has to be treated as a tensor.

Generally, in time varying fields, the electric and magnetic fields are coupled, as demonstrated by eqns (1.1) and (1.2). Both sources ρ and **J** are present and are linked through the equation of continuity (1.5). Similarly the electric scalar potential ϕ and the magnetic vector potential **A** can be combined using the *Lorentz gauge*

$$\nabla \cdot \mathbf{A} + \frac{1}{c^2}\frac{\partial \phi}{\partial t} = 0 \tag{1.55}$$

where c is the velocity of light. The Lorentz gauge simplifies the relationship between the potentials and the sources, which may be shown to be

$$\nabla^2\mathbf{A} - \frac{1}{c^2}\frac{\partial^2\mathbf{A}}{\partial t^2} = -\mu\mathbf{J} \tag{1.56}$$

and

$$\nabla^2\phi - \frac{1}{c^2}\frac{\partial^2\phi}{\partial t^2} = -\frac{\rho}{\varepsilon} \tag{1.57}$$

where μ and ε have been taken as constants. Thus $\mathbf{A}$ depends on $\mathbf{J}$ only, whereas ϕ depends on ρ only. The electric and magnetic fields are obtained from the potentials by the relationships

$$\mathbf{E} = -\frac{\partial\mathbf{A}}{\partial t} - \nabla\phi \tag{1.58}$$

and

$$\mathbf{B} = \nabla \times \mathbf{A} \tag{1.59}$$

Other conditions on the potentials are possible through the use of the general gauge transformation

$$\mathbf{A}' = \mathbf{A} + \nabla\psi \tag{1.60}$$

$$\phi' = \phi - \frac{\partial\psi}{\partial t} \tag{1.61}$$

with a subsidiary condition

$$\nabla^2\psi - \frac{1}{c^2}\frac{\partial^2\psi}{\partial t^2} = 0 \tag{1.62}$$

Thus ψ obeys the wave equation and the electric and magnetic fields of eqns (1.58) and (1.59) are not affected by this transformation. Examples of possible useful transformations include choosing ϕ' to be zero or making $\mathbf{A}'$ solenoidal irrespective of the sources. A comprehensive survey of various possible formulations may be found in the papers by Hammond [1.21] and Carpenter [1.7].

Finally, we turn our attention again to sinusoidally varying fields. It is often convenient to separate the *imposed* current density from the conduction current by writing

$$\underline{\mathbf{J}} = \sigma\underline{\mathbf{E}} + \underline{\mathbf{J}}_s \tag{1.63}$$

Using eqns (1.12), (1.13), (1.6) and (1.7) we can now write

$$\nabla \times \underline{\mathbf{H}} = (\sigma + \mathrm{j}\omega\varepsilon)\underline{\mathbf{E}} + \underline{\mathbf{J}}_s \tag{1.64}$$

$$\nabla \times \underline{\mathbf{E}} = -\mathrm{j}\omega\mu\underline{\mathbf{H}} \tag{1.65}$$

Moreover,

$$\underline{\mathbf{B}} = \nabla \times \underline{\mathbf{A}} \tag{1.66}$$

and

$$\underline{\mathbf{E}} = -j\omega\underline{\mathbf{A}} - \nabla\underline{\phi} \tag{1.67}$$

Taking the *curl* of eqn (1.66) in combination with eqns (1.64) and (1.7) yields

$$\nabla \times \nabla \times \underline{\mathbf{A}} = \mu(\sigma + j\omega\varepsilon)\underline{\mathbf{E}} + \mu\underline{\mathbf{J}}_s \tag{1.68}$$

and thus

$$\nabla\nabla \cdot \underline{\mathbf{A}} - \nabla^2\underline{\mathbf{A}} = -j\omega\mu(\sigma + j\omega\varepsilon)\underline{\mathbf{A}} - \mu(\sigma + j\omega\varepsilon)\nabla\underline{\phi} + \mu\underline{\mathbf{J}}_s \tag{1.69}$$

The Lorentz gauge may now be imposed in the form

$$\nabla \cdot \underline{\mathbf{A}} = -\mu(\sigma + j\omega\varepsilon)\underline{\phi} \tag{1.70}$$

and with the aid of the following substitution

$$k^2 = j\omega\mu(\sigma + j\omega\varepsilon) \tag{1.71}$$

we arrive at two non-homogeneous Helmholtz equations linking current and charge sources with the vector potential and the scalar potential, respectively, as

$$\nabla^2\underline{\mathbf{A}} - k^2\underline{\mathbf{A}} = -\mu\underline{\mathbf{J}}_s \tag{1.72}$$

$$\nabla^2\underline{\phi} - k^2\underline{\phi} = -\frac{\rho}{\varepsilon} \tag{1.73}$$

The choice of an appropriate gauge for the vector potential is an important matter in most computational schemes, and in particular in the finite-element formulations, where some gauges lead to equations which may become badly conditioned at very low frequencies and some loss in accuracy has also been reported at high frequencies. For a comprehensive discussion of magnetic vector formulations and gauge transformations in finite-element methods the reader is referred to reference [1.5], pp. 365–379.

1.3 CLASSIFICATION OF FIELDS

The following provides a list of the most common types of equations which are frequently used to describe particular types of fields under static and time varying conditions. Most chapters in this book describe ways of solving such equations under particular conditions. Every equation has its *scalar* form and a *vector* form, because they may be applied to a scalar potential or a vector potential.

Laplace's equation:

$$\nabla^2\phi = 0 \tag{1.74}$$

$$\nabla^2\mathbf{A} = 0 \tag{1.75}$$

Poisson's equation:

$$\nabla^2\phi = f \tag{1.76}$$

$$\nabla^2\mathbf{A} = \mathbf{F} \tag{1.77}$$

Helmholtz equation:

$$\nabla^2\phi + k^2\phi = f \tag{1.78}$$

$$\nabla^2\mathbf{A} + k^2\mathbf{A} = \mathbf{F} \tag{1.79}$$

Diffusion equation:

$$\nabla^2\phi = \alpha\frac{\partial\phi}{\partial t} \tag{1.80}$$

$$\nabla^2\mathbf{A} = \alpha\frac{\partial\mathbf{A}}{\partial t} \tag{1.81}$$

Wave equation:

$$\nabla^2\phi - \beta^2\frac{\partial^2\phi}{\partial t^2} = f \tag{1.82}$$

$$\nabla^2\mathbf{A} - \beta^2\frac{\partial^2\mathbf{A}}{\partial t^2} = \mathbf{F} \tag{1.83}$$

The above relationships fall into various classes of partial differential equations. Thus the equation may be

(1) elliptical {eqns (1.74), (1.75), (1.76), (1.77), (1.78) and (1.79)},
(2) parabolic {eqns (1.80) and (1.81)},
(3) hyperbolic {eqns (1.82) and (1.83)}.

1.4 BOUNDARY CONDITIONS

Two requirements are implied by the term *boundary conditions*. First, it is necessary to contain the region in which a second-order partial differential equation is to be solved by specifying on the boundary of the region sufficient information to make the solution unique. Essentially, it is a matter of conveying to the interior solution the effect of any field sources outside the region of interest.

Secondly, when two sub-regions having different magnetic or electric properties meet, it is necessary to be able to link the solutions on both sides of this interface. Although included under the generic name 'boundary conditions', the latter are perhaps more appropriately called *interface conditions*.

Boundary conditions at an interface

We shall derive the interface conditions at a magnetic boundary (and, by analogy, a dielectric boundary), where in the more general case, two materials of different relative permeability μ_{r1} and μ_{r2} meet. Applying Gauss's theorem to the surface of a coin-shaped volume that just squeezes a small piece of the interface δs between its two circular faces, we have

$$\oint \mathbf{B} \cdot d\mathbf{s} = \mathbf{B}_2 \cdot \delta\mathbf{s} - \mathbf{B}_1 \cdot \delta\mathbf{s} = 0 \tag{1.84}$$

where the contribution of the ribbon edge of the coin-shaped volume is negligible, and we are assuming that the positive direction of the magnetic field is from region 1 to region 2. Thus

$$B_{n2} = B_{n1} \tag{1.85}$$

where the suffix n denotes the component of $\mathbf{B}$ normal to the interface.

In the analogous case of a dielectric interface, because of the possibility of a free surface charge of density q (C m^{-2}), the normal component of the electric flux density is discontinuous and given by

$$D_{n2} - D_{n1} = q \tag{1.86}$$

An application of Ampère's law to a small rectangular path that just squeezes the interface between its two longer sides δl yields

$$\mathbf{H}_2 \cdot \delta\mathbf{l} - \mathbf{H}_1 \cdot \delta\mathbf{l} = 0 \tag{1.87}$$

Whilst we can state that the tangential component of $\mathbf{H}$ is usually continuous across the boundary, i.e. $H_{t2} = H_{t1}$, the tangential component is actually a two-dimensional vector. Not only this, but if there is a surface current density $\mathbf{K}$ (A m^{-1}) on the boundary, eqn (1.87) can be usefully generalized as

$$(\mathbf{H}_2 - \mathbf{H}_1) \times \mathbf{n} = \mathbf{K} \tag{1.88}$$

where, because $\mathbf{K}$ and the discontinuity in $\mathbf{H}_t$ must be orthogonal, the vector product with the unit vector $\mathbf{n}$ gives the direction of $\mathbf{K}$.

In a similar way we can write, for the electric field $\mathbf{E}$,

$$(\mathbf{E}_2 - \mathbf{E}_1) \times \mathbf{n} = 0 \tag{1.89}$$

Considering again the magnetic interface, if a flux line in region 1 enters the surface at an angle θ_1 to the normal and exits into region 2 at an angle θ_2, then eqns (1.85) and (1.88) can be written

$$B_2 \cos\theta_2 = B_1 \cos\theta_1 \tag{1.90}$$

$$H_2 \sin\theta_2 = H_1 \sin\theta_1 \tag{1.91}$$

Dividing eqn (1.91) by eqn (1.90), and using eqn (1.7), gives

$$\frac{\tan\theta_2}{\tan\theta_1} = \frac{\mu_0\mu_{r2}}{\mu_0\mu_{r1}} = \frac{\mu_{r2}}{\mu_{r1}} \tag{1.92}$$

Suppose now that region 1 is very highly permeable and region 2 is air. The right-hand side of eqn (1.92) tends to zero and it therefore appears that θ_2 must be very close to 0°, i.e. the flux enters or leaves a very highly permeable surface at right angles. This is indeed so under most conditions, but there are circumstances where θ_2 can be considerably greater than zero even with very high values of μ_{r1}.

If, for example, we consider the field inside a cylindrical tube of high permeability situated in a uniform field transverse to its axis, the field in the screened region is always parallel to the applied field and so cannot leave and enter the curved interior surface everywhere at right angles. The reason is the dominant tangential field within the tube wall which may cause θ_1 to be very close to 90° over large portions of the surface. If $\mu_{r1} = 1000$ and $\theta_1 = 89.9°$, for example, eqn (1.92) yields $\theta_2 = 29.8°$, i.e. considerably greater than would normally be expected on the air side of a highly permeable surface.

Surface polarity

Although the source of a magnetic field is electric current, it is sometimes convenient to introduce magnetic poles or polarity density q_m to represent the effect of a magnetic surface [1.40; 1.5]. q_m is analogous to the bound electric charge density that can be used to model a dielectric surface. Note that q in eqn (1.86) is the free electric charge density which has no counterpart in magnetism; the bound charge is hidden in the definition of electric flux density **D**. If q_m is the local value of surface magnetic polarity, then, by Gauss's theorem,

$$q_m = 2\mu_0 H_n' \tag{1.93}$$

where H_n' is the normal component of the magnetic field due to q_m. If the normal component of the applied field due to all other sources is H_n, eqns (1.85) and (1.7) yield

$$\mu_0\mu_{r2}(H_n - H_n') = \mu_0\mu_{r1}(H_n + H_n')$$

or

$$H_n' = \frac{\mu_{r2} - \mu_{r1}}{\mu_{r2} + \mu_{r1}} H_n \tag{1.94}$$

from which q_m follows in terms of H_n. Region 1 is assumed here to be the region of lower permeability, usually air, and it may be more convenient to express q_m in terms of the total normal component $H_{n1} = H_n + H_n'$. Thus from eqns (1.93) and (1.94)

$$q_m = (\mu_{r2} - \mu_{r1}) \frac{\mu_0}{\mu_{r2}} H_{n1} \tag{1.95}$$

It is important to take care with the use of eqn (1.95) because it is an alternative way of representing the magnetized surface and restricts us to using the **H** field or the scalar potential related directly to it (eqn 1.52).

External boundary conditions

Some problems have open boundaries that extend to infinity in a mathematical sense. Analytically this is not usually difficult, but to deal with such problems numerically may require special techniques. More usually a problem will have a natural physical boundary on which one of the components of the magnetic field may be zero or can be assumed to be a well defined function of position. We note that, from eqn (1.95), a surface layer of magnetic poles can cause a discontinuity in the normal component of **H**, and, from eqn (1.88), a surface current can set up a discontinuity in the tangential component of **H**.

We are thus able to convey information about magnetic fields and field sources external to the region of interest via specification of either $\mathbf{H}_t$ or $\mathbf{H}_n$ over parts or the whole of the selected boundary surface. However, specification of $\mathbf{H}_n$ is not sufficient without imposing additionally the satisfaction of Ampère's law for the volume concerned. The three-dimensional generalization of Ampère's law is [1.38]:

$$\iint_S (\mathbf{n} \times \mathbf{H})\, \mathrm{d}S = \iiint_v \mathit{curl}\, \mathbf{H}\, \mathrm{d}v = \iiint_v \mathbf{J}\, \mathrm{d}v \tag{1.96}$$

In a two-dimensional problem with the current flow perpendicular to the plane of the region eqn (1.96) reduces to

$$\oint H_t\, \mathrm{d}l = \iint J\, \mathrm{d}S \tag{1.97}$$

In terms of the scalar magnetic potential ϕ_m (eqn 1.52), H_t corresponds to specification of ϕ_m. This potential specified boundary condition is known as a Dirichlet condition. Alternatively, H_n leads to $\partial\phi_m/\partial n$, the Neumann condition.

In terms of the magnetic vector potential **A** (eqn 1.43), we require $\mathbf{n} \times \mathit{curl}\, \mathbf{A}$ or $\mathbf{n} \times \mathbf{A}$ respectively, which reduce, in the two-dimensional case where the only component of **A** (A_p say) is perpendicular to the two-dimensional plane, to the Dirichlet specification A_p (for H_n) and the Neumann specification $\partial A_p/\partial n$ (for H_t).

1.5 MAGNETIC MATERIALS

Equation (1.7) gave the linear relationship between **B** and **H**. More precisely we have

$$\mathbf{B} = \mu_0 \mathbf{H} + \mathbf{M} \tag{1.98}$$

where **M** is the magnetization vector which is the magnetic moment per unit volume. Only in linear isotropic materials is **M** both parallel to and proportional to **H**, i.e.

$$\mathbf{M} = \mu_0 \chi_m \mathbf{H} \tag{1.99}$$

where χ_m is a constant known as the *magnetic susceptibility*. Equation (1.98) then becomes

$$\mathbf{B} = \mu_0 (1 + \chi_m) \mathbf{H} = \mu_0 \mu_r \mathbf{H} \tag{1.100}$$

where the relative permeability

$$\mu_r = 1 + \chi_m \tag{1.101}$$

Although eqn (1.100) is an approximation for many magnetic materials, and it is often the practice to allow μ_r to vary as a function of $\mathbf{H}$, it does allow us to classify magnetic materials into three categories:

Diamagnetic χ_m is a very small negative number ($\mu_r < 1$),
Paramagnetic χ_m is a very small positive number,
Ferromagnetic χ_m is a large positive number ($\mu_r \gg 1$).

Diamagnetism and paramagnetism are very small effects, with χ_m typically of the order of 10^{-5}, and have to do with the orbital motion and spin of the electrons within the material, but ferromagnetism is very strong and a vital feature in a wide range of power apparatus. Ferromagnetic materials are made up of domains which are already magnetized in the sense that each domain contains aligned magnetic dipoles formed by electron spin. When the sample of material is in an unmagnetized state the domains are so aligned that their net magnetization cancels. For example, in an iron crystal there are six directions of 'easy' magnetization: the positive and negative directions along three orthogonal axes. If we assume, for the purpose of the argument only, that all domains are the same size, then in each of the 'easy' directions there will be as many domains pointing in the positive as in the negative directions. Now if a small magnetic field is gradually applied to the unmagnetized crystal nearly parallel to one of the easy directions, the domains with polarity in the same direction, i.e. assisting the applied field external to themselves, begin to grow at the expense of adjacent parallel domains with opposite polarity, by virtue of small movements of the domain walls. This process is virtually loss free if it takes place slowly.

Let us move now to consideration of a larger sample made up of many crystals or crystal fragments, each containing many domains, in the form of a ring on which there is a uniformly wound toroidal magnetizing winding producing a uniform applied field H parallel to the internal axis of the ring cross-section. This configuration gets over the problem of possible differences between the applied and the local field because there is no surface magnetic polarity present. Thus, with a simple search coil to measure the corresponding flux density B, a curve of B against H, starting with an unmagnetized ring, can be drawn as shown in Fig. 1.1. The previously described process with a low value of H applied, takes us to point P_1. If this field is removed the operating point drops back to the origin because the process is reversible. To reach the higher point P_2, the domains already aligned with the field direction continue to grow, other opposing domains completely reverse, and those at other angles reverse polarity if this brings them more nearly into line with the applied field. All this activity involves extensive domain wall movement and the expenditure of energy. As a result when the field is reduced to zero the flux density only drops to the point represented by Q_2, and if the applied

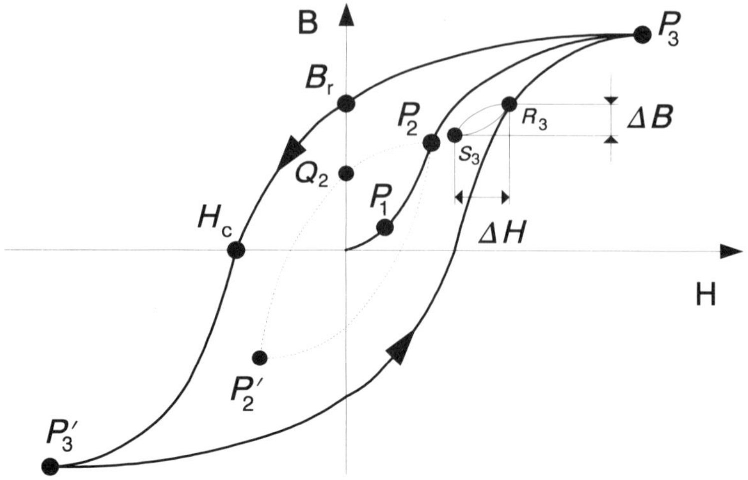

Fig. 1.1 *Hysteresis loops for a ferromagnetic material*

field is then cycled between equal negative and positive maximum values the complete loop P_2P_2' is repeatedly traversed. This is known as a *hysteresis loop*, derived from the Greek word meaning 'to lag', and it can be shown that the expenditure of energy per cycle is given by the area of the loop. Alternatively, rising up the initial magnetization curve (also called the normal magnetization curve) to the higher point P_3 takes the material into magnetic saturation whereby almost complete alignment of the domains in the field direction has taken place. For a full discussion of the processes involved see Bozorth [1.6] or Chikazumi [1.12].

Symmetrical hysteresis loops, know as major loops, can be formed from any points such as P_2 or P_3. The value of flux density at which $H = 0$ on any loop is known as the *remanent density* B_r and the value of H at which $B = 0$ is the *coercive force* H_c. The so-called *permanent magnetic materials* operate in the second quadrant of Fig. 1.1 and require a very high coercive force, or coercivity, to prevent demagnetization during operation.

Hysteresis loops do not have to be symmetrical. An important example occurs if we are on a major loop such as that represented by P_3P_3' in Fig. 1.1 but stop at point R_3 and begin to reduce H again but only by a small amount to point S_3. By causing H to oscillate by ΔH produces the minor hysteresis loop shown. The mean slope of this loop, i.e. $\Delta B/\Delta H$, is known as the *incremental permeability* μ_i. Minor loops can also be generated from points on the initial magnetization curve. The incremental permeability at a given point decreases with ΔH and the limiting value of μ_i as ΔH tends to zero is the *reversible permeability* μ_{rev}. It is necessary to consider the incremental or reversible permeability in situations where a small a.c. perturbation exists in the presence of a much larger d.c. field, because, for example, the eddy-current loss can be much greater than it would be if only the absolute permeability ($\mu_0\mu_r$) or *differential permeability* ($\mathrm{d}B/\mathrm{d}H$) is taken into account.

1.6 ENERGY STORAGE AND FORCES IN MAGNETIC FIELD SYSTEMS

In the magnetic field model, the energy associated with the system is distributed throughout the space occupied by the field. Assuming no losses, energy supplied per unit volume (the energy density) when increasing the flux density from zero to B is

$$w_m = \int_0^B H\,\mathrm{d}B \;\; (\mathrm{J\,m^{-3}}) \tag{1.102}$$

The total magnetic energy stored is

$$W_m = \iiint_v \left(\int_0^B H\,\mathrm{d}B \right) \mathrm{d}v \tag{1.103}$$

If, for simplicity, we assume a uniform distribution of field in the region concerned of volume v, then

$$W_m = v \int_0^B H\,\mathrm{d}B \tag{1.104}$$

If, in addition, we have a linear magnetic material

$$W_m = v \frac{1}{2\mu} B^2 \tag{1.105}$$

Clearly the higher μ, the lower the total energy stored. For the same flux density, the energy stored in a given volume of air is orders of magnitude higher than in the same volume of iron.

For a singly excited system, eqn (1.104) can be written in the form

$$W_m = \int_0^B (H \times \mathit{length})(\mathit{area} \times \mathrm{d}B) = \int_0^\phi \mathit{mmf}(\phi)\,\mathrm{d}\phi \tag{1.106}$$

where ϕ is the total flux passing through the region. But $mmf = ni$ where n is the number of turns on the excitation winding and $\phi = \lambda/n$ where λ is the total flux linkage. Thus

$$W_m = \int_0^i i(\lambda)\,\mathrm{d}\lambda \tag{1.107}$$

Note that the current is a function of the flux linkages and the *mmf* is a function of the flux; their relations depend on the geometry of the magnetic circuit and windings, as well as the magnetic properties of the core material. Equations (1.104), (1.106) and (1.107) may be graphically interpreted as the area shown in Fig. 1.2 labelled as *energy*. The other area, labelled as *coenergy*, can be expressed as

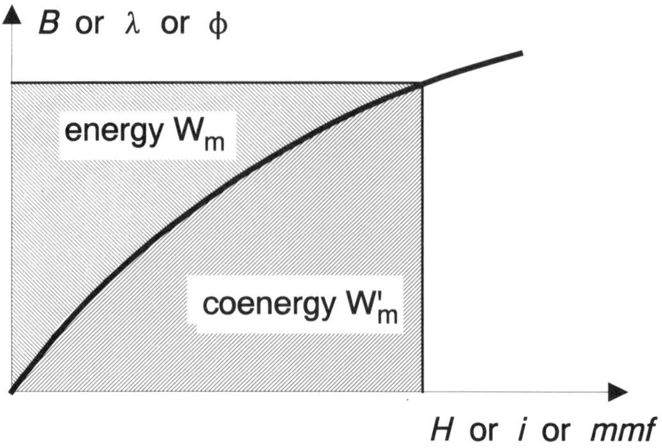

Fig. 1.2 *Energy and coenergy*

$$W'_m = v\int_0^H B\,\mathrm{d}H = \int_0^i \lambda(i)\,\mathrm{d}i = \int_0^{mmf} \phi(mmf)\,\mathrm{d}mmf \qquad (1.108)$$

For a linear system in which B and H or λ and i or ϕ and mmf are proportional, the energy and coenergy are equal. However, for a nonlinear system the energy and coenergy will differ but their sum for a singly excited system is given by

$$W'_m + W_m = vBH = \lambda i = mmf\,\phi \qquad (1.109)$$

Consider a simple electromechanical system where movement is allowed in one direction only, say x, and electrical and mechanical losses are neglected. An energy balance equation may be written

electrical energy input = *mechanical energy output*
+ *increase in stored energy*

In the simple situation of a change in x at constant λ there is no *emf* induced and therefore no electrical input so that the force is given by

$$F = -\left(\frac{\partial W_m}{\partial x}\right)_{\lambda=constant} \qquad (1.110)$$

Consider now Fig. 1.3 which shows a change of W_m at constant current and the associated electrical input due to the changing flux. We can now write

$$F = i\left(\frac{\partial \lambda}{\partial x}\right)_{i=constant} - \left(\frac{\partial W_m}{\partial x}\right)_{i=constant} \qquad (1.111)$$

or, in terms of the coenergy

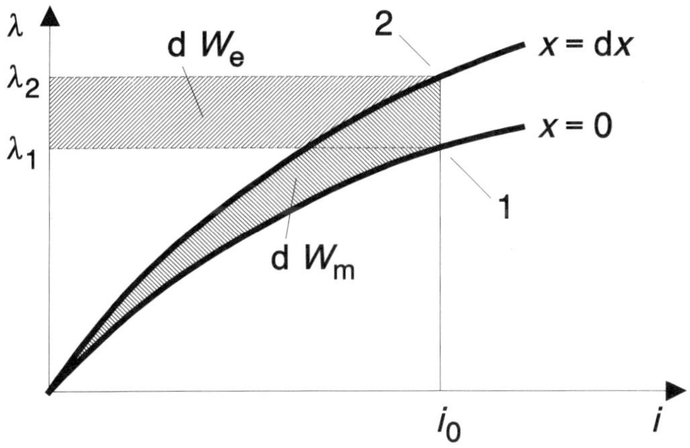

Fig. 1.3 *Change of energy at constant current*

$$F = \left(\frac{\partial(i\lambda)}{\partial x}\right)_{i=constant} - \left(\frac{\partial W_m}{\partial x}\right)_{i=constant} = \left(\frac{\partial W_m'}{\partial x}\right)_{i=constant} \tag{1.112}$$

It is interesting to notice the change in sign between eqns (1.110) and (1.112). If the relationship between λ and i is a straight line the magnitude of the two expressions for the force is the same but the sign is reversed because of the electrical input.

Calculations of forces and torques based on the *energy method* described briefly above are usually quite convenient. They involve volume integration of energy or coenergy and such routines are usually provided by post-processors of commercial software. However, at least two field solutions are required in order to determine the space derivatives. For improved accuracy several solutions may be necessary and thus computation times may be long.

Another approach, known as the *stress method*, looks at tensile and compressive stresses along iron surfaces and relies on surface integration of appropriate expressions. We have already seen that it is possible to represent surface magnetization in terms of a distributed magnetic polarity density q_m. If, in eqn (1.94), $\mu_{r1} = 1$, i.e. region 1 is air, then from eqn (1.93)

$$q_m = 2\mu_0 \frac{\mu_r - 1}{\mu_r + 1} H_n \tag{1.113}$$

Combining eqns (1.95) and (1.113) we also obtain

$$H_n = \frac{\mu_r + 1}{2\mu_r} H_{n1} \tag{1.114}$$

Now, since the force on a single pole Q is given by $Q\mathbf{H}_a$, where $\mathbf{H}_a$ is the applied field, the force per unit surface area on q_m is given by the stress components

$$F_n = q_m H_n \quad \text{and} \quad F_t = q_m H_t \tag{1.115}$$

Thus, after substitution, we have

$$F_n = 2\mu_0 \frac{\mu_r - 1}{\mu_r + 1} H_n^2 = \frac{1}{2} \mu_0 H_{n1}^2 (1 - \mu_r^{-2}) \tag{1.116}$$

$$F_t = 2\mu_0 \frac{\mu_r - 1}{\mu_r + 1} H_n H_t = \mu_0 H_{n1} H_{t1} (1 - \mu_r^{-1}) \tag{1.117}$$

since the contribution of q_m to the local tangential field is zero, and so the applied tangential component of field H_t is equal to the total H_{t1}. In the common case where $\mu_r \gg 1$, $H_{t1} \to 0$ and $F_n \to \mu_0 H_{n1}^2/2$ and $F_t \to 0$. These stresses are not unique, but since the surface polarization correctly reproduces the magnetic field conditions at all points outside the iron body, the total force on the iron obtained by surface integration is unique.

The full equations for F_n and F_t only apply to the surface of a homogeneous iron region with a uniform value of μ_r: otherwise there is a volume distribution of polarity present. It would be useful to have more general expressions for F_n and F_t that could be applied to any closed surface in air that surrounds a composite device consisting of nonlinear steel, current carrying conductors etc., in the confidence that the resultant force on whatever is within the enclosed region is correctly obtained by surface integration of the two stress components. Such expressions are *Maxwell's stresses* in air (vacuum) [1.16]:

$$F_n = \tfrac{1}{2} \mu_0 (H_n^2 - H_t^2) \tag{1.118}$$

$$F_t = \mu_0 H_n H_t \tag{1.119}$$

where F_n is directed outwards from the surface. The suffix 1 is no longer required to distinguish the region exterior to the iron from the interior: H_n and H_t now represent the components of the magnetic field at the chosen, and, to some extent, arbitrary surface. Whereas the earlier expressions can be thought of as stresses acting on the iron surface, the much more general Maxwell forms represent stresses transmitted through the chosen computation surface by the magnetic field.

1.7 EDDY-CURRENT SKIN EFFECT

If a very large flat conducting surface on the plane $y = 0$, say, is subjected to a uniform alternating field parallel to the surface, $H_z = H_s \cos \omega t$, then the solution of the complex one-dimensional differential equation (Helmholtz equation) for H_z inside the conducting slab is

$$\underline{H}_z = H_s \mathrm{e}^{-\alpha y} \tag{1.120}$$

where $\alpha = (1 + \mathrm{j})/\delta$ and $\delta = (\omega\mu\sigma/2)^{-1/2}$. The corresponding current density distribution is given by

$$\underline{J}_x = \frac{\mathrm{d}\underline{H}_z}{\mathrm{d}y} = -\alpha H_s \mathrm{e}^{-\alpha y} = -J_s \mathrm{e}^{-\alpha y} \tag{1.121}$$

and the eddy-current loss per unit surface area of the slab is

$$P_e = \frac{1}{2\sigma}\int_0^\infty |\underline{J}_x|^2 \,\mathrm{d}y = \frac{H_s^2}{\sigma\delta^2}\int_0^\infty \mathrm{e}^{-2y/\delta}\,\mathrm{d}y = \frac{H_s^2}{2\sigma\delta} \tag{1.122}$$

But, using Ampère's law, $H_s = I$, the peak current per unit width. Thus the loss can be written in the form

$$P_e = \left(\frac{I}{\sqrt{2}}\right)^2 \frac{1}{\sigma\delta} = I_{rms}^2 R' \tag{1.123}$$

where $R' = (\sigma\delta)^{-1}$ is the resistance of unit length of a layer of thickness δ and unit width. For this reason the parameter δ is known as the eddy-current *skin depth* because the correct value of loss is obtained if the current is assumed to flow with uniform density in a layer of depth δ.

In practice, the magnetic field and current density fall exponentially in magnitude away from the surface. Not only this, but there is a phase change with increasing depth. For example, from eqn (1.121), the instantaneous value of the current density is given by

$$J_x = \mathrm{Re}\,\{\underline{J}_x \mathrm{e}^{\mathrm{j}\omega t}\} = -J_s \mathrm{e}^{-y/\delta}\cos\left(\omega t - \frac{y}{\delta}\right) \tag{1.124}$$

so that, when $y = \pi\delta$, J_x is in anti-phase with its surface value $(-J_s)$. This is typical of a diffusion process. At any depth, the local peak value does not occur until $y/\omega\delta$ seconds after it occurs on the surface, and, from one point of view, it is as if a damped wave of current (or flux) is moving into the material at velocity $\omega\delta$.

The relative smallness of the total penetration is particularly important. If we regard anything less than 5% of the surface density as negligible, then the effective penetration is given by 3δ. For copper at 50 Hz, δ is typically about 9 mm so that the penetration, on this criterion, is 27 mm. But δ is inversely proportional to the square root of frequency so that, at 500 kHz, the effective penetration has reduced to 0.27 mm. The same sort of reduction occurs even at power frequencies if the material has a high relative permeability. The result is that this simple one-dimensional approach can be used to model, at least approximately, the surface of conductors of any shape if the major dimensions of the conductor are much greater than δ, because locally the surface effect is almost one-dimensional. We only require to know the distribution of the surface field H_s. In numerical finite-difference or finite-element solutions this can be particularly important because the alternative will be the need to discretise the grid or mesh sufficiently finely to follow the skin phenomenon correctly.

The simpler numerical technique is to render the conducting region impermeable to flux by setting $\mu_r = 0$ and therefore excluding this region from the computation by specifying $H_n = 0$ on its surface. A practical example, although most often solved analytically, is the wall of a waveguide.

In situations where the skin effect is rather less pronounced, but the penetration nevertheless still small, the conducting surface can be modelled as a surface impedance. This is the equivalent of a local linear one-dimensional solution because the surface impedance is defined as

$$Z_s = \frac{E_s}{H_s} \tag{1.125}$$

where E_s is the peak value of the surface electric field. From Ohm's law, eqn (1.8), and eqn (1.121),

$$E_s = \frac{J_s}{\sigma} = \frac{\alpha H_s}{\sigma} = (1 + \mathrm{j})\frac{H_s}{\sigma\delta} \tag{1.126}$$

Thus

$$Z_s = \frac{1 + \mathrm{j}}{\sigma\delta} \tag{1.127}$$

The above situation is one in which the eddy currents are limited by their own field. The currents are said to be *inductance limited*. There is an opposite extreme when the currents are *resistance limited* owing to lack of space. This is precisely what is required in a transformer lamination, for example. The full linear solution for the eddy-current loss per unit surface area of a plate or lamination of thickness $2b$ is [1.39]

$$P_e = \frac{H_s^2}{\sigma\delta}\,\frac{\sinh\gamma - \sin\gamma}{\cosh\gamma + \cos\gamma} \tag{1.128}$$

where $\gamma = 2b/\delta$. It is clear that as γ tends to infinity, the loss tends to $H_s^2/\sigma\delta$, i.e. twice the loss per unit surface area of a massive slab, because we have two surface effects. However, when $\gamma \ll 1$, the loss tends to

$$P_e' = \frac{H_s^2}{\sigma\delta}\,\frac{\gamma^3}{6} \tag{1.129}$$

which is the result that would have been obtained by ignoring the reaction field of the eddy currents altogether.

The condition for this resistance-limited regime is that the thickness of the conductor is less than δ. If such a conductor is present in a field problem it can be modelled as having zero conductivity, i.e. no eddy currents, as far as the external system is concerned. Clearly, if it is a load carrier, an imposed uniform current density must be specified.

1.8 GEOMETRICAL STRUCTURE OF ELECTROMAGNETIC FIELDS AND DIFFERENTIAL FORMS

Vector algebra provides a familiar notation for describing electromagnetic relations and is used throughout this book. However, there is much to be said for using *differential forms* instead of vectors, because these forms underline the geometrical structure of the equations. Whereas the vector calculus cannot extend to four dimensions, both three- and four-dimensional problems can be readily treated by differential forms. Other advantages of such treatment include the unification of static and dynamic fields, the clarification of the difference between gradient (polar) and flux (axial) vectors and the replacement of the operators *grad*, *div* and *curl* by a single exterior derivative. In this section we shall introduce very briefly this powerful notation, in which physical and geometrical information is combined. For full treatment the reader is referred to specialist literature, in particular works by Hammond and Baldomir, for example references [1.2; 1.3; 1.4; and 1.22].

In terms of differential forms, Maxwell's equations can be expressed as [1.1]

$$\mathrm{d}D = \rho, \quad \mathrm{d}H = J + \partial_t D, \quad \mathrm{d}B = 0, \quad \mathrm{d}E = -\partial_t B \tag{1.130}$$

together with the constitutive relationships

$$D = \varepsilon * E, \quad B = \mu * H \tag{1.131}$$

The variables are now known as forms: a 0-form is a scalar, a 1-form has a short length associated with it, a 2-form an area and a 3-form a volume. For example, the 2-form B is equivalent to $\mathbf{B} \cdot \delta\mathbf{s}$ in vector notation. Hence we can write

ϕ	0-form				scalar potential
A		1-form			magnetic vector potential
E		1-form			electric field strength
H		1-form			magnetic field strength
D			2-form		electric flux density
B			2-form		magnetic flux density
J			2-form		electric current density
ρ				3-form	electric charge density

In the notation of eqns (1.130) and (1.131), d is the exterior differential operator (which corresponds to divergence and curl in vector analysis), ∂_t is the differential operator with respect to time and $*$ denotes the Hodge operator linking a differential 2-form (flux) having area with a differential 1-form having length. This linkage defines a volume element involving the product of two field quantities. The products of D and E and of B and H have the dimensions of energy, and the constitutive equations therefore define the local energy density associated with a local volume.

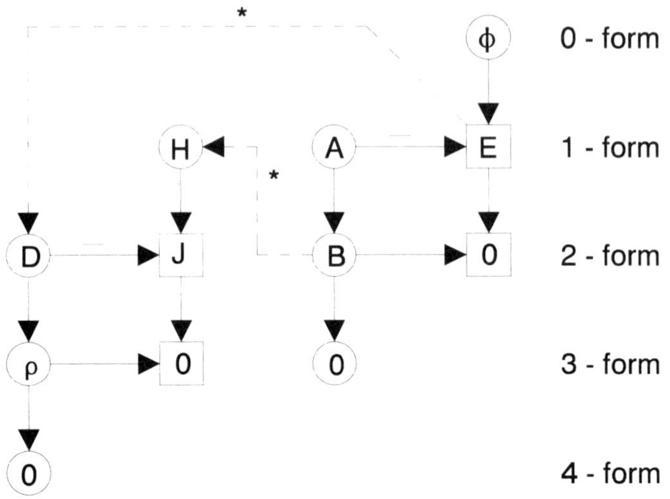

Fig. 1.4 *Maxwell's equations in differential forms*

It is interesting to note that the double exterior differentiation is always zero, i.e. $d \cdot d\phi = 0$, $d \cdot dE = 0$, $d \cdot dD = 0$. This is a general property of differential forms and is known as Poincaré's lemma. It corresponds to *curl grad* $\phi = 0$ and *div curl* $\mathbf{E} = 0$ in vector algebra. Notice also that a 1-form is always associated with an element of length and a 2-form with an element of area. This avoids the confusion which often arises between axial and polar vectors in vector analysis.

Equations (1.130) and (1.131) can be simplified by introducing the potentials A and ϕ, together with a gauge condition between them. We obtain

$$\Box A = -\mu * J \tag{1.132}$$

$$\Box \phi = -\frac{1}{\varepsilon} * \rho \tag{1.133}$$

where the Lorentz gauge has been used. This gauge is given by

$$\delta A + \mu\varepsilon \partial_t \phi = 0 \tag{1.134}$$

Here the operator δ is the co-derivative of the exterior differential operator d and is defined as $\delta = *d*(-1)^{p(n-p+1)+s}$, where n is the dimension of space, p is the order of the differential form and s is the number of negative terms in the metric. The Laplace–Beltrami operator $\Box$ is defined as $\Box = d\delta + \delta d$. In three dimensions, $\Box \mathbf{A}$ is equivalent to *grad div – curl curl* in vector notation, and $\Box \phi$ is equivalent to *div grad*.

Maxwell's equations can be represented diagrammatically as in Fig. 1.4 [1.1]. Each level of the diagram corresponds to a different form of a particular order. Horizontal arrows with their signs indicate the time derivative ∂_t, and the vertical arrows indicate the exterior derivative d. Convergence of two arrows at a node implies summation. The dotted lines show the constitutive relations. Quantities, obtained by summing terms in which both d and ∂_t are involved, are shown in squares. The scalar potential ϕ is a 0-form and the vector potential A is a 1-form. The introduction of the potentials makes the right-hand side of the diagram complete.

The scalar and vector potentials are related to the field vectors via the following relationships

$$B = \mathrm{d}A \tag{1.135}$$

and

$$E = -\partial_t A - \mathrm{d}\phi \tag{1.136}$$

Figure 1.4 displays a particular type of symmetry, where two blocks are displaced with respect to each other, as regards the order of the differential forms involved. The two blocks are linked through the constitutive relationships, eqn (1.131). The symmetry between the blocks allows the correspondences [1.1]

$$\mathrm{d}\rho = 0 \simeq \mathrm{d}B = 0 \tag{1.137}$$

$$\mathrm{d}D = \rho \simeq \mathrm{d}A = B \tag{1.138}$$

$$\mathrm{d}H = J + \partial_t D \simeq -\mathrm{d}\phi = E + \partial_t A \tag{1.139}$$

$$\mathrm{d}J + \partial_t \rho = 0 \simeq \mathrm{d}E + \partial_t B = 0 \tag{1.140}$$

in which each of Maxwell's equations is paired with a relationship contained in the other block. For example, eqn (1.137) states that the electric charge density ρ is locally conserved, like the flux associated with the magnetic induction B. Equation (1.138) parallels Gauss's law to the existence of vector potential. Equations (1.139) and (1.140) embody Ampère's law and Faraday's law, as well as the local conservation of sources.

Electrostatic and magnetostatic fields may be retrieved from Fig. 1.4 by prolonging the dotted paths joining the fields with their duals as far as possible, on either side, without introducing time-varying quantities. The relevant equations are found for electrostatics as

$$\mathrm{d} * E = \frac{\rho}{\varepsilon} \quad \text{and} \quad \mathrm{d}E = 0 \tag{1.141}$$

and for magnetostatics

$$\mathrm{d} * H = 0 \quad \text{and} \quad \mathrm{d}H = J \tag{1.142}$$

Both pairs of relationships lead to Poisson's equations. The corresponding diagrams are depicted in Fig. 1.5.

The product of E and D (and H and B) in the notation of differential forms is a 3-form and is, therefore, associated with a small volume of geometrical space. Hence the division of space into 'cells' implies the division of the energy. Another 3-form is associated with the 'assembly work' of, say, charges described by the product of ϕ and ρ. The integrals of the field energy and

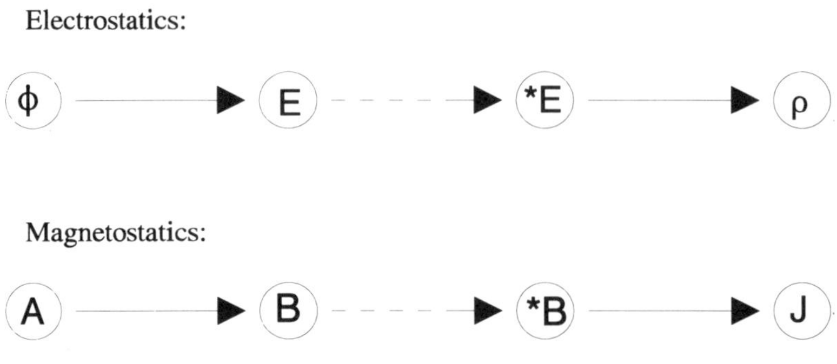

Fig. 1.5 *Paths for electrostatics and magnetostatics*

the assembly work are equal for the system, but locally the product $\phi\rho$ can be positive, negative or zero, whereas the product ED is non-negative. We can describe the field in two different ways. The first starts with the 1-form (E or H) and its associated increments of potential difference, whereas the second begins with the 2-form (D or B) which gives a flux through an area. Application of the Hodge operator and identification of relevant area or line segment produce a volume and an associated energy. These elementary volumes are then joined in parallel or series to form 'slices' or 'tubes', which are then connected in series or parallel to fill the entire space of the system. The implementation involves simple multiplication, followed by addition in series and parallel; it does not require the solution of simultaneous equations. The technique is known as the method of *tubes and slices* [1.22; 1.43] and is briefly described in the next section.

1.9 THE METHOD OF TUBES AND SLICES

The method of *tubes and slices* is based on a dual energy formulation. Foundations of this approach may be traced back to Maxwell, who in his famous treatise on electricity and magnetism [1.30], describes a variational method applied to the calculation of resistance of conductors of varying cross-section. The method relies on subdivision of the conductor into *slices* formed by equipotential surfaces and *tubes* separated by very thin insulating sheets. The two calculations yield lower and upper bounds of the resistance, respectively. The approach is applicable to other types of vector fields as demonstrated by Hammond [1.20; 1.22]. In electrostatics Maxwell's method of 'slices' is identical to the application of the variational principle

$$\langle(\rho - \nabla \cdot \mathbf{D}), \delta\phi\rangle = \langle\{\rho + \nabla \cdot (\varepsilon\nabla\phi)\}, \delta\phi\rangle = 0 \tag{1.143}$$

where the brackets $\langle\ \rangle$ indicate integration through the region of interest, and his method of 'tubes' is identical to the application of

$$\langle(\mathbf{E} + \nabla\phi), \delta\mathbf{D}\rangle = \left\langle\left(\frac{1}{\varepsilon}\mathbf{D} + \nabla\phi\right), \delta\mathbf{D}\right\rangle = 0 \tag{1.144}$$

The two functionals are equivalent to introducing additional fictitious sources, of *divergence* or *curl* type respectively. In capacitance problems where the energy can be written as $\frac{1}{2}\phi^2 C$ or $\frac{1}{2}Q^2/C$, upper bounds of the capacitance are produced by the variational statement of equation 1 (slices) and lower bounds result from the use of equation 2 (tubes). Thus any potential map produces an upper bound of capacitance and any flux map produces a lower bound.

In magnetostatics the equilibrium conditions can be described by two variational principles

$$\langle (\nabla \times \mathbf{H} - \mathbf{J}'), \delta \mathbf{A} \rangle = 0 \tag{1.145}$$

and

$$\langle (\nabla \times \mathbf{A} - \mathbf{B}), \delta \mathbf{H} \rangle = 0 \tag{1.146}$$

where $\mathbf{J}'$ is the assigned current density. The first variational principle implies that $\mathbf{B} = \nabla \times \mathbf{A}$, so that $\nabla \cdot \mathbf{B} = 0$ and thus there are no *divergence* sources for the magnetic field. However, the expression $\nabla \times \mathbf{H} - \mathbf{J}'$ allows a small variation in $\nabla \times \mathbf{H}$ from its correct value, so that the variation allows a small additional distribution of *curl* sources. The product of this small fictitious current multiplied by the small variation of $\mathbf{A}$ gives an energy variation of the second order of small quantities which can be put to zero.

The second variational principle assumes that $\nabla \times \mathbf{H} = \mathbf{J}'$, so that the *curl* sources of the magnetic field are correct. However, the expression $\nabla \times \mathbf{A} - \mathbf{B}$ allows a small variation in the *divergence* sources. The product of this small polarity distribution multiplied by the small variation of $\mathbf{H}$ gives an energy variation of the second order of small quantities which can be put to zero. The field energy can be expressed either in terms of the field vectors $\mathbf{H}$ and $\mathbf{B}$ by

$$U = \tfrac{1}{2} \langle \mathbf{B}, \mathbf{H} \rangle \tag{1.147}$$

or in terms of the interaction of the current sources with the vector potential $\mathbf{A}$ by

$$U = \tfrac{1}{2} \langle \mathbf{J}', \mathbf{A} \rangle + \tfrac{1}{2} [\mathbf{I}', \mathbf{A}] \tag{1.148}$$

where $\mathbf{I}'$ is the assigned line density of current on the surface, and the brackets [] represent integration over the closed boundary surface. $\mathbf{I}'$ is related to $\mathbf{J}'$ so as to make the total current in the system zero. This isolates the system and gives it a unique energy.

The first variational principle is applied to the energy in terms of $\mathbf{A}$ and $\mathbf{B} = \nabla \times \mathbf{A}$ by writing

$$\delta U(\mathbf{A}) = \delta \left(\langle \mathbf{J}', \mathbf{A} \rangle + [\mathbf{I}', \mathbf{A}] - \frac{1}{2} \left\langle \mathbf{B}, \frac{\mathbf{B}}{\mu} \right\rangle \right) = 0 \tag{1.149}$$

The second variation is therefore negative

$$\delta^2 U(\mathbf{A}) \leqslant 0 \tag{1.150}$$

The second variational principle is applied to the energy in terms of **H** by writing

$$\delta U(\mathbf{H}) = \delta(\tfrac{1}{2}\langle \mathbf{H}, \mu\mathbf{H}\rangle) = 0 \tag{1.151}$$

Hence

$$\delta^2 U(\mathbf{H}) \geqslant 0 \tag{1.152}$$

For simplicity μ has been assumed constant and this gives the factor 1/2. However, the method is applicable to permeabilities which are single-valued functions of the field strength [1.20]. The second variations show the possibility of obtaining both upper and lower bounds for the energy. The first variational principle treats the field as a collection of tubes and the second one as a collection of slices.

Similar arguments may be presented for a steady current flow and calculation of resistance, where the analogue of permittivity ε or magnetic permeability μ is electric conductivity σ. Extension of the method to time-varying problems is possible and some interesting suggestions may be found in references [1.31; 1.4; and 1.20].

The implementation of the method is relatively straightforward. For example, in electrostatic problems, for a given field solution, the flux map and the potential map may be represented using connections of simple parallel-plate capacitors, as demonstrated in Fig. 1.6.

Tubes:

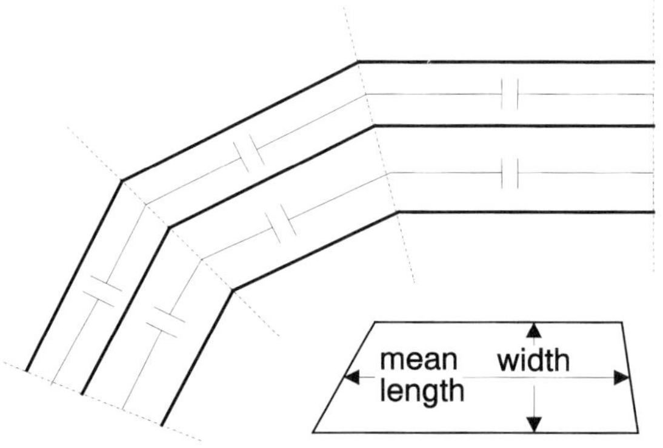

(a)

$$C_{ij} = \frac{\varepsilon \times width}{mean\ length}\ (\text{F/m})$$

Slices

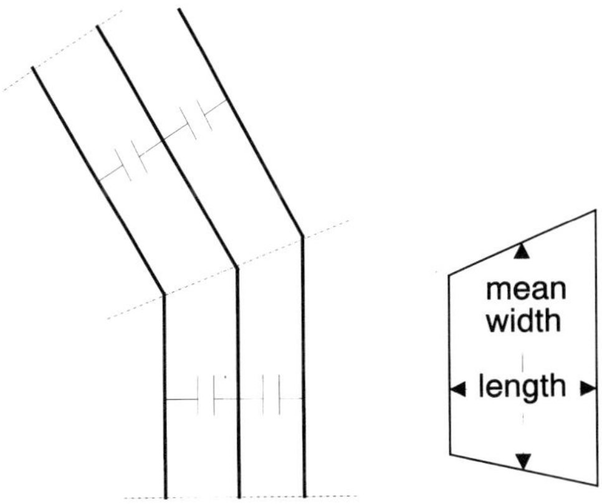

(b)

$$C_{ij} = \frac{\varepsilon \times mean\ width}{length}\ (\mathrm{F/m})$$

Fig. 1.6 *Circuit representation of tubes and slices in electrostatics*

These component capacitors are connected in parallel and in series, as shown in Fig. 1.7, to form an equivalent circuit of the field map.

Tubes:

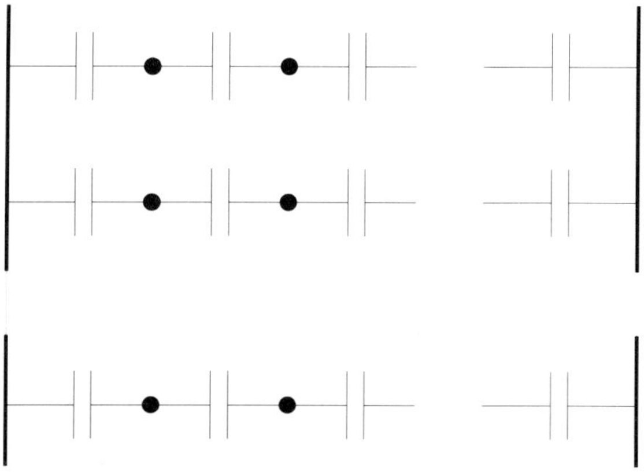

(a)

Slices:

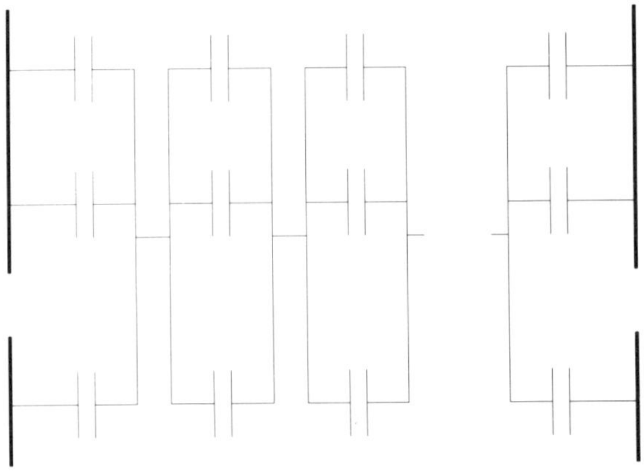

(b)

Fig. 1.7 *Series/parallel connections of component capacitors*

Neither of the two representations is exact owing to the approximations introduced. First, the number of subdivisions will necessarily be finite. Secondly, the flux or potential lines may not be in the correct position so that the orthogonality of the two field maps is violated. This second consideration is very important and leads to the following equation for tubes

$$C^{-} = \sum_{i=1}^{n} \frac{1}{\sum_{j=1}^{m} \frac{1}{\left(\frac{\varepsilon S_{ij}}{l_{ij}}\right)}} \tag{1.153}$$

whereas for slices

$$C^{+} = \frac{1}{\sum_{i=1}^{m} \frac{1}{\sum_{j=1}^{n} \left(\frac{\varepsilon S_{ij}}{l_{ij}}\right)}} \tag{1.154}$$

where S_{ij} and l_{ij} are dimensions of the component capacitors. Thus dual bounds for the capacitance are established.

For the steady current flow problems the resistance may be calculated in an analogous way. Physically, subdivision into tubes can be achieved by inserting thin insulating sleeves between the tubes, which must always increase the resistance unless the flow of current is undisturbed. Very thin sleeves in the correct direction everywhere will have negligible effect, but if such sleeves are not strictly in the direction of current flow they will increase the resistance. The undisturbed resistance is therefore a minimum. Equally, the insertion of infinitely conducting sheets for slices will reduce the resistance if they disturb the flow. The undisturbed resistance is therefore a maximum. Thus a division into tubes and slices enables one to calculate upper and lower bounds for the unknown resistance.

Finally, in many electrical devices the magnetic circuit is designed in such a way that very little *mmf* is absorbed in the iron core and attention is focused on the shape and dimensions of the air-gap. An unsaturated iron surface may be assumed to have a constant magnetic potential and thus becomes a slice. At the same time the flux distribution may be described in terms of tubes. These tubes terminate on iron surfaces. We can work in terms of *permeance* which is the analogue of conductance in the electric circuit. However, for problems outside a current region the calculation of inductance is reduced to a calculation of permeance. Thus, for calculating inductance, equations analogous to (1.153) and (1.154) may be used with μ substituted for ε. We now have a system of equations for calculating circuit parameters R, C or L for many practical problems under static conditions.

Calculation of an internal inductance, on the other hand, involves working in terms of energy instead of the circuit parameters. A possible approximation technique is described for example in reference [1.43].

The simplicity of the final expressions (eqns (1.153) to (1.154)) is very striking and should be contrasted with the complexity of other numerical formulations, such as finite elements. Although the TAS method, like indeed many other methods, usually calls for more than one calculation to achieve the desired accuracy, this should hardly matter in view of this simplicity of computation.

1.10 FIELD CALCULATIONS USING TUBES AND SLICES

The *tubes and slices* method is essentially a geometrical approach, based not on solving equations but using sketching of fields as a means of finding the solution. Full advantage can be taken of interactive graphics capabilities of modern computers. An appropriate computer package called TAS is described in reference [1.43]. In particular, the program has been found extremely useful in teaching, as discussed in reference [1.17] and fully implemented in a recent book [1.16].

In the TAS program the user is in constant interaction with the field solution provided in the form of approximate distributions of tubes and slices. These distributions may be easily modified by moving appropriate construction points and lines on the screen and the system responds almost instantly with a pair of bounded solutions given by simple calculations of the type given by eqns (1.153) and (1.154). The bounded values give confidence limits to the solution, whereas the orthogonality of the two field patterns, or lack of it, sets a criterion for further modifications.

In order to demonstrate the application of the method, let us estimate the inductance per unit length of the coaxial line, with a strip centre conductor, as shown in Fig. 1.8. Owing to symmetry only a quarter of the system needs to be investigated (see Fig. 1.9). A possible set of 'construction lines' and

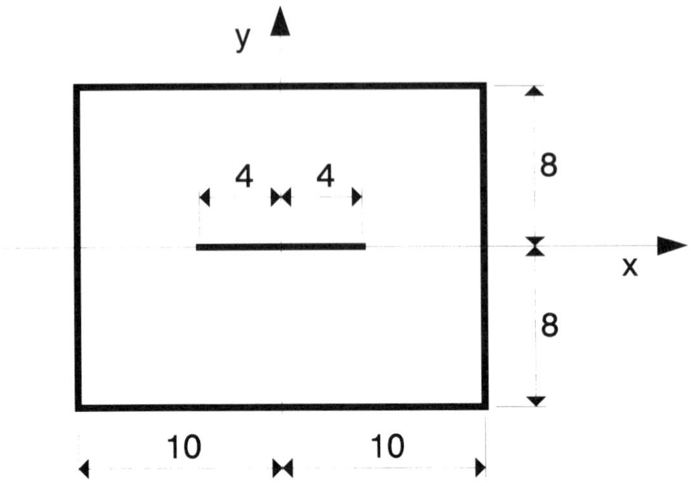

Fig. 1.8 *A rectangular coaxial line with a strip centre conductor*

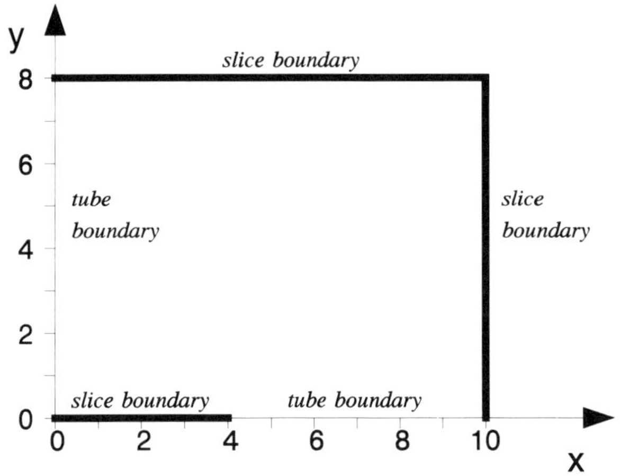

Fig. 1.9 *Computational model of the coaxial line*

'quadrilaterals' is demonstrated in Fig. 1.10. Such quadrilaterals (each with a diagonal line) are used to match a particular shape of system boundaries.

These construction lines and quadrilaterals are at the same time used to generate tubes and slices of Fig. 1.11. The whole process is fully automated, although the user may wish to change manually some distributions. An improved subdivision may be achieved by adding more construction lines to the basic distribution of Fig. 1.11 and repositioning them so that a more orthogonal system is obtained. An example of such an improved distribution is shown in Fig. 1.12. The results are summarized in Table 1.1 and compared with a finite-element solution. It is interesting to note that although the

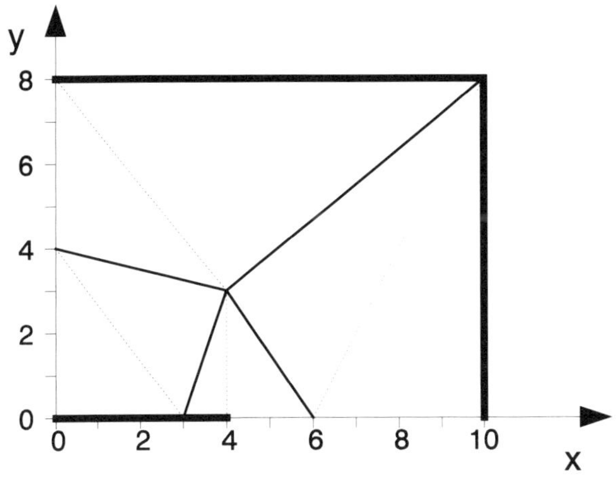

Fig. 1.10 *Construction lines forming a set of quadrilaterals*

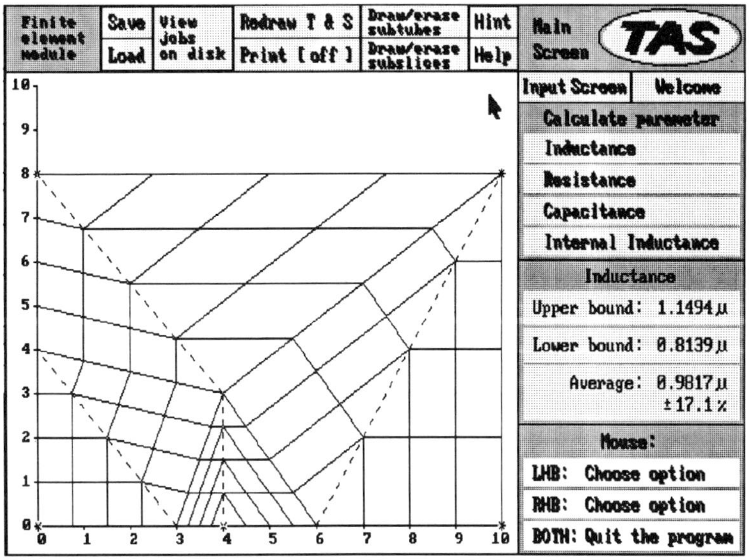

Fig. 1.11 *Distribution of tubes and slices*

Table 1.1 Summary of results

	Coarse distribution	Improved distribution	Finite element
L H m^{-1} (per unit length)	0.9817 × μ	0.9968 × μ	0.9993 × μ
Confidence limits	±17.1%	±7.4%	±0.7%
Actual error (against FE)	−1.76%	−0.25%	—

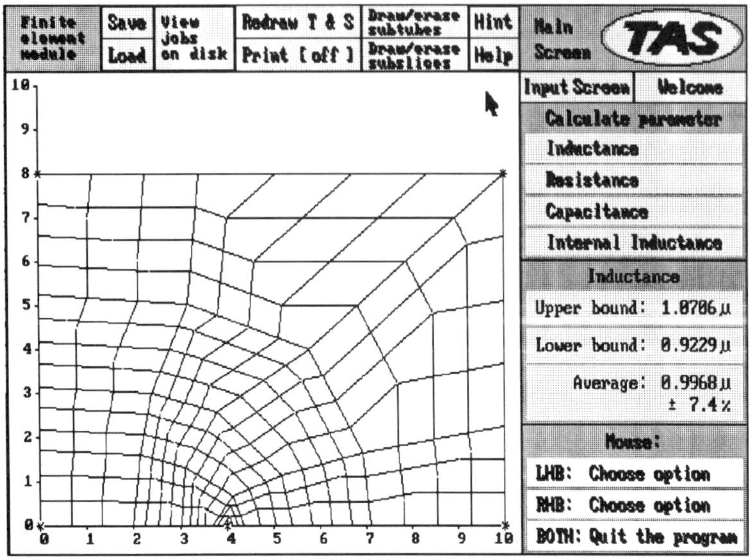

Fig. 1.12 *Improved distribution of tubes and slices*

improved solution exhibits improved orthogonality, the average value of the capacitance is hardly changed. Nevertheless, the smaller error band of the improved solution gives additional confidence to the user.

The TAS program has a built-in finite-element module. This combination of the two techniques has been found to be very useful. First, an approximate field distribution is calculated by the tube/slice process and then the accuracy is improved by means of a finite-element procedure. A much smoother distribution is obtained as shown in Fig. 1.13. It should also be noted that the finite-element method has been adapted to produce upper and lower bounds to the unknown exact values.

The two methods complement each other. The construction lines and quadrilaterals from the TAS calculations are now used as a basis of a FE mesh. This mesh will typically be refined automatically to improve accuracy. The values at all nodes are first calculated from the TAS solution so that a relatively small number of iterations is required for the solution to converge. It is also worth noting that the mesh is very regular and that most mesh lines follow the direction of the field or the direction of equipotential surfaces. Further information about this powerful combination of the two techniques may be found in references [1.17] and [1.44].

1.11 PREPARATORY AND FURTHER READING

Before proceeding to the next chapters which discuss particular methods and techniques of computational magnetics, the reader may have decided to study further the fundamentals and background theory. If the short review provided

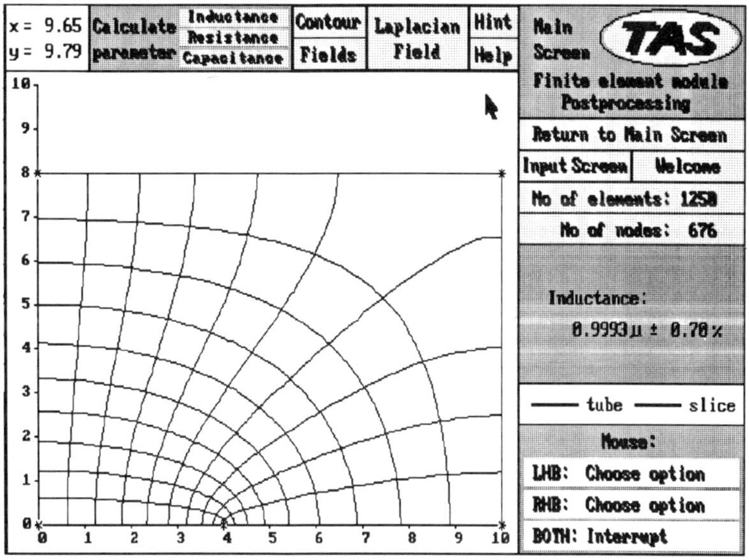

Fig. 1.13 *Finite-element solution for the coaxial line*

in this chapter has not been sufficient, we would like to make some recommendations to aid such studies. A very comprehensive treatment of both the theory and computational methods at the undergraduate level is offered in the book by Hammond and Sykulski [1.16], where the reader will find a full derivation of most of the equations reviewed in this chapter, complemented by thorough discussion of underlying physical processes and details of the numerical implementation of methods.

For a more introductory treatment of the subject the reader may wish to refer to the textbooks by Carter [1.10], Christopulos [1.13], Hayt [1.24] or Lancaster [1.27]. It is also appropriate to mention some more advanced texts by Carter [1.11], Dobbs [1.14], Harrington [1.23], Moon and Spencer [1.31], Smythe [1.36] and Stratton [1.41]. A recent significant book on analytical and numerical solution of electric and magnetic fields is by Binns, Lawrenson and Trowbridge [1.5].

Eddy currents have always been of particular interest to both researchers and practising engineers, and the following three books may be recommended for full treatment in various aspects and from a slightly different point of view: Lammeraner and Stafl [1.26], Stoll [1.39] and the most recent by Krawczyk and Tegopoulos [1.25].

There is a vast literature on implementation of numerical methods to solutions of magnetic field problems. It is not practical to offer a comprehensive bibliography within the compass of such a book as this one. A wide range of papers is published in professional journals, such as the *IEEE Transactions on Magnetics*, the *IEE Proceedings A* and *B*, the *Journal of Applied Physics*, the *Journal of Physics D* (*Applied Physics*), and the *International*

Journal for Computation and Mathematics in Electrical and Electronic Engineering, *COMPEL*. Relevant publications may also be found in the *IEEE Transactions on Power Apparatus and Systems*, the *International Journal for Numerical Methods in Engineering*, the *Computer Methods in Applied Mechanics and Engineering*, and *Computer Aided Design*.

Much of the current material in any rapidly developing area is presented at professional conferences. Those relevant to computational magnetics are COMPUMAG, INTERMAG, MMM, CEFC, CEM, ISEF and ELECTROSOFT. Selected and refereed papers from many of the above conferences often appear in one of the professional journals, see for example proceedings of ISEF'91 [1.42].

2
Finite-Difference Methods

Richard L. Stoll and Kazimierz Zakrzewski

2.1 INTRODUCTION

Finite-difference methods are important for two reasons. First, they form the background to almost all later developments. Secondly, a finite-difference method is relatively easy to construct and program to solve a particular problem, or class of problems, that may not be suitable for an existing general purpose software package using, say, finite elements, and so is ideal for the relatively rapid development of special purpose computer programs.

The finite-difference approach to the numerical solution of field problems is basically simple in concept, but considerably harder to examine theoretically in terms of accuracy and stability. To solve any field problem we must first select the dependent variable (field component or components, potential etc.). This variable is a continuous function within the space and time domain of the problem, and subject to appropriate boundary conditions (see Chapter 1). The application of the finite-difference technique is then essentially one of overlaying the problem domain with a *net*, *grid* or *mesh* that is uniform in size but may alter from one sub-domain to another. The variable now assumes a set of discrete values at the junctions or *nodes* of the mesh, but remains undefined at intermediate points. There is, of course, a lack of preciseness in this definition. For example, if the function $f(x)$ is defined by its values at the discrete points $x = 0$, Δx, $2\Delta x$, ..., $n\Delta x$, the function $h(x) = f(x) + g(x) \sin(\pi x/\Delta x)$ is similarly defined but has different values elsewhere. However, we aim to deal with smooth functions in the sense that Δx must be sufficiently small to allow $\Delta f/\Delta x$ to be a good approximation to $\mathrm{d}f/\mathrm{d}x$; something that would not be true for $h(x)$.

The equations to be solved by the finite-difference techniques described here are the second-order partial differential equations classified under their mathematical types: elliptic and parabolic, in Section 1.3. The actual elliptic equations of interest are Laplace, Poisson and Helmholtz which describe boundary value problems. Parabolic equations on the other hand are, in our context, the diffusion equations ((1.80) and (1.81)), which describe initial boundary value or marching problems in the sense that the solution marches out from the initial state, at time $t = 0$ say, guided and modified in transit by the surrounding boundary conditions.

2.2 DISCRETIZATION, FINITE-DIFFERENCE OPERATORS AND TRUNCATION ERRORS

Consider a function $f(x)$ which is a continuous function of x in the domain $\langle a, b \rangle$. The domain $X = b - a$ is now divided into M elements of equal length Δx, thus defining a set of $M + 1$ nodes at discrete values of x given by

$$x_i = a + i\Delta x, \quad 0 \leqslant i \leqslant M \tag{2.1}$$

The function $f(x)$ is now represented by its values at these nodes,

$$f_i = f(x_i) \tag{2.2}$$

Between, nodes $f(x)$ can only be found by interpolation. For example, using linear interpolation in the ith element to find an approximation f' to $f(x')$, where $x_i \leqslant x' \leqslant x_{i+1}$, we have

$$f' = \varepsilon f_{i+1} + (1 - \varepsilon)f_i \tag{2.3}$$

where

$$\varepsilon = \frac{x' - x_i}{\Delta x} \tag{2.4}$$

The error in f' is given by the difference $f(x') - f'$ and is dependent on the interpolation technique used. However, in practice, in the computation schemes of interest here, the values of f_i are not the exact analytically determined values given by eqn (2.2) but are themselves found using a finite-difference process. Thus a study of errors and their possible propagation is important. To this end it is sometimes helpful to represent $f(x)$ or its error in terms of a Fourier series in complex exponential form

$$f(x) = \sum_{n=-\infty}^{\infty} g_n \mathrm{e}^{\mathrm{j}2\pi nx/X} \tag{2.5}$$

where $\mathrm{j} = \sqrt{-1}$, and

$$g_n = \frac{1}{X}\int f(x)\, \mathrm{e}^{-\mathrm{j}2\pi nx/X}\, \mathrm{d}x \tag{2.6}$$

Equations (2.5) and (2.6) can be expressed in discrete form as

$$f_i = \sum_{n=1}^{M} G_n \mathrm{e}^{\mathrm{j}2\pi ni/M} \tag{2.7}$$

$$G_n = \frac{1}{M}\sum_{i=1}^{M} f_i \mathrm{e}^{-\mathrm{j}2\pi ni/M} \tag{2.8}$$

Once discretization has taken place the number of harmonics are clearly limited. The wavelength of the smallest harmonic is equal to Δx.

We not only need to discretize functions but also to derive suitable finite-difference approximations to the first and second derivatives appearing in the field equations. The Taylor series expansion is one method that can be used, whereby, writing $h = \Delta x$ for brevity

$$f(x+h) = f(x) + hf'(x) + \tfrac{1}{2}h^2 f''(x) + \tfrac{1}{6}h^3 f'''(x) + \cdots \quad (2.9)$$

$$f(x-h) = f(x) - hf'(x) + \tfrac{1}{2}h^2 f''(x) - \tfrac{1}{6}h^3 f'''(x) + \cdots \quad (2.10)$$

Subtraction of these expansions yields

$$f(x+h) - f(x-h) = 2hf'(x) + O(h^3) \quad (2.11)$$

where $O(h^3)$ indicates terms containing third and higher powers of the small length h. Thus

$$f'(x) = \frac{f(x+h) - f(x-h)}{2h} + O(h^2) \quad (2.12)$$

If $O(h^2)$ is omitted from eqn (2.12), we say that the remaining difference formula for $f'(x)$ has a *truncation* error of order h^2. A difference formula, with similar error, can be obtained for the second derivative $f''(x)$ by adding eqns (2.9) and (2.10):

$$f''(x) = \frac{f(x+h) - 2f(x) + f(x-h)}{h^2} + O(h^2) \quad (2.13)$$

It is sometimes necessary, e.g. when dealing with boundary conditions, to use eqn (2.9) to obtain a first derivative of the form

$$f'(x) = \frac{f(x+h) - f(x)}{h} + O(h) \quad (2.14)$$

This is the so-called forward-difference formula (a backward-difference equivalent comes from eqn (2.10)), but the penalty is the first-order error involved. However, formulas of the type given in eqn (2.14) are frequently used for time-stepping diffusion equations.

2.3 A ONE-DIMENSIONAL PARABOLIC FINITE-DIFFERENCE EQUATION

At this stage it is probably helpful to examine some possible finite-difference formulations of a simple one-dimensional magnetic field problem that will be solved in more detail later. One-dimensional problems are only significant from the numerical point of view if either the excitation is of too complicated a form to be easily handled analytically, or, the excitation is simple (e.g. varies sinusoidally with time) but the response is nonlinear. We will consider the latter as being the more general and take the example of a large plate of ferromagnetic material immersed in a uniform time-varying magnetic field, e.g. $H_z = H_s \cos \omega t$, applied parallel to the surface of the plate. Let the plane of symmetry of the plate or lamination be on the plane $y = 0$ as shown in Fig. 2.1. If the plate thickness $2b$ is much less than the other dimensions,

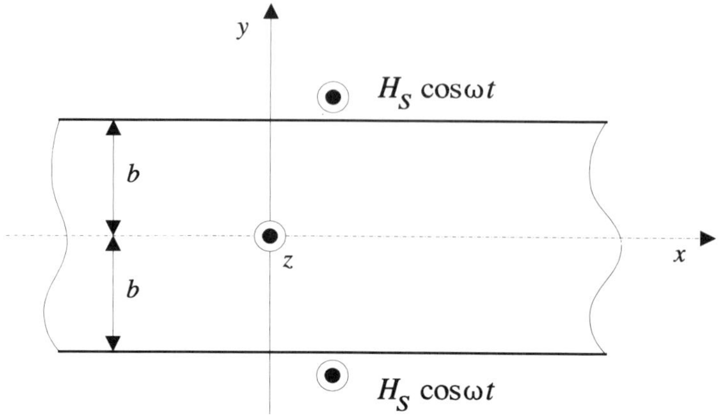

(a)

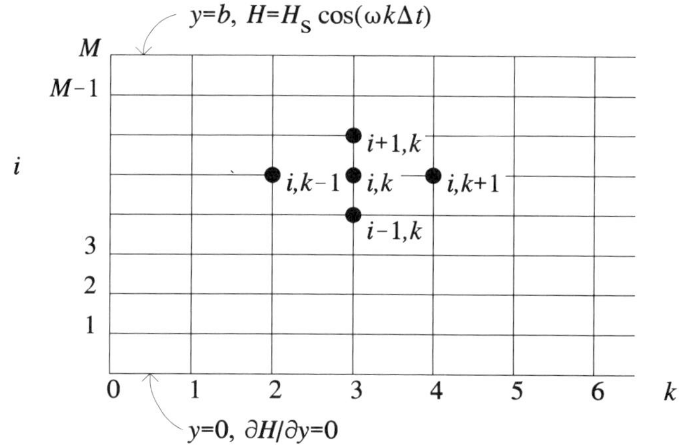

(b)

Fig. 2.1 *(a) Ferromagnetic plate immersed in uniform magnetic field* $H_s \cos \omega t$, *(b) half thickness of plate modelled in the space-time domain*

the problem can be modelled by a one-dimensional diffusion equation of the form

$$\frac{\partial^2 H}{\partial y^2} = \sigma \frac{dB}{dH} \frac{\partial H}{\partial t} \tag{2.15}$$

where H is an abbreviation for $H_z(y, t)$ inside the plate. Equation (2.15) can be obtained from eqns (1.1), (1.2) and (1.8) by neglecting displacement current (negligible in conducting materials) and assuming the conductivity σ to be a constant. Note that the constitutive equation (eqn (1.7)) has not been used because B and H are related by the magnetization curve, or a hysteresis loop, as illustrated in Fig. 1.1. For brevity, however, let us define a parameter β as

$$\beta = \sigma \frac{dB}{dH} = \sigma\mu_d \tag{2.16}$$

where μ_d is the differential permeability and a function of both y and t. Thus eqn (2.15) becomes

$$\frac{\partial^2 H}{\partial y^2} = \beta \frac{\partial H}{\partial t} \tag{2.17}$$

By symmetry the condition at the centre of the plate, $y = 0$, is that $\partial H/\partial y = 0$, and only half the plate $(0 \leqslant y \leqslant b)$ need be considered and divided into M mesh elements, say, where $M = b/\Delta y$. H is thus defined at $M + 1$ nodes in space at the points $y = i\Delta y$, where $i = 0, 1, 2, \ldots, M$ and this space 'ladder' is marched forward in time t taking discrete steps Δt so that $t = k\Delta t$, where $k = 0, 1, 2, \ldots$. A given node in this discretized space-time domain can be addressed using the integers i and k, so that H at node (i, k) is denoted by $H_{i,k}$.

From the brief introduction to finite-difference formulae for first- and second-order derivatives at the end of Section 2.2, we might suggest the following centre difference algorithm to represent eqn (2.17)

$$\frac{H_{i+1,k} - 2H_{i,k} + H_{i-1,k}}{\Delta y^2} = \beta \frac{H_{i,k+1} - H_{i,k-1}}{2\Delta t} \tag{2.18}$$

However, this deceptively logical equation, first proposed by Richardson [2.41] before the days of high-speed computers, is unconditionally unstable and serves as a warning that the matter of convergence and stability cannot be taken lightly. These topics are briefly considered in the next Section.

The next obvious alternative is a simple explicit two-time step algorithm based on the forward-difference formula in time (see eqn (2.14)). Thus

$$\frac{H_{i+1,k} - 2H_{i,k} + H_{i-1,k}}{\Delta y^2} = \beta \frac{H_{i,k+1} - H_{i,k}}{\Delta t} \tag{2.19}$$

giving

$$H_{i,k+1} = rH_{i+1,k} + (1 - 2r)H_{i,k} + rH_{i-1,k} \tag{2.20}$$

where

$$r = \frac{\Delta t}{\beta \Delta y^2} \tag{2.21}$$

It can be shown (see Section 2.4) that $r \leqslant 0.5$ to ensure convergence and stability in the *linear* case. In practice this means using very small values of time step. In complete contrast, the backward-difference formulation in time, yields

$$-rH_{i+1,k+1} + (1 + 2r)H_{i,k+1} - rH_{i-1,k+1} = H_{i,k} \tag{2.22}$$

which is stable for all values of constant r, i.e. in the linear case. The penalty to be paid for the likelihood, even in a nonlinear system, of good stability, is that eqn (2.22) must be solved simultaneously with all M equations that apply to time step $k+1$. The coefficient matrix is tri-diagonal and a simple form of Gaussian elimination can be used [2.41] given in a very neat form by the Thomas algorithm [2.51].

The truncation error of both eqns (2.20) and (2.22) is $O(\Delta y^2 + \Delta t)$, i.e. they suffer from a first-order error in Δt as we might have anticipated from eqn (2.14). In the search for a smaller error consider a weighted average of the second-order space derivatives on time rows k and $k+1$, so that

$$r\{\Theta(H_{i+1,k+1} - 2H_{i,k+1} + H_{i-1,k+1}) + (1-\Theta)(H_{i+1,k} - 2H_{i,k} + H_{i-1,k})\} = H_{i,k+1} - H_{i,k} \qquad (2.23)$$

Clearly if $\Theta = 0$ we have the simple explicit equation, and if $\Theta = 1$, the simple implicit equation. Equation (2.23) is unconditionally convergent and stable for $\tfrac{1}{2} \leqslant \Theta \leqslant 1$, but for $0 \leqslant \Theta < \tfrac{1}{2}$, we have the condition [2.41]

$$r \leqslant \frac{1}{2(1-2\Theta)} \qquad (2.24)$$

However, for constant r, the minimum local truncation error occurs if [2.2]

$$\Theta = \frac{1}{2}\left(1 - \frac{1}{6r}\right) \qquad (2.25)$$

Equation (2.25) suggests that, for fairly large values of r or values that are variable, the sensible choice of Θ is 1/2, i.e. on the boundary of unconditional stability. Equation (2.23) then reduces to the Crank–Nicolson form

$$-rH_{i+1,k+1} + (2+2r)H_{i,k+1} - rH_{i-1,k+1} = rH_{i+1,k} + (2-2r)H_{i,k} + rH_{i-1,k} \qquad (2.26)$$

with error $O(\Delta y^2 + \Delta t^2)$. This algorithm is a popular one because of its stability and higher accuracy, which effectively means the ability to use a larger time step for the same accuracy.

A relatively large time step in a nonlinear problem with r a function of H requires at least one step of an iteration process, and possibly more, to modify r_i before moving forward by Δt. Bearing in mind that the Crank–Nicolson algorithm for $H_{i,k+1}$ (eqn (2.26)) is symmetrically centred about the intermediate node $(i, k+\tfrac{1}{2})$, the value of $H_{i,k+\frac{1}{2}}$ is first found by a half-step Crank–Nicolson process using the existing value β of the nonlinear parameter at each node, so that

$$r_{i,k} = \frac{\Delta t}{2\beta\,\Delta y^2} \qquad (2.27)$$

Once $H_{i,k+\frac{1}{2}}$ is known, the new parameter to be used in the main step forward is

$$\beta' = \beta(H_{i,k+1/2}) \qquad (2.28)$$

and

$$r_{i,k} = \frac{\Delta t}{\beta' \Delta y^2} \tag{2.29}$$

Implicit equations have the disadvantage of needing simultaneous solution of all M unknowns at each time step, and each half time step in the Crank–Nicolson scheme with nonlinear coefficients. Explicit algorithms, if convergent and stable for relatively large values of r, would be an obvious advantage. One well-known explicit method is that of DuFort and Frankel [2.15] which simply replaces $H_{i,k}$ in the basic explicit scheme of eqn (2.20) by the average of the $i, k-1$ and $i, k+1$ values, i.e.

$$H_{i,k} = \frac{H_{i,k+1} + H_{i,k-1}}{2} \tag{2.30}$$

The DuFort–Frankel algorithm is thus a 3-step one of the form

$$H_{i,k+1} = H_{i,k-1} + \frac{2r}{2r+1}(H_{i+1,k} - 2H_{i,k-1} + H_{i-1,k}) \tag{2.31}$$

where $r = r_{i,k}$ is still the value of the parameter at node i, k. The local truncation error is $O(\Delta y^2 + \Delta t^2 + \Delta t^2/\Delta y^2)$ where $\Delta t/\Delta y$ must tend to zero as Δy and Δt separately tend to zero. In practice this means that

$$\frac{\Delta t}{\Delta y} \ll \sqrt{\beta} \tag{2.32}$$

otherwise the term $(\Delta t/\Delta y)^2 \partial^2 H/\partial t^2$ is effectively being added to the right-hand side of the original diffusion equation (eqn (2.17)) making it an hyperbolic equation.

A brief investigation will show that the Crank–Nicolson scheme can be derived from the application of the simple explicit and implicit eqns (2.20) and (2.22) on alternative time rows and then shrinking this double time-step process to one step. In a similar way, the DuFort–Frankel method is equivalent to the application of eqns (2.20) and (2.22) to alternate nodes. Thus if the explicit equation is first used to step forward from alternate nodes on row k to row $k+1$, the application of the implicit equation to the remaining nodes on row $k+1$ is an explicit process because at node $(i, k+1)$, for example, the flanking values at nodes $(i-1, k+1)$ and $(i+1, k+1)$ have already been computed. Not only this, but immediate substitution of $H_{i,k+1}$ into the explicit equation for the next step forward to time $k+2$ yields

$$H_{i,k+2} = 2H_{i,k+1} - H_{i,k} \tag{2.33}$$

and, after treatment of all nodes in this way, we are ready to use the implicit equation again on row $k+2$. Thus, provided the intermediate nodal values are not required, there need only be the cyclic application of eqn (2.22) (in explicit form) and eqn (2.33), and the computation involved is reduced by about 30%. It is clear that, regarding the space-time domain as a chess board, only nodes of one colour need be recorded and this led to the name Hopscotch for this algorithm when originally proposed by Gourlay [2.20].

Unfortunately, some circumstances may make it necessary to return to the original DuFort–Frankel form of the algorithm but the separation of the nodes into two interlaced sets, black and white, is an important concept because it is a warning not to mix the sets when calculating the coefficients β. Because β at the black node (i, k), for example, is used to obtain H at the white node $(i, k+1)$, it must be a function of the average of $H_{i+1,k}$ and $H_{i-1,k}$, not of $H_{i,k}$.

It is worth commenting on this, at first, surprising fact that two methods (Crank–Nicolson and DuFort–Frankel) so closely based on the dual application of simple algorithms with truncation error $O(\Delta y^2 + \Delta t)$ should yield an increased second-order accuracy in time (Δt^2). The reason is that the simple explicit method approaches the true solution from below and the simple implicit method from above [2.38], i.e. the fundamental components of the error are opposite in sign.

2.4 CONVERGENCE, COMPATIBILITY AND STABILITY

Sufficient reference has been made to the local truncation error which is defined as the difference between the partial differential equation and its approximating difference equation at any given point. The *discretization* error on the other hand is the difference between the *exact solution u* of the differential equation and the *exact solution* of the difference equation U. The scheme is said to converge if U tends to u at all nodes as the space and time steps tend to zero, sometimes subject to a link between the latter (e.g. as in the case of the DuFort–Frankel algorithm).

Because convergence is difficult to investigate, the simpler concept of *compatibility* (or *consistency*) is often used. The difference equation is usually said to be compatible with the differential equation if the limiting value of the local truncation error is zero as the space and time steps tend to zero. A very important theory propounded by Lax [2.59] states that, given a properly posed linear initial-value problem and a linear finite-difference approximation to it satisfying the compatibility condition, *stability* is the necessary and sufficient condition for convergence. Thus it may not be necessary to conduct the harder examination of convergence directly.

Stability has to do with the growth, or strictly the lack of growth, of errors. All numerical processes generate errors, e.g. rounding errors, and the actual computed solution of the difference equation N differs from the exact solution U, so that the total error at node (i, k) say is

$$u_{i,k} - N_{i,k} = (u_{i,k} - U_{i,k}) + (U_{i,k} - N_{i,k}) \tag{2.34}$$

where the first term on the right-hand side is the discretization error and the second term is the global rounding error R. In contrast to the discretization error, the rounding error does not decrease as the mesh lengths decrease, rather it will get worse because more arithmetic operations are involved. Digital computers hold numbers to many significant figures and so rounding errors are basically small. Nevertheless, for fixed values of the small mesh lengths, it is possible for the time-stepping process to allow the errors to swamp the true solution.

One technique for investigating the behaviour of errors is the Fourier stability method where a space row (or space grid if the problem is two-dimensional) of errors is inserted in the form of a Fourier series with as many terms as there are nodes in the row. In a linear system it is sufficient to consider one harmonic only (because superposition can be used) and consider its propagation as time t increases. Consider the nth harmonic, of unit magnitude for convenience, so that the error E at $t = 0$ at node $(i, 0)$ is, from eqn (2.7),

$$E_{i,0} = \mathrm{e}^{\mathrm{j}2\pi ni/M} = \mathrm{e}^{\mathrm{j}\lambda ih} \tag{2.35}$$

where $\lambda = 2\pi n/Mh$ and $h = \Delta x$ for brevity. The errors that are now carried forward in time are subject to the same difference equation as the exact solution U. If we suppose that the method of separation of variables can be applied to the difference equation, it is possible to write E at any node (i, k) as

$$E_{i,k} = \mathrm{e}^{\mathrm{j}\lambda ih}\,\mathrm{e}^{\gamma kp} = \mathrm{e}^{\mathrm{j}\lambda ih}\,[\mathrm{e}^{\gamma p}]^{k} \tag{2.36}$$

where $p = \Delta t$ for brevity and γ is a constant. The error will not increase with time, i.e. k, provided

$$|\mathrm{e}^{\gamma p}| \leqslant 1 \tag{2.37}$$

As an example, let us submit these error functions to the simple explicit equation. Substituting from eqn (2.36) into eqn (2.20) and dividing by $E_{i,k}$ we obtain

$$\begin{aligned} \mathrm{e}^{\gamma p} &= r\mathrm{e}^{\mathrm{j}\lambda h} + (1 - 2r) + r\mathrm{e}^{-\mathrm{j}\lambda h} \\ &= 2r\cos\lambda h + (1 - 2r) \\ &= 1 - 2r(1 - \cos\lambda h) = 1 - 4r\sin^2\left(\frac{\lambda h}{2}\right) \end{aligned} \tag{2.38}$$

Since r is positive, $\mathrm{e}^{\gamma p}$ is always less than unity. There are three conditions possible for the error solution (eqn 2.36):

(1) if $0 \leqslant \mathrm{e}^{\gamma p} < 1$, the error decays steadily as $k \to \infty$;
(2) if $-1 < \mathrm{e}^{\gamma p} < 0$, the error decays in magnitude but oscillates in sign;
(3) if $\mathrm{e}^{\gamma p} < -1$, the error oscillates with increasing magnitude, and the solution is unstable.

The explicit equation is thus stable if $\mathrm{e}^{\gamma p} \geqslant -1$, or $1 - 4r\sin^2(\lambda h/2) \geqslant -1$ so that

$$r \leqslant [2\sin^2(\lambda h/2)]^{-1} \tag{2.39}$$

The sine term cannot exceed unity, at which point $r \leqslant \tfrac{1}{2}$ as stated in Section 2.3.

Consider now the DuFort–Frankel algorithm. Writing eqn (2.31) in terms of the error functions (eqn (2.36)), dividing by $E_{i,k}$ and putting the factor $2r(2r+1)^{-1}$ as α for brevity, we obtain

$$e^{\gamma p} - e^{-\gamma p} = \alpha\left(e^{j\lambda h} - 2e^{-\gamma h} + e^{-j\lambda h}\right) \tag{2.40}$$

Let $K = e^{\gamma p}$ so that eqn (2.40) reduces to the quadratic

$$K^2 - 2\alpha K \cos \lambda h + (2\alpha - 1) = 0 \tag{2.41}$$

with the solution

$$K = \alpha \cos \lambda h \pm \left[\alpha^2 \cos^2 \lambda h - (2\alpha - 1)\right]^{1/2} \tag{2.42}$$

If, as stated in Section 2.3, the algorithm is unconditionally stable, r can take any positive value. When r is large α tends to unity from below, and in the limit

$$K = \cos \lambda h \pm j \sin \lambda h \tag{2.43}$$

Thus $K \leqslant 1$ as required (see eqn (2.37)).

The Fourier stability method is the least rigorous of several techniques of stability analysis, not least because boundary conditions are ignored. For a more detailed survey the reader is referred to excellent books by Smith [2.41] and Ames [2.2]. In particular, the stability and convergence of the DuFort–Frankel algorithm has been investigated in detail for some one- and two-dimensional nonlinear magnetic field problems by Wiak [2.53–2.56].

2.5 CONVERGENCE OF NONLINEAR DIFFUSION EQUATIONS

The detailed analysis of Wiak on the DuFort–Frankel scheme mentioned at the end of Section 2.4 is beyond the scope of this book, but it is worth attempting two things briefly here. First to list various forms of the one-dimensional diffusion equation in terms of three selected dependent variables H_z, E_x and A_x (using the same model configuration as in Section 2.3) with the magnetic or electric material parameter as a variable (but excluding dielectric materials). Secondly, to examine the convergence of a nonlinear version of the basic explicit algorithm (eqn (2.20)) as the simplest example of the difficulties to be faced if the question of convergence is to be addressed theoretically.

From eqns (1.1), (1.2), (1.7), (1.8) and (1.43) we have the one-dimensional first-order equations,

$$\frac{\partial H_z}{\partial y} = J_x = \sigma(E) E_x \tag{2.44}$$

$$\frac{\partial E_x}{\partial y} = \frac{\partial B_z}{\partial t} = \frac{dB}{dH}\frac{\partial H_z}{\partial t} = \mu_d(H)\frac{\partial H_z}{\partial t} \tag{2.45}$$

$$\frac{\partial A_x}{\partial y} = -B_z = -\mu(H)H_z \tag{2.46}$$

The permeability expressed as some suitable function of H can represent either a single-valued magnetisation curve or an hysteresis loop. The conductivity σ expressed as a function of E models one form of nonlinear conductor. Other possibilities include variation with temperature, which would require the coupling of the thermal and electromagnetic systems in some form of interactive hybrid scheme, or the addition of H to give $\sigma(E, H)$ as proposed in a recent model of a high-temperature ceramic superconductor [2.34].

Let us now assume σ to be a constant and differentiate eqns (2.44) to (2.46) in turn with respect to y and isolate the corresponding dependent variable H_z, E_x and A_x by suitable combination and manipulation, to yield

$$\frac{\partial^2 H_z}{\partial y^2} = \sigma\mu_d(H)\frac{\partial H_z}{\partial t} \tag{2.47}$$

$$\frac{\partial^2 E_x}{\partial y^2} = \sigma\mu_d(H)\frac{\partial E_x}{\partial t} + \frac{\sigma}{\mu_d(H)}\frac{\partial \mu_d(H)}{\partial H_z}E_x\frac{\partial E_x}{\partial y} \tag{2.48}$$

$$\frac{\partial}{\partial y}\left[\frac{1}{\mu(H)}\frac{\partial A_x}{\partial y}\right] = -\sigma E_x = \sigma\frac{\partial A_x}{\partial t} \tag{2.49}$$

since, from eqns (1.2) and (1.43), $E_x = -\partial A_x/\partial t$. Equation (2.47) is identical to eqn (2.15) already introduced in the simple model problem of Section 2.3.

Alternatively with $\mu = \mu_d = \text{constant}$, and σ a variable:

$$\frac{\partial}{\partial y}\left[\frac{1}{\sigma(E)}\frac{\partial H_z}{\partial y}\right] = \mu\frac{\partial H_z}{\partial t} \tag{2.50}$$

$$\frac{\partial^2 E_x}{\partial y^2} = \sigma(E)\mu\frac{\partial E_x}{\partial t} + \mu E_x\frac{\partial \sigma(E)}{\partial t} \tag{2.51}$$

$$\frac{\partial^2 A_x}{\partial y^2} = \sigma(E)\mu\frac{\partial A_x}{\partial t} \tag{2.52}$$

The mathematical similarity between eqns (2.47) and (2.52), and eqns (2.49) and (2.50) is clearly seen. The type of boundary condition will be an additional consideration and may suggest one particular dependent variable, but, apart from this, there is an obvious advantage in using equations of the simplest form (eqns (2.47) and (2.52)).

Consider now the general form of eqns (2.47)

$$u_t = \Phi(u, u_{yy}) \tag{2.53}$$

where the suffix denotes differentiation with respect to t or y, and we assume

$$\frac{\mathrm{d}\Phi}{\partial u_{yy}} \geqslant c > 0 \tag{2.54}$$

$$\left|\frac{\partial \Phi}{\partial u}\right| + \frac{\partial \Phi}{\partial u_{yy}} \leqslant d \tag{2.55}$$

What follows is a simplified version of a proof of convergence based on the maximum principle analysis and given in Ames [2.2]. The simple explicit algorithm for eqn (2.53) is

$$U_{i,k+1} = U_{i,k} + \Delta t \Phi\left(U_{i,k}, \frac{\partial_y^2 U_{i,k}}{\Delta y^2}\right) \tag{2.56}$$

where

$$\partial_y^2 U_{i,k} = U_{i+1,k} - 2U_{i,k} + U_{i-1,k} \tag{2.57}$$

On the other hand, from eqn (2.53)

$$\begin{aligned} u_{i,k+1} &= u_{i,k} + \Delta t\ u_t|_{i,k} + \mathrm{O}(\Delta t^2) \\ &= u_{i,k} + \Delta t \Phi\left(u_{i,k}, \frac{\delta_y^2 u_{i,k}}{\Delta y^2}\right) + \mathrm{O}(\Delta t^2 + \Delta t \Delta y^2) \end{aligned} \tag{2.58}$$

Subtracting eqn (2.56) from eqn (2.58) the error function is given by

$$E_{i,k+1} = E_{i,k} + \Delta t \Delta \Phi + \mathrm{O}(\Delta t^2 + \Delta t \Delta y^2) \tag{2.59}$$

where

$$\Delta \Phi = \frac{\partial \Phi}{\partial u} E_{i,k} + \frac{\partial \Phi}{\partial u_{yy}}\left(\frac{\delta_y^2 E_{i,k}}{\Delta y^2}\right) \tag{2.60}$$

Thus, using eqn (2.57)

$$\begin{aligned} E_{i,k+1} = {} & \frac{\Delta t}{\Delta y^2}\frac{\partial \Phi}{\partial u_{yy}} E_{i+1,k} + \left(1 + \Delta t \frac{\partial \Phi}{\partial u} - \frac{2\Delta t}{\Delta y^2}\frac{\partial \Phi}{\partial u_{yy}}\right) E_{i,k} \\ & + \frac{\Delta t}{\Delta y^2}\frac{\partial \Phi}{\partial u_{yy}} E_{i-1,k} + \mathrm{O}(\Delta t^2 + \Delta t \Delta y^2) \end{aligned} \tag{2.61}$$

All the coefficients of eqn (2.61) are positive if

$$1 + \Delta t \frac{\partial \Phi}{\partial u} - \frac{2\Delta t}{\Delta y^2}\frac{\partial \Phi}{\partial u_{yy}} \geqslant 0 \tag{2.62}$$

or, using eqn (2.55), if

$$1 - \Delta t\ d - \frac{2\Delta t}{\Delta y^2} d \geqslant 0 \tag{2.63}$$

giving the condition,

$$\frac{\Delta t}{\Delta y^2} \leqslant \frac{1 - \Delta t\ d}{2d} = \frac{1}{2}\left(\frac{1}{d} - \Delta t\right) \approx \frac{1}{2d} \tag{2.64}$$

Subject to the condition expressed in eqn (2.64), eqn (2.61) can be written in terms of the absolute values of the error function with $|E_{i,k+1}|$ less than or equal to the right-hand side. If $\|E_k\|$ is the maximum of the absolute values of $E_{i,k}$ on time row k, then

$$\begin{aligned}\|E_{k+1}\| &\leqslant \left(1 + \Delta t \frac{\partial \Phi}{\partial u}\right) \|E_k\| + \Delta t A(\Delta t + \Delta y^2) \\ &\leqslant (1 + \Delta t d)\|E_k\| + \Delta t A \end{aligned} \tag{2.65}$$

where $A(\Delta t + \Delta y^2)$ is some small quantity of order Δt and Δy^2. But $\|E_0\| = 0$ and so, from eqn (2.65)

$$\begin{aligned} \|E_1\| &\leqslant \Delta t A \\ \|E_2\| &\leqslant (1 + \Delta t d)\Delta t A + \Delta t A \\ \|E_{k+1}\| &\leqslant [1 + (1 + \Delta t d) + (1 + \Delta t d)^2 + \cdots + (1 + \Delta t d)^k]\Delta t A \\ &= \frac{(1 + \Delta t d)^k - 1}{(1 + \Delta t d) - 1}\Delta t A = [(1 + \Delta t d)^k - 1]\frac{A}{d} \end{aligned} \tag{2.66}$$

The error function thus tends to zero as Δt and Δy tend to zero. Given eqns (2.54) and (2.55), this nonlinear form of the simple explicit equation is convergent if condition (2.64) is satisfied.

It may be helpful to look at a specific case of nonlinearity. The modified Frolich representation of the magnetization curve is sometimes used, whereby

$$B = \mu_0 H + \frac{H}{a + b|H|} \tag{2.67}$$

giving, from eqn (2.16),

$$\beta = \sigma \frac{\mathrm{d}B}{\mathrm{d}H} = \sigma\mu_0 + \frac{\sigma a}{(a + b|H|)^2} \tag{2.68}$$

Writing eqn (2.47) in the form of eqn (2.53)

$$\frac{\partial H}{\partial t} = \Phi = \frac{1}{\beta}\frac{\partial^2 H}{\partial y^2} \tag{2.69}$$

so that

$$\frac{\partial \Phi}{\partial H_{yy}} = \frac{1}{\beta} \tag{2.70}$$

and, with $\alpha = a + b|H|$ for brevity,

$$\frac{\partial \Phi}{\partial H} = \frac{\partial^2 H}{\partial y^2}\frac{\partial}{\partial H}\left(\frac{1}{\beta}\right) = \frac{\partial^2 H}{\partial y^2}\frac{2ab\alpha}{\sigma(a + \mu_0\alpha^2)^2} = \frac{\partial H}{\partial t}\frac{2ab}{\alpha(a + \mu_0\alpha^2)} \tag{2.71}$$

we have, from eqn (2.55),

$$d \geqslant \frac{1}{(a + \mu_0\alpha^2)}\left\{\left|\frac{\partial H}{\partial t}\right|\frac{2ab}{\alpha} + \frac{\alpha^2}{\sigma}\right\} \tag{2.72}$$

If we assume that the maximum value of $|\partial H/\partial t|$ is ωH_s (as for sinusoidal variation with time with peak value H_s) and take typical values of the constants in eqn (2.67) as $a = 600$ and $b = 0.6$, we can examine the condition on the parameter d for different values of magnetic field. For very low H, $\alpha \approx a$, and with a good conductor ($\sigma > 10^6$) the term α^2/σ in eqn (2.72) is negligible. Also $a + \mu_0\alpha^2 \approx a$, so that $d \geqslant 2\omega|H|b/a$. With 50 Hz excitation this means that $d \geqslant 0.628|H|$. The maximum value of H to reasonably satisfy the condition $\alpha \approx a$ is $50\,\mathrm{A\,m^{-1}}$, giving $d_{min} = 31.4$, so that eqn (2.64) becomes $\Delta t \leqslant \Delta y^2/62.8$ and Δt may have to be extremely small, depending on the space discretization necessary. For values of H higher than about $50\,\mathrm{A\,m^{-1}}$, eqn (2.72) must be examined in more detail. However, for very high magnetic field strength, $\alpha \approx b|H|$ so that $(a + \mu_0\alpha^2) = (a + \mu_0 b^2|H|^2)$ and both terms in the factor may be important. As an example, suppose $|H|$ is such that $\mu_0 b^2|H|^2 = a$, then $d \geqslant [|\partial H/\partial t|/|H| + (2\sigma\mu_0)^{-1}]$. One problem is evident; namely that of finding a representative value of the ratio $|\partial H/\partial t|/|H|$. Although both maximum values do not occur simultaneously or everywhere in the plate, it is probably reasonable to take $|\partial H/\partial t|/|H| = \omega H_s/H_s = \omega$, and, since the factor containing σ now becomes negligible in contrast, we have $d \geqslant \omega = 100\pi$. As a result, the modified Frolich curve does not appear to be very promising when used with the simple explicit equation. However, it must be remembered that the maximum analysis of convergence is very pessimistic in nature. The form of the nonlinearity is also important. For example, taking the curve represented by $B = aH^{1/n}$, where n is an integer, a wide range of H from low values upwards can be shown to require $d \geqslant \omega(1 - n^{-1})$.

2.6 FINITE-DIFFERENCE METHODS APPLIED TO MAGNETIC FIELD PROBLEMS

There are numerous examples in the literature of the application of finite-difference methods to magnetic field problems. The categories of problems can be listed as follows:

(1) one- and two-dimensional (in space) solutions of the diffusion equation in various forms, with nonlinear coefficients and/or transient or periodic nonsinusoidal excitation;
(2) two-dimensional solutions of the Helmholtz equation, where the dependent variable is complex. By assumption this class is linear with sinusoidal excitation;
(3) two- or three-dimensional solutions of Laplace's or Poisson's equations.

We will consider some of these possibilities in turn with examples.

2.6.1 Steady-state eddy currents in ferromagnetic plates, laminations or cylinders

This type of problem was introduced in Section 2.3 where the applied magnetic field is both uniform and parallel to the surfaces of a flat plate. With an assumed constant permeability, the only solutions of interest involve transient excitation or periodic excitation with some complex non-sinusoidal form. Otherwise an analytical solution is straightforward.

Here we are primarily concerned with the nonlinearity introduced by the magnetization curve of the material. Most examples in the literature consider a simple sinusoidally varying field, $H_s \sin \omega t$, applied parallel to the surface of a plate or parallel to the axis of a cylinder. Abrams and Gillott [2.1], for example, consider the case of a long steel rod, and include the effect of hysteresis.

The one-dimensional cylindrical form of the diffusion equation in H_z is

$$\frac{\partial^2 H_z}{\partial r^2} + \frac{1}{r}\frac{\partial H_z}{\partial r} = \sigma\mu_d \frac{\partial H_z}{\partial t} \tag{2.73}$$

The simple explicit finite-difference form of eqn (2.73) is

$$\frac{H_{i+1,k} - 2H_{i,k} + H_{i-1,k}}{\Delta r^2} + \frac{H_{i+1,k} - H_{i-1,k}}{2r\Delta r} = \sigma\mu_d \frac{H_{i,k+1} - H_{i,k}}{\Delta t} \tag{2.74}$$

At the centre of the rod, there is a problem with the second term on the left-hand side of eqns (2.73) and (2.74) as r tends to zero. However, from Maclaurin's expansion

$$H'(r) = H'(0) + rH''(0) + \cdots \tag{2.75}$$

and, therefore, since $H'(0) = 0$ (i.e. $\partial H/\partial r = 0$ at $r = 0$), we have $H'(r)/r$ tending to $H''(0)$ as r tends to zero. Thus eqn (2.73) becomes, for $r = 0$

$$2\frac{\partial^2 H_z}{\partial r^2} = \sigma\mu_d \frac{\partial H_z}{\partial t} \tag{2.76}$$

and eqn (2.74) is

$$\frac{4(H_{i+1,k} - H_{i,k})}{\Delta r^2} = \sigma\mu_d \frac{H_{i,k+1} - H_{i,k}}{\Delta t} \tag{2.77}$$

It is interesting that Abrams and Gillott [2.1] appear to have succeeded in using the simple explicit method. At about the same time Poritsky and Butler [2.32], considering both thick flat slabs and thin plates, found some instabilities and turned to the simple implicit scheme (see eqn (2.22)), but with an iterative solution at each time step based on an initial set of new values extrapolated from the two previous time rows.

The most general case of periodic H boundary conditions at the surfaces of a ferromagnetic plate is:

at $y = 0$

$$H_z = H_0 + \sum_{l=1}^{q} H_l \sin(\omega l t + \phi_l) \tag{2.78}$$

at $y = 2b$

$$H_z = H_0' + \sum_{l=1}^{q} H_l' \sin(\omega l t + \phi_l') \tag{2.79}$$

where H_0 and H_0' are constant d.c. field components. At $t = 0$, $0 < y < 2b$, the initial value of H_z is usually set to zero.

The lack of symmetry indicated by the prime notation allows a component of imposed transport current to be present, e.g. if $H_l' = -H_l$ we have the situation of imposed current only and no ambient magnetic field. In finite-difference form the equivalent of eqn (2.78) for the surface $y = 0$ is

$$H_{0,k} = H_0 + \sum_{l=1}^{q} H_l \sin(\omega l k \Delta t + \phi_l) \tag{2.80}$$

and similarly for $H_{2M,k}$ on the surface $y = 2b$, where $M = b/\Delta y$. Only in the two symmetrical situations can the finite-difference solution be confined to half the plate. For odd symmetry ($H_l' = -H_l$) we have $H_z = 0$, i.e. $H_{M,k} = 0$ at $y = b$, and for even symmetry ($H_l' = H_l$) $\partial H_z/\partial y = 0$, giving $H_{M+1,k} = H_{M-1,k}$ so that the equation for $H_{M,k+1}$ can be rewritten with node $M + 1$ eliminated. For example, the DuFort–Frankel equation (eqn (2.31)) becomes

$$H_{M,k+1} = H_{M,k-1} + \frac{2r}{2r+1}(2H_{M-1,k} - 2H_{M,k-1}) \tag{2.81}$$

The alternative to an imposed surface magnetisation is an imposed flux per unit width of the plate. If the excitation winding producing the magnetic flux in the plate or stack of laminations is connected to a sinusoidally varying *emf* source, Faraday's law indicates that the total flux in the plate and the electric field on the surface of the plate are also sinusoidal. Since, by symmetry,

$$2E_s = -\frac{\partial \Phi}{\partial t} = -\frac{\partial}{\partial t}(\hat{\Phi} \sin \omega t) \tag{2.82}$$

where E_s is the instantaneous value of the surface electric field, we have

$$E_s = -\tfrac{1}{2}\omega \hat{\Phi} \cos \omega t \tag{2.83}$$

But

$$J_x = \sigma E_x = \frac{\partial H_z}{\partial y} \tag{2.84}$$

and so we have the surface condition on $y = 0$ in finite-difference form

$$\frac{H_{1,k} - H_{-1,k}}{2\Delta y} = -\frac{1}{2}\omega\sigma\hat{\Phi}\cos(\omega k\Delta t) \tag{2.85}$$

The fictitious node $(-1, k)$ in the finite-difference equation for $H_{0,k+1}$ can therefore be eliminated. At the centre of the plate, node $(M, k+1)$, an equation of the form of eqn (2.81) still holds because $E_x = 0$.

The surface, and possibly the interior, values of the electric field are also of interest when surface H is specified. Three-point polynomial fits, with error $O(\Delta y^2)$, for the surface nodes are given by [2.11]

$$E_{0,k} = \frac{-3H_{0,k} + 4H_{1,k} - H_{2,k}}{2\sigma\Delta y} \tag{2.86}$$

$$E_{2M,k} = \frac{H_{2M-2,k} - 4H_{2M-1,k} + 3H_{2M,k}}{2\sigma\Delta y} \tag{2.87}$$

In the symmetrical cases, with only half the plate computed, $E_{2M,k} = \pm E_{0,k}$ as appropriate.

The most important use of the surface electric field is in the calculation of the total loss, both ohmic and hysteresis, per unit surface area of the plate. From Poynting's theorem, the loss is obtained by integrating the product $(\mathbf{E} \times \mathbf{H}) \cdot \mathbf{n}$ on both surfaces of the plate over a complete period T of the time cycle, where $\mathbf{n}$ is the unit vector normal to the surface of the plate and outwardly directed. Thus $\mathbf{n} = -\mathbf{j}$ on the surface $y = 0$ and $\mathbf{n} = \mathbf{j}$ on $y = 2b$, giving

$$\begin{aligned} P &= \frac{1}{T}\int_t^{t+T} \{[\hat{\mathbf{i}}E_x(0) \times \hat{\mathbf{k}}H_z(0)] \cdot (-\hat{\mathbf{j}}) + [\hat{\mathbf{i}}E_x(2b) \times \hat{\mathbf{k}}H_z(2b)] \cdot \hat{\mathbf{j}}\}\, \mathrm{d}t \\ &= \frac{1}{T}\int_t^{t+T} [E_x(0)H_z(0) - E_x(2b)H_z(2b)]\, \mathrm{d}t \end{aligned} \tag{2.88}$$

Once the solution, in terms of the nodal values of H has converged, which will take several cycles, one final cycle can be computed during which time the surface nodal values of E can be obtained from eqns (2.86) and (2.87), and the loss obtained by numerical integration using either the trapezoidal rule or Simpson's rule. In terms of the latter, the first part of eqn (2.88) becomes, with $T = N\Delta t$,

$$\begin{aligned} P_1 = \frac{1}{3N}(&E_{0,n}H_{0,n} + 4E_{0,n+1}H_{0,n+1} + 2E_{0,n+2}H_{0,n+2} \\ &+ 4E_{0,n+3}H_{0,n+3} + \cdots + 4E_{0,N+n-1}H_{0,N+n-1} + E_{0,N+n}H_{0,N+n}) \end{aligned} \tag{2.89}$$

where n is the initial value of the time integer k for the final cycle.

The solution marches forward from the initial nodal values of $H_{i,0} = 0$ at $k = 0$ and progresses through a numerical transient that may also be close to the real transient if the initial conditions are realizable. However, the important requirement is convergence to the steady state, which can be checked at the end of each cycle by a simple test comparing $H_{i,k}$ with $H_{i,k-N}$. For example if

$$|H_{i,k} - H_{i,k-N}| < \varepsilon \qquad (2.90)$$

for all values of i, where ε is some small positive quantity, the solution has converged. ε can be found by numerical experiment and will depend on the results being sought.

2.6.2 Treatment of magnetic nonlinearity and hysteresis

The main, indeed the only, difficult problem with these one-dimensional solutions concerns the variable coefficients $r_{i,k}$ that contain the differential permeability μ_d. Ferromagnetic materials are subject to hysteresis (see Section 1.5). Strictly speaking therefore, the hysteresis loops need to be modelled in some way, although the single valued magnetization curve is often used when modelling hard materials such as silicon steel laminations. The additional problem is that, with cyclic magnetization, the operating loop is a function of the peak value of B or H. Clearly it is not possible to store a large number of loops and so some means of interpolating between loops is required.

There are two forms of major hysteresis loop. The first is the overall loop formed by plotting the average flux density in a plate or lamination against the surface value of magnetic field strength for a given fundamental frequency. This is known as the dynamic loop because it contains the effect of eddy currents, and can only be obtained once the full numerical solution is known. The second type is the static hysteresis loop relating local values of B and H and obtained by measuring successive small changes in B produced by small and slow changes in H (see Section 1.5). Clearly, since domain wall movements are involved, the peak to peak swings in H must be slow enough not to produce significant additional microscopic eddy currents induced by rapid domain wall movement. In the numerical solution it is necessary to neglect the latter and assume that the static loops apply within the material. As the result of normal macroscopic eddy-current screening the peak value of H will decrease with depth from the surface and a different loop will apply at each level.

Zakrzewski and Pietras [2.62] describe a method in which, starting with the initial condition $H_z = 0$ at all nodes, the solution moves up the peak magnetization curve from the demagnetized state. This curve is found experimentally by joining the tips of a set of major hysteresis loops, and is almost identical to the normal initial magnetization curve. Once the local maximum value of $H_{i,k}$ has been reached, the solution moves back down the upper branch of the appropriate major hysteresis loop. When the maximum negative value of $H_{i,k}$ occurs, corresponding to some value B_m of flux density, the operating point is moved horizontally across to the reversal peak magnetization curve (i.e. selecting the corresponding value B_m, H_m on the latter curve)

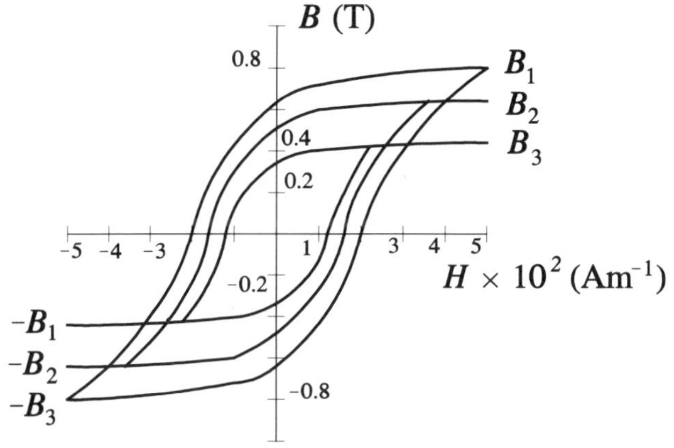

Fig. 2.2 *Illustration of the upper and lower branches used in the computation process to form a set of major hysteresis loops*

in order to join the lower branch of another major hysteresis loop. The same process repeats when the maximum positive value of $H_{i,k}$ is obtained. After a few cycles the value of H at each node will be found to be following the symmetrical major loop appropriate to the peak field strength achieved at that node. This numerical transient is similar to the actual transient process that might occur in practice.

As already mentioned, only a few major hysteresis loop branches can be stored as tables or files of H_p, B_p number pairs, where p represents the entry number. However, the approximate profile of any intermediate branch can be obtained by linearly interpolating according to the ratio

$$R = \frac{B_m - B_u}{B_{u-1} - B_u} \tag{2.91}$$

where B_u and B_{u-1} are the peak values of flux density of the branches on either side of B_m. In order to find R the stored H, B entries in each branch file must be artificially extended to the maximum value of H for the largest hysteresis loop of the complete set, as illustrated in Fig. 2.2. Assuming that the current value of $H_{i,k}$ corresponds to an entry, say H_p, in the files, the required value of B is given by

$$B = B_{u,p} + R(B_{u-1,p} - B_{u,p}) \tag{2.92}$$

where $B_{u,p}$ is the pth flux density entry in the file for branch u. In practice, $H_{i,k}$ will not match a value H_p exactly, but to obtain $\mu_d(=\Delta B/\Delta H)$ we actually require the values of B and H on either side of the point corresponding to $H_{i,k}$.

Further details of the technique are given by Zakrzewski and Pietras [2.62] who also show two examples; one for a thick slab in which hysteresis is important at low excitation levels and one for high excitation in a thin lamination where hysteresis loss may be comparatively important because of the

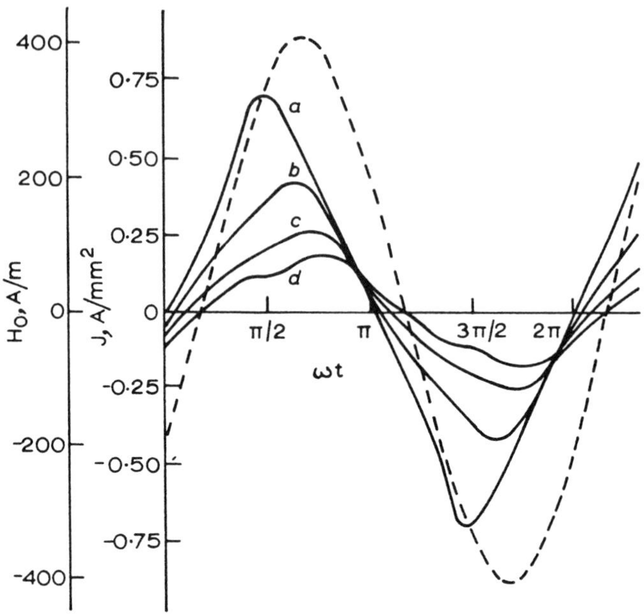

Fig. 2.3 *Eddy-current density waveforms at different depths below the surface of a large flat solid iron slab with a peak surface field of 410 A m^{-1}: (a) 0 mm (b) 0·25 mm (c) 0·50 mm (d) 0·75 mm*

reduced eddy-current flow (and loss) due to saturation. Figure 2.3 illustrates the eddy-current density as a function of depth from the surface of a slab of solid iron on which the peak applied field is 410 A m^{-1}. The frequency is 50 Hz and conductivity 5.4×10^6 S m^{-1}. $\Delta y = 0.125$ mm and time steps in the region of 80 μs were used. The computed surface density agrees well with measurement on the surface of a mild steel cylinder inside a solenoid. The computed total loss per unit surface area is 20.2 W m^{-2}, whereas without hysteresis the loss is about 25% less.

Other attempts to represent hysteresis include semi-analytical models, e.g. Coulson, Slater and Simpson [2.10] where flux density errors less than 10% are claimed, and Del Vecchio [2.14], both based on the Preisach domain model. On the other hand, Rivas [2.36] is typical of a good fully analytical model using a rational fraction of the form of eqn (2.97) (see below) which has also been employed for single-valued magnetization curves.

In fact it is more usual to neglect hysteresis and work with the peak magnetization curve. The latter can be represented as a single table of B, H points. Any value of B can be found by linear interpolation, i.e. if H lies between the table entries H_p and H_{p+1}, we have

$$B = B_p + \frac{H - H_p}{H_{p+1} - H_p}(B_{p+1} - B_p) \qquad (2.93)$$

Clearly, it may be more convenient to list the pairs (μ, H), (μ_d, H), (μ_d, B),

(ν, B^2), where ν is the reluctivity, depending on the form of the finite-difference equation to be solved. The alternative is an algebraic function expressing B in terms of H. Fischer and Moser [2.17] have surveyed a large number of possibilities, and Trutt, Erdelyi and Hopkins [2.49] examine some in terms of the goodness of fit to actual measured curves. A simple equation that is recommended for materials with a relatively smooth knee to the curve, i.e. transition into saturation, is the modified Frolich equation

$$B = \mu_0 H + \frac{H}{a + b|H|} \tag{2.94}$$

which leads to

$$\mu = \mu_0 + (a + b|H|)^{-1} \tag{2.95}$$

and

$$\mu_d = \mu_0 + a(a + b|H|)^{-2} \tag{2.96}$$

A rather more complicated expression is the rational fraction approximation of Widger [2.58]

$$B = \frac{a_0 + a_1|H| + a_2|H|^2}{1 + b_1|H| + b_2|H|^2} H \tag{2.97}$$

where higher order terms, i.e. in H^3 etc., can be added. However, the preferred form is that proposed by Rivas [2.36] where eqn (2.97) is taken to represent the magnetization vector **M** not **B** (see eqn (1.98)) so that

$$B = \mu_0 H + \frac{a_0 + a_1|H| + a_2|H|^2}{1 + b_1|H| + b_2|H|^2} H \tag{2.98}$$

A short but useful discussion of the treatment of nonlinearity is also given by Krawczyk and Tegopoulos [2.25]. The clear advantage of eqns (2.94) to (2.98) is the limited computation involved. Lim and Hammond [2.28; 2.29], for example, have used the original Frolich formulation (without the term $\mu_0 H$ in eqn (2.94)), together with the DuFort–Frankel algorithm, for extensive studies of the effect of saturation on the eddy-current losses in steel plates.

A much more complex expression which gives an excellent fit for silicon steel laminations with a sharp knee in the B/H curve is

$$B = \mu_0 H + S_1 \tan^{-1}(A_1 H) + S_2 \tan^{-1}(A_2 H) \tag{2.99}$$

where, as a guide, typical values of the constants are $S_1 = 0.987$, $S_2 = 0.323$, $A_1 = 0.030\,8$, and $A_2 = 0.000\,110$.

2.6.3 Transient one-dimensional solutions

Transient solutions of the one-dimensional plate problem are considered by Wiak [2.53; 2.54] who, as already mentioned, pays detailed attention to the DuFort–Frankel algorithm. In reference [2.53] he formulates the problem in terms of B, having chosen to specify the surface value of B in the form $B_m\{1-\exp(-t/\tau)\}$ where B_m is the final constant value and τ the time constant. A small value $\tau = 1$ ms gives something close to a step-function impact excitation, and, especially with a small space step of 0.1 mm for a fairly thin 5 mm plate, requires a very small time step of 1 μs to follow the transient solution accurately. In fact, Wiak specifies the inequalities

$$\left(\frac{\Delta t}{\Delta y}\right) \leqslant 0.01 \qquad \text{and} \qquad \left(\frac{\tau}{\Delta t}\right) \geqslant 1000 \tag{2.100}$$

Not surprisingly, the requirements for a transient solution are more severe than with periodic steady-state excitation where the initial transient is not of interest.

It is important to model the start of a real transient correctly; something that is not possible using a three time-step scheme. Wiak proposes the two-step Saulev algorithm [2.53] which has an asymmetric form with the second derivative approximated by (for the first step from $k=0$ to $k=1$)

$$\frac{\partial^2 B}{\partial y^2} \approx \frac{B_{i+1,0} - B_{i,0} - B_{i,1} + B_{i-1,1}}{\Delta y^2} \tag{2.101}$$

The Saulev scheme is normally used with opposite skew on alternate time rows, but here, from $k=2$ onwards, the DuFort–Frankel equation takes over.

It is not clear why Wiak uses the B formulation, which, although the basic one-dimensional diffusion equation is not listed in Section 2.5, can easily be derived and shown to be more complicated than the equivalent eqn (2.47) for H. After all, in one class of one-dimensional diffusion problem (current driven), surface H is the natural boundary condition. However, in reference [2.54], Wiak examines the same transient solution in the E formulation (see eqn (2.48)). His conclusions about Δt and Δx are the same as in eqn (2.100). For the situation where a plate is immersed in a magnetic field provided by a voltage-driven excitation winding, the specification of E follows from the fact that the surface electric field is related to $\mathrm{d}\Phi/\mathrm{d}t$, so that we have the transient equivalent of eqn (2.82). This fact does not necessarily make the more complicated E equation preferable to the H equation, but it is likely to be so because of the Dirichlet boundary conditions. As already noted in Section 2.6.1, the H formulation here requires the surface Neumann condition $\partial H/\partial y$ $(=\sigma E_x)$. Alternatively, the vector potential formulation (eqn (2.49)) can be used because, from *curl* $\mathbf{A} = \mathbf{B}$,

$$\Phi(t) = \int_0^b B_z \,\mathrm{d}y = -\int_0^b \frac{\partial A_x}{\partial y}\,\mathrm{d}y = A_s(t) \tag{2.102}$$

where A_s is the instantaneous surface value of A_x, and A_x at the centre $y=b$ is assumed to be zero.

Stoll [2.43] gives another example of transient current density patterns at three levels inside a 5 mm steel plate when the surface H field is suddenly

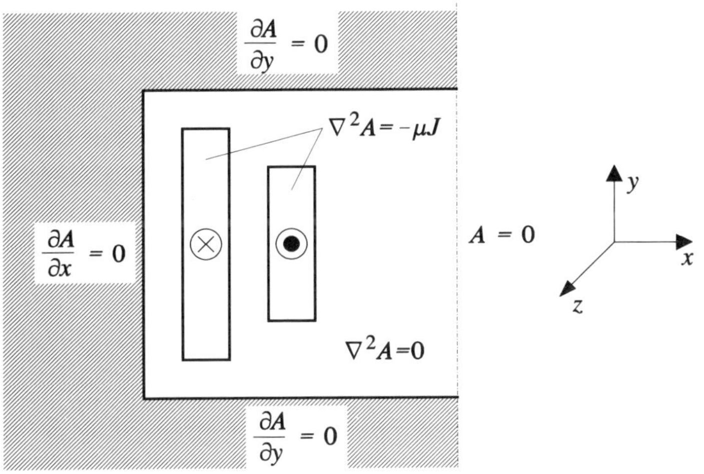

Fig. 2.4 *Low and high voltage coil cross-sections in half the window of a single-phase transformer*

switched on or off. There is an interesting difference in the magnitude of the current response under the two regimes, not due to hysteresis because this is not included, but to magnetic saturation. The current density is always greater at or near the beginning of the transient, but, due to saturation, the rate of fall of flux is less than the unsaturated rate of rise.

2.6.4 Two-dimensional static solutions in multi-region problems

Two-dimensional problems involving Laplace's or Poisson's equations are often amenable to analytical solution (see, for example, the excellent survey by Binns, Lawrenson, and Trowbridge [2.4]) unless the boundary is of complicated shape or there are two or more regions involved, particularly if the latter involve magnetic materials.

The most complicated Poissonian type of multi-region problem solved analytically uses a method devised by Roth, and one transformer example is outlined by Binns *et al.* [2.3; 2.4]. The disadvantage of the method is that the field quantities, and resulting parameters such as the force on a conductor, appear as double Fourier series. Thus the computation involved is substantial. In addition, the conductor cross-sections must be rectangular and the current density uniform. The alternative of a finite-difference or finite-element solution soon becomes attractive.

A typical configuration is given in Fig. 2.4 which shows half the window of a transformer and the cross-section of the low and high voltage coils. The plane of symmetry is shown as a dotted line. The two-dimensional Poissonian equation to be solved is

$$\frac{\partial^2 A_z}{\partial x^2} + \frac{\partial^2 A_z}{\partial y^2} = -\mu_0 J_z(x, y) \tag{2.103}$$

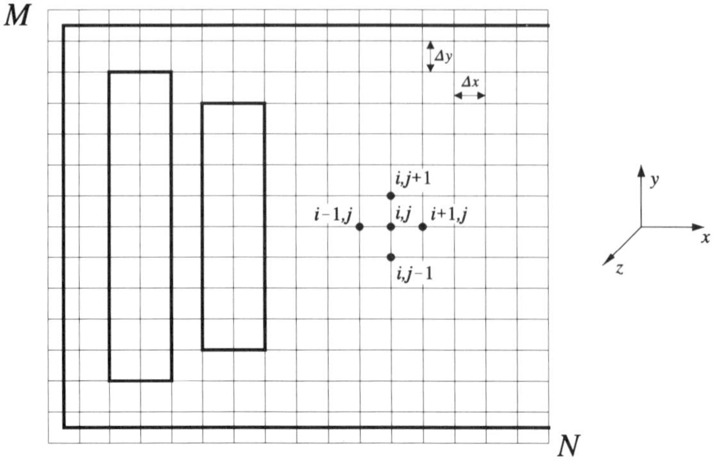

Fig. 2.5 *Finite-difference mesh representing the problem in Fig. 2.4*

the vector potential being an obvious choice of dependent variable because A and J are parallel. From eqn (1.43)

$$B_x = \frac{\partial A_z}{\partial y} \quad \text{and} \quad B_y = -\frac{\partial A_z}{\partial x} \tag{2.104}$$

and the boundary conditions in Fig. 2.4 follow because the transformer core is assumed to be infinitely permeable so that there is no tangential component of magnetic field at the window surfaces. On the plane of symmetry, $B_x = 0$, so that $\partial A/\partial y = 0$ or $A = \text{constant}$. Only the derivatives of A are important and the constant can be taken as zero for convenience.

The finite-difference grid representing Fig. 2.4 is shown in Fig. 2.5. A rectangular mesh is used with a uniform cell of dimensions $\Delta x \times \Delta y$. Referring to eqn (2.13), a finite-difference equivalent of eqn (2.103) can be written immediately as

$$\frac{A_{i+1,j} - 2A_{i,j} + A_{i-1,j}}{\Delta x^2} + \frac{A_{i,j+1} - 2A_{i,j} + A_{i,j-1}}{\Delta y^2} = -\mu_0 J_{i,j} \tag{2.105}$$

where $J_{i,j}$ is the current density associated with node i, j. The nodal current is assumed to flow with uniform density $J_{i,j}$ in a cell of dimension $\Delta x \times \Delta y$ having node i, j as its centre. It is therefore a simple matter to write difference equations for nodes on the flat surface of a conductor, at a corner, or even at an inverted (concave) corner such as occurs with an L-shape, by using the weighted averages $0.5J$, $0.25J$, or $0.75J$, respectively in place of J in eqn (2.105).

With the Neumann boundaries, where $\partial A/\partial n$ is specified, the boundary can either bisect the outer cells as illustrated in Fig. 2.5, or coincide with the outer nodes. Treatment of the latter case is as given in the one-dimensional problem of Section 2.6.1 where the fictitious outside node is eliminated from the equation for the boundary nodes. In Fig. 2.5, a node outside the boundary has the same potential as the corresponding node just inside, so that, for example, in eqn (2.105) written for a node adjacent to the left-hand boundary, we replace $A_{i-1,j}$ by A_{ij}, since $\partial A/\partial x = 0$.

Apart from the nodes adjacent to the boundaries, all internal equations have the form, from eqn (2.105),

$$A_{i,j} = \frac{\Delta y^2}{2(\Delta x^2 + \Delta y^2)}(A_{i+1,j} + A_{i-1,j}) + \frac{\Delta x^2}{2(\Delta x^2 + \Delta y^2)}(A_{i,j+1} + A_{i,j-1}) + \frac{\Delta x^2 \Delta y^2}{2(\Delta x^2 + \Delta y^2)}\mu_0 J_{i,j} \tag{2.106}$$

It is clear that if the complete set of equations for all the unknown nodal values was written in matrix form, the coefficient matrix would be extremely sparse, having at most five non-zero entries per row, whereas the dimension of the matrix even in the simple illustration of Fig. 2.5 is 195. The orderly addressing system in Fig. 2.5 also means that the matrix entries form a compact band surrounding the main diagonal. Normal matrix elimination methods (e.g. Gaussian elimination) are not efficient in this case and it is preferable to use an iterative method, or, for a simple uniformly meshed rectangular region such as Fig. 2.5 with boundary conditions that are uniform on each boundary, one of the so-called direct methods. These can be expected to give at least one order of magnitude improvement in solution time over a good iterative method.

Interestingly, these direct methods closely resemble the original Roth double-Fourier series analysis of a continuous variable but applied now to a set of nodal potential and current density values, where the number of harmonics is limited to the number of nodes in each direction. In 1965, Hockney [2.21] described the application of Fourier expansions in one or other, or both, coordinate directions, but advocated series expansions in one direction only due to the time taken to determine the Fourier coefficients. By substituting the series, in x say, into the original Poisson's equation, an ordinary differential equation (in y) is obtained for each harmonic coefficient. The resulting set of equations can be written in tri-diagonal matrix form and solved numerically. A process known as cyclic reduction lessens the computation considerably [2.21]. Since the work by Hockney, several developments have been reported [2.6; 2.26; 2.39; 2.40; 2.48], but still restricted to basically rectangular and uniformly bounded problems.

A further method, applicable to both Poisson's and Helmholtz equation in less restrictive geometries with graded meshes, is the Capacitance Matrix technique (CMT) described by Proskurowski and Widlund [2.33]. A combination of CMT and the earlier methods has been used by Lukaniszyn [2.30] to produce the magnetic leakage field surrounding the windings of a current transformer. Figure 2.6 shows the contours of constant vector potential obtained for two two-dimensional solutions, the first (a) on the centre plane of the core and window, and the second (b) perpendicular to the first and on

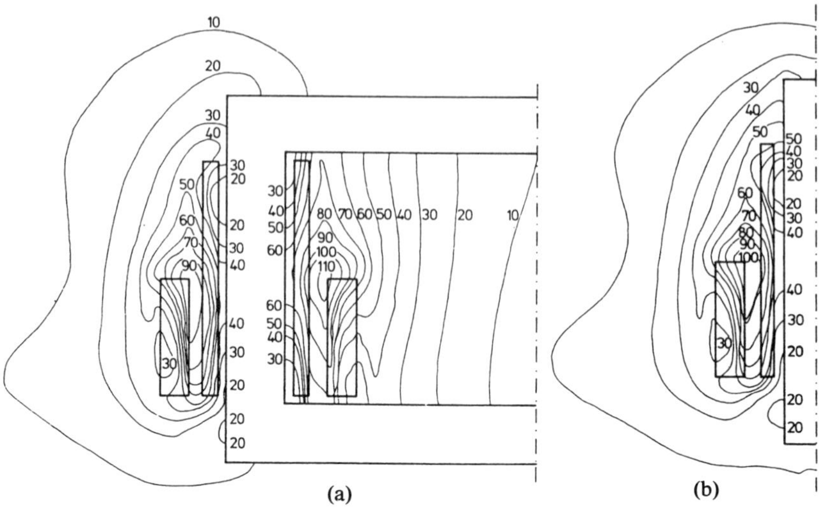

Fig. 2.6 *Contours of constant vector potential in and around two transformer coils: (a) on the centre plane of the window, and (b) perpendicular to the window and on the centre plane of the core leg*

the centre plane of a vertical core leg. Although to treat a three-dimensional problem as a pair of orthogonal two-dimensional solutions in this way is to take considerable liberties with the geometry, it is possible to obtain the leakage reactance within a few percent of the measured value.

Considering just one of the two-dimensional fields, e.g. Fig. 2.6a, the self inductance per unit length of one coil with N uniformly wound turns is

$$L = \frac{N}{IS} \iint (A - A') \, \mathrm{d}s \tag{2.107}$$

where I is the coil current, S the cross-section, and A and A' are the values of vector potential at corresponding points in the two sections of the coil, when only that coil is excited. The mutual inductance between two coils is given by

$$M = \frac{N_2}{I_1 S_2} \iint (A_1 - A_1') \, \mathrm{d}s_2 \tag{2.108}$$

where I_1 is the current in one coil, A_1 is the vector potential produced by that coil, and the integration is carried out over the cross-section S_2 of the other coil. If L_1 and L_2 are the self inductances of the primary and secondary coils respectively, then the leakage inductance referred to the primary side is usually defined as $[L_1 - M(L_1/L_2)^{1/2}]$.

In a regular mesh, where the conductor or coil side boundaries coincide with the grid lines, the numerical integration is particularly simple, so that

$$\iint A \, \mathrm{d}s \approx \sum \frac{1}{4} (A_{i,j} + A_{i+1,j} + A_{i+1,j+1} + A_{i,j+1}) \Delta x \Delta y \qquad (2.109)$$

where the summation takes place over all the cells in the cross-section of one coil side.

The force on a conductor or coil can be obtained in a similar way. In general

$$\mathbf{F} = \iiint (\mathbf{J} \times \mathbf{B}) \, \mathrm{d}v \qquad (2.110)$$

Now in x,y coordinates,

$$\mathbf{B} = B_x \hat{\mathbf{i}} + B_y \hat{\mathbf{j}} = \frac{\partial A_z}{\partial y} \hat{\mathbf{i}} - \frac{\partial A_z}{\partial x} \hat{\mathbf{j}} \qquad (2.111)$$

so that the two components of force on one of the coils in Fig. 2.6a are

$$F_x = \iint J_z \frac{\partial A_z}{\partial x} \mathrm{d}x \, \mathrm{d}y \qquad (2.112)$$

$$F_y = \iint J_z \frac{\partial A_z}{\partial y} \mathrm{d}x \, \mathrm{d}y \qquad (2.113)$$

In the example considered here J_z is a constant and so we are left with the need to integrate $\partial A / \partial x$ etc. over the coil cross-section

$$\iint \frac{\partial A}{\partial x} \mathrm{d}x \, \mathrm{d}y \approx \sum \frac{1}{2} [(A_{i+1,j} - A_{i,j}) + (A_{i+1,j+1} - A_{i,j+1})] \, \Delta y \qquad (2.114)$$

The alternative to direct methods of solving the sets of nodal equations of the type found in eqns (2.105) and (2.106), and essential in more complicated multi-region problems, are iterative methods of which successive over-relaxation (SOR) is the most common. The potentials of all the unspecified nodes are initially set to zero and we begin to scan the grid repetitively in an orderly and systematic manner, forcing eqn (2.105) to be satisfied at each node. One possibility is to scan the nodes of Fig. 2.5 by columns bottom to top and from left to right, so that, during the $(k + 1)$th scan or iteration, the form of eqn (2.106) becomes

$$A_{i,j}^{(k+1)} = C_1 A_{i+1,j}^{(k)} + C_2 A_{i,j+1}^{(k)} + C_3 A_{i-1,j}^{(k+1)} + C_4 A_{i,j-1}^{(k+1)} + C_5 J_{i,j} \qquad (2.115)$$

where C_1, C_2 etc. are the known coefficients in terms of Δx and Δy etc. We note that the new values of the potential are used as soon as they are available and so only one array of values need be stored at any one time. Equation (2.115) expresses the Gauss–Seidel iterative algorithm. The extension to SOR is simply to obtain the change that is about to be made to $A_{i,j}$ (known as the displacement or residual) i.e.

$$R_{i,j}^{(k+1)} = A_{i,j}^{(k+1)} - A_{i,j}^{(k)} \qquad (2.116)$$

and amplify it by multiplying by a relaxation factor α that is greater than unity, to give

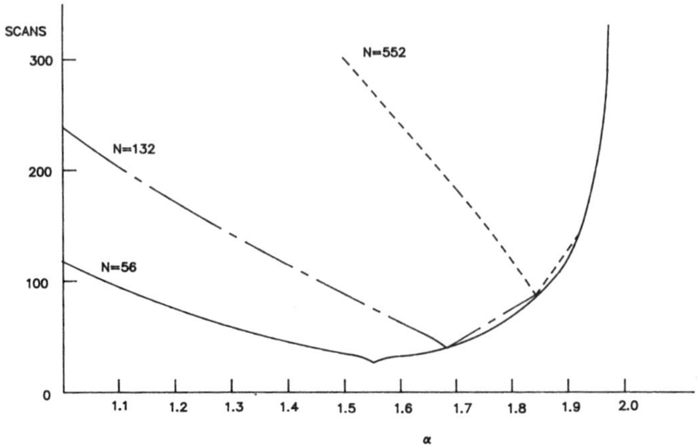

Fig. 2.7 *The number of SOR iterations needed for a given minimum residual as a function of the number of internal nodes (N) in a simple square mesh and the value of the relaxation factor a*

$$A_{i,j}^{(k+1)} = A_{i,j}^{(k)} + \alpha R_{i,j}^{(k+1)} \tag{2.117}$$

If $\alpha = 1$, eqns (2.116) and (2.117) reduce to eqn (2.115) again. The SOR iterations will converge if $1 \leqslant \alpha < 2$, but the solution is very significantly speeded up in terms of the number of scans required, for a given error, if α is close to its optimum value α_o for the size and type of problem being solved. Typically the number of scan will be an order of magnitude less than that needed by the original Gauss–Seidel scheme.

For a uniform rectangular region $a \times b$, Garabedian [2.19] has shown that

$$\alpha_o = \frac{2K}{K + \Delta y \sqrt{K}\,[(\pi/a)^2 + (\pi/b)^2]^{1/2}} \tag{2.118}$$

where $K = 1 + (\Delta y/\Delta x)^2$. However, in more complicated problems, the method of Carré [2.9], that deduces successive improvements to α during the iterative process, approaching α_o from below, can be very effective. Figure 2.7 illustrates the way in which the number of scans required decreases rapidly near the optimum value and also the penalty of exceeding α_o by even a small amount.

The work of this Section has been limited to the numerical solution of Laplace's and Poisson's equations, with uniform meshing, in non-magnetic current carrying regions. Once we have mixed magnetic and non-magnetic regions special finite-difference equations will apply at nodes on the interfaces. Although techniques exist to formulate these equations by applying the magnetic boundary conditions (see, for example, Binns and Lawrenson [2.3]), a more general approach is outlined in the following Section.

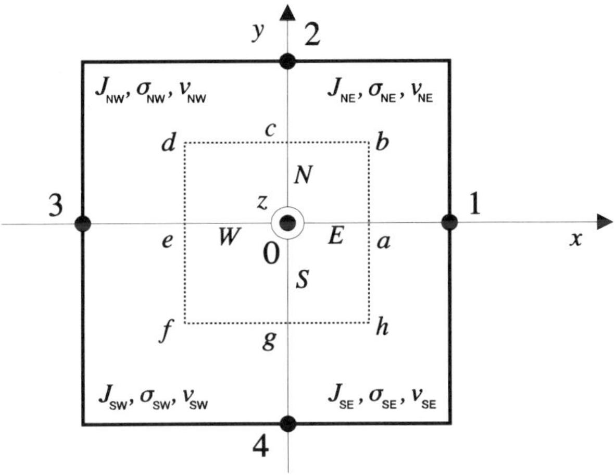

Fig. 2.8 *A node in an irregular grid with the surrounding cells*

2.6.5 *More two-dimensional static and steady-state solutions*

In developing a finite-difference equation for a general node in a graded mesh that is also on a conductor or magnetic region boundary, it is helpful to start with Helmholtz equation for the vector potential component A_z. This is the complex form of Poisson's equation and describes sinusoidal steady-state behaviour, with ordinary Poisson's equation as the special zero-frequency case. We now have the possibility of nonlinear permeability, or reluctivity ν (or strictly, with Helmholtz equation, constant μ or ν that change from one cell to the next), and the symmetrical form of Helmholtz equation in x, y coordinates is

$$\frac{\partial}{\partial x}\left(\nu \frac{\partial \underline{A}}{\partial x}\right) + \frac{\partial}{\partial y}\left(\nu \frac{\partial \underline{A}}{\partial y}\right) = \mathrm{j}\omega\sigma\underline{A} - J_s \tag{2.119}$$

where the current density is given the suffix s to denote an imposed current (which we assume to be a real quantity for convenience) to distinguish it from an induced current in the time-dependent system. Consider the four graded cells shown in Fig. 2.8, each cell having specified dimensions h_1 etc., and current density, conductivity and reluctivity. For ease of description the compass point notation is used, e.g. σ_{NE} is the conductivity of the cell to the North-East of the central node 0 under consideration. The most convenient method of obtaining a finite-difference equation is to apply Stoke's theorem

$$\iint (curl\,\mathbf{A}) \cdot \mathrm{d}\mathbf{s} = \oint \mathbf{A} \cdot \mathrm{d}\mathbf{l} \tag{2.120}$$

to an auxiliary cell *bdfh* centred on node 0 [2.43]. Thus we require to perform a surface integration of the *curl* form of eqn (2.119),

$$curl_z\left(\hat{\mathbf{i}}\nu\frac{\partial \underline{A}}{\partial y} - \hat{\mathbf{j}}\nu\frac{\partial \underline{A}}{\partial x}\right) = J_s - \mathrm{j}\omega\sigma\underline{A} \tag{2.121}$$

over the cell $bdfh$, to give the line integral equation,

$$\oint\left(\hat{\mathbf{i}}\nu\frac{\partial \underline{A}}{\partial y} - \hat{\mathbf{j}}\nu\frac{\partial \underline{A}}{\partial x}\right)\cdot \mathrm{d}\mathbf{l} = \int_{-S}^{N}\int_{-W}^{E}(J_s - \mathrm{j}\omega\sigma\underline{A})\,\mathrm{d}x\,\mathrm{d}y \tag{2.122}$$

where $E = h_1/2$, $N = h_2/2$, $W = h_3/2$, and $S = h_4/2$. On ab the contribution to the line integral is

$$\int_0^N\left(\hat{\mathbf{i}}\nu\frac{\partial \underline{A}}{\partial y} - \hat{\mathbf{j}}\nu\frac{\partial \underline{A}}{\partial x}\right)\cdot \hat{\mathbf{j}}\,\mathrm{d}y = -\int_0^N \nu_{NE}\frac{\partial \underline{A}}{\partial x}\bigg|_{x=E}\mathrm{d}y \tag{2.123}$$

Now $\partial A/\partial x$ at $x = E$ can be expanded as a Taylor series about point a, at which $A = A_a$,

$$\frac{\partial A}{\partial x}\bigg|_{x=E} = \frac{\partial A}{\partial x}\bigg|_a + y\frac{\partial^2 A}{\partial x\partial y}\bigg|_a + \mathrm{O}(N^2) \tag{2.124}$$

But

$$A_1 = A_a + E\frac{\partial A}{\partial x}\bigg|_a + \frac{1}{2}E^2\frac{\partial^2 A}{\partial x^2}\bigg|_a + \mathrm{O}(E^3) \tag{2.125}$$

and

$$A_0 = A_a - E\frac{\partial A}{\partial x}\bigg|_a + \frac{1}{2}E^2\frac{\partial^2 A}{\partial x^2}\bigg|_a - \mathrm{O}(E^3) \tag{2.126}$$

Subtracting eqn (2.126) from eqn (2.125) we have

$$\frac{\partial A}{\partial x}\bigg|_a = \frac{A_1 - A_0}{h_1} - \mathrm{O}(E^2) \tag{2.127}$$

and substituting eqn (2.127) in eqn (2.124)

$$\frac{\partial A}{\partial x}\bigg|_{x=E} = \frac{A_1 - A_0}{h_1} - \mathrm{O}(N) \tag{2.128}$$

The approximate value of the line integral along ab is therefore $-N\nu_{NE}(A_1 - A_0)/h_1$ with an error of $\mathrm{O}(N^2)$, and, following similar arguments, the complete line integral of eqn (2.122) is given by

$$(\alpha_1 + \alpha_2 + \alpha_3 + \alpha_4)A_0 - (\alpha_1 A_1 + \alpha_2 A_2 + \alpha_3 A_3 + \alpha_4 A_4)$$

where

$$\alpha_1 = \frac{N\nu_{NE} + S\nu_{SE}}{h_1} \tag{2.129}$$

$$\alpha_2 = \frac{E\nu_{NE} + W\nu_{NW}}{h_2} \tag{2.130}$$

$$\alpha_3 = \frac{N\nu_{NW} + S\nu_{SW}}{h_3} \tag{2.131}$$

$$\alpha_4 = \frac{E\nu_{SE} + W\nu_{SW}}{h_4} \tag{2.132}$$

The error is $O(K^2)$, where K is the larger of the half mesh dimensions N, S, E and W.

The surface integral on the right-hand side of eqn (2.122) is the sum of the integral over each quadrant of Fig. 2.8. Consider the first quadrant:

$$\iint_{0\,0}^{NE} (J_s - j\omega\sigma\underline{A})\,dx\,dy = NEJ_{NE} - j\omega\sigma_{NE}\iint_{0\,0}^{NE} \underline{A}\,dx\,dy \tag{2.133}$$

A double Taylor expansion about node 0 yields

$$A = A_0 + x\frac{\partial A}{\partial x}\bigg|_0 + y\frac{\partial A}{\partial y}\bigg|_0 + O(K^2) \tag{2.134}$$

so that

$$\iint_{0\,0}^{NE} A\,dx\,dy = NEA_0 + \frac{1}{2}NE^2\frac{\partial A}{\partial x}\bigg|_0 + \frac{1}{2}N^2E\frac{\partial A}{\partial y}\bigg|_0 + O(K^4) \tag{2.135}$$

If we ignore terms $O(K^3)$, bearing in mind that the approximation to the line integral has an error $O(K^2)$, the complete surface integral has the form $I_0 - j\omega Q_0$, so that eqn (2.122) finally becomes

$$\underline{A}_0 = \frac{(\alpha_1\underline{A}_1 + \alpha_2\underline{A}_2 + \alpha_3\underline{A}_3 + \alpha_4\underline{A}_4) + I_0}{(\alpha_1 + \alpha_2 + \alpha_3 + \alpha_4) + j\omega Q_0} \tag{2.136}$$

where I_0 is the total imposed current associated with node 0, i.e.

$$I_0 = NEJ_{NE} + NWJ_{NW} + SEJ_{SE} + SWJ_{SW} \tag{2.137}$$

and similarly,

$$Q_0 = NE\sigma_{NE} + NW\sigma_{NW} + SE\sigma_{SE} + SW\sigma_{SW} \tag{2.138}$$

The simple notation of eqn (2.136) can easily be generalized to apply to any node (i, j), with the cell parameters being given the same address as the node in the bottom left-hand corner of the cell, e.g. the cell in the second quadrant of Fig. 2.8, at present labelled NW, is now cell $(i-1, j)$ and $\nu_{NW} = \nu_{i-1,j}$ etc. The important point is that eqn (2.136) is completely general and can be applied to a node in any position including all types of internal boundary. It is also highly suitable, in its static or diffusion form, for nonlinear problems where the reluctivity will be different in each cell, and a function of the average values of the derivatives of A in the cell after each scan of the iterative solution, e.g. since

$$B^2 = \left(\frac{\partial A}{\partial x}\right)^2 + \left(\frac{\partial A}{\partial y}\right)^2 \tag{2.139}$$

we have, for cell (i, j)

$$B_{av}^2 = \left\{\frac{1}{2}\left(\frac{A_{i+1,j} - A_{i,j}}{\Delta x}\right) + \frac{1}{2}\left(\frac{A_{i+1,j+1} - A_{i,j+1}}{\Delta x}\right)\right\}^2 + \left\{\frac{1}{2}\left(\frac{A_{i,j+1} - A_{i,j}}{\Delta y}\right) + \frac{1}{2}\left(\frac{A_{i+1,j+1} - A_{i+1,j}}{\Delta y}\right)\right\}^2 \tag{2.140}$$

This is why it was suggested earlier that one useful way of storing the magnetization details of a magnetic material is in the form of a B^2/ν file.

Equation (2.136) can be solved by successive over-relaxation (SOR) as described in Section 2.6.4. However, for a nonlinear solution, solved in terms of A, it is essential to under-relax the cell reluctivity ν obtained from the value of B^2, so that, for cell (i, j),

$$\nu_{i,j}^{(k+1)} = \nu_{i,j}^{(k)} + w(\nu - \nu_{i,j}^{(k)}) \tag{2.141}$$

where w is less than one (typically 0.2, but problem dependent).

Clearly, for a ferromagnetic region, a reluctivity map in the form of a separate array of cell values is required. However, where J and σ are constant in large conducting regions two more full arrays would be extremely wasteful in space and so some form of rationalization is necessary, even to the extent of listing special versions of eqn (2.136) applicable to each region and boundary node type. In the extreme case of a node in the interior of a region of constant conductivity, reluctivity, and current density, that has been meshed with a grid of simple square cells, eqn (2.136) reduces to

$$A_0 = \frac{A_1 + A_2 + A_3 + A_4 + \mu_0\mu_r I_0}{4 + j\omega\mu_0\mu_r Q_0} \tag{2.142}$$

where $I_0 = J\Delta x^2$ and $Q_0 = \sigma\Delta x^2$. It can be seen that eqn (2.106) will reduce to the same equation if $\Delta y = \Delta x$ and $\sigma = 0$. The SOR solution of a linear complex equation such as eqn (2.142), or eqn (2.136) with constant values of α, ideally involves a complex relaxation factor. A near optimum value of the latter can be derived during the solution using a modified version of the Carré method [2.9; 2.42; 2.43] mentioned in Section 2.6.4.

It is worth noting that in the alternative two-dimensional problem where the magnetic field is perpendicular to the x, y plane, i.e. we have H_z only, the diffusion equation in H is

$$\frac{\partial}{\partial x}\left(\rho\frac{\partial H}{\partial x}\right) + \frac{\partial}{\partial y}\left(\rho\frac{\partial H}{\partial y}\right) = j\omega\mu H \tag{2.143}$$

and is analogous to eqn (2.119) with H in place of A, resistivity ρ in place of reluctivity ν, and permeability μ in place of conductivity σ. The only difference is the absence of the current density term that appears in eqn (2.119). Thus, since entirely mathematical arguments were used in the deriva-

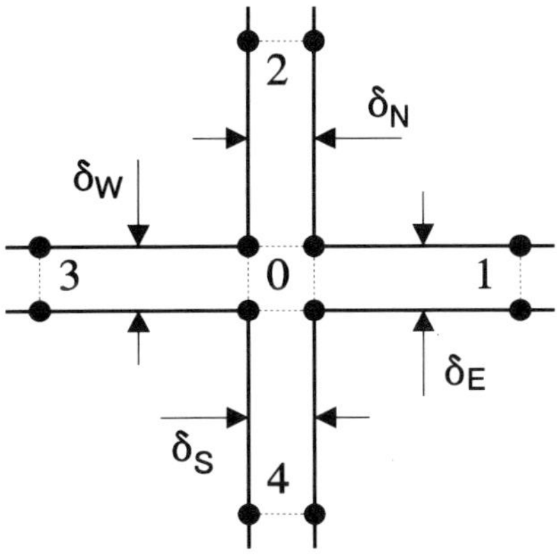

Fig. 2.9 *Small air gaps inserted into a rectangular grid*

tion of eqn (2.136), a finite-difference approximation to eqn (2.143) has the same form as eqn (2.136) but with I_0 zero, σ_{NE} etc. in eqn (2.138) for Q_0 replaced by μ_{NE} etc., and similarly ν_{NE} etc. in eqns (2.129) to (2.132) for the α coefficients replaced by ρ_{NE} etc.

A very useful feature of the way in which eqn (2.136) was derived is that the presence of very small air gaps between highly magnetic surfaces can be incorporated in the line integral formulation. To actually mesh a very small air gap is expensive in nodes because even the cell dimension parallel to the gap is restricted by the aspect ratio of the cell, which should not be greater than 10 to 1. Suppose the gap is 0.5 mm and we require at least two cells to straddle it. Typical cell dimensions then become 0.25 mm × 2 mm, say, which in an extensive air gap, will result in a large number of cells.

Inclusion of the very small air gaps illustrated in Fig. 2.9 can be achieved by adding their *mmf* contributions to the line integral of eqn (2.122). In air, we have

$$H_x = \frac{1}{\mu_0}\left(\frac{\partial A}{\partial y}\right) \quad \text{and} \quad H_y = -\frac{1}{\mu_0}\left(\frac{\partial A}{\partial x}\right) \tag{2.144}$$

If the relative permeability is sufficiently high for the flux to cross the small gap perpendicular to the iron surfaces, then the contribution of gap δ_E, for example, is simply

$$\delta_E H_y = \frac{\delta_E}{\mu_0}\left(-\frac{\partial A}{\partial x}\right) \approx \frac{\delta_E}{\mu_0}\frac{A_0 - A_1}{h_1} \tag{2.145}$$

We note that points on opposite sides of the air gap both share the same node number under this assumption. The total contribution of all four air gaps to the *mmf*, bearing in mind that the line integration is anti-clockwise, is

$$\frac{1}{\mu_0}\left\{\delta_E\frac{(A_0 - A_1)}{h_1} + \delta_N\frac{(A_0 - A_2)}{h_2} + \delta_W\frac{(A_0 - A_3)}{h_3} + \delta_S\frac{(A_0 - A_4)}{h_4}\right\}$$

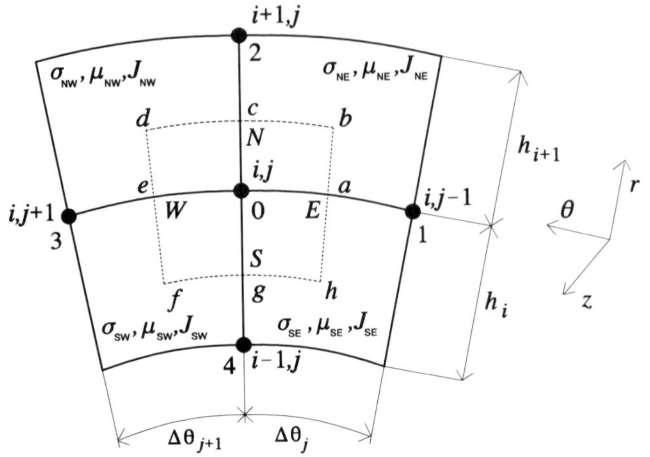

Fig. 2.10 *A node in a polar grid with the surrounding cells*

From the form of eqn (2.136), we thus see that a term of the form $\delta/(\mu_0 h)$ must be added to each α coefficient, so that, for example, $\beta_3 = \alpha_3 + \delta_w/(\mu_0 h_3)$, and the new finite-difference equation is identical in form to eqn (2.136) but with the new coefficients β_i in place of $\alpha_i (i = 1\text{–}4)$. This is a general form of the equation and does not imply that gaps must exist in all four branches at once.

Stoll [2.44] has applied the above technique to the problem of calculating the transverse force on a magnetic filler block in a flexibility slot in the solid steel pole face of a two-pole turbogenerator rotor where the likely clearance gaps are less than 0.25 mm, but not uniform.

The alternative two-dimensional geometry is the polar coordinate system illustrated in Fig. 2.10. The Helmholtz equation equivalent to eqn (2.121) now becomes

$$curl_z \left(\hat{\mathbf{a}}_r \frac{\nu}{r} \frac{\partial \underline{A}}{\partial \theta} - \hat{\mathbf{a}}_\theta \nu \frac{\partial \underline{A}}{\partial r} \right) = J_s - \mathrm{j}\omega\sigma\underline{A} \tag{2.146}$$

The diffusion form has $\sigma\,\partial A/\partial t$ in place of $\mathrm{j}\omega\sigma\underline{A}$.

Applying Stoke's theorem in a similar way to that for the rectangular system, Wiak [2.55] shows that the finite-difference equation is

$$\begin{aligned}(\alpha_1 + \alpha_2 + \alpha_3 + \alpha_4)\underline{A}_{i,j} &- (\alpha_1\underline{A}_{i,j-1} + \alpha_2\underline{A}_{i+1,j} + \alpha_3\underline{A}_{i,j+1} + \alpha_4\underline{A}_{i-1,j}) \\ &= \frac{r_i}{4}[(h_{i+1}\Delta\theta_j J_{NE} + h_{i+1}\Delta\theta_{j+1}J_{NW} + h_i\Delta\theta_{j+1}J_{SW} + h_i\Delta\theta_j J_{SE}) \\ &\quad - (h_{i+1}\Delta\theta_j\sigma_{NE} + h_{i+1}\Delta\theta_{j+1}\sigma_{NW} + h_i\Delta\theta_{j+1}\sigma_{SW} + h_i\Delta\theta_j\sigma_{SE})\mathrm{j}\omega\underline{A}_{i,j}]\end{aligned} \tag{2.147}$$

where

$$\alpha_1 = \frac{(h_{i+1}\nu_{NE} + h_i\nu_{SE})}{2r_i\Delta\theta_j} \tag{2.148}$$

$$\alpha_2 = \frac{(\Delta\theta_{j+1}\nu_{NW} + \Delta\theta_j\nu_{NE})(r_i + \frac{1}{2}h_{i+1})}{2h_{i+1}} \tag{2.149}$$

$$\alpha_3 = \frac{(h_i\nu_{SW} + h_{i+1}\nu_{NW})}{2r_i\Delta\theta_{j+1}} \tag{2.150}$$

$$\alpha_4 = \frac{(\Delta\theta_{j+1}\nu_{SW} + \Delta\theta_j\nu_{SE})(r_i - \frac{1}{2}h_i)}{2h_i} \tag{2.151}$$

Equation (2.147) has been presented in this way, with $\underline{A}_{i,j}$ on both sides, so that the modification to the diffusion equation is easily accomplished by replacing $j\omega\underline{A}$ on the right-hand side by $\partial A/\partial t$ in the desired numerical form.

The components of flux density in a given cell can be obtained by averaging the derivatives of A on opposite sides of the cell, e.g. in the *North West* cell in Fig. 2.10 (cell (i, j) bearing in mind the directions of increasing i and j) we have

$$B_r = \frac{1}{r}\frac{\partial A}{\partial \theta} \approx \frac{1}{2}\left(\frac{A_{i+1,j+1} - A_{i+1,j}}{r_{i+1}\Delta\theta_{j+1}} + \frac{A_{i,j+1} - A_{i,j}}{r_i\Delta\theta_{j+1}}\right) \tag{2.152}$$

Wiak and Pelikant [2.57] interestingly solve the static Poissonian version of eqn (2.147), i.e. without the last induced eddy-current term, for a switched reluctance motor. However, the earliest and major application of the above forms of nonlinear Poisson's equation (modified versions of eqns (2.136) and (2.147)) was to the analysis of d.c. machines by Erdélyi and Fuchs [2.16; 2.18]. Although computer programs of this complexity are unlikely to be written today, in view of the present widespread availability of commercial and in-house general purpose finite-element packages, one interesting feature of the work in the 1970s was the development of ways of improving the convergence of steady-state iterative methods such as SOR. In this respect, the integral correction method of De la Vallée Poussin [2.13], being based on Ampère's law (eqn (1.27)), has been found to be helpful.

Consider a contour such as that shown dotted in Fig. 2.11 which is close to the boundary of the problem and bisects a series of mesh links that are locally perpendicular to the boundaries. The contour encloses a known total current I. Let Δn denote the length of these perpendicular links, that may vary along the contour, and Δt the tangential distances between adjacent nodes. Since $\mathbf{B} = curl\,\mathbf{A}$ or $\mathbf{H} = \nu\, curl\,\mathbf{A}$, a numerical form of Ampère's law applied to the contour becomes, in terms of A_z:

$$\sum \nu (A_i - A_0)_{av}\left(\frac{\Delta t}{\Delta n}\right) = I \tag{2.153}$$

where Δn will be constant on the straight sections of the contour and $(A_i - A_0)_{av}$ represents the average difference between the values A_i on the

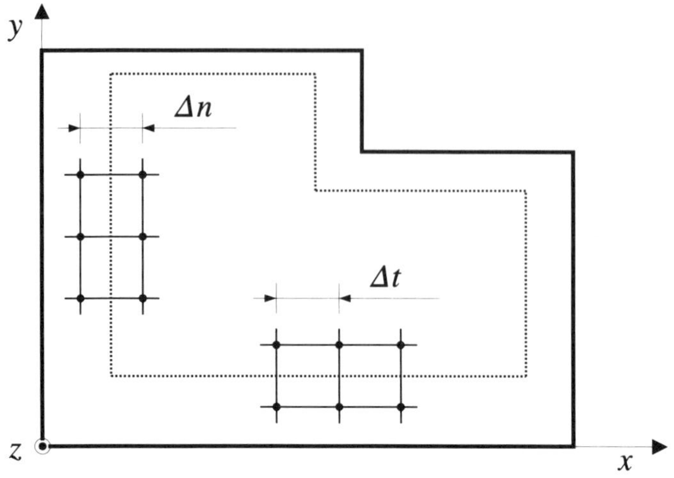

Fig. 2.11 *Integration contour for integral correction method of improving convergence*

corners of a given cell (of reluctance ν) inside the contour and the values A_0 on the corners outside the contour. However, at a particular stage in the iterative process eqn (2.153) will not hold, because the solution has not converged. Let a constant value ΔA be added to all nodes inside the contour, such that eqn (2.153) is satisfied, i.e.

$$\sum \nu (A_i + \Delta A - A_0)_{av} \left(\frac{\Delta t}{\Delta n} \right) = I \tag{2.154}$$

Thus, solving for the constant ΔA, we have

$$\Delta A = \frac{\left[I - \sum \nu (A_i - A_0)_{av} \left(\frac{\Delta t}{\Delta n} \right) \right]}{\sum \nu \left(\frac{\Delta t}{\Delta n} \right)} \tag{2.155}$$

Although this technique is not known to have been applied to steady-state solutions of Helmholtz equation, where the vector potential component is complex, there appears to be no reason why it should not be as effective as it is for static solutions.

2.6.6 Two-dimensional transient or nonlinear steady-state problems

With time-stepping problems in two space dimensions it is almost essential to employ an explicit method such as DuFort–Frankel, or a semi-implicit method that reduces or simplifies what would otherwise be a full matrix solution for a large number of nodes at each time step. If $j\omega A$ on the right-hand side of eqn (2.147) is replaced by

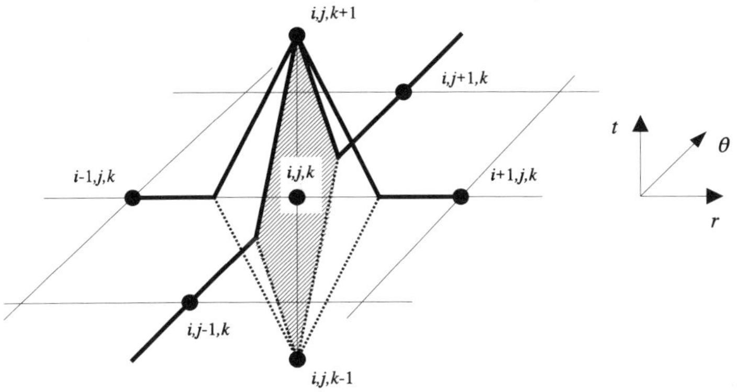

Fig. 2.12 *Finite-difference molecule in two space dimensions (i, j) and time (k)*

$$\frac{\partial A}{\partial t} \approx \frac{A_{i,j,k+1} - A_{i,j,k}}{\Delta t} \tag{2.156}$$

and $A_{i,j,k}$ is then replaced on both sides by $(A_{i,j,k+1} + A_{i,j,k-1})/2$, we have the DuFort–Frankel form of the diffusion equation for A_z in polar coordinates. Omitting the imposed current density term, and writing

$$Q_{i,j} = \tfrac{1}{4} r_i (h_{i+1}\Delta\theta_j \sigma_{NE} + h_{i+1}\Delta\theta_{j+1}\sigma_{NW} + h_i \Delta\theta_{j+1}\sigma_{SW} + h_i\Delta\theta_j\sigma_{SE}) \tag{2.157}$$

for brevity, eqn (2.147) becomes

$$A_{i,j,k+1}\left(\alpha_1 + \alpha_2 + \alpha_3 + \alpha_4 + \frac{Q_{i,j}}{\Delta t}\right) = 2\alpha_1 A_{i,j-1,k} + 2\alpha_2 A_{i+1,j,k} + 2\alpha_3 A_{i,j+1,k} + 2\alpha_4 A_{i-1,j,k} - A_{i,j,k-1}\left(\alpha_1 + \alpha_2 + \alpha_3 + \alpha_4 - \frac{Q_{i,j}}{\Delta t}\right) \tag{2.158}$$

The difference molecule is shown in Fig. 2.12.

Extending the work of Richmyer and Morton [2.35], who examined the stability of three-time-layer difference schemes using the method of energy inequality, Wiak [2.55] shows that the DuFort–Frankel scheme must satisfy the condition

$$1 + \frac{2\Delta t(\Delta r^2 + r^2\Delta\theta^2)}{r^2\Delta r^2\Delta\theta^2\sigma}(\nu_{min} - \nu_{max}) \geqslant 0 \tag{2.159}$$

where the radial mesh length, Δr, and angular displacement, $\Delta\theta$, are now assumed to be constants. With strong nonlinearity, $\nu_{max} \gg \nu_{min}$, so that eqn (2.159) reduces to a condition on the time step,

$$\Delta t \leqslant \frac{r^2\Delta r^2\Delta\theta^2\sigma}{2(\Delta r^2 + r^2\Delta\theta^2)\nu_{max}} \tag{2.160}$$

Wiak [2.55] also shows that eqn (2.160) holds in the nonlinear case, and goes on to solve the problem of the eddy currents induced in a ferromagnetic rotor screen of a synchronous generator with a superconducting d.c. field winding when the rotor oscillates (hunts) about synchronous speed. The onset of numerical instability is shown as the time step increases above the value of 33 μs set by eqn (2.160) for the parameters involved.

The above example actually solves the DuFort–Frankel algorithm inside the conducting screen but uses analytical solutions outside, in terms of the Fourier series that result from the normal separation of variables method applied to Laplace's equation. The analytical and numerical solutions are joined at the boundaries of the conducting region. A more general concept is that of linking the numerical diffusion algorithm with one suitable for Laplace's equation, e.g. SOR. This boundary matching can be illustrated as follows. At each new time step the nodal values of A_z on the surface of the conducting region are used as Dirichlet boundary conditions for the adjacent air space. The updated nodal values within this Laplacian region are found by performing a few SOR iterations. For example, in the air gap of an electrical machine about five iterations are sufficient, not only because of the limited air gap length but also because the change in the interface values from one time step to the next is small. The row of nodal values just inside the air region and parallel to the interface are then used as the boundary values for the next DuFort–Frankel step in the conducting region, so that the two regions overlap by one layer of nodes. In the process of moving from conductor to air the normal components of B, in terms of the surface nodal values of A_z, are supplied. Returning from air to conductor, the tangential components of B are effectively specified by the values of A_z at each pair of nodes perpendicular to the air side of the interface. Jack and Stoll [2.22] have used this technique to find the negative sequence induced currents and ohmic losses in the solid rotor of a turbogenerator.

A popular semi-implicit method is the Alternating-Direction Implicit method (ADI) first proposed by Peaceman and Rachford [2.31], and examined in great detail in the book by Wachspress [2.52] for elliptic equations. For the purpose of ease of demonstration we will take the simple implicit form of the diffusion version of eqn (2.105) in x, y coordinates for a uniform rectangular cell size, $\Delta x \times \Delta y$, and constant permeability and conductivity, i.e.

$$\frac{1}{(\Delta x)^2}(A_{i+1,j,k+1} - 2A_{i,j,k+1} + A_{i-1,j,k+1}) + \frac{1}{(\Delta y)^2}(A_{i,j+1,k+1} - 2A_{i,j,k+1} + A_{i,j-1,k+1}) = \frac{\sigma\mu}{\Delta t}(A_{i,j,k+1} - A_{i,j,k}) \tag{2.161}$$

The ADI algorithm moves forward in time with a repetitive two-step cycle. In moving from time k to $k+1$, eqn (2.161) becomes half explicit in the y direction, to give

$$\frac{1}{(\Delta x)^2}(A_{i+1,j,k+1} - 2A_{i,j,k+1} + A_{i-1,j,k+1}) + \frac{1}{(\Delta y)^2}(A_{i,j+1,k} - 2A_{i,j,k} + A_{i,j-1,k}) = \frac{\sigma\mu}{\Delta t}(A_{i,j,k+1} - A_{i,j,k}) \tag{2.162}$$

whereas, in moving from time $k+1$ to $k+2$, eqn (2.161) becomes half explicit in the x direction,

$$\frac{1}{(\Delta x)^2}(A_{i+1,j,k+1} - 2A_{i,j,k+1} + A_{i-1,j,k+1}) + \frac{1}{(\Delta y)^2}(A_{i,j+1,k+2} - 2A_{i,j,k+2} + A_{i,j-1,k+2}) = \frac{\sigma\mu}{\Delta t}(A_{i,j,k+2} - A_{i,j,k+1}) \tag{2.163}$$

At each time step, therefore, sets of one-dimensional implicit equations must be solved, each set having a simple tri-diagonal coefficient matrix. In the linear case, the process is simple and efficient because the coefficient matrix remains constant throughout and can be decomposed into the product of lower (L) and upper (U) triangular matrices, where, because of symmetry, U is equal to the transpose of L. L can be found by the Choleski method for the larger of the row or column dimensions. Brief details are given by Stoll [2.43], including the use of ADI for steady-state complex solutions. The main difficulty with the ADI method is that its successful performance is very problem dependent. This follows from the substantial explicit component. Thus it is worth investigating any alternative that increases the implicit content, but without going to the computational burden of the two-dimensional form of the simple implicit equation (eqn (2.161)) or the Crank–Nicolson algorithm. One possibility is the interesting strongly-implicit iterative method of Stone [2.46], outlined in some detail by Stoll [2.43].

2.6.7 Three-dimensional solution of Laplacian and Poissonian problems

Laplacian problems can be solved in terms of a magnetic scalar potential such that $\mathbf{H} = -grad\, V$. The magnetic flux density satisfies $div\,\mathbf{B} = 0$, and so,

$$div\,(\mu\mathbf{H}) = div\,(\mu\, grad\, V) = 0 \tag{2.164}$$

Applying Gauss' theorem (eqn (1.22)) to the small volume surrounding node 0 of Fig. 2.13, where the faces of the volume bisect the mesh links, the volume integral of the divergence becomes the surface integral of the vector $\mu\, grad\, V$, so that

$$\oint \mu\, grad\, V \cdot \mathrm{ds} = 0 \tag{2.165}$$

The derivation of a three-dimensional finite-difference equation now follows a similar process to that used for the development of the Helmholtz (or diffusion) equation equivalent in two-dimensions, where a line integral in

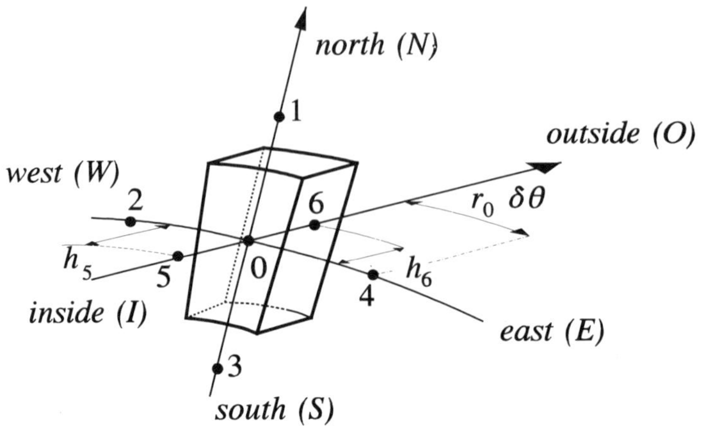

Fig. 2.13 *Three-dimensional finite-difference molecule showing surface of integration*

terms of A_z was involved (see Section 2.6.5). Each of eight three-dimensional cells, not fully shown in Fig. 2.13 but all sharing node 0, is defined by a constant angular increment $\Delta\theta$ but varying linear dimensions h_1, h_3, h_5 and h_6 in the radial and axial directions. For example, h_1 is not shown to avoid confusing the diagram but is the mesh length between nodes 0 and 1 in the positive radial direction. In the more general case, each cell can have a different permeability, so that μ_{NEI} stands for the permeability of the cell to the North East and inside (i.e. decreasing z) of node 0. Leurs and Stoll [2.27] give the finite-difference equation that results from approximating the surface integral over the auxiliary cell of Fig. 2.13

$$
\begin{aligned}
&\frac{V_1 - V_0}{4h_1}\left(r_0 + \frac{1}{2}h_1\right)\Delta\theta[h_5(\mu_{NEI} + \mu_{NWI}) + h_6(\mu_{NEO} + \mu_{NWO})] \\
&+ \frac{V_2 - V_0}{4r_0\Delta\theta}[h_5(h_1\mu_{NWI} + h_3\mu_{SWI}) + h_6(h_1\mu_{NWO} + h_3\mu_{SWO})] \\
&+ \frac{V_3 - V_0}{4h_3}\left(r_0 - \frac{1}{2}h_3\right)\Delta\theta[h_5(\mu_{SWI} + \mu_{SEI}) + h_6(\mu_{SWO} + \mu_{SEO})] \\
&+ \frac{V_4 - V_0}{4r_0\Delta\theta}[h_5(h_1\mu_{NEI} + h_3\mu_{SEI}) + h_6(h_1\mu_{NEO} + h_3\mu_{SEO})] \\
&+ \frac{V_5 - V_0}{4h_5}\Delta\theta\left[h_1\left(r_0 + \frac{1}{4}h_1\right)(\mu_{NWI} + \mu_{NEI}) + h_3\left(r_0 - \frac{1}{4}h_3\right)(\mu_{SWI} + \mu_{SEI})\right] \\
&+ \frac{V_6 - V_0}{4h_6}\Delta\theta\left[h_1\left(r_0 + \frac{1}{4}h_1\right)(\mu_{NWO} + \mu_{NEO}) + h_3\left(r_0 - \frac{1}{4}h_3\right)(\mu_{SWO} + \mu_{SEO})\right] \\
&= 0
\end{aligned}
\tag{2.166}
$$

The resulting symmetric set of linear equations can be solved by block successive over-relaxation as an efficient alternative to normal SOR in three dimensions. The unknowns that form a block are the potentials on an r, θ

plane ($z =$ constant). To minimize the band width, the nodes are numbered in the circumferential direction first. Each block is solved by Cholesky decomposition. In any block containing nodes that are all in air, the block matrix only needs to be decomposed once. For example, with $h_1 = h_3 = \Delta r$, and $\mu = \mu_0$, eqn (2.166) reduces to

$$\frac{V_1 - V_0}{2\Delta r}\left(r_0 + \frac{1}{2}\Delta r\right)\Delta\theta(h_5 + h_6) + \frac{V_3 - V_0}{2\Delta r}\left(r_0 - \frac{1}{2}\Delta r\right)\Delta\theta(h_5 + h_6)$$
$$+ \frac{V_2 - V_0}{2r_0\Delta\theta}\Delta r(h_5 + h_6) + \frac{V_4 - V_0}{2r_0\Delta\theta}\Delta r(h_5 + h_6)$$
$$+ \frac{V_5 - V_0}{h_5} r_0\Delta\theta\Delta r + \frac{V_6 - V_0}{h_6} r_0\,\Delta\theta\,\Delta r = 0 \tag{2.167}$$

and the relative ease with which a set of equations of this form could be handled is clear.

Suppose we now have a current sheet in the plane $r = r_0$, carrying a surface current of density

$$\mathbf{K} = \hat{\mathbf{a}}_\theta K_\theta + \hat{\mathbf{a}}_z K_z \tag{2.168}$$

subject to the condition

$$div\,\mathbf{K} = \frac{1}{r}\frac{\partial K_\theta}{\partial\theta} + \frac{\partial K_z}{\partial z} = 0 \tag{2.169}$$

The nodes in Fig. 2.13 can now be considered to be split by the infinitely thin sheet as shown in Fig. 2.14. Equation (2.165), with $\mu = \mu_0$, must still be satisfied over the auxiliary cell surface (not shown in Fig. 2.14), so that eqn (2.167) becomes

$$\alpha_1(V_1 - V_0) + \alpha_3(V_3 - V_0') + \tfrac{1}{2}\alpha_2[(V_2 - V_0) + (V_2' - V_0')]$$
$$+ \tfrac{1}{2}\alpha_4[(V_4 - V_0) + (V_4' - V_0')]$$
$$+ \tfrac{1}{2}\alpha_5[(V_5 - V_0) + (V_5' - V_0')]$$
$$+ \tfrac{1}{2}\alpha_6[(V_6 - V_0) + (V_6' - V_0')] = 0 \tag{2.170}$$

where the α coefficients follow from eqn (2.167). The terms of the form $V_i' - V_0'$ can be found by applying Ampère's law to the vanishingly thin rectangular paths formed by the corner nodes $0, i, i', 0'$. For example, consider the path $044'0'$ in Fig. 2.14, for which $\oint \mathbf{H}\cdot d\mathbf{l} = I$ becomes

$$r_0\Delta\theta H_{\theta 04} - r_0\Delta\theta H'_{\theta 04} \approx K_{z04} r_0 \Delta\theta \tag{2.171}$$

where the positive directions of the components of current sheet density normal to the paths are shown in Fig. 2.14. Now $H_\theta = -(1/r)(\partial V/\partial\theta)$ so that eqn (2.171) finally becomes

$$(V_4' - V_0') = (V_4 - V_0) + r_0\Delta\theta K_{z04} \tag{2.172}$$

and $V_4' - V_0'$ can be eliminated from eqn (2.170). Similarly for the other three paths we have

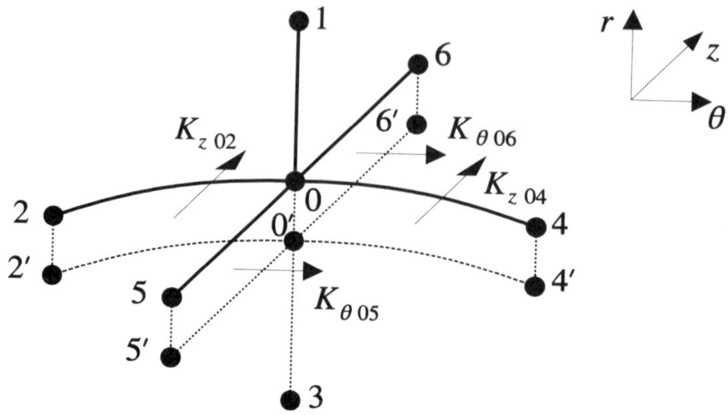

Fig. 2.14 *Three-dimensional finite-difference molecule with split nodes for current sheet insertion*

$$(V_2' - V_0') = (V_2 - V_0) - r_0 \Delta\theta K_{z02} \tag{2.173}$$

$$(V_5' - V_0') = (V_5 - V_0) + h_5 K_{\theta 05} \tag{2.174}$$

$$(V_6' - V_0') = (V_6 - V_0) - h_6 K_{\theta 06} \tag{2.175}$$

The only unknown primed term that remains in eqn (2.170) is $(V_3 - V_0')$. This can be determined by considering the total current i_0 flowing in the sheet in one of the coordinate directions (either θ or z) between node 0 and some specified reference point where $V = 0$. Thus $V_0' = V_0 + i_0$ and

$$(V_3 - V_0') = (V_3 - V_0) - i_0 \tag{2.176}$$

Thus the terms in the original finite-difference equation (2.167) can be simply modified to produce the equation that applies to nodes on the surface of a defined current sheet. We also note that, if it is inconvenient to specify both K_θ and K_z everywhere, one component can be generated from the other using eqn (2.169).

If it is necessary to enter a current carrying region, i.e. the current sheet simplification is unacceptable, Carpenter's electric vector potential (**T** vector), see for example [2.7], can be used. Here the magnetic field vector is defined as

$$\mathbf{H} = \mathbf{T} - grad\, V \tag{2.177}$$

where $curl\,\mathbf{T} = \mathbf{J}$ to satisfy Ampère's law. The advantage of the **T** vector in problems in which the current density vector is two-dimensional, or approximately so, is that **T** can be represented in terms of only one component, orthogonal to the current directions. Reference [2.8] shows an example of the use of this technique. The **T** vector is also applicable to current sheets as an alternative to the previous development entirely in terms of V [2.27].

The alternative for general current carrying Poissonian regions is a full development in terms of **A** or **H**. For example, Sarma [2.37] gives finite-difference equations in terms of A_x, A_y and A_z. However, it is unlikely that complicated three-dimensional finite-difference approaches will be a productive development in situations where good finite-element software already exists. Notwithstanding, some interesting and quite powerful uses have been reported in the literature over the last three decades, e.g. nonlinear cartesian Laplace in V [2.23], linear cartesian Poisson in **A** [2.12; 2.60; 2.61] (the first reference being an example of the use of electrical networks in field solutions), linear cartesian Helmholtz in **H** [2.45] with multiple zoning of resistivity.

3
Finite-Element and Boundary-Element Methods

Ryszard Sikora

3.1 INTRODUCTION

Finite-elements and boundary-elements methods are widely applied in electromagnetic field analysis and synthesis [3.2; 3.3; 3.4; 3.5; 3.7; 3.10; 3.17; 3.22; 3.23; 3.25; 3.28; 3.29; 3.30; 3.31; 3.32; 3.33; 3.34; 3.35; 3.36; 3.38].

The development of computers made it possible to solve many problems using numerical methods. The combination of the availability of computers and the Finite-Elements Method (FEM) and Boundary-Elements Method (BEM) contributed to establishing the methods and enabled them to be applied to many engineering problems. In the past 30 years a large number of computer programs based on FEM and BEM have been developed. Application of FEM can be dated back to 1967 when O.C. Zienkiewicz used the FEM for calculations of mechanical structures. The FEM was introduced by P.P. Silvester and M.V.K. Chari for electrical engineering calculations [3.7; 3.30]. This method can be used in calculations of stationary and transient, linear and nonlinear, isotropic and anisotropic, homogeneous and non-homogeneous electromagnetic problems. FEM finds application in calculation of quantities describing electromagnetic field distribution (e.g. $\mathbf{E}$, $\mathbf{H}$, $\mathbf{D}$, $\mathbf{B}$, $\mathbf{J}$, $\mathbf{A}$, $\mathbf{T}$, V, Ω) and integral quantities (e.g., P, W, L, R). The method is quite useful for computation in bounded regions, but is inconvenient when the region is unbounded. There are several useful methods for electromagnetic field calculation in unbounded regions, e.g. the Infinite-Elements Method (IEM) [3.18; 3.19; 3.27].

Application of the boundary-element method can be dated back to 1978 when C.A. Brebbia published his book about BEM [3.5]. This method can be used for electromagnetic field calculation in bounded and unbounded regions in linear media.

3.2 WEIGHTED RESIDUALS [3.15; 3.35; 3.38; 3.40]

The weighted residual method can be used as a basis for the finite-elements method and boundary-elements method [3.15]. It is helpful in unifying software design for both FEM and BEM. Consider a linear, operational equation:

$$L(u_0) = 0 \quad \text{in space } \Omega \text{ (Fig. 3.1)} \tag{3.1}$$

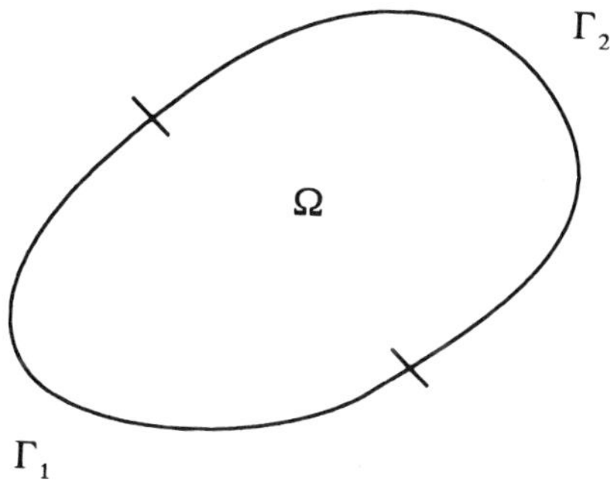

Fig. 3.1 *Bounded region*

with boundary conditions:

$$S(u_0) = s \qquad \text{on } \Gamma_1 \tag{3.2}$$

$$G(u_0) = g \qquad \text{on } \Gamma_2 \tag{3.3}$$

where:

$\Gamma = \Gamma_1 + \Gamma_2$
L N-order differential operator,
G $N-1$-order differential operator,
S $N-2$-order differential operator,
u_0 unknown function.

Equations (3.2) define the essential and natural boundary conditions. In this section differential equations of the second order ($N=2$) will only be taken into account so the essential boundary conditions mean Dirichlet and the natural boundary conditions mean Neumann conditions.

The exact solution u_0 is approximated by a complete series of functions

$$[\varphi_j]^{\mathrm{T}}, j = 1, 2, \ldots,$$

$$u = \alpha_1\varphi_1 + \alpha_2\varphi_2 + \cdots + \alpha_n\varphi_n, \tag{3.4}$$

where:

u an approximating function,
α_i unknown coefficients,
φ_i linearly independent functions.

An error inside the domain Ω is determined by a relation:

$$\varepsilon = L(u) - L(u_0) \tag{3.5}$$

Similarly on the Γ_1 border

$$\varepsilon_1 = S(u) - S(u_0) \tag{3.6}$$

and on the Γ_2 border

$$\varepsilon_2 = G(u) - G(u_0) \tag{3.7}$$

Depending on the type of approximating solution of eqns (3.1)–(3.3), part of the errors is minimized and part is neglected. Let us assume that an error is equal to zero on the Γ border. This assumption is used in the spatial methods. The error is minimized by the formula

$$\langle \varepsilon, w \rangle \equiv \langle L(u), w \rangle \equiv \int_{\Omega} L(u) w \, d\Omega = 0 \tag{3.8}$$

where:

$$w = \beta_1 \psi_1 + \beta_2 \psi_2 + \cdots + \beta_i \psi_i + \cdots + \beta_m \psi_m, \tag{3.9}$$

β_i any coefficient,
ψ_i linearly independent functions.

The application of the weighted residual method depends on the selection of appropriate function series $[\psi_i]^T$. Some possible choices are discussed below which leads to the following particular methods.

Point collocation method [3.8]
If

$$\psi_i = \delta_i$$

then

$$\langle L(u), \delta_i \rangle = 0 \qquad \text{for } i = 1, 2, \ldots, m \tag{3.8a}$$

where δ_i is the Dirac delta function in point i.

A set of equations (3.8a) may be over-determined when $m > n$ (n was defined by eqn (3.4)).

Bubnov–Galerkin method [3.14]
It is very easy to introduce FEM on the basis of the Bubnov–Galerkin method. In expressions (3.4) and (3.9) we assume that $n = m$. Thus

$$u = \alpha_1 \varphi_1 + \alpha_2 \varphi_2 + \cdots + \alpha_i \varphi_i + \cdots + \alpha_m \varphi_m \tag{3.10}$$

$$w = \beta_1 \varphi_1 + \beta_2 \varphi_2 + \cdots + \beta_i \varphi_i + \cdots + \beta_m \varphi_m \tag{3.11}$$

If

$$\varphi_i = \begin{cases} \neq 0 & \text{in } \Omega_i \\ = 0 & \text{in } \Omega \setminus \Omega_i \end{cases}$$

Where Ω_i is a finite element, we obtain a numerical implementation of the Bubnov–Galerkin method which is the FEM.

The least-squares method [3.8, 3.14]
This method is based on minimizing the square of error $\varepsilon^2 = (L(u) - L(u_0))^2$, which is equivalent to the assumption that $\psi_i = L(\varphi_i)$.

If the operator L (3.1) is symmetric, integration by parts is usually used, and instead of the scalar product $\langle L(u), w \rangle$ the following relation is assumed

$$\langle L(u), w \rangle = -\langle D(u), D(w) \rangle + \langle G(u), S(w) \rangle_{\Gamma}, \tag{3.12}$$

where

$$\langle G(u), S(w)\rangle_{\Gamma} \equiv \int_{\Gamma} G(u)S(w)\,\mathrm{d}\Gamma \tag{3.13}$$

For the Laplace's operator ($N = 2, L = \nabla^2$) eqn (3.12) is similar to Green's formula

$$\int_{\Omega} \nabla^2 uw\,\mathrm{d}\Omega = -\int_{\Omega} \mathit{grad}\,u\,\mathit{grad}\,w\,\mathrm{d}\Omega + \int_{\Gamma} \frac{\partial u}{\partial n}\,\mathrm{d}\Gamma \tag{3.14}$$

L is an N-order operator and D is of order $N-1$. In this case it is possible to use basic functions $[\varphi_i]^{\mathrm{T}}$ belonging to the space $C^{N-2}(\Omega)$. For the case ($N = 2$) functions $[\varphi_i]^{\mathrm{T}}$ are continuous and their derivatives are discontinuous. Moreover, if the operator L is positive definite the equations obtained from the Bubnov–Galerkin method are the same as equations obtained from Ritz's method.

The boundary methods belong to the next group. They are based on the assumption that functions $[\varphi_i]^{\mathrm{T}}$ or $[\psi_i]^{\mathrm{T}}$ satisfy eqn (3.1). According to this $\varepsilon = 0$ whereas ε_1 and ε_2 are minimized. Let us present boundary methods for the cases where L is a symmetric operator.

Boundary collocation method [3.8]

If functions $[\varphi_i]^{\mathrm{T}}$ satisfy eqn (3.1) but do not fulfil the boundary conditions, the error is minimized on the boundary

$$\langle \varepsilon_1, w\rangle_{\Gamma_1} \equiv \langle S(u) - s, G(w)\rangle_{\Gamma_1} = 0 \tag{3.15}$$

$$\langle \varepsilon_2, w\rangle_{\Gamma_2} \equiv \langle G(u) - g, S(w)\rangle_{\Gamma_2} = 0 \tag{3.16}$$

Trefftz method [3.8]

If functions ψ_i satisfy eqn (3.1), integrating by parts the relationship (3.8), we obtain

$$\langle L(u), w\rangle \equiv \langle u, L^*(w)\rangle + \langle S(w), G(w)\rangle_{\Gamma} = \langle S(u), G^*(w)\rangle_{\Gamma} = 0 \tag{3.8c}$$

where L^*, G^* are the operators adjoint to L, G.

Substituting eqn (3.1) into eqn (3.8c) the following boundary equation is obtained

$$\langle S(w), G(u)\rangle_{\Gamma} = \langle S(u), G^*(w)\rangle_{\Gamma} \tag{3.17}$$

Method based on the singular solution

Assuming that the weight functions w_i are solutions of the equation

$$L^*(w) = \delta_i \tag{3.18}$$

in the domain Ω_i the following equation is obtained

$$u_i + \langle S(w), G(w)\rangle_{\Gamma} = \langle S(u), G^*(w)\rangle_{\Gamma} \tag{3.19}$$

and on the border Γ surrounding domain Ω_i

$$C_i u_i + \langle S(w), G(u) \rangle_\Gamma = \langle S(u), G^*(w) \rangle_\Gamma \tag{3.20}$$

Equation (3.20) is the basis of the BEM.

3.3 THE FINITE-ELEMENT APPROACH [3.4; 3.26; 3.31; 3.38]

The error in the Bubnov–Galerkin (Ritz) method can be minimized by:

- using a large number (m) of elements in the approximating formula (3.10);
- dividing the domain Ω into many subdomains Ω_i (elements).

In the latter case a simple approximating function (small m) is applied within the element Ω_i. The classical Bubnov–Galerkin (Ritz) method is based on the first approach. More complicated approximating functions (larger m) must be used for more exact solutions. Complicated functions need sophisticated programming and time-consuming calculations. The complexity of the functions depends on the shape of the domain border.

FEM is based on the second approach. The FEM may be implemented in several ways. For two-dimensional problems the simplest subdomain (element) is a triangle with three vertices (nodes). The required function u^e in this kind of element is linear with respect to coordinates. Derivatives of u^e are constant inside element ($[\varphi_i]^T$ are of C^0 class). Between the elements, the functions u^e are continuous. In order to have a nearly exact solution ($u \approx u_0$) and fairly smooth derivatives of u we have to use a sufficiently dense triangular mesh. Alternative FEM formulations will be presented in Section 3.4. FEM will be presented as a subdomain technique based on the Ritz's method.

Consider the operational equation

$$L(u) = f \tag{3.21}$$

which has a different right-hand side from eqn (3.1). Operator L is positive definite. The solution of eqn (3.21) is provided by minimizing the functional $\Phi(u)$. The shape of functional $\Phi(u)$ depends on: the operator L, f and the boundary conditions. A few functionals will be shown later. In applying FEM, the domain, in which functional $\Phi(u)$ is minimized, is divided into subdomains Ω_i (finite elements). This is shown in Fig. 3.2.

Elements are defined by nodes (vertices) (i, j, k). For simplicity only triangular elements will be considered for two-dimensional problems. Inside any triangular element Ω_i the solution u^e fulfils the following relationship

$$u^e(x, y) = [N_i(x, y), N_j(x, y), N_k(x, y)] \begin{bmatrix} u_i \\ u_j \\ u_k \end{bmatrix} = [N]^T [u]^e \tag{3.22}$$

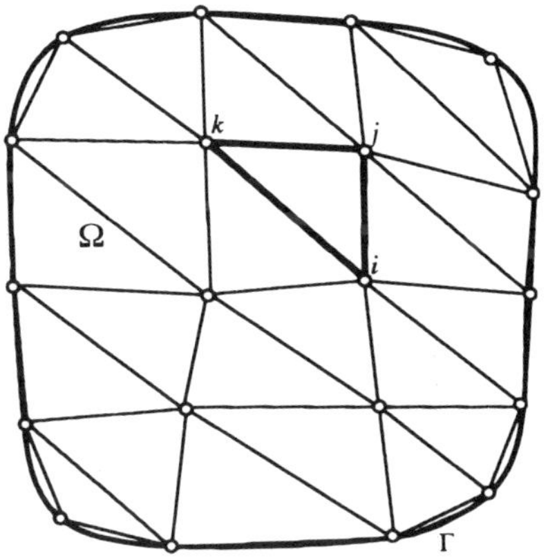

Fig. 3.2 *Division of domain into elements*

where:

$[N]^{\mathrm{T}}$	row vector of the shape functions,
$[u]^e$	column vector,
u_i, u_j, u_k	set of the function u with values at nodes i, j, k, respectively,
$N_i(x, y)$, $N_j(x, y)$, $N_k(x, y)$	must be unity at the ith, jth, kth nodes respectively and zero at all other nodes (e.g. $N_i(x_i, y_i) = 1$, $N_i(x_j, y_j) = 0$).

The functional $\Phi(u)$ is minimized with respect to values of the function u in all the nodes (vertices) of domain Ω (together with border Γ), which means with respect to a vector:

$$[u] = \begin{bmatrix} u_1 \\ u_2 \\ \vdots \\ u_r \end{bmatrix} \tag{3.23}$$

where r is the total number of nodes after division of domain Ω into elements. In order to minimize the functional $\Phi(u)$, $\Phi(u)$ must be differentiated with respect to column vector $[u]$ and equated to zero

$$\frac{\partial \Phi}{\partial [u]} = \begin{bmatrix} \dfrac{\partial \Phi}{\partial u_1} \\ \dfrac{\partial \Phi}{\partial u_2} \\ \vdots \\ \dfrac{\partial \Phi}{\partial u_r} \end{bmatrix} = 0 \tag{3.24}$$

Taking into account that

$$\Phi = \sum_{e=1}^{t} \Phi^e \tag{3.25}$$

where:

- Φ total functional in domain Ω,
- Φ^e functional in subdomain Ω^e,
- e element number,
- t total number of elements,

we obtain for any equation belonging to a set (3.24)

$$\frac{\partial \Phi}{\partial u_k} = \frac{\partial \sum_{e=1}^{t} \Phi^e}{\partial u_k} = \sum_{e=1}^{t} \frac{\partial \Phi^e}{\partial u_k} = 0 \tag{3.26}$$

For any element, where we take into account an elliptical equation, we obtain

$$\frac{\partial \Phi^e}{\partial [u]^e} = [h]^e[u]^e + [S]^e \tag{3.27}$$

or

$$\begin{bmatrix} \dfrac{\partial \Phi^e}{\partial u_i} \\ \dfrac{\partial \Phi^e}{\partial u_j} \\ \dfrac{\partial \Phi^e}{\partial u_k} \end{bmatrix} = \begin{bmatrix} h^e_{ii} & h^e_{ij} & h^e_{ik} \\ h^e_{ji} & h^e_{jj} & h^e_{jk} \\ h^e_{ki} & h^e_{kj} & h^e_{kk} \end{bmatrix} \begin{bmatrix} u_i \\ u_j \\ u_k \end{bmatrix} + \begin{bmatrix} S_i \\ S_j \\ S_k \end{bmatrix} \tag{3.27a}$$

where:

- $[h]$ stiffness matrix,
- $[S]$ load vector.

Finally the set of equations transforms to

$$\frac{\partial \Phi}{\partial [u]} = [H][u] + [S] = [0] \tag{3.24a}$$

where

$$[S] = \begin{bmatrix} S_1 \\ S_2 \\ \cdot \\ \cdot \\ \cdot \\ S_r \end{bmatrix}; \qquad [H] = \begin{bmatrix} H_{11} & \cdot & \cdot & \cdot & \cdot & H_{1r} \\ \cdot & \cdot & \cdot & \cdot & \cdot & \cdot \\ H_{i1} & \cdot & H_{ii} & H_{ij} & \cdot & H_{ir} \\ \cdot & \cdot & \cdot & \cdot & \cdot & \cdot \\ H_{r1} & \cdot & \cdot & \cdot & \cdot & H_{rr} \end{bmatrix} \tag{3.24b}$$

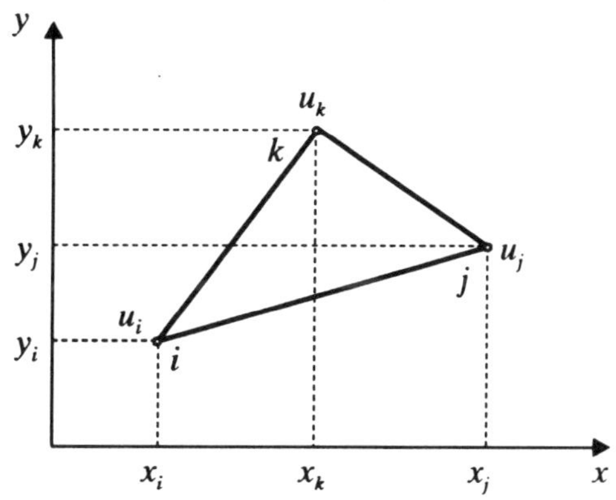

Fig. 3.3 *Triangular elements in cartesian coordinates*

$$H_{ij} = \sum_{e}^{t} h_{ij}^{e}, \qquad S_i = \sum_{e}^{t} S_i^e \tag{3.24c}$$

Vector $[u]$ is a solution of equation (3.24) or (3.24a). Inside the triangular elements (Fig. 3.3) function u is linear with respect to coordinates x, y:

$$u^e(x, y) = \alpha_1 + \alpha_2 x + \alpha_3 y \tag{3.28}$$

At the nodes (vertices of triangles) function u satisfies the following set of equations

$$\begin{aligned} u_i &= \alpha_1 + \alpha_2 x_i + \alpha_3 y_i \\ u_j &= \alpha_1 + \alpha_2 x_j + \alpha_3 y_j \\ u_k &= \alpha_1 + \alpha_2 x_k + \alpha_3 y_k \end{aligned} \tag{3.29}$$

or in matrix notation

$$\begin{bmatrix} u_i \\ u_j \\ u_k \end{bmatrix} = \begin{bmatrix} 1 & x_i & y_i \\ 1 & x_j & y_j \\ 1 & x_k & y_k \end{bmatrix} \begin{bmatrix} \alpha_1 \\ \alpha_2 \\ \alpha_3 \end{bmatrix}, \tag{3.29a}$$

or

$$[u] = [\Delta][\alpha] \tag{3.29b}$$

Solving eqn (3.29) we obtain

$$\begin{aligned} \alpha_1 &= \frac{1}{2\Delta}(a_i u_i + a_j u_j + a_k u_k) \\ \alpha_2 &= \frac{1}{2\Delta}(b_i u_i + b_j u_j + b_k u_k) \\ \alpha_3 &= \frac{1}{2\Delta}(c_i u_i + c_j u_j + c_k u_k) \end{aligned} \tag{3.30}$$

where

$$2\Delta = \det \begin{bmatrix} 1 & x_i & y_i \\ 1 & x_j & y_j \\ 1 & x_k & y_k \end{bmatrix} \tag{3.31}$$

Δ is the area of a triangle i, j, k, and

$$\begin{aligned} a_i &= x_j y_k - x_k y_j \\ b_i &= y_j - y_k = y_{jk} \\ c_i &= x_k - x_j = x_{kj} \end{aligned} \tag{3.32}$$

Coefficients $a_j, b_j, \ldots$ are calculated by cyclic permutation of suffices i, j, k:

$$\begin{aligned} a_j &= x_k y_i - x_i y_k \\ b_j &= y_k - y_i \\ c_j &= x_i - x_k, \ldots \end{aligned} \tag{3.33}$$

Substituting eqn (3.30) into eqn (3.28) we obtain eqn (3.22), where

$$N_m = \frac{1}{2\Delta}(a_m + b_m x + c_m y) \qquad m = i, j, k \tag{3.34}$$

The method is applied to the solution of a two-dimensional problem, described by Helmholtz's equation:

$$-\frac{\partial}{\partial x}\left(p_x \frac{\partial u}{\partial x}\right) - \frac{\partial}{\partial y}\left(p_y \frac{\partial u}{\partial y}\right) + k^2 u = f \tag{3.35}$$

If $p_x = p_y = 1$ the first part of eqn (3.35) is simplified to

$$\frac{\partial}{\partial x}\left(\frac{\partial u}{\partial x}\right) + \frac{\partial}{\partial y}\left(\frac{\partial u}{\partial y}\right) = \frac{\partial^2 u}{\partial x^2} + \frac{\partial^2 u}{\partial y^2} = \Delta u \tag{3.36}$$

Equation (3.35) is valid in anisotropic ($p_x \neq p_y$) and non-homogeneous ($p_x(x, y)$, $p_y(x, y)$) media. In isotropic and homogeneous media, function $u(x, y)$ satisfies the simplier eqn (3.36). Coefficients p_x, p_y are physical parameters e.g. ε_x, ε_y, μ_x, μ_y, or σ_x, σ_y, according to the prescribed field (electrostatic, magnetostatic, or electromagnetic). Coefficients p_x and p_y must be differentiable whereas k^2 and f must be continuous. The shape of the function depends on a differential equation and boundary conditions. For Dirichlet's conditions

$$u|_\Gamma = u(B) = u_0 \tag{3.37a}$$

where B is any point on the border Γ, the functional is as follows

$$\Phi(u) = \int_\Omega \left[p_x \left(\frac{\partial u}{\partial x}\right)^2 + p_y \left(\frac{\partial u}{\partial y}\right)^2 + k^2 u^2 - 2fu\right] \mathrm{d}x\,\mathrm{d}y. \tag{3.38}$$

Minimizing the functional (3.38) is equivalent to a boundary value problem (3.37a) for eqn (3.36). For Neumann's conditions

$$\left.\frac{\mathrm{d}u}{\mathrm{d}n}\right|_{\Gamma} = \varphi(B) = -\varphi_0 \tag{3.37b}$$

The following functional is now obtained

$$\Phi(u) = \int_{\Omega} \left[p_x \left(\frac{\partial u}{\partial x}\right)^2 + p_y \left(\frac{\partial u}{\partial y}\right)^2 + k^2 u^2 - 2fu \right] \mathrm{d}x\,\mathrm{d}y + 2\int_{\Gamma} \varphi_0 u \mathrm{d}\Gamma \tag{3.39}$$

For Poisson's or Laplace's equations in functional (3.38) or (3.39) the relevant components $(k^2u^2, 2fu)$ are neglected.

Equation (3.24a) is resolved in the following manner. At first the functional must be differentiated with respect to a nodal value of u_m

$$\frac{\partial \Phi^e}{\partial u_m} = 2\int_{\Omega^e} \left[p_x \frac{\partial u^e}{\partial x} \frac{\partial}{\partial u_m}\left(\frac{\partial u^e}{\partial x}\right) + p_y \frac{\partial u^e}{\partial y} \frac{\partial}{\partial u_m}\left(\frac{\partial u^e}{\partial y}\right) + k^2 u^e \frac{\partial u^e}{\partial u_m} - f\frac{\partial u^e}{\partial u_m} \right] \mathrm{d}x\,\mathrm{d}y$$

$$+ 2\int_{\Gamma^e} \varphi^0 \frac{\partial u^e}{\partial u_m} \mathrm{d}\Gamma, \qquad m = i, j, k \tag{3.40}$$

where Γ^e is the boundary of the element Ω^e.

Taking into account that

$$\frac{\partial u^e}{\partial x} = \left[\frac{\partial N_i}{\partial x}, \frac{\partial N_j}{\partial x}, \frac{\partial N_k}{\partial x}\right] \begin{bmatrix} u_i \\ u_j \\ u_k \end{bmatrix} = \frac{1}{2\Delta}[b_i, b_j, b_k][u]^e$$

$$\frac{\partial u^e}{\partial y} = \left[\frac{\partial N_i}{\partial y}, \frac{\partial N_j}{\partial y}, \frac{\partial N_k}{\partial y}\right] \begin{bmatrix} u_i \\ u_j \\ u_k \end{bmatrix} = \frac{1}{2\Delta}[c_i, c_j, c_k][u]^e \tag{3.41}$$

$$\frac{\partial}{\partial u_m}\left(\frac{\partial u^e}{\partial x}\right) = \frac{b_m}{2\Delta} \qquad \frac{\partial}{\partial u_m}\left(\frac{\partial u^e}{\partial y}\right) = \frac{c_m}{2\Delta}$$

$$\frac{\partial u^e}{\partial u_m} = N_m = \frac{1}{2\Delta}(a_m + b_m x + c_m y)$$

eqn (3.40) becomes

$$\frac{\partial \Phi^e}{\partial u_m} = 2\int_{\Omega^e} \left(\frac{1}{4\Delta^2} b_m p_x [b_i, b_j, b_k][u]^e + \frac{1}{4\Delta^2} c_m p_y [c_i, c_j, c_k][u]^e + \right.$$

$$\left. + k^2 [N]^{\mathrm{T}} [u]^e N_m - f N_m \right) \mathrm{d}x\,\mathrm{d}y + 2\int_{\Gamma^e} \varphi_0 N_m \,\mathrm{d}\Gamma \tag{3.42a}$$

If domain Ω is discretized into a large number of elements Ω^e, then p_x, p_y and f may be assumed constant inside Ω^e. Additionally taking into account that

$$\int_{\Omega^e} N_i \, \mathrm{d}x \, \mathrm{d}y = \int_{\Omega^e} N_j \, \mathrm{d}x \, \mathrm{d}y = \int_{\Omega^e} N_k \, \mathrm{d}x \, \mathrm{d}y = \frac{1}{3} \Delta \tag{3.42b}$$

we obtain

$$\frac{\partial \Phi^e}{\partial u_m} = \frac{1}{2\Delta} [p_x b_i b_m + p_y c_i c_m, p_x b_j b_m + p_y c_j c_m, p_x b_k b_m + p_y c_k c_m] [u]^e$$
$$+ \int_{\Omega^e} k^2 [N]^{\mathrm{T}} [u] N_m \, \mathrm{d}x \, \mathrm{d}y - \frac{2}{3} \Delta f + 2 \int_{\Gamma^e} \varphi_0 N_m \, \mathrm{d}\Gamma^e \tag{3.43}$$

or

$$\begin{bmatrix} \dfrac{\partial \Phi^e}{\partial u_i} \\ \dfrac{\partial \Phi^e}{\partial u_j} \\ \dfrac{\partial \Phi^e}{\partial u_k} \end{bmatrix} = \begin{bmatrix} h^e_{ii} & h^e_{ij} & h^e_{ik} \\ h^e_{ji} & h^e_{jj} & h^e_{jk} \\ h^e_{ki} & h^e_{kj} & h^e_{kk} \end{bmatrix} \begin{bmatrix} u_i \\ u_j \\ u_k \end{bmatrix} + \begin{bmatrix} S^e_i \\ S^e_j \\ S^e_k \end{bmatrix} \tag{3.44}$$

where, for example

$$h_{ij} = \frac{p_x b_i b_j + p_x c_i c_j}{2\Delta} + 2 \int_{\Omega^e} k^e N_i N_j \, \mathrm{d}x \, \mathrm{d}y \tag{3.45}$$

$$S^e_m = -\frac{2}{3} \Delta f + 2 \int_{\Gamma^e}^{\varphi_0} \varphi_0 N_m \mathrm{d}\Gamma$$

$$m = i, j, k \tag{3.45a}$$

For the full Ω domain the following system of equations is obtained

$$\frac{\partial \Phi}{\partial [u_i]} = \sum_{e=1}^{t} \frac{\partial \Phi^e}{\partial [u_i]} = [H][u] + [S] = [0] \tag{3.46}$$

Inside Ω a function $[u]$ satisfies the equation

$$[H][u] + [S] = [0] \tag{3.47}$$

To illustrate the method mentioned above the problem shown in Fig. 3.4 will be solved. After discretization of Ω, the total number of nodes is equal to 5. In Fig. 3.5a–e the procedure of formulating eqn (3.46) is shown. Figures 3.5a, b, c and d present equations for successive elements Ω_e. The first element is defined by nodes 1, 2, 3. According to this matrices $[H]^1$, $[u]^1$ and $[S]^1$ were formed. Non-zero coefficients are marked by lines in cells. Empty cells mean zero coefficients (Fig. 3.5a). Using this procedure eqns (3.46) are formed for elements 2, 3 and 4 (Fig. 3.5b, c and d). In the next stage the global matrix is written (Fig. 3.5e). It is easy to observe (Fig. 3.5e) that the global stiffness matrix $[H]$ is symmetric. Only nodes 1 and 4, 2 and 5 are not directly connected in this example (Fig. 3.4). Thus $h_{14} = h_{41} = 0$ and $h_{25} = h_{52} = 0$ (Fig. 3.5e). If a domain were divided into many elements (e.g. Fig. 3.6) the matrix $[H]$ would include more coefficients equal to zero because only a few

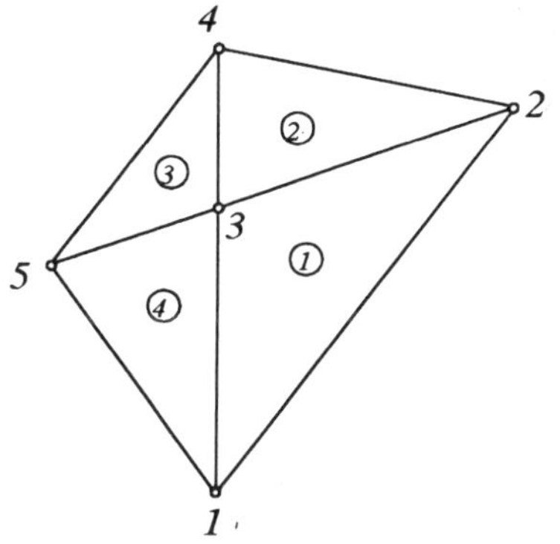

Fig. 3.4 *Division of domain into four triangular elements*

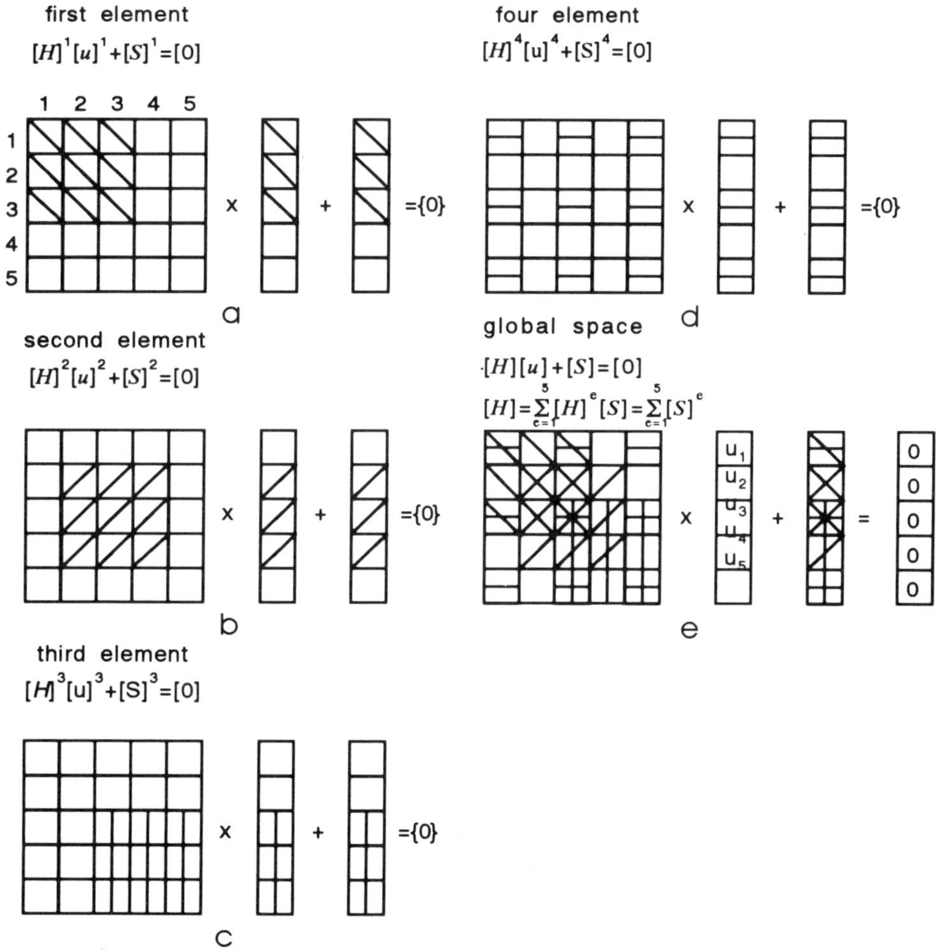

Fig. 3.5 *Assembling of FEM equations for the example from Fig. 3.4*

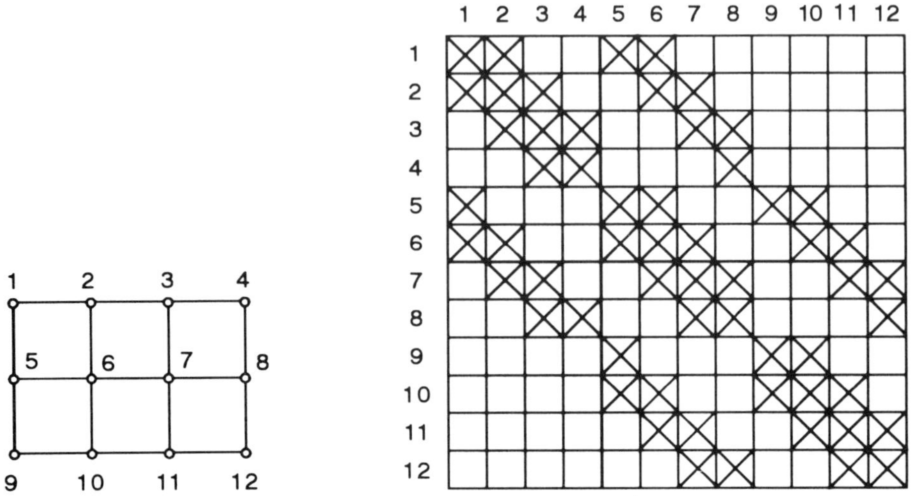

Fig. 3.6 *12-node mesh and its [H] matrix*

nodes are connected directly. Thus in this case the matrix becomes a sparse matrix. A sparse matrix may be banded if non-zero coefficients are placed in a band. The band width is equal to the maximum difference in number of nodes plus one. Thus in the problem presented in Fig. 3.6, the band width is equal to 6 (e.g. 1st-element $6 - 1 + 1 = 6$, 6th-element $12 - 7 + 1 = 6$). The matrix $[H]$ is symmetric and banded. Only the upper triangular part of the matrix, including the main diagonal, needs to be stored in a computer memory.

3.4 ISOPARAMETRIC ELEMENTS [3.38; 3.40]

The simple three node triangular element was described in Section 3.3. Sometimes the boundary Γ of a domain is strongly curvilinear. In these cases an accurate representation of Γ by triangular elements requires a very dense mesh. A problem like this may be solved by using curvilinear elements. It is then possible to obtain accurate representation of the boundary Γ by a small number of elements. In this section four elements will be considered, i.e. (Fig. 3.7):

- four-nodal quadrilateral,
- six-nodal isoparametric triangle,
- six-nodal rectilinear triangle,
- eight-nodal isoparametric quadrilateral.

Within a three-node triangular element, the solution is given by three terms (3.4). Generally a number of function terms is equal to a number of element nodes. Thus for a four-nodal element

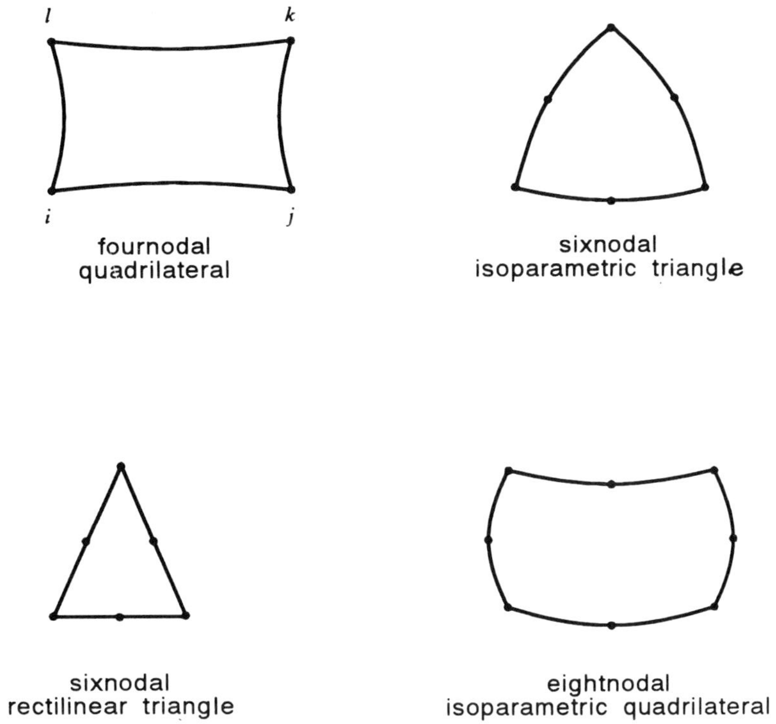

Fig. 3.7 *Elements including more than three nodes*

$$u = \alpha_1 + \alpha_2 x + \alpha_3 y + \alpha_4 xy \tag{3.48}$$

and for six nodes

$$u = \alpha_1 + \alpha_2 x + \alpha_3 y + \alpha_4 xy + \alpha_5 x^2 + \alpha_6 y^2 \tag{3.49}$$

It is very convenient to use local normalized coordinates (η, ξ) in deriving the isoparametric stiffness matrix. To achieve compatibility with Gaussian's quadrature, coordinates (η, ξ) are defined in the interval $(-1, 1)$. For the two-dimensional case, application of local coordinates leads to the equation

$$u^e(\eta, \xi) = [N]^{\mathrm{T}}[u]^e \tag{3.50}$$

The local coordinates (η, ξ) of a point in the rectilinear quadrilateral (Fig. 3.8a) correspond to the point (x, y) in a curvilinear quadrilateral (Fig. 3.8b) defined by the relationships

$$x(\eta, \xi) = [N]^{\mathrm{T}}[x]^e \tag{3.51}$$

$$y(\eta, \xi) = [N]^{\mathrm{T}}[y]^e$$

where

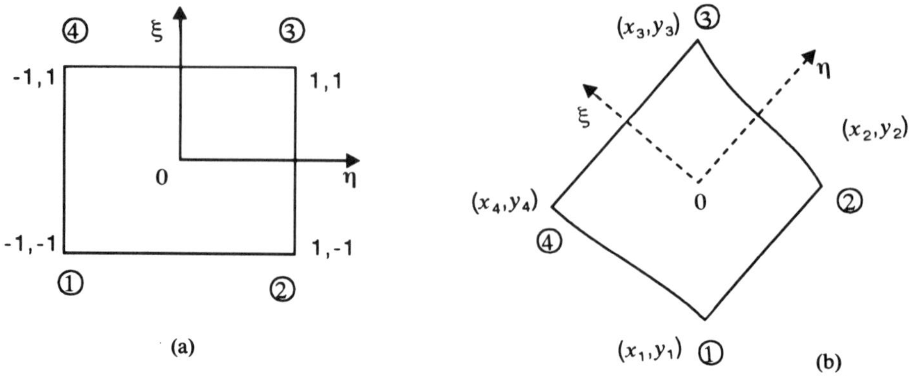

Fig. 3.8 *Four-nodal isoparametric elements*

$$[x]^e = \begin{bmatrix} x_i \\ x_j \\ x_k \\ x_l \end{bmatrix}, \qquad [y]^e = \begin{bmatrix} y_i \\ y_j \\ y_k \\ y_l \end{bmatrix}$$

For the quadrilateral (Fig. 3.8) the shape functions are defined by

$$N_i(\eta, \xi) = \tfrac{1}{4}(1 + \eta\eta_i)(1 + \xi\xi_i), \qquad 1 \leqslant i \leqslant 4 \tag{3.52}$$

where

$$\begin{aligned} \eta_1 &= \eta_4 = -1 \\ \eta_2 &= \eta_3 = 1 \\ \xi_1 &= \xi_2 = -1 \\ \xi_3 &= \xi_4 = 1 \end{aligned} \tag{3.53}$$

On the sides of elements ($\eta = \pm 1$ or $\xi = \pm 1$) shape functions $N_m(\eta, \xi)$ are linear. They satisfy the following conditions:

$$\sum_{m=1}^{M} N_m(\eta, \xi) = 1 \tag{3.54}$$

$$N_m(\eta_j, \xi_j) = \begin{cases} 1, & j = m \\ 0, & j \neq m \end{cases}$$

Global coordinates x, y are explicitly defined by local coordinates η, ξ. The inverse is not always true. Fortunately as a rule there is no need to calculate η, ξ but only to have the relationship between derivatives:

$$\begin{bmatrix} \dfrac{\partial u}{\partial \eta} \\ \dfrac{\partial u}{\partial \xi} \end{bmatrix} = \begin{bmatrix} \dfrac{\partial x}{\partial \eta} & \dfrac{\partial y}{\partial \eta} \\ \dfrac{\partial x}{\partial \xi} & \dfrac{\partial y}{\partial \xi} \end{bmatrix} \begin{bmatrix} \dfrac{\partial u}{\partial x} \\ \dfrac{\partial u}{\partial y} \end{bmatrix} \tag{3.55}$$

Derivatives are calculated by the following formulae

$$\frac{\partial x}{\partial \eta} = \frac{\partial}{\partial \eta}[N]^{\mathrm{T}}[x]^e, \qquad \frac{\partial y}{\partial \eta} = \frac{\partial}{\partial \eta}[N]^{\mathrm{T}}[y]^e$$
$$\frac{\partial x}{\partial \xi} = \frac{\partial}{\partial \xi}[N]^{\mathrm{T}}[x]^e, \qquad \frac{\partial y}{\partial \xi} = \frac{\partial}{\partial \xi}[N]^{\mathrm{T}}[y]^e \tag{3.56}$$

The matrix belonging to an element e, e.g. the stiffness matrix, may be expressed as

$$[h]^e = \int_{\Omega^e} [B]^e \, \mathrm{d}x \, \mathrm{d}y \tag{3.57}$$

where

$$B_{ij} = \frac{\partial N_i}{\partial x}\frac{\partial N_j}{\partial x} + \frac{\partial N_i}{\partial y}\frac{\partial N_j}{\partial y}$$

$$\begin{bmatrix} \dfrac{\partial N_i}{\partial x} \\ \dfrac{\partial N_i}{\partial y} \end{bmatrix} = [J]^{-1} \begin{bmatrix} \dfrac{\partial N_i}{\partial \eta} \\ \dfrac{\partial N_i}{\partial \xi} \end{bmatrix}$$

$$[J] = \begin{bmatrix} \sum_{i=1}^{M} \dfrac{\partial N_i}{\partial \eta} x_i & \sum_{i=1}^{M} \dfrac{\partial N_i}{\partial \eta} y_i \\ \sum_{i=1}^{M} \dfrac{\partial N_i}{\partial \xi} x_i & \sum_{i=1}^{M} \dfrac{\partial N_i}{\partial \xi} y_i \end{bmatrix} \tag{3.58}$$

where $[J]$ is the Jacobian matrix.

Taking into account the equations for isoparametric elements, formula (3.57) is transformed to

$$[h]^e = \int_{-1}^{1} \int_{-1}^{1} [B]^e |J| \, \mathrm{d}\eta \, \mathrm{d}\xi \tag{3.59a}$$

and for numerical calculation

$$[h]^e = \sum_{l=1}^{NIP} w_l [B_l]^e |J_l| \tag{3.59b}$$

where $[B_l]^e$ and $|J_l|$ are calculated at points and multiplied by weight coefficients w_l, and NIP is a number of integration points.

3.5 NONLINEARITY IN STATICS [3.14; 3.32; 3.38; 3.43; 3.44]

In this Chapter FEM will be applied to solve nonlinear stationary problems on the basis of:

- Newton–Raphson method, and
- direct iterative method.

Newton–Raphson's method
For nonlinear media, the stiffness matrix coefficients are functions of u. u itself satisfies a system of nonlinear algebraic equations, i.e.

$$[H(u)][u] - [S] = [0] \tag{3.60}$$

The left-hand side of eqn (3.60) may be written as

$$[F(u)] = [H(u)][u] - [S] \tag{3.61}$$

Equation (3.61) is solved by Newton–Raphson's iterative method. After n iterations the $(n+1)$th solution of eqn (3.60) is given by

$$[u^{n+1}] = [u^n] + [\Delta u^n] \tag{3.62}$$

The function $[F(u^{n+1})]$, defined by eqn (3.61), may be expanded into a Taylor's series:

$$[F(u^{n+1})] = [F(u^n)] + \left.\frac{\partial[F(\cdot)]}{\partial[u]}\right|_n [\Delta u^n] + \left.\frac{\partial^2[F(\cdot)]}{\partial[u^2]}\right|_n \frac{[\Delta u^n]^2}{2} + \ldots \tag{3.63}$$

Truncating series (3.63) to two terms we obtain Newton–Raphson's method

$$\begin{aligned} &[H_T(u^n)][\Delta u^n] = [F(u^n)] \\ &[u^{n+1}] = [u^n] + [\Delta u^n] \end{aligned} \tag{3.64}$$

where

$$[H_T(u^n)] = \left.\frac{\partial[F(\cdot)]}{\partial[u]}\right|_n \tag{3.65}$$

is a Jacobian (tangential) matrix. The basic version of Newton–Raphson's algorithm is shown in Fig. 3.9.

Direct iterative method
The direct iterative method is based on successively solving a system of linear equations. Equation (3.60) is tranformed to

$$[H(u^n)][u^{n+1}] = [S] \tag{3.66}$$

and solved by the algorithm shown in Fig. 3.10.

The Newton–Raphson's method and the direct iterative methods will be explained by using examples of magnetic field calculation (thus $u = A$). The Newton–Raphson's example uses a magnetic field calculation, excited in non-linear media by direct current.

Let us assume, that

$$\mathbf{A} = A_z\hat{\mathbf{k}} = A\hat{\mathbf{k}}, \qquad \mathbf{J} = J_z\hat{\mathbf{k}} = J\hat{\mathbf{k}} \tag{3.67}$$

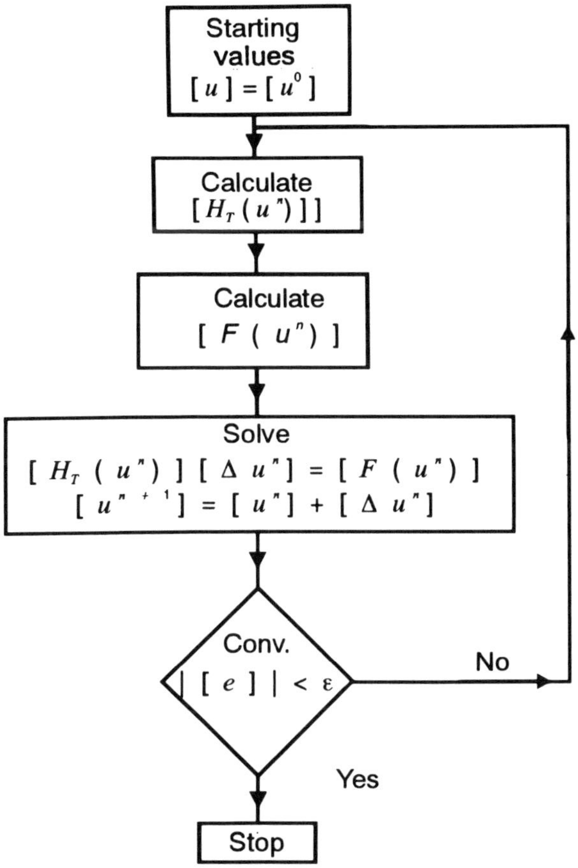

Fig. 3.9 *Flow chart for Newton–Raphson's method*

where:

$\hat{\mathbf{k}}$ unit vector in the Z-direction,
A magnetic vector potential,
J current density vector.

The starting point for solving the assumed example is the first Maxwell's equation

$$curl\,\mathbf{H} = \mathbf{J} \tag{3.68}$$

Taking into account that

$$curl\,\mathbf{A} = \mathbf{B} \tag{3.69}$$

$$\mathbf{B} = \mu(B)\mathbf{H} \tag{3.70}$$

$$\nu = \frac{1}{\mu}$$

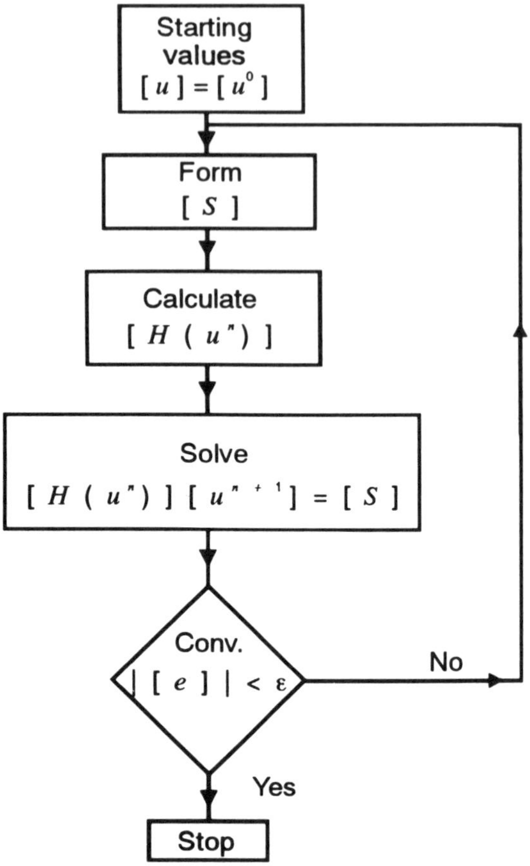

Fig. 3.10 *Flow chart for direct iterative method*

where $\nu(B)$ is an explicit function of B or H (known as reluctivity), we obtain a nonlinear Poisson's equation:

$$curl\,(\nu(B)\,curl\,\mathbf{A}) = -\mathbf{J} \tag{3.71}$$

Taking into account eqn (3.67), eqn (3.71) is transformed to

$$div(\nu(\cdot)\,grad\,A) = -J \tag{3.71a}$$

or

$$\nabla\cdot(\nu(\cdot)\nabla A) = -J$$

The boundary Γ, surrounding the domain Ω, will be divided into two parts. Two boundary conditions (Dirichlet or Neumann's) are applied in magnetostatics. For simplicity it is assumed that the boundary conditions are homogeneous

$$A = 0 \qquad \text{on } \Gamma_1 \tag{3.72}$$

$$\frac{\partial A}{\partial n} = 0 \qquad \text{on } \Gamma_2 \tag{3.73}$$

where n is the normal to the boundary Γ_2.

In Newton–Raphson's method the fundamental problem is to find the Jacobian matrix (3.65). Thus

$$\begin{aligned}[H_T(u)] &= \frac{\partial [F]}{\partial [u]} = \frac{\partial}{\partial [u]}([H(u)][u]) = \frac{\partial}{\partial [u]}[H(u)][u] + [H(u)] \\ &= [T(u)] + [H(u)]\end{aligned} \tag{3.74}$$

because

$$\frac{\partial}{\partial [u]}[H(u)][u] = [T(u)] \tag{3.75}$$

The shape of matrix $[T]$ depends on how reluctivity is defined. In the case when reluctivity is

$$\nu(\cdot) = \nu(B) = \nu(|\nabla \times A|) \tag{3.76}$$

the coefficients of matrix $[T]$ are given by

$$t_{ij}^e = \int_{\Omega^e} [\nabla N_i]^{\mathrm{T}} \frac{\partial v}{\partial B} \sqrt{[(\nabla [N][u]^{\mathrm{T}})(\nabla [N][u])]}\, \nabla N_j \, \mathrm{d}\Omega \tag{3.77}$$

For the direct iteration method, coefficients of the stiffness matrix and load matrix must be calculated

$$h_{ij}^e = \int_{\Omega^e} [\nabla N_i]^{\mathrm{T}} \nu(\cdot) \nabla Nj \, \mathrm{d}\Omega \tag{3.78}$$

$$S_i^e = \int_{\Omega^e} J N_i \, \mathrm{d}\Omega \tag{3.79}$$

The next step is the reluctivity formulation. For this purpose the magnetization curve is approximated by a multisegment method. The magnetization curve is given by points, and successive points define the beginning and end of successive segments. Reluctivity is constant inside intervals. Starting with the first point, in defining the curve, we obtain

$$\begin{aligned} &\nu_1 = \nu_2 \\ (H_i, B_i); \qquad &\nu_i = \frac{H_i}{B_i}, \quad i = 2, 3, \ldots, N+1 \end{aligned} \tag{3.80}$$

Reluctivity inside respective segments is calculated by the formula

$$\nu(B) = \frac{\nu_{i+1} - \nu_i}{B_{i+1} - B_i}(B - B_i) + \nu_i \tag{3.81}$$

where B belongs to $[B_i, B_{i+1}]$.

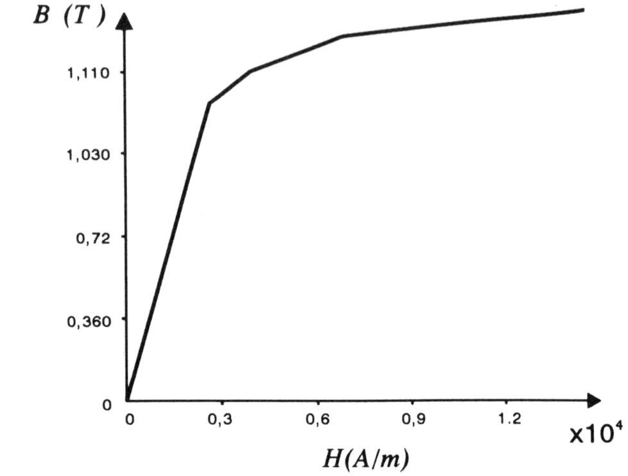
B (T)
1,110
1,030
0,72
0,360
0
0
0,3
0,6
0,9
1.2
x10⁴
H(A/m)

(a)

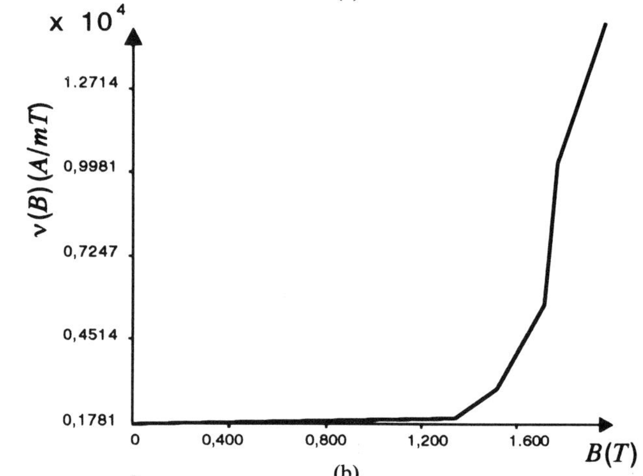
x 10⁴
1.2714
0,9981
0,7247
0,4514
0,1781
ν(B) (A/mT)
0
0,400
0,800
1,200
1.600
B(T)

(b)

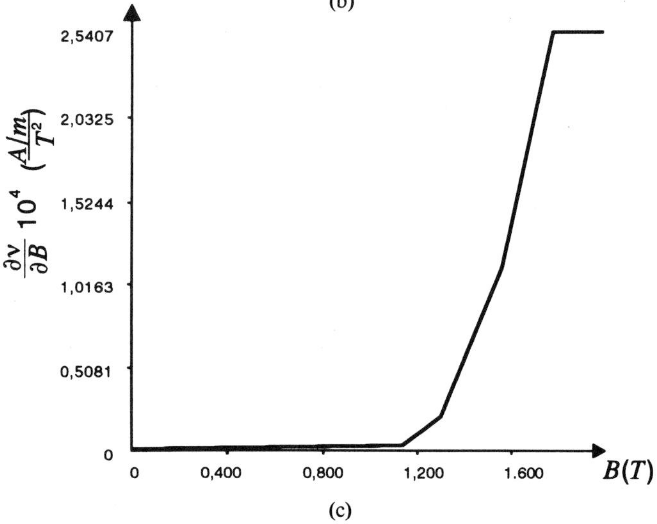
2,5407
2,0325
1,5244
1,0163
0,5081
0
∂ν/∂B 10⁴ (A/m / T²)
0
0,400
0,800
1,200
1.600
B(T)

(c)

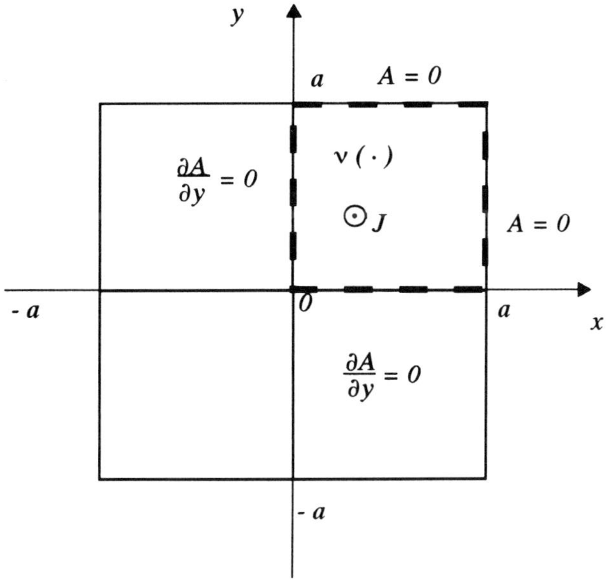

(d)

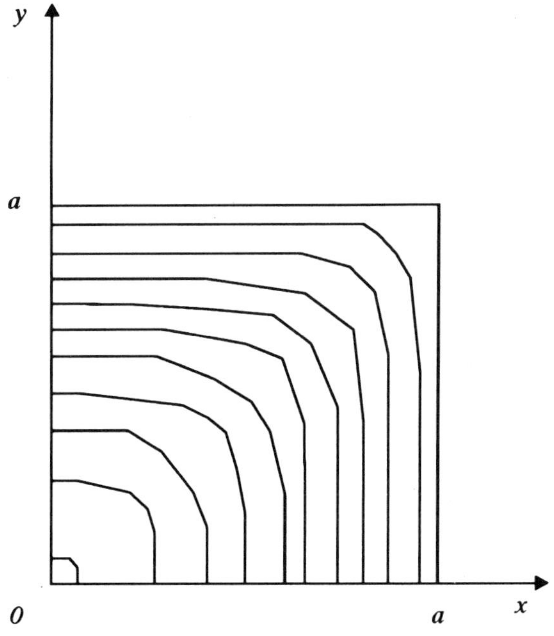

(e)

Fig. 3.11 *(a) Magnetization curve, (b) reluctivity curve, (c) reluctivity curve gradient, (d) infinitely long ferromagnetic conductor, (e) magnetic field distribution*

The gradient of the reluctivity curve is defined by coefficients

$$\alpha_i = \frac{\nu_{i+1} - \nu_i}{B_{i+1} - B_i} \tag{3.82}$$

with the following assumptions:

- if $B = 0$ then $\dfrac{\partial \nu}{\partial B} = 0$,
- $\alpha_i = \left(\dfrac{\partial \nu}{\partial B}\right)_i$ is the gradient for $\tilde{B}_i = \dfrac{B_{i+1} + B_i}{2}$,
- the gradient becomes linear inside successive segments defined by $\tilde{B}$,
- for $B > B_{saturation}$ the gradient is equal to zero.

Thus for flux density B

$$\tilde{B}_i \leqslant B \leqslant \tilde{B}_{i+1}$$

the gradient of the reluctivity curve is given by the formula

$$\frac{\partial \nu}{\partial B} = \frac{\alpha_{i+1} - \alpha_i}{\tilde{B}_{i+1} - \tilde{B}_i}(B - \tilde{B})_i + \alpha_i \tag{3.83}$$

In Fig. 3.11a–e the magnetization curve [3.11], reluctivity curve and reluctivity curve gradient are presented for the steel containing 0.3% C. The method is illustrated by a magnetic field calculation in an infinitely long ferromagnetic conductor (Fig. 3.11d). As the conductor is symmetrical, the magnetic field was calculated for a quarter of the cross-section (Fig. 3.11e).

3.6 LINEAR TIME-DEPENDENT PROBLEMS [3.14; 3.32; 3.38; 3.40]

The FEM described in this Section is applied to time-varying electromagnetic field. As there is a conducting media the problem can be described by a non-homogeneous partial differential equation

$$-\frac{\partial}{\partial x}\left(p_x \frac{\partial u}{\partial x}\right) - \frac{\partial}{\partial y}\left(p_y \frac{\partial u}{\partial y}\right) + k^2 \frac{\partial u}{\partial t} = f \tag{3.36a}$$

with the following boundary conditions

$$p_n \frac{\partial u}{\partial n} = u_1(t) \quad \text{on } \Gamma_1 \tag{3.84}$$

$$u = 0 \quad \text{on } \Gamma_2$$

$$\frac{\partial u}{\partial n} = 0 \quad \text{on } \Gamma_3$$

and the initial condition

$$u_{(t=0)} = u_0(x, y) \qquad \text{for } (x, y) \text{ belonging to } \Omega \tag{3.85}$$

where

$$\Gamma_1 + \Gamma_2 + \Gamma_3 = \Gamma$$

As in Section 3.5, a solution for time-varying problems will be presented for the magnetic field calculation described by the equation

$$\nabla \cdot (\nu \nabla A) = \sigma \frac{\partial A}{\partial t} - J \tag{3.86}$$

with boundary conditions

$$\nu \frac{\partial A}{\partial n} = h(t) \quad \text{on } \Gamma_1 \tag{3.87}$$

$$A = 0 \quad \text{on } \Gamma_2$$

$$\frac{\partial A}{\partial n} = 0 \quad \text{on } \Gamma_3$$

and the initial condition

$$A_{t=0} = A_0(x, y), \qquad (x, y) \text{ belongs to } \Omega \tag{3.88}$$

The domain under consideration is presented in Fig. 3.12. Equations (3.86)–(3.88) are substituted according to the weighted residual method, by the following integral equation

$$\int_\Omega w_1 \nabla \cdot (\nu \nabla A)\, \mathrm{d}\Omega - \int_\Omega w_1 \sigma \frac{\partial A}{\partial t} \mathrm{d}\Omega + \int_\Omega w_1 J_0 \,\mathrm{d}\Omega + \int_{\Gamma_1} \left(\nu \frac{\partial A}{\partial n} - h \right) w_2 \,\mathrm{d}\Gamma = 0 \tag{3.89}$$

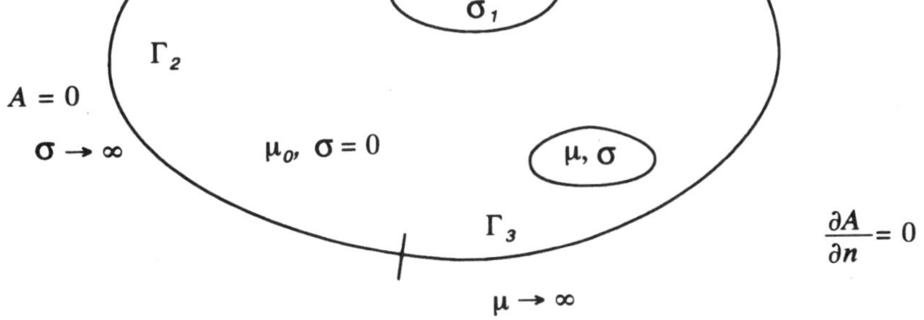

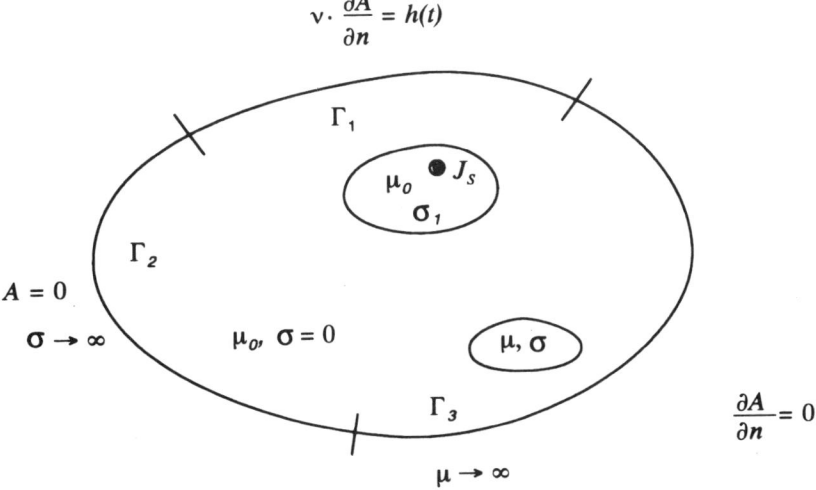

Fig. 3.12 *Domain under investigation for time-dependent problem*

After assuming, that

$$A = [N]^{\mathrm{T}}[u] \tag{3.90}$$
$$w_1 = N_i$$
$$w_2 = -\sigma N_i$$

applying the second Green's identity and Galerkin's method, eqn (3.89) is transformed to

$$\int_{\Omega} \nabla N_i^{\mathrm{T}} \nu \nabla([N]^{\mathrm{T}}[u])\,\mathrm{d}\Omega + \int_{\Omega} \sigma N_i \left([N]^{\mathrm{T}} \frac{\mathrm{d}[u]}{\mathrm{d}t}\right) \mathrm{d}\Omega - \int_{\Omega} N_i J_0\,\mathrm{d}\Omega$$
$$- \int_{\Gamma_1} h N_i\,\mathrm{d}\Gamma = 0, \qquad i = 1, 2, 3, \tag{3.91}$$

or

$$[H][u] + [C]\frac{\mathrm{d}[u]}{\mathrm{d}t} = [S] \tag{3.92}$$

where

$$h_{ij}^e = \int_{\Omega^e} \nabla N_i^{\mathrm{T}} \nu \nabla N_j\,\mathrm{d}\Omega \tag{3.93}$$
$$S_i^e = \int_{\Omega^e} N_i J_0\,\mathrm{d}\Omega + \int_{\Gamma_1} h N_i\,\mathrm{d}\Gamma$$

and $[C]$ is the conductivity (mass) matrix.

The boundary value problem (3.36a), (3.84), (3.85) will be solved by the one-step Θ (theta) method. After the assumption

$$[u^n] = [u(n\Delta t)]$$
$$[u^{n+\Theta}] = (1-\Theta)[u^n] + \Theta[u^{n+1}] \tag{3.94}$$
$$\frac{\mathrm{d}[u]}{\mathrm{d}t} \approx \frac{[u^{n+1}] - [u^n]}{\Delta t} \quad \text{for } t = (n+\Theta)\Delta t$$

equation (3.92) for the instant $t = (m + \Theta)\Delta t$ is written as

$$\left(\frac{1}{\Delta t}[C] + \Theta[H]\right)[u^{n+1}] = [S^{n+\Theta}] + \left(\frac{1}{\Delta t}[C] - (1-\Theta)[H]\right)[u^n] \tag{3.95}$$

with the initial condition (3.85). The Θ method is absolutely stable for $\frac{1}{2} < \Theta < 1$. For $\Theta = \frac{1}{2}$, the accuracy is of second order (in time), while for $\Theta \neq \frac{1}{2}$ the order of accuracy is one.

The kind of Θ method to be used may be selected according to the values of Θ:

$\Theta = 0$ explicit Euler's method,
$\Theta = \frac{1}{2}$ Crank–Nicolson's (trapezoidal) method,

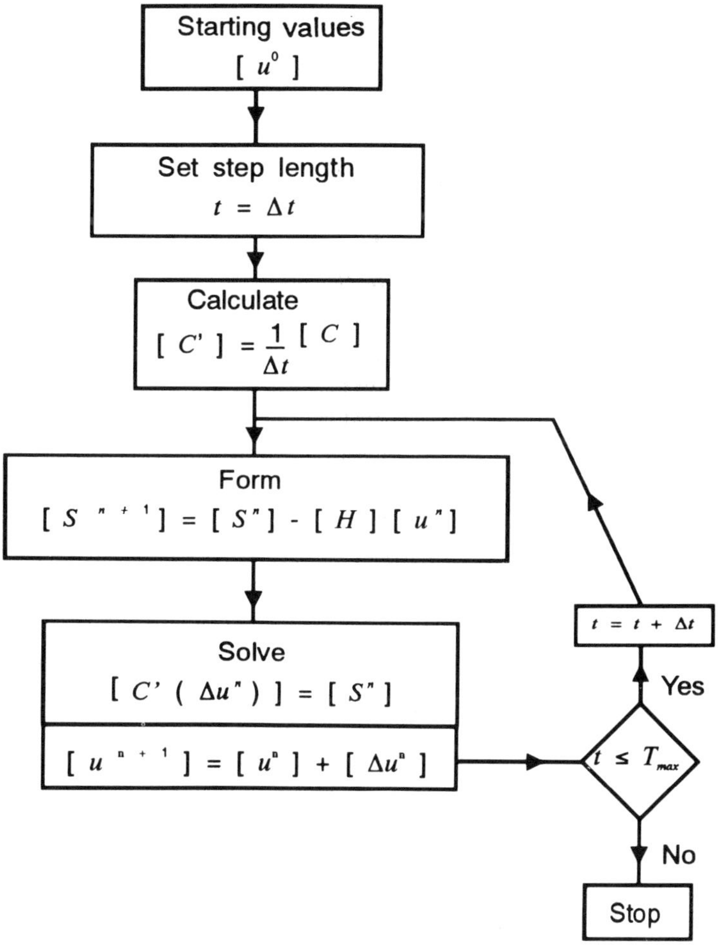

Fig. 3.13 *Flow chart for explicit Euler's method*

$\Theta = \frac{2}{3}$ Galerkin's implicit scheme,
$\Theta = 1$ implicit Euler's method.

Explicit Euler's method

The explicit Euler's method will now be used for the solution of the problem described by eqns (3.86)–(3.88). In this case eqn (3.95) takes the form:

$$\frac{1}{\Delta t}[C][u^{n+1}] = [S] + \left(\frac{1}{\Delta t}[C] - [H]\right)[u^n] \tag{3.96}$$

Equation (3.96) may be solved by a *consistent mass matrix* or by *diagonal mass matrix*. The flow chart for the algorithm of consistent mass matrix is presented in Fig. 3.13. The above algorithm is based on the following set of equations

$$\frac{1}{\Delta}[C][\Delta u^n] = [S^n] - [H][u^n] \tag{3.97}$$

$$[u^{n+1}] = [u^n] + [\Delta u^n] \tag{3.98}$$

where

$$\begin{aligned} C_{ij} &= \int_{\Omega^e} \sigma N_i N_j \, d\Omega \\ h_{ij} &= \int_{\Omega^e} \nabla N_i^T \nu \nabla N_j \, d\Omega \\ S_i^n &= \int_{\Omega^e} N_i J^n \, d\Omega + \int_{\Gamma^1} h^n N_i \, d\Gamma \end{aligned} \tag{3.99}$$

In order to use the algorithm of the diagonal mass matrix, the matrix:

$$[C] = \begin{bmatrix} C_{11}\, C_{12} \cdots\cdots \\ \cdots\cdots\cdots\cdots \\ \cdots C_{ii}\, C_{ij} \cdots \\ \cdots\cdots\cdots\cdots \end{bmatrix}$$

in eqn (3.95) is replaced by a diagonal matrix

$$[D] = \begin{bmatrix} d_{11} \cdots\cdots\cdots \\ \cdots\cdots d_{22} \cdots\cdots \\ \cdots\cdots\cdots d_{ii} \cdots \\ \cdots\cdots\cdots\cdots \end{bmatrix},$$

Matrix coefficients are calculated from relationships

$$d_{ii} = C_{ii}\frac{T}{\alpha} \tag{3.100}$$

$$d_{ij} = \begin{cases} \neq 0 & \text{for } i = j \\ = 0 & \text{for } i \neq j \end{cases}$$

where

$$T = \sum_i \sum_j C_{ij}, \qquad \alpha = \sum_i C_{ii}$$

For the majority of problems (except axisymmetric) $T = 1$. This mass matrix formulation simplifies calculation because it does not require a solution of a system of equations.

3.7 NONLINEAR TIME-DEPENDENT PROBLEMS [3.7; 3.12; 3.38; 3.44]

As in Sections 3.5 and 3.6, magnetic field calculations will be used to explain the solution of nonlinear time-dependent problems. For the domain, shown in Fig. 3.14, the following space-time description is applied

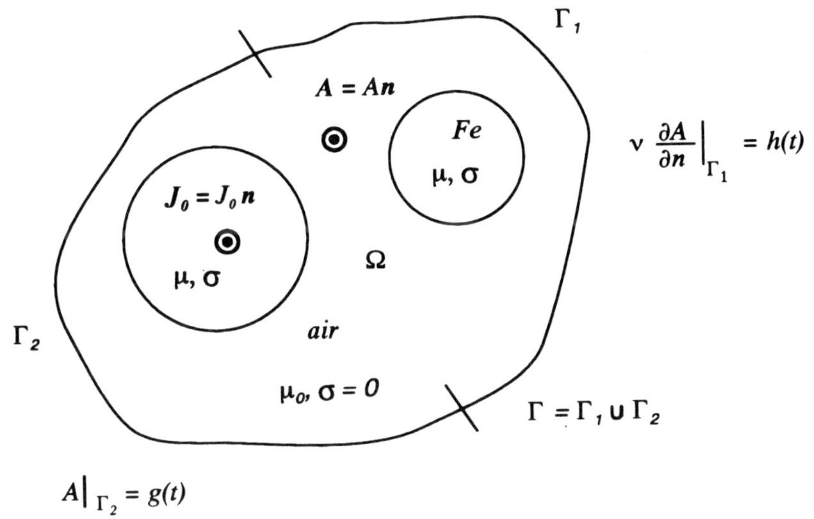

Fig. 3.14 *Nonlinear time-dependent problem (γ-conductivity)*

$$\nabla\cdot(\nu(\cdot)\nabla A) = \sigma\frac{\partial A}{\partial t} - J_0 \tag{3.101}$$

$$\nu\frac{\partial A}{\partial n} = h(t) \quad \text{on } \Gamma_1$$

$$A = g(t) \quad \text{on } \Gamma_2$$

$$A(x, y, t = 0) = A_0(x, y)$$

For any weight functions w_1 and w_2 the following equation is satisfied

$$\int_\Omega \left\{(\nabla\cdot(\nu(\cdot)\nabla A)) - \sigma\frac{\partial A}{\partial t} + J_0\right\} w_1\,\mathrm{d}\Omega + \int_\Gamma \left(\nu(\cdot)\frac{\partial A}{\partial n} - h(t)\right) w_2\,\mathrm{d}\Gamma = 0 \tag{3.102}$$

After applying the Green's first theorem

$$\int_\Omega \nabla\cdot(\nu(\cdot)\nabla A)w_1\,\mathrm{d}\Omega = -\int_\Omega \nabla A\nu(\cdot)\nabla w_1\,\mathrm{d}\Omega + \oint_\Gamma \nu(\cdot)\frac{\partial A}{\partial n}w_1\,\mathrm{d}\Gamma \tag{3.103}$$

eqn (3.102) becomes

$$\int_\Omega \left(-\nabla A\nu(\cdot)\nabla w_1 - \sigma\frac{\partial A}{\partial t}w_1 + J_0 w_1\right)\mathrm{d}\Omega + \oint_\Gamma \nu(\cdot)\frac{\partial A}{\partial n}w_1\,\mathrm{d}\Gamma + {}$$
$$+ \int_{\Gamma_1}\left(\nu(\cdot)\frac{\partial A}{\partial n} - h(t)\right) w_2\,\mathrm{d}\Gamma = 0 \tag{3.104}$$

Assuming, that $w_2 = -w_1$ and applying trial function w_1 which satisfies the condition

$$w_1 \big|_{\Gamma_2} = 0$$

we finally obtain

$$\int_{\Omega} \left(\nabla A \nu(\cdot) \nabla w_1 + \sigma \frac{\partial A}{\partial t} w_1 - J_0 w_1 \right) d\Omega - \int_{\Gamma_1} h(t) w_1 \, d\Gamma = 0 \quad (3.105)$$

After assuming that $w_1 = N_i$, eqn (3.105) transforms to

$$\int_{\Omega} \left(\nabla N_i \nu(\cdot) \nabla([N]^{\mathrm{T}}[u]) + \sigma N_i \frac{\partial([N]^{\mathrm{T}}[u])}{\partial t} - N_i J_0 \right) d\Omega - \int_{\Gamma_1} N_i h(t) \, d\Gamma = 0 \quad (3.106)$$

for FEM eqn (3.103) is transformed to

$$\sigma[T(\cdot)][u] + [H(\cdot)][u] = [R] \quad (3.107)$$

where

$$[\dot{u}] = \frac{d[u]}{dt} \quad (3.108)$$

$$t_{ij}^e = \int_{\Omega^e} N_i N_j \, d\Omega \quad (3.109)$$

$$h_{ij}^e = \int_{\Omega^e} \nabla N_i \nu(\cdot) \nabla N_j \, d\Omega \quad (3.110)$$

$$r_i^e = \int_{\Omega^e} N_i J_0 \, d\Omega + \int_{\Gamma_1^e} N_i h(t) \, d\Gamma \quad (3.111)$$

Equation (3.106) is solved by the Θ method. In the calculation the following relationships are used

$$t_m = m\Delta t, \qquad \Delta t = \frac{T}{M} \quad (3.112)$$

$$u_i^m = u_i(t_m) = u_i(m\Delta t) \quad (3.113)$$

$$[\dot{u}] \approx \frac{[u^{m+1}] - [u^m]}{\Delta t} \quad (3.114)$$

$$[u^{m+\Theta}] = (1-\Theta)[u^m] + \Theta[u^{m+1}], \qquad \tfrac{1}{2} \leqslant \Theta \leqslant 1 \quad (3.115)$$

Hence the set of equations (3.107) becomes

$$\sigma[T] \frac{[u^{m+1}] - [u^m]}{\Delta t} + [H(\cdot)^{m+\Theta}][u^{m+\Theta}] = [R^{m+\Theta}], \qquad m \geqslant 0 \quad (3.116)$$

The set of equations (3.116) is nonlinear. These equations will be solved by predictor–corrector (P–C) method. In step P eqns (3.116) are written as

$$\sigma[T]\,\frac{[\tilde{u}^{m+1}]-[u^m]}{\Delta t}+[H^m(\cdot)]\,[\tilde{u}^{m+\Theta}]=[R^{m+\Theta}] \tag{3.117}$$

and in step C:

$$\sigma[T]\,\frac{[u^{m+1}]-[u^m]}{\Delta t}+[\tilde{H}^{m+1}(\cdot)][u^{m+\Theta}]=[R^{m+\Theta}] \tag{3.118}$$

where

$$[u^{m+\Theta}]=(1-\Theta)[u^m]+\Theta[u^{m+1}] \tag{3.119}$$

$$[\tilde{u}^{m+\Theta}]=(1-\Theta)[u^m]+\Theta[\tilde{u}^{m+1}] \tag{3.120}$$

$$[H^m(\cdot)]=[H(u^m)] \tag{3.121}$$

$$[\tilde{H}^{m+1}(\cdot)]=[H(\tilde{u}^{m+1})] \tag{3.122}$$

The problem formulated in this way requires the solution of two sets of equations at the same time: one for extrapolation (P) and one for interpolation (C). For $\Theta=1/2$, Crank–Nicolson's method is used. This method is based on the Richardson's extrapolation [3.12]. Let E be the mean extrapolation operator

$$E[u^m]=\tfrac{3}{2}[u^m]-\tfrac{1}{2}[u^{m-1}]+\Theta(\Delta t^2)\qquad m\geqslant 1 \tag{3.123}$$

Taking into account eqn (3.115), eqn (3.116) is transformed to

$$\sigma[T]\,\frac{[u^{m+1}]-[u^m]}{\Delta t}+[H(Eu^m)][u^{m+1/2}]=[R^{m+1/2}] \tag{3.124}$$

where

$$[u^{m+1/2}]=\tfrac{1}{2}([u^{m+1}]+[u^m]) \tag{3.125}$$

For this formulation eqn (3.124) is solved in one step only, but the solutions in the two previous steps have to be known.

One of the best methods for solving eqn (3.116) is the Newton–Raphson's method [3.33].

Newton–Raphson's method

This set of eqns (3.117) is modified before applying Newton–Raphson's method. For this purpose the following formulation is used

$$[u^{m+1}]-[u^m]=\frac{[u^{m+\Theta}]-[u^m]}{\Theta},\qquad \frac{1}{2}\leqslant\Theta\leqslant 1 \tag{3.126}$$

$$[\dot{u}]\approx\frac{[u^{m+\Theta}]-[u^m]}{\Theta\Delta t} \tag{3.127}$$

$$[u^{m+1}] = [u^m] + \frac{1}{\Theta}([u^{m+\Theta}] - [u^m]), \tag{3.128}$$

Taking into account eqn (3.127), the set of eqns (3.116) is transformed to

$$\left([H(\cdot)^{m+\Theta}] + \frac{\sigma}{\Delta t\Theta}[T]\right)[u^{m+\Theta}] = \frac{\sigma}{\Delta t\Theta}[T][u^m] + [R^{m+\Theta}] \tag{3.129}$$

or for brevity

$$[K^{m+\Theta}(\cdot)][u^{m+\Theta}] = [S^{m+\Theta}] \tag{3.130}$$

Denoting $[F(\cdot)]$ as

$$[F(\cdot)] = [K^{m+\Theta}(\cdot)][u^{m+\Theta}] - [S^{m+\Theta}] \tag{3.131}$$

and developing $[F(\cdot)]$ in $k+1$ iterative steps into a Taylor's series

$$[F(\cdot)]|_{k+1} = [F(\cdot)]|_k + \left.\frac{\partial[F(\cdot)]}{\partial[u]}\right|_k [\Delta u_k] + \cdots = 0 \tag{3.132}$$

the Newton–Raphson's method is formulated as

$$\begin{aligned}[K_T^{m+\Theta}(\cdot)]_k[\Delta u_k] &= [S^{m+\Theta}] - [K^{m+\Theta}(\cdot)]_k[u_k^{m+\Theta}] \\ [u_{k+1}^{m+\Theta}] &= [u_k^{m+\Theta}] + [\Delta u_k]\end{aligned} \tag{3.133}$$

where

$$[K_T^{m+\Theta}(\cdot)] = \left.\frac{\partial[F(\cdot)]}{\partial[u]}\right|_k^{m+\Theta} \tag{3.134}$$

is the tangential matrix.

After solving in time $(m+\Theta)\Delta t$, solution in time $(m+1)\Delta t$ is calculated from eqn (3.120). The tangential matrix is calculated in the following way

$$[K_T^{m+\Theta}(\cdot)] = \frac{\partial}{\partial[u_k^{m+\Theta}]}\left(\left([H^{m+\Theta}(\cdot)]_k + \frac{\sigma}{\Delta t\Theta}[T]\right)[u_k^{m+\Theta}]\right) \tag{3.135}$$

For simplicity, indices $m+\Theta$ and k are ignored. In order to calculate coefficients k_{Tij}^e of the matrix $[K_T(\cdot)]$ the following derivatives are calculated

$$\begin{aligned}&\frac{\partial}{\partial u_j}\left\{\int_{\Omega^e} \nabla N_i \nu(\cdot)\nabla([N]^{\mathrm{T}}[u])\,\mathrm{d}\Omega + \frac{\sigma}{\Delta t + \Theta}\int_{\Omega^e} N_i([N]^{\mathrm{T}}[u])\,\mathrm{d}\Omega\right\} \\ &= \int_{\Omega^e} \nabla N_i \nu(\cdot)\nabla N_j[u]\,\mathrm{d}\Omega + \int_{\Omega^e} \nabla N_i \frac{\partial \nu(\cdot)}{\partial u_j}\nabla([N]^{\mathrm{T}}[u])\,\mathrm{d}\Omega \\ &\quad + \frac{\sigma}{\Delta t\Theta}\int_{\Omega^e} N_i N_j\,\mathrm{d}\Omega\end{aligned} \tag{3.136}$$

At the start of the slope of the reluctivity curve $\partial\nu(\cdot)/\partial u_j$ is calculated. If the reluctivity curve is given as a function of the flux density B the tangential matrix $[K_T(\cdot)]$ is symmetric. This matrix is also symmetric if $\nu = \nu(B^2)$. For calculation it is easier to use $\nu = \nu(B^2)$ than $\nu = \nu(B)$:

$$B^2 = B_x^2 + B_y^2 \tag{3.137}$$

$$B_x = \frac{\partial([N]^{\mathrm{T}}[u])}{\partial y}; \qquad B_y = \frac{\partial([N]^{\mathrm{T}}[u])}{\partial x} \tag{3.138}$$

$$\frac{\partial \nu(\cdot)}{\partial u_j} = \frac{\partial \nu}{\partial B^2} \cdot \frac{\partial B^2}{\partial u_j} = \frac{\mathrm{d}\nu}{\mathrm{d}B^2} \left(2\frac{\partial([N]^{\mathrm{T}}[u])}{\partial y} \frac{\partial N_j}{\partial y} + 2\frac{\partial([N]^{\mathrm{T}}[u])}{\partial x} \frac{\partial N_j}{\partial x} \right)$$
$$= 2\frac{\mathrm{d}\nu}{\mathrm{d}B^2} \nabla([N]^{\mathrm{T}}[u]) \nabla N_j \tag{3.139}$$

After substituting eqn (3.139) into eqn (3.136) we obtain

$$k_{Tij}^e = \int_{\Omega} \nabla N_i \nu(\cdot) \nabla N_j \, \mathrm{d}\Omega + \frac{\sigma}{\Delta t \Theta} \int_{\Omega^e} N_i N_j \, \mathrm{d}\Omega$$
$$+ 2 \int_{\Omega^e} \nabla N_i \nabla([N]^{\mathrm{T}}[u]) \frac{\mathrm{d}\nu}{\mathrm{d}B^2} \nabla([N]^{\mathrm{T}}[u]) \nabla N_j \, \mathrm{d}\Omega \tag{3.140}$$

In Fig. 3.15 the flow chart for the predictor–corrector method is presented.

3.8 BOUNDARY-ELEMENT METHOD [3.4; 3.5; 3.6; 3.24; 3.25; 3.35]

The bases for the BEM formulation are Green's second theorem:

$$\int_{\Omega} (u\Delta G - G\Delta u) \, \mathrm{d}\Omega = \oint_{\Gamma} \left(u\frac{\partial G}{\partial n} - G\frac{\partial u}{\partial n} \right) \mathrm{d}\Gamma \tag{3.141}$$

and the Green's function G. For the Laplace operation Δ in two-dimensional space:

$$G = \frac{1}{2\pi} \ln \frac{1}{r} \tag{3.142}$$

and in three-dimensions:

$$G = \frac{1}{4\pi r} \tag{3.143}$$

where:
r distance between the source and point under consideration.

Let us derive the boundary-integral equations. Figure 3.16 shows space Ω surrounded by a boundary surface Γ. Inside the domain Ω, Poisson's equation $\nabla^2 u = -f$ is satisfied. Let us assume that on Γ_1 Dirichlet's conditions hold $u = u_0$ and on Γ_2 Neumann's conditions apply $\partial u/\partial n = \varphi_0$. The Green's function satisfies the following equation

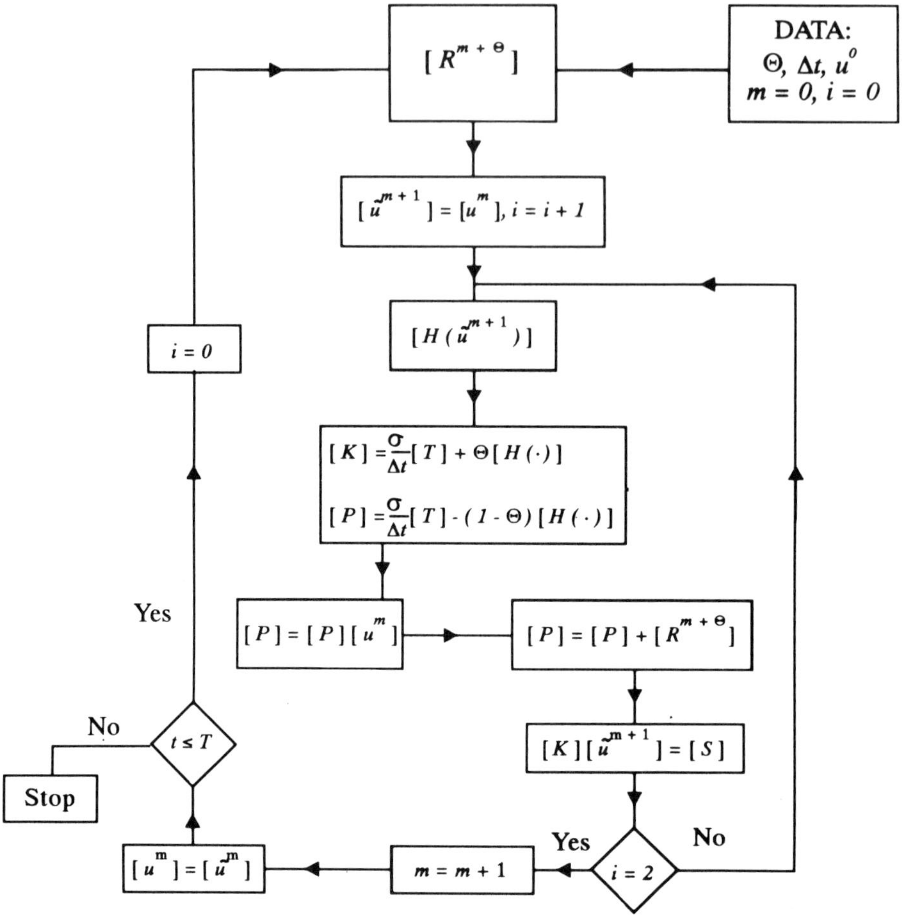

Fig. 3.15 *Flow chart for predictor–corrector method*

$$\nabla^2 G = -\delta(P) \tag{3.144}$$

where

$$\delta(P) = \text{Dirac function}$$

$$\delta(P) = \begin{cases} \neq 0 & \text{for } r = \rho \\ = 0 & \text{for } r \neq \rho. \end{cases}$$

Formula (3.141) takes the form

$$u(P) = \int_\Omega G f \, \mathrm{d}\Omega - \oint_\Gamma u \frac{\partial G}{\partial n} \mathrm{d}\Gamma + \oint_\Gamma G \frac{\partial u}{\partial n} \mathrm{d}\Gamma \tag{3.145}$$

which allows calculation of $u(P)$ inside domain Ω on the basis of u and $\partial u / \partial n$ on the boundary Γ.

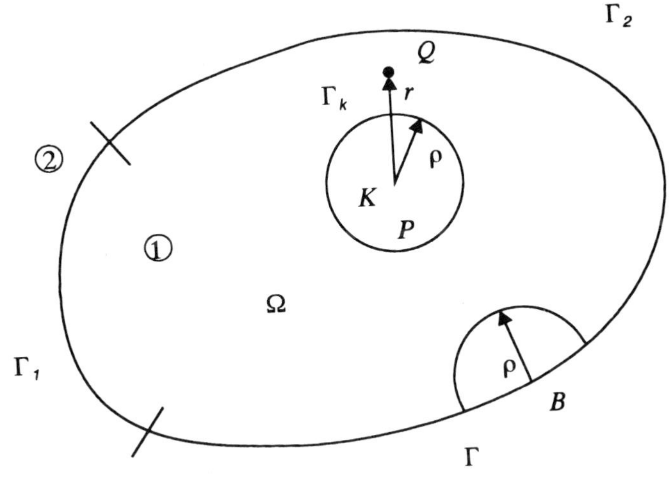

Fig. 3.16 *Introducing BEM*

Equation (3.145) will now be derived. For this, circle K with radius ρ is removed from domain Ω. In consequence of this for eqn (3.141) an integral along the boundary Γ_k integration must be added

$$\int_{\Omega-K} (u\Delta G - G\Delta u)\,\mathrm{d}\Omega = \oint_{\Gamma}\left(u\frac{\partial G}{\partial n} - G\frac{\partial u}{\partial n}\right)\mathrm{d}\Gamma + \oint_{\Gamma_k}\left(u\frac{\partial G}{\partial n} - G\frac{\partial u}{\partial n}\right)\mathrm{d}\Gamma_k \tag{3.146}$$

After taking into account that in the region $\Omega - K$:

$$\nabla^2 G = 0 \tag{3.147}$$

we obtain:

$$\int_{\Omega-K} Gf\,\mathrm{d}\Omega = \oint_{\Gamma}\left(u\frac{\partial G}{\partial n} - G\frac{\partial u}{\partial n}\right)\mathrm{d}\Gamma + \oint_{\Gamma_k}\left(u\frac{\partial G}{\partial n} - G\frac{\partial u}{\partial n}\right)\mathrm{d}\Gamma_k \tag{3.148}$$

Now an integral along the boundary Γ_k, when $\rho \to 0$, will be calculated:

$$\lim_{\rho\to 0}\int_{\Gamma_k} u(P)\frac{\partial G}{\partial n}\,\mathrm{d}\Gamma_k = \lim_{\rho\to 0}\int_0^{2\pi} u(P)\frac{1}{2\pi}\left(-\frac{1}{\rho}\right)\rho\,\mathrm{d}\varphi = -u(P)$$

where

$$\frac{\partial G}{\partial n} = \frac{\partial}{\partial \rho}\left(\frac{1}{2\pi}\ln\frac{1}{\rho}\right) = \frac{1}{2\pi}\left(-\frac{1}{\rho}\right)$$

$$\mathrm{d}\Gamma_k = \rho\,\mathrm{d}\varphi$$

and

$$\lim_{\rho\to 0}\int_{\Gamma_k}\frac{\partial u}{\partial n}\bigg|_P G\,\mathrm{d}\Gamma_k = \lim_{\rho\to 0}\int_0^{2\pi}\frac{\partial u}{\partial n}\bigg|_P \frac{1}{2\pi}\ln\frac{1}{\rho}\,\rho\,\mathrm{d}\varphi = 0$$

because

$$\lim_{\rho\to 0}\rho\ln\frac{1}{\rho} = -\lim_{\rho\to 0}\rho\ln\rho = -\lim_{\rho\to 0}\frac{\ln\rho}{1/\rho}$$

and after application of d'Hospital's principle

$$-\lim_{\rho\to 0}\frac{\ln\rho}{1/\rho} = \lim_{\rho\to 0}\frac{1/\rho}{1/\rho^2} = \lim_{\rho\to 0}\rho = 0$$

In the surroundings of point B (Fig. 3.16) integration follows the semicircle (φ belongs to $(0, \pi)$), then

$$\lim_{\rho\to 0}\int_{\Gamma_k} u(B)\frac{\partial G}{\partial n}\mathrm{d}\Gamma_k = -\frac{1}{2}u(B) \tag{3.149}$$

and

$$\int_{\Gamma_k} G\frac{\partial u}{\partial n}\mathrm{d}\Gamma_k = 0$$

Thus eqn (3.146) reduces to eqn (3.145) inside Ω, whereas on the boundary

$$\frac{1}{2}u(B) - \int_\Omega Gf\,\mathrm{d}\Omega = \oint_\Gamma\left(u\frac{\partial G}{\partial n} - G\frac{\partial u}{\partial n}\right)\mathrm{d}\Gamma_1 \tag{3.150}$$

Finally, by applying boundary conditions, we find

$$\frac{1}{2}u(B) - \int_\Omega Gf\,\mathrm{d}\Omega = \int_{\Gamma_1} u\frac{\partial G}{\partial n}\mathrm{d}\Gamma_1 - \int_{\Gamma_1} G\frac{\partial u}{\partial n}\mathrm{d}\Gamma_1 + \int_{\Gamma_2} u\frac{\partial G}{\partial n}\mathrm{d}\Gamma_2 - \int_{\Gamma_2} G\varphi_0\,\mathrm{d}\Gamma_2 \tag{3.151}$$

The boundary-element method is a numerical procedure for solving boundary-integral equations. For this, boundary Γ is divided into elements: Γ_1 into N and Γ_2 into M elements. Various approximations may be used on these elements. We will obtain the simplest formulae when it is assumed that the function u or its derivatives are constant along the element.

Then (Fig. 3.17)

$$\int_{\Gamma_1} G\frac{\partial u}{\partial n}\mathrm{d}\Gamma_1 = \sum_{n=1}^{N} G\frac{\partial u}{\partial n}\bigg|_n \Delta l_n \tag{3.152}$$

$$\int_{\Gamma_2} u\frac{\partial G}{\partial n}\mathrm{d}\Gamma_2 = \sum_{m=1}^{M}\frac{\partial G}{\partial n}u_m\Delta l_m \tag{3.153}$$

Finally formula (3.151) transforms to

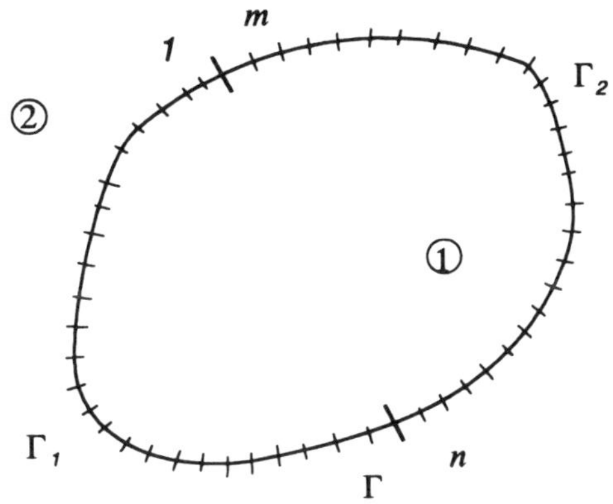

Fig. 3.17 *Boundary discretization*

$$\frac{1}{2}u_i + \sum_{n=1}^{N} G\frac{\partial u}{\partial n}\bigg|_n \Delta l_n - \sum_{m=1}^{M} \frac{\partial G}{\partial n} u_m \Delta l_m = \int_{\Gamma_1} u_0 \frac{\partial G}{\partial n} \mathrm{d}\Gamma_1 - \int_{\Gamma_2} G\varphi_0 \,\mathrm{d}\Gamma_2 + \int_{\Omega} Gf\,\mathrm{d}\Omega \tag{3.154}$$

where i is any point on the boundary Γ.

For $N+M$ unknown values of function or their normal derivatives, eqns (3.154) are formulated. For calculation of coefficients occurring in eqn (3.154), Green's function (3.152) or its derivatives (3.153) and three integrals occurring on the right-hand side of eqn (3.154) must be calculated. The given quantities are

$$G, \frac{\partial G}{\partial n}, u_0, \varphi_0, f, \Delta l_n, \Delta l_m$$

3.9 OPEN-BOUNDARY PROBLEMS [3.2; 3.3; 3.27; 3.38]

Sometimes it is necessary to calculate the electromagnetic field in an unbounded region (Fig. 3.18). In a bounded region (1) Laplace's, Poisson's, Helmholtz's or Fourier's equations may be satisfied. Media existing in region (1) may be linear or nonlinear. All the sources are in the bounded region. In an unbounded (2) region Laplace's equation is satisfied.

Let us solve the problem shown in Fig. 3.18 by interfacing FEM and BEM. The bounded region (1) is divided into finite elements, whereas in the exterior region (2) BEM is used. On the interface Γ the boundary conditions are unknown. For region (1) FE equations are formulated and BE (nodes $i = 1, 2, \ldots, M$) are added to them. Two sets of equations (FE and BE) are combined together and solved on a computer.

For elements which do not have nodes on the border Γ, the stiffness matrix $[H]$ is calculated in the way shown above. Numbering of border elements is clockwise. For this kind of numbering the normal unit vector (**n**) is directed from region (2) to (1). Let us discuss elements placed on the border Γ. A

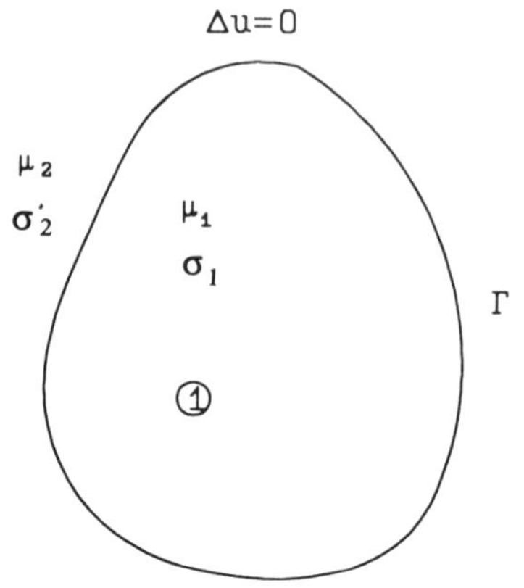

Fig. 3.18 *Electromagnetic field calculation in unbounded region*

segment connecting nodes i, $i+1$ placed on the border Γ is the ith boundary element Γ_i (Fig. 3.19).

First, let us assume that Neumann's boundary condition is known on the boundary. For this element

$$[K^e][u^e] = -[F^e] \tag{3.155}$$

Coefficients of the vector $[F^e]$ are calculated from

$$F_i^e = \int_{\Gamma_i} N_i q \, d\Gamma_i \tag{3.156}$$

$$F_{i+1}^e = \int_{\Gamma_i} N_{i+1} q \, d\Gamma_i \tag{3.157}$$

where:

q normal derivative,
N_i, N_{i+1} shape functions on the boundary Γ_i.

$$N_i = \tfrac{1}{2}(1-\zeta) \tag{3.158}$$

$$N_{i+1} = \tfrac{1}{2}(1+\zeta) \tag{3.158a}$$

$$\zeta \text{ belongs to } (-1, 1), \qquad d\Gamma = \frac{L_i}{2} d\zeta$$

where L_i is the element length.

The minus sign in eqn (3.155) results from the opposing direction of the unit normal vector. After assuming a linear approximation of the derivative on the border

$$q|_{\Gamma_i} = q_i N_i + q_{i+1} N_{i+1} \tag{3.159}$$

we obtain

$$F_i^e = \int_{\Gamma_i} N_i^2 \, d\Gamma_i q_i + \int_{\Gamma_i} N_i N_{i+1} \, d\Gamma_i q_{i+1} \tag{3.160}$$

$$F_{i+1}^e = \int_{\Gamma_i} N_i N_{i+1} \, d\Gamma_i q_i + \int_{\Gamma_i} N_{i+1}^2 \, d\Gamma_i q_{i+1} \tag{3.161}$$

After integration

$$\begin{bmatrix} F_i^e \\ F_{i+1}^e \end{bmatrix} = \frac{L_i}{6} \begin{bmatrix} 2 & 1 \\ 1 & 2 \end{bmatrix} \begin{bmatrix} q_i \\ q_{i+1} \end{bmatrix} \tag{3.162}$$

Global matrix coefficients are calculated from the formula

$$F_i = F_i^{e_i} + F_i^{e_{i-1}} = \frac{L_{i-1}}{6} q_{i-1} + \frac{L_{i-1} + L_i}{3} q_i + \frac{L_i}{6} q_{i+1} \tag{3.163}$$

where

$$L_0 = L_M, \quad q_0 = q_M, \quad q_{m+1} = q_1$$

The global matrix is defined by a formula

$$[F] = [A][Q] \tag{3.164}$$

Matrix $[A]$ has a dimension $M \times M$ and reads

$$[A] = \begin{bmatrix} \frac{L_1 + L_M}{3} & \frac{L_1}{6} & 0 & 0 & \cdots & 0 & \frac{L_M}{6} \\ \frac{L_1}{6} & \frac{L_1 + L_2}{3} & \frac{L_2}{6} & 0 & \cdots & 0 & 0 \\ 0 & \frac{L_2}{6} & \frac{L_2 + L_3}{3} & \frac{L_3}{6} & \cdots & 0 & 0 \\ \cdot & \cdot & \cdot & \cdot & \cdots & \cdot & \cdot \\ \frac{L_M}{6} & 0 & 0 & 0 & \cdots & \frac{L_{M-1}}{6} & \frac{L_{M-1} + L_M}{3} \end{bmatrix}. \tag{3.165}$$

Vector $[Q]$ has dimension M and contains normal derivatives. The following boundary integral equation can be written in the region (2)

$$C_i u_i + \sum_{j=1}^{M} \int_{\Gamma_j} u \frac{\partial G}{\partial n} \, d\Gamma_j = \sum_{j=1}^{M} \int_{\Gamma_j} qG \, d\Gamma_j + \lim_{R \to \infty} \int_0^{2\pi} \left[qG - u \frac{\partial G}{\partial n} \right] R \, d\varphi \tag{3.166}$$

where:

$$G = \frac{1}{2\pi} \ln \frac{1}{r}$$

i any point on the border Γ,
j boundary element number.

The last integral of the right-hand side of eqn (3.166) is equal to zero. A similar situation exists for the application of the BEM to bounded regions. Coefficients C_i do not have to be calculated explicitly. However, there is a difference in the way that the elements of the matrix $[H]$ are determined. For this purpose we will assume that potential u is constant in domain (2), hence the right-hand side of eqn (3.166) will not be equal to zero as in the interior domain. This follows from

$$\lim_{R\to\infty} \int_0^{2\pi} u \frac{\partial G}{\partial n} R \, d\varphi = \lim_{R\to\infty} u \frac{-1}{2\pi} \int_0^{2\pi} \frac{1}{R} R \, d\varphi = -u \tag{3.167}$$

Thus

$$[H] \begin{bmatrix} u \\ u \\ \vdots \\ u \end{bmatrix} = [u]$$

$$H_{ii} = -\sum_{\substack{j=1 \\ j\neq i}}^{M} H_{ij} + 1 \tag{3.168}$$

where $i = 1, 2, \ldots, M$.

The determination of H_{ij} and G_{ij} is the same as in the classical version of the BEM. Vector $[F]$ can be determined by the boundary Γ potential $[u_\Gamma]$:

$$[F] = [A][G]^{-1}[H][u_\Gamma] \tag{3.169}$$

It will be noticed that the product of $[A]$, $[G]^{-1}$ and $[K]$ is a stiffness matrix for the macroelement with M nodes placed on the boundary Γ (Fig. 3.19). This matrix is asymmetrical.

In recent years a number of special elements have been introduced in order to extend the scope of the standard finite-element method. 'Infinite finite-elements' have become increasingly used and accepted as a means of solving unbounded field problems [3.18; 3.19; 3.27]. The interior region (1) is modelled by finite elements and the exterior region (2) is represented by infinite elements. A very simple infinite-element for a two-dimensional problem was proposed by Pissanetzky [3.27]. This element has two nodes and conforms with linear triangles and linear quadrilaterals.

In Fig. 3.20a subspace Ω_i is divided into finite elements and in Fig. 3.20b the infinite element is shown. This element has two points 1 and 2 defined by coordinates (X_1, Y_1) and (X_2, Y_2). The auxiliary coordinates system (v, w) is introduced. Axis v is parallel to segment 1–2. For calculation, the following functions are used

$$l = \tfrac{1}{2}\sqrt{[(X_2 - X_1)^2 + (Y_2 - Y_1)^2}$$

$$d = \frac{1}{2l}(X_1 Y_2 - X_2 Y_1)$$

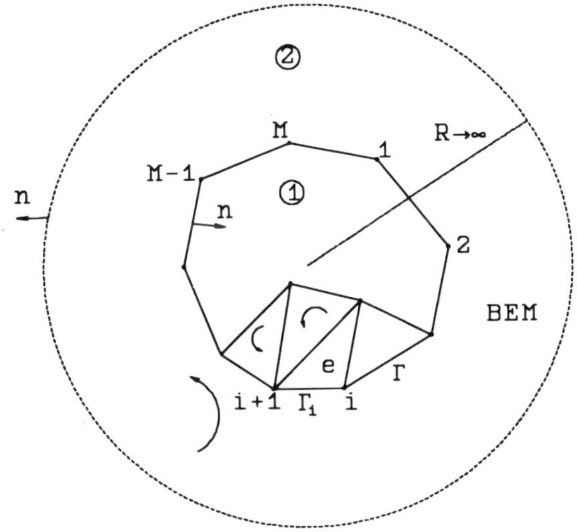

Fig. 3.19 *FEM and BEM connection*

$$b = \frac{1}{4l}(X_2^2 - X_1^2 + Y_2^2 - Y_1^2) \tag{3.170}$$

$$r_m = \sqrt{(d^2 + b^2)}$$

$$\tan\varphi_1 = \frac{1}{d}(b - l), \qquad \tan\varphi_2 = \frac{1}{d}(b + l)$$

Additional axis η is introduced by the formula:

$$\eta = \frac{1}{l}(d\tan\varphi - b) \tag{3.171}$$

where

$$\tan\varphi = \frac{1}{d}(b + \eta l) \tag{3.171a}$$

The function u is linear between points 1 and 2:

$$u_{1-2}(\eta) = \tfrac{1}{2}[(1-\eta)u_1 + (1+\eta)u_2]. \tag{3.172}$$

In terms of the polar coordinates, points belonging to segment 1–2 have coordinates $(d/\cos\varphi, \varphi)$. If one wants u to decay as a given function $f(r)$ towards infinity for any $\varphi = constant$, it is enough to assume in the infinite element

$$u(r, \varphi) = \frac{f(r)}{2f(d/\cos\varphi)}[(1-\eta)u_1 + (1+\eta)u_2]$$

Thus the shape functions for the infinite elements are

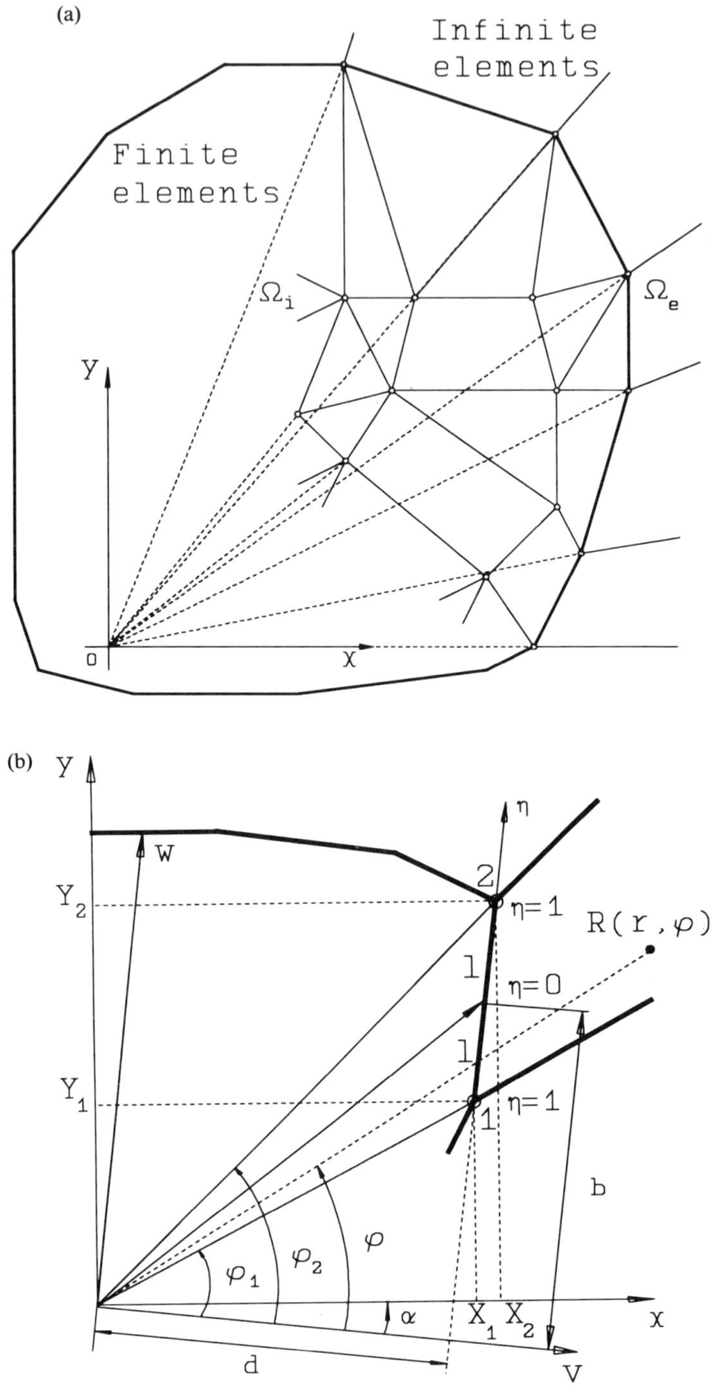

Fig. 3.20 *(a) Division of domain into finite and infinite elements, (b) two-nodal infinite element*

$$N_1(r, \varphi) = \frac{f(r)}{2f(d/\cos\varphi)}(1 - \eta)$$
$$N_2(r, \varphi) = \frac{f(x)}{2f(d/\cos\varphi)}(1 + \eta) \quad (3.173)$$

Let us consider the particular case when $f(r) = 1/r^n$, where $n \geqslant 1$. Taking into account eqns (3.171) and (3.173) and the relationship

$$\tan\varphi = (r\sin\varphi)/(r\cos\varphi) = w/v$$

we obtain

$$N_1(v, w) = \frac{d^n}{2v^n}\left(1 + \frac{b}{l} - \frac{dw}{lv}\right)$$
$$N_2(v, w) = \frac{d^n}{2v^n}\left(1 - \frac{b}{l} - \frac{dw}{lv}\right) \quad (3.174)$$

For solution of the above mentioned problem, the global functional Φ has to be minimized in subspaces (1) + (2). In subspace (2) Laplace's equation is satisfied:

$$\nabla^2 u = 0 \quad (3.175)$$

The functional for Laplace's equation is the particular case for functional (3.38) and is expressed by the formula

$$\Phi^e = \int_{IE}\left[\left(\frac{\partial u}{\partial x}\right)^2 + \left(\frac{\partial u}{\partial y}\right)^2\right] \mathrm{d}x\,\mathrm{d}y \quad (3.176)$$

We will write this functional in the $v0w$ coordinates system. The relationship between $v0w$ and $x0y$ systems is the following

$$\begin{bmatrix} x \\ y \end{bmatrix} = \begin{bmatrix} \cos\alpha & \sin\alpha \\ -\sin\alpha & \cos\alpha \end{bmatrix}\begin{bmatrix} v \\ w \end{bmatrix} \quad (3.177)$$

$$\cos\alpha = \frac{1}{2l}(Y_2 - Y_1), \qquad \sin\alpha = \frac{1}{2l}(X_2 - X_1)$$

where α is the angle between coordinates $v0w$ and $x0y$, and the inverse relation is as follows

$$\begin{bmatrix} v \\ w \end{bmatrix} = \begin{bmatrix} \cos\alpha & -\sin\alpha \\ \sin\alpha & \cos\alpha \end{bmatrix}\begin{bmatrix} x \\ y \end{bmatrix} \quad (3.178)$$

Derivatives are expressed by relationships

$$\frac{\partial u}{\partial x} = \frac{\partial u}{\partial v}\frac{\partial v}{\partial x} + \frac{\partial u}{\partial w}\frac{\partial w}{\partial x}$$
$$\frac{\partial u}{\partial y} = \frac{\partial u}{\partial v}\frac{\partial v}{\partial y} + \frac{\partial u}{\partial w}\frac{\partial w}{\partial y} \quad (3.179)$$

or in a matrix form, after taking into account eqn (3.178)

$$\begin{bmatrix} \dfrac{\partial u}{\partial x} \\ \dfrac{\partial u}{\partial y} \end{bmatrix} = \begin{bmatrix} \cos\alpha & \sin\alpha \\ -\sin\alpha & \cos\alpha \end{bmatrix} \begin{bmatrix} \dfrac{\partial u}{\partial v} \\ \dfrac{\partial u}{\partial w} \end{bmatrix} \tag{3.180}$$

From eqn (1.179) it follows, that

$$\left(\frac{\partial u}{\partial x}\right)^2 + \left(\frac{\partial u}{\partial y}\right)^2 = \left(\frac{\partial u}{\partial v}\right)^2 + \left(\frac{\partial u}{\partial w}\right)^2 \tag{3.181}$$

Because

$$\int_\Omega f(x, y)\, \mathrm{d}x\, \mathrm{d}y = \int_\Omega g(v, w)\, |J|\, \mathrm{d}v\, \mathrm{d}w \tag{3.182}$$

where $f(x, y)$ and $g(v, w)$ are alternative expressions for functions in coordinates $x0y$ or $v0x$ and $|J|$ is the Jacobian matrix,

$$\begin{aligned} \mathrm{d}x\, \mathrm{d}y = |J|\, \mathrm{d}v\, \mathrm{d}w &= \left|\frac{\partial(x, y)}{\partial(v, w)}\right| \mathrm{d}v\, \mathrm{d}w = \begin{vmatrix} \partial x/\partial v & \partial x/\partial w \\ \partial y/\partial v & \partial y/\partial w \end{vmatrix} \mathrm{d}v\, \mathrm{d}w \\ &= \begin{vmatrix} \cos\alpha & \sin\alpha \\ -\sin\alpha & \cos\alpha \end{vmatrix} \mathrm{d}v\, \mathrm{d}w = \mathrm{d}v\, \mathrm{d}w \end{aligned} \tag{3.183}$$

and finally the functional (3.176) can be written as

$$\Phi^e = \int_{IE} \left[\left(\frac{\partial u}{\partial v}\right)^2 + \left(\frac{\partial u}{\partial w}\right)^2\right] \mathrm{d}v\, \mathrm{d}w \tag{3.184}$$

For nodes 1 and 2, after differentiation of the functional (3.184) we obtain

$$\frac{\partial \Phi^e}{\partial u_i} = 2\int_e \left[\frac{\partial u}{\partial v}\frac{\partial}{\partial u_i}\left(\frac{\partial u}{\partial v}\right) + \frac{\partial u}{\partial w}\frac{\partial}{\partial u_i}\left(\frac{\partial u}{\partial w}\right)\right] \mathrm{d}v\, \mathrm{d}w, \qquad i = 1, 2 \tag{3.185}$$

where

$$\frac{\partial u}{\partial v} = \left[\frac{\partial N_1}{\partial v}, \frac{\partial N_2}{\partial v}\right] \begin{bmatrix} u_1 \\ u_2 \end{bmatrix}; \qquad \frac{\partial u}{\partial w} = \left[\frac{\partial N_1}{\partial w}, \frac{\partial N_2}{\partial w}\right] \begin{bmatrix} u_1 \\ u_2 \end{bmatrix} \tag{3.186}$$

$$\frac{\partial}{\partial u_i}\left(\frac{\partial u}{\partial v}\right) = \frac{\partial N_i}{\partial v}, \qquad \frac{\partial}{\partial u_i}\left(\frac{\partial u}{\partial w}\right) = \frac{\partial N_i}{\partial w} \tag{3.187}$$

Taking into account the integration limits for variables v and w

$$v\begin{cases} \infty \\ d \end{cases} \qquad w\begin{cases} (b + l)v/d \\ (b - l)v/d \end{cases} \tag{3.188}$$

for the whole IE we obtain

$$\begin{bmatrix} \dfrac{\partial \Phi^e}{\partial u_1} \\ \dfrac{\partial \Phi^e}{\partial u_2} \end{bmatrix} = 2 \int_{\frac{b-l}{d}v}^{\frac{b+l}{d}v} \int_d^{\infty} \begin{bmatrix} \left(\dfrac{\partial N_1}{\partial v}\right)^2 + \left(\dfrac{\partial N_1}{\partial w}\right)^2; & \left(\dfrac{\partial N_2}{\partial v}\dfrac{\partial N_1}{\partial v} + \dfrac{\partial N_2}{\partial w}\dfrac{\partial N_1}{\partial w}\right) \\ \dfrac{\partial N_1}{\partial v}\dfrac{\partial N_2}{\partial v} + \dfrac{\partial N_1}{\partial w}\dfrac{\partial N_2}{\partial w}\Bigg); & \left(\dfrac{\partial N_2}{\partial v}\right)^2 + \left(\dfrac{\partial N_2}{\partial w}\right)^2 \end{bmatrix} \mathrm{d}v\,\mathrm{d}w \begin{bmatrix} u_1 \\ u_2 \end{bmatrix} \tag{3.189}$$

Finally, the stiffness matrix is found as

$$[H]^e = \frac{1}{6nd} \left[\begin{array}{c|c} (4n^2 + 2n + 1)l - 6nb & (2n^2 - 2n - 1)l \\ + 3r_m^2/l & - 3r_m^2/l \\ \hline \text{sym} & (4n^2 + 2n + 1)l \\ & + 6nb + 3r_m^2/l \end{array} \right] \tag{3.190}$$

The closed-form expressions for the element matrix can also be obtained for the infinite element which has three nodes and conforms with 'quadratic quadrilaterals' [3.18].

Figure 3.21 shows an example internal subdivision into finite elements and external subdivision into infinite elements.

The geometrical relationships for 3-node infinite elements are the same as those for 2-node elements. In the middle of segment 1–2 a third node (marked as 3) is added with appropriate description. Other notation is the same as in Fig. 3.20. The unknown function u has values u_1, u_2, u_3 in nodes 1, 2, 3, respectively and varies parabolically along the segment 1–2–3

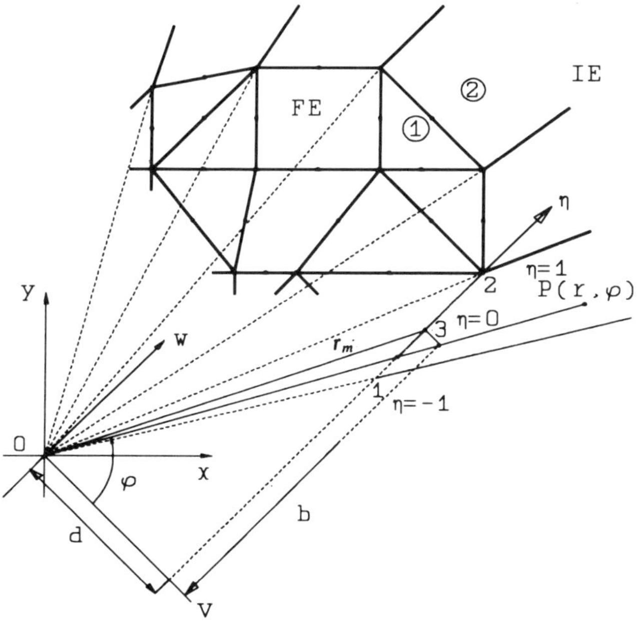

Fig. 3.21 *Internal subdivision into finite elements and external subdivision into infinite elements*

$$u_{1\text{-}3\text{-}2}(\eta) = -\eta\frac{1-\eta}{2}u_1 + \eta\frac{1+\eta}{2}u_2 + (1-\eta^2)u_3 \tag{3.191}$$

The shape functions for the three nodal infinite elements, in terms of their polar coordinates r, φ are

$$\begin{aligned} N_1(r,\varphi) &= \frac{-f(r)}{f(d/\cos\varphi)}\eta\frac{1-\eta}{2} \\ N_2(r,\varphi) &= \frac{f(r)}{f(d/\cos\varphi)}\eta\frac{1+\eta}{2} \\ N_3(r,\varphi) &= \frac{f(r)}{f(d/\cos\varphi)}(1-\eta^2) \end{aligned} \tag{3.192}$$

Coordinate η is given by formula (3.171). Let us assume that as for 2-nodal infinite elements

$$f(r) = \frac{1}{r^n}, \; n \geqslant 1.$$

We will write the shape functions in local coordinates

$$\begin{aligned} N_1(v,w) &= \frac{d^n}{2v^n}\left(\frac{b}{l} - \frac{d}{l}\frac{w}{v}\right)\left(1 - \frac{d}{l}\frac{w}{v} + \frac{b}{l}\right) \\ N_2(v,w) &= \frac{d^n}{2v^n}\left(\frac{d}{l}\frac{w}{v} - \frac{b}{l}\right)\left(1 + \frac{d}{l}\frac{w}{v} - \frac{b}{l}\right) \\ N_3(v,w) &= \frac{d^n}{v^n}\left[1 - \left(\frac{d}{l}\frac{w}{v} - \frac{b}{l}\right)^2\right] \end{aligned} \tag{3.193}$$

Thus, after calculation, we obtain stiffness matrix elements for 3-nodal infinite elements

$$H^e_{ij} = 2\int\limits_{(b-l)\frac{v}{d}}^{(b+l)\frac{v}{d}}\int\limits_{d}^{\infty}\left(\frac{\partial N_i}{\partial v}\frac{\partial N_j}{\partial v} + \frac{\partial N_i}{\partial w}\frac{\partial N_j}{\partial w}\right)dv\,dw, \qquad i = 1,2,3; \quad j = 1,2,3 \tag{3.194}$$

After a simple but very laborious calculation the following result is obtained

$$\begin{aligned} H^e_{11} &= \frac{1}{30nd}[(8n^2 + 22n + 17)l - 10b(3n+4) + 35r_m^2/l] \\ H^e_{12} &= \frac{1}{30nd}[(-2n^2 + 2n + 7)l + 5r_m^2/l] \\ H^e_{13} &= \frac{2}{15nd}[(n^2 - n - 6)l + 10b - 10r_m^2/l], \\ H^e_{21} &= H^e_{12}, \\ H^e_{22} &= \frac{1}{30nd}[(8n^2 + 22n + 17)l + 10b(3n+4) + 35r_m^2/l] \end{aligned} \tag{3.195}$$

$$H_{23}^{e} = \frac{2}{15nd}\left[(n^2 - n - 6)l - 10b - 10r_m^2/l\right]$$

$$H_{31}^{e} = H_{13}^{e}$$

$$H_{32}^{e} = H^{e}{}_{23}$$

$$H_{33}^{e} = \frac{8}{15nd}\left[(2n^2 - 2n + 3)l + 5r_m^2/l\right]$$

3.10 HIERARCHICAL ELEMENTS [3.1; 3.23; 3.38; 3.39]

The hierarchical elements are used in computation of electromagnetic fields by the finite-element and boundary-element methods. They are very useful in local improvement of the accuracy of calculations without changing the finite elements mesh. Local improvement can be achieved by:

(1) local high discretization of the finite elements mesh,
(2) application of hierarchical elements.

Let us discuss hierarchical elements. In this kind of element shape functions are applied in which increasing the order of the polynomial does not break the continuity of the interface between elements. Accuracy of calculations increases by increasing the order of the polynomial without increasing mesh density. One polynomial for this type of application is the Hermitian polynomial. These are used in the solution of the following differential equation [3.20]

$$\frac{d^2H_n(x)}{dx^2} - 2x\frac{dH_n(x)}{dx} + 2nx = 0 \tag{3.196}$$

where $n = 0, 1, 2, \ldots$.

$$\begin{aligned} H_n(x) &= (-1)^n e^{x^2}\frac{d^n}{dx^n}(x^{e^{-x^2}}) \\ &= 2^n x^n - 2^{n-1}\binom{n}{2}x^{n-2} + 2^{n-2}\cdot 1\cdot 3\cdot\binom{n}{4}x^{n-4} - \cdots \end{aligned} \tag{3.197}$$

For $n = 0, 1, 2, 3, 4$ and 5 Hermitian polynomials have the following forms

$$H_0(x) = 1;\quad H_1(x) = 2x;\quad H_2(x) = 4x^2 - 2$$
$$H_3(x) = 8x^3 - 12x;\quad H_4(x) = 16x^4 - 48x^2 + 12$$
$$H_5(x) = 32x^5 - 160x^3 + 120$$

Hermitian polynomials satisfy the following recursive relationships

$$\begin{aligned} H_{n-1}(x) &= 2xH(x) - 2nH_{n-1}(x) \\ \frac{dH_n(x)}{dx} &= 2nH_{n-1}(x) \end{aligned} \tag{3.198}$$

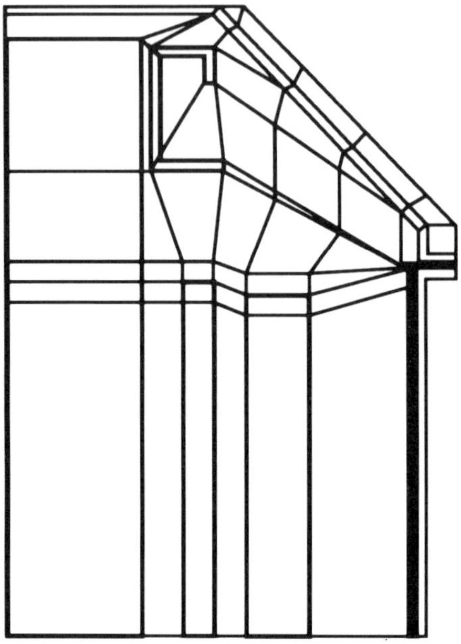

Fig. 3.22 *The discretization of a model of a transformer leakage area into elements*

In reference [3.23] Hermitian polynomials are used for a transformer leakage magnetic field calculation. In Fig. 3.22 one quarter of a transformer cross-section with finite elements is shown.

The vector potential is used for calculations. In the model, potential A has only one component

$$\mathbf{A} = A_k \hat{\mathbf{k}} = A\hat{\mathbf{k}} \qquad A = A(x, y) \tag{3.199}$$

satisfies the following equation

$$\begin{aligned} \frac{\partial^2 A}{\partial x^2} + \frac{\partial^2 A}{\partial y^2} &= \mathrm{j}\sigma\omega\mu A - \mu J && \text{in coils} \\ \frac{\partial^2 A}{\partial x^2} + \frac{\partial^2 A}{\partial y^2} &= \mathrm{j}\sigma\omega\mu A && \text{in conductors} \\ \frac{\partial^2 A}{\partial x^2} + \frac{\partial^2 A}{\partial y^2} &= 0 && \text{in dielectrics} \end{aligned} \tag{3.200}$$

where J is the current density in the coil, and σ is the conductivity.

Vector potential $\mathbf{A}$ was interpolated by the following expression:

$$A(\eta, \xi) = H_{00}A + H_{10}\frac{\partial A}{\partial \eta} + H_{01}\frac{\partial A}{\partial \xi} + H_{11}\frac{\partial^2 A}{\partial \eta \partial \xi} + H_{20}\frac{\partial^2 A}{\partial \eta^2} + H_{02}\frac{\partial^2 A}{\partial \xi^2} \tag{3.201}$$

where:

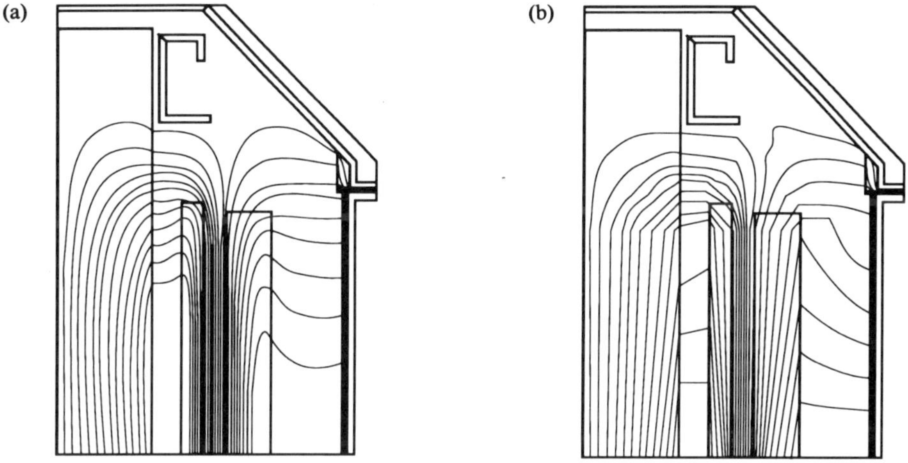

Fig. 3.23 *Field lines distribution for (a) zero, (b) first and second order Hermitian polynomials*

η, ξ local coordinates,
$H_{m,n}$ Hermitian polynomials,
m, n 0, 1, 2.

In Figs 3.23a, b distributions of vector potential A are shown which are obtained from FEM calculations using zero, first- and second-order Hermitian elements.

3.11 THREE-DIMENSIONAL PROBLEMS [3.4; 3.10; 3.19; 3.25; 3.38]

A more realistic image of electromagnetic phenomena is obtained using three-dimensional analysis. In a number of practical applications a two-dimensional approximation may be sufficient without causing significant error in results. An application of FEM to two-dimensional calculations was presented in Section 3.3. In this Section FEM is applied to three-dimensional calculations. Thus in eqn (3.38) describing a two-dimensional scalar field, we will add a derivative with respect to the third coordinate and the equation becomes

$$-\frac{\partial}{\partial x}\left(p_x \frac{\partial u}{\partial x}\right) - \frac{\partial}{\partial y}\left(p_y \frac{\partial u}{\partial y}\right) - \frac{\partial}{\partial z}\left(p_z \frac{\partial u}{\partial z}\right) + k^2 u = f \tag{3.202}$$

The functional, equivalent to the non-homogeneous Neumann's boundary value problem for eqn (3.202) has the following form

$$\Phi(u) = \int_\Omega \left[p_x \left(\frac{\partial u}{\partial x}\right)^2 - p_y \left(\frac{\partial u}{\partial y}\right)^2 - p_z \left(\frac{\partial u}{\partial z}\right)^2 + k^2 u^2 - 2fu\right] \mathrm{d}x\, \mathrm{d}y\, \mathrm{d}z$$

$$-2\int_\Gamma \varphi(B) u \, \mathrm{d}\Gamma \tag{3.203}$$

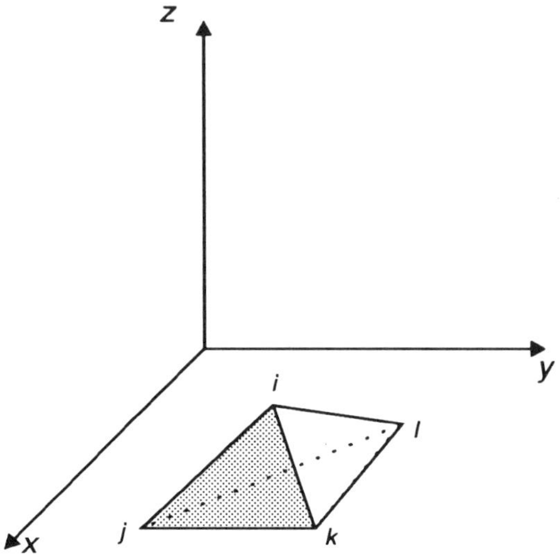

Fig. 3.24 *Tetrahedral element*

The derivative of functional (3.203) is as follows

$$\frac{\partial \Phi}{\partial u_m} = 2\int_{\Omega^e}\left[p_x \frac{\partial u^e}{\partial x}\frac{\partial}{\partial u_m}\left(\frac{\partial u^e}{\partial x}\right) + p_y \frac{\partial u^e}{\partial y}\frac{\partial}{\partial u_m}\left(\frac{\partial u^e}{\partial y}\right) + p_z \frac{\partial u^e}{\partial z}\frac{\partial}{\partial u_m}\left(\frac{\partial u^e}{\partial z}\right) + k^2 u^e \frac{\partial u^e}{\partial u_m} - f\frac{\partial u^e}{\partial u_m}\right] \mathrm{d}x\,\mathrm{d}y\,\mathrm{d}z + 2\int_{\Gamma^e} \varphi_0 \frac{\partial u^e}{\partial u_m}\,\mathrm{d}\Gamma \tag{3.204}$$

In Fig. 3.24 the simplest three-dimensional tetrahedral element is presented. For this element $m = i, j, k, l$ and

$$u^e(x,y,z) = [N_i(x,y,z),\ N_j(x,y,z),\ N_k(x,y,z),\ N_l(x,y,z)]\begin{bmatrix} u_i \\ u_j \\ u_k \\ u_l \end{bmatrix} \tag{3.205}$$

Equation (3.29a) for the tetrahedral element has the following form

$$\begin{bmatrix} \dfrac{\partial \Phi^e}{\partial u_i} \\ \dfrac{\partial \Phi^e}{\partial u_j} \\ \dfrac{\partial \Phi^e}{\partial u_k} \\ \dfrac{\partial \Phi^e}{\partial u_l} \end{bmatrix} = \begin{bmatrix} h^e_{ii} & h^e_{ij} & h^e_{ik} & h^e_{il} \\ h^e_{ji} & h^e_{jj} & h^e_{jk} & h^e_{jl} \\ h^e_{ki} & h^e_{kj} & h^e_{kk} & h^e_{kl} \\ h^e_{li} & h^e_{lj} & h^e_{lk} & h^e_{ll} \end{bmatrix}\begin{bmatrix} u_i \\ u_j \\ u_k \\ u_l \end{bmatrix} + \begin{bmatrix} S_i \\ S_j \\ S_k \\ S_l \end{bmatrix} \tag{3.206}$$

Instead of eqn (3.28) we have

$$u_m = \alpha_1 + \alpha_2 x_m + \alpha_3 y_m + \alpha_4 z_m \tag{3.207}$$

where $m = i, j, k, l$, and instead of eqn (3.30):

$$\begin{aligned}
\alpha_1 &= \frac{1}{6V}(a_i u_i + a_j u_j + a_k u_k + a_l u_l) \\
\alpha_2 &= \frac{1}{6V}(b_i u_i + b_j u_j + b_k u_k + b_l u_l) \\
\alpha_3 &= \frac{1}{6V}(c_i u_i + c_j u_j + c_k u_k + c_l u_l) \\
\alpha_4 &= \frac{1}{6V}(d_i u_i + d_j u_j + d_k u_k + d_l u_l)
\end{aligned} \tag{3.208}$$

where

$$6V = \det\begin{bmatrix} 1 & x_i & y_i & z_i \\ 1 & x_j & y_j & z_j \\ 1 & x_k & y_k & z_k \\ 1 & x_l & y_l & z_l \end{bmatrix} \tag{3.209}$$

and V is the volume of tetrahedral element i, j, k, l.

$$\begin{aligned}
a_i &= \det\begin{bmatrix} x_j & y_j & z_j \\ x_l & y_l & z_l \\ x_k & y_k & z_k \end{bmatrix} \qquad
b_i = \det\begin{bmatrix} 1 & y_j & z_j \\ 1 & y_l & z_l \\ 1 & y_k & z_k \end{bmatrix} \\
c_i &= \det\begin{bmatrix} x_j & 1 & z_j \\ x_l & 1 & z_l \\ x_k & 1 & z_k \end{bmatrix} \qquad
d_i = \det\begin{bmatrix} x_j & y_j & 1 \\ x_l & y_l & 1 \\ x_k & y_k & 1 \end{bmatrix}
\end{aligned} \tag{3.210}$$

The other coefficients are defined by right-hand cyclic permutation of the subscripts in the order k, i, j, l.

Formula (3.36) will transform into

$$u^e(x, y, z) = [N_i, N_j, N_k, N_l]\begin{bmatrix} u_i \\ u_j \\ u_k \\ u_l \end{bmatrix} = [N]^T[u]^e \tag{3.211}$$

where

$$N_m = \frac{1}{6V}(a_m + b_m x + c_m y + d_m z) \tag{3.212}$$

Formulae (3.41), (3.43), (3.44) and (3.45) will be transformed similarly. In the same way, as shown in Fig. 3.5, the full equation for the three-dimensional case is assembled. More complicated elements are also used.

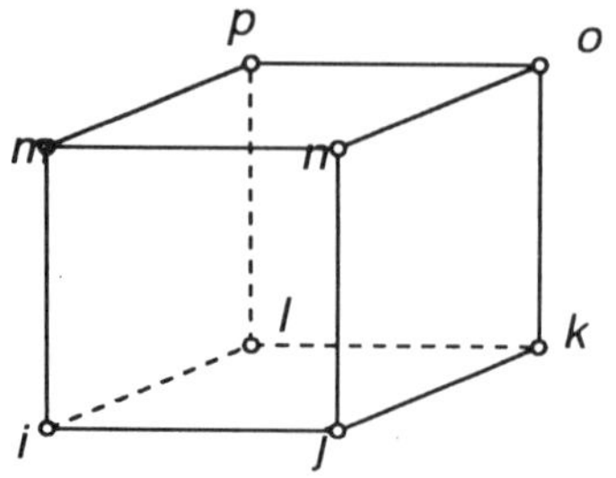

Fig. 3.25 *Linear hexahedral element*

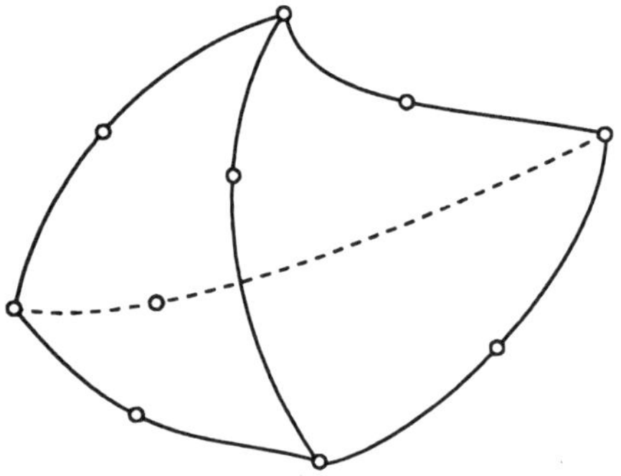

Fig. 3.26 *Ten-nodal tetrahedral element*

Figure 3.25 shows a linear hexahedral and Fig. 3.26 demonstrates a 10-node tetrahedral element. For the linear hexahedral element eqn (3.28) becomes:

$$u^e = \alpha_1 + \alpha_2 x + \alpha_3 y + \alpha_4 z + \alpha_5 xy + \alpha_6 yz + \alpha_7 xz + \alpha_8 xyz \quad (3.213)$$

The calculations are similar to those which were done for tetrahedral elements. In three-dimensional calculations isoparametric elements are applied in the same way as calculations which were done in Section 3.4.

The simple example [3.17] of scalar magnetic field calculation will now be presented. The model is shown in Fig. 3.27. Inside the ferromagnetic material ($\mu \gg \mu_0$) there is an empty hexahedron ($\mu = \mu_0$). On the surface between regions with $\mu \gg \mu_0$ and $\mu = \mu_0$ electric current flows with linear density K_0. We will calculate the magnetic field inside the air region. In this region the following equations are satisfied

$$\begin{aligned} curl\,\mathbf{H} &= 0 \\ div\,\mathbf{B} &= 0 \\ \mathbf{B} &= \mu_0 \mathbf{H} \end{aligned} \quad (3.214)$$

Magnetic scalar potential is determined by the relationships

$$\mathbf{H} = -grad\,\varphi \quad (3.215)$$

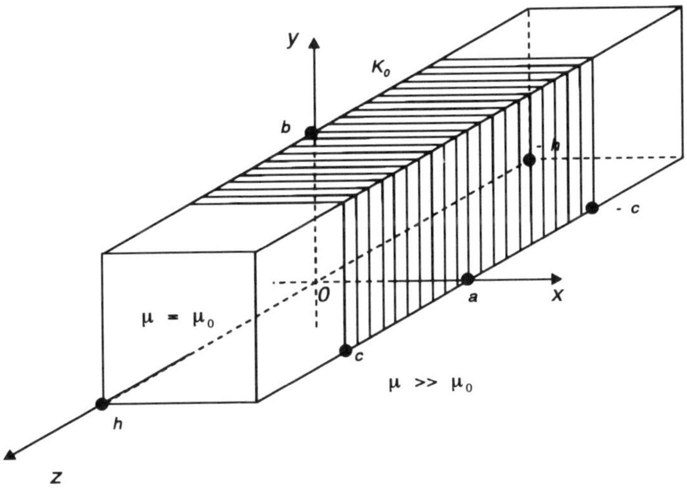

Fig. 3.27 *Model for scalar magnetic field calculation*

and fulfils Laplace's equation

$$\frac{\partial^2\varphi}{\partial x^2} + \frac{\partial^2\varphi}{\partial y^2} + \frac{\partial^2\varphi}{\partial z^2} = 0 \qquad (3.216)$$

Taking symmetry of the model into account, calculations were done for half of the model presented in Fig. 3.27.

The following boundary conditions are prescribed:

$$\mathbf{n} \cdot \mathbf{B} = 0 \quad \text{for} \begin{cases} x = 0, y \in (0, b), \ z \in (0, h) \\ y = 0, x \in (0, a), \ z \in (0, h) \end{cases} \qquad (3.217)$$

$$\mathbf{H} \times \mathbf{n} = \mathbf{K} \qquad (3.218)$$

on the rest of the walls

$$\mathbf{K} = \begin{cases} K_0\hat{\mathbf{j}} & x = a, y \in (0, b), \ z \in (0, c) \\ 0 & x = a, y \in (0, b), \ z \in (c, h) \\ -K_0\hat{\mathbf{i}} & y = b, x \in (0, a), \ z \in (0, c) \\ 0 & y = b, x \in (0, a), \ z \in (c, h) \\ 0 & z = 0, x \in (0, a), \ y \in (0, b) \\ 0 & z = h, x \in (0, a), \ y \in (0, b) \end{cases}$$

Taking into account eqn (3.215) boundary conditions (3.217) and (3.218) are expressed in terms of scalar potential. Thus condition (3.217) reads

$$\frac{\partial\varphi}{\partial n} = 0 \quad \text{for} \begin{cases} x = 0, y \in (0, b), z \in (0, h) \\ y = 0, x \in (0, a), z \in (0, h) \end{cases} \qquad (3.217a)$$

and eqn (3.217) becomes

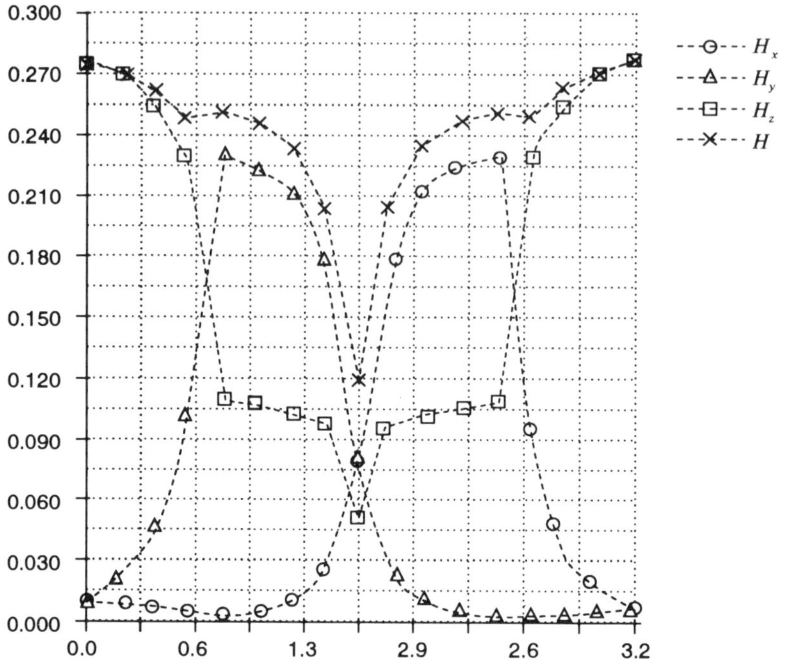

Fig. 3.28 *Numerical solution*

$$\begin{aligned}
\varphi &= 0 && \text{for } z = 0, x \in (0, a), y \in (0, b) \\
\varphi &= -K_0 z && \text{for } x = a, y \in (0, b), z \in (0, c) \\
\varphi &= -K_0 c && \text{for } x = a, y \in (0, b), z \in (c, h) \\
\varphi &= -K_0 z && \text{for } y = b, x \in (0, a), z \in (0, c) \\
\varphi &= -K_0 c && \text{for } y = b, x \in (0, a), z \in (c, h)
\end{aligned} \tag{3.218a}$$

The results of the numerical calculations are presented in Fig. 3.28. The problem has an analytical solution. For comparison, in Fig. 3.29, results of the analytical solution are also presented. More information on the example may be found in reference [3.17].

3.12 NUMERICAL SOLUTION OF LARGE SYSTEMS OF EQUATIONS [3.4; 3.21; 3.37; 3.40]

Electromagnetic field calculations by finite element and boundary element methods require the solution of large systems of algebraic equations. The methods for solving such large systems of equations may be divided into two basic groups:

- direct methods, and
- iterative methods.

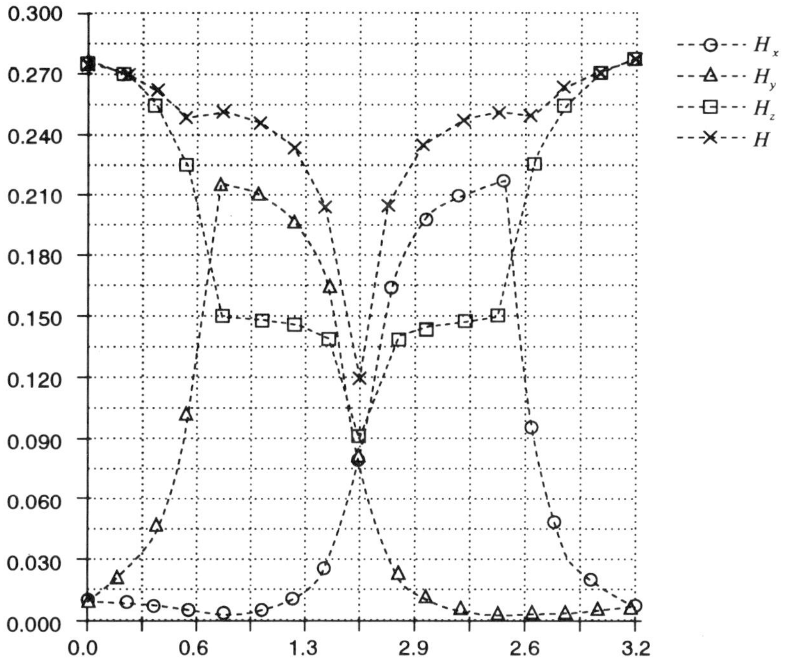

Fig. 3.29 *Analytical solution*

Direct methods

Let us consider an algebraic system of equations:

$$[k][a] = [r] \tag{3.219}$$

where:

$[k]$ square matrix of coefficients,
$[a]$ vector of unknown values,
$[r]$ vector of right-hand sides.

Let the coefficient matrix $[k]$ not require changing rows or columns. This will be the case if $[k]$ is symmetric and positive (or negative) definite. It may or may not be true when a matrix is asymmetric or not definite. If row or column changes are necessary, then the following procedure has to be modified.

Let us additionally assume that $[k]$ can be written as a product of two matrices

$$[k] = [L][U] \tag{3.220}$$

where:

$[L]$ a lower triangular matrix with a unity main diagonal,
$[U]$ an upper triangular matrix,

$$
[L] = \begin{bmatrix} 1 & 0 & ---- & 0 \\ L_{21} & 1 & ---- & 0 \\ - & - & ---- & - \\ L_{n1} & L_{n2} & ---- & 1 \end{bmatrix}
\qquad
[U] = \begin{bmatrix} U_{11} & U_{12} & ---- & U_{1n} \\ 0 & U_{22} & ---- & U_{2n} \\ - & - & ---- & - \\ 0 & 0 & ---- & U_{nn} \end{bmatrix}
\tag{3.221}
$$

The above process is known as *triangular decomposition* of the matrix $[k]$. Thus the solution of eqn (3.219) may be expressed in terms of the solution of two equations:

$$
\begin{aligned} [L][y] &= [r] \\ [U][a] &= [y] \end{aligned}
\tag{3.222}
$$

The solution is very easy in this case and can be written as

$$
\begin{aligned} y_1 &= r_1 \\ y_1 &= r_i - \sum_{j=1}^{i-1} L_{ij} y_j \qquad i = 2, 3, \ldots, n \end{aligned}
\tag{3.223a}
$$

$$
\begin{aligned} a_n &= y_n / U_{nn} \\ a_i &= \left(y_i - \sum_{j=i+1}^{n} U_{ij} a_{ij} \right) \Big/ U_{ii} \end{aligned}
\tag{3.223b}
$$

The procedures described by eqns (3.223a) and (3.223b) may be referred to as forward elimination and backward substitution, respectively. The first three steps of the decomposition process are shown in Table 3.1.

Table 3.1 Triangular division of matrix $[k]$

Step 1:

$$
\begin{bmatrix} k_{11} & k_{12} & k_{13} \\ k_{21} & k_{22} & k_{23} \\ k_{31} & k_{32} & k_{33} \end{bmatrix}
\begin{bmatrix} L_{11} = 1 & & \\ & & \\ & & \end{bmatrix}
\begin{bmatrix} U_{11} = k_{11} & & \\ & & \\ & & \end{bmatrix}
$$

Step 2:

$$
\begin{bmatrix} k_{11} & k_{12} & k_{13} \\ k_{21} & k_{22} & k_{23} \\ k_{31} & k_{32} & k_{33} \end{bmatrix}
\begin{bmatrix} 1 & 0 & \\ L_{21} = k_{21}/U_{11} & L_{22} = 1 & \\ & & \end{bmatrix}
\begin{bmatrix} U_{11} & U_{12} = k_{12} & \\ 0 & U_{22} = k_{22} - L_{22} U_{12} & \\ & & \end{bmatrix}
$$

Step 3:

$$
\begin{bmatrix} & & k_{13} \\ & & k_{23} \\ k_{31} & k_{32} & k_{33} \end{bmatrix}
\begin{bmatrix} 1 & 0 & 0 \\ L_{21} & 1 & 0 \\ L_{31} = k_{31}/U_{11} & & L_{33} = 1 \\ L_{32} = (k_{32} - L_{31} U_{12})/U_{22} & & \end{bmatrix}
\begin{bmatrix} U_{11} & U_{12} & U_{13} = k_{13} \\ 0 & U_{22} & U_{23} = k_{23} - L_{21} U_{13} \\ 0 & 0 & U_{33} \\ & & = k_{33} - L_{31} U_{13} - L_{32} U_{23} \end{bmatrix}
$$

The algorithm of decomposition of a square matrix of order $n \times n$ is as follows

$$U_{11} = k_{11}; \qquad L_{11} = 1 \tag{3.223}$$

then for each active area (from 2 to n)

$$\begin{aligned} L_{j1} &= k_{j1} U_{11} \\ U_{1j} &= k_{1j} \end{aligned} \tag{3.224}$$

followed by

$$L_{ji} = \left(k_{ji} - \sum_{m=1}^{i-1} L_{jm} U_{mi} \right) \Big/ U_{ii} \tag{3.225}$$

$$U_{ij} = k_{ij} - \sum_{m=1}^{i-1} L_{im} U_{mi} \tag{3.226}$$

and finally

$$L_{jj} = 1 \tag{3.225a}$$

$$U_{jj} = k_{jj} - \sum_{m=1}^{j-1} L_{jm} U_{mj} \tag{3.226a}$$

The decomposition process is shown in Fig. 3.30. After the single matrix decomposition eqn (3.185) can be solved repeatedly for several right-hand side vector $[r]$ values. For a very large matrix $[k]$ the process of decomposition may be very time-consuming; thus repeated calculations without a need to perform the decomposition is a valuable property. The above analysis was done for general types of matrix $[k]$. In the case of FEM, coefficients of matrices have special properties. Very often the matrix is symmetric ($k_{ij} = k_{ji}$), thus the following relation is true

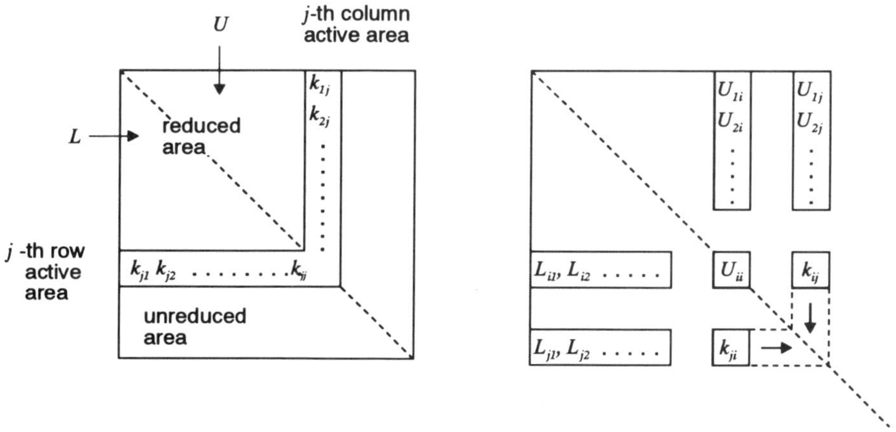

Fig. 3.30 *Matrix decomposition*

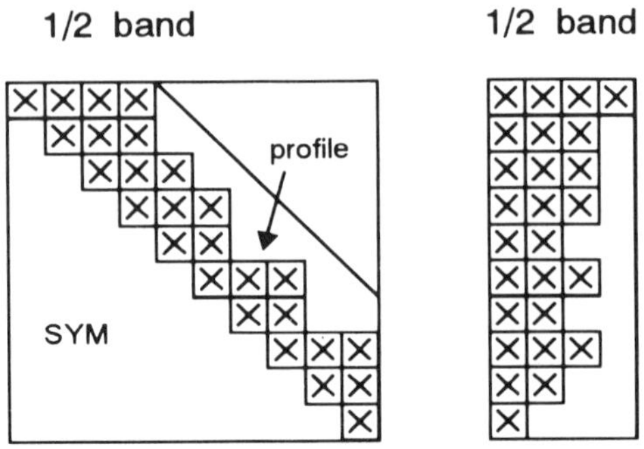

Fig. 3.31 *Typical storage scheme for a profile matrix*

$$U_{ij} = L_{ji} U_{ii}$$

Hence, for such problems, it is not necessary to store both matrices [L] and [U] but only one of them (upper or lower triangular matrix). This causes reduction by half of the required computer memory. We obtain the further savings of memory if only non-zero coefficients are stored. In FEM the maximum band width is usually equal to 10–20% of the number of unknowns. Figure 3.31 shows a typical storage scheme for a profile matrix. Further simplification is demonstrated by Fig. 3.32. More details about efficient storage systems may be found in Section 7.4 of Chapter 7.

Iterative methods [3.37]

The two simplest methods are those of Gauss–Seidel and over-relaxation: in Gauss–Seidel's method, in the first step, matrix [k] is divided into two triangular matrices [L] and [U]

$$[k] = [L] + [U] \tag{3.227}$$

where [L] is the lower triangular matrix defined by

$$L_{ij} = k_{ij} \begin{cases} i = 1, 2, \ldots, n \\ j = 1, 2, \ldots, i \end{cases}$$

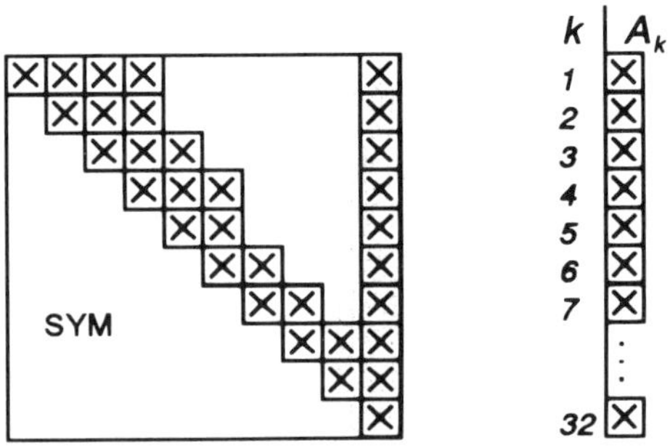

Fig. 3.32 *Reducing computer memory*

$[U]$ is the upper triangular matrix defined by:

$$U_{ij} = k_{ij} \qquad i = 1, 2, \ldots, n-1$$
$$j = i+1, i+2, \ldots, n$$

All the other elements of matrices $[L]$ and $[U]$ are equal to zero.

The basic algorithm of Gauss–Seidel's method is described by the following equations

$$[a^0] = [v]$$
$$[L][a^{n+1}] = [r] - [U][a^n] \tag{3.228}$$

where $[v]$ is the initial vector and indices correspond to successive iterations.

If matrix $[k]$ is symmetrical and positive definite then the Gauss–Seidel's method is convergent, even though the process may be very slow.

In order to improve the rate of convergence over-relaxation may be applied. Subtracting $[L][a^n]$ from both sides of eqn (3.228) yields

$$[L][\Delta a] = [r] - [k][a^n] \tag{3.229}$$

and

$$[a^{n+1}] = [a^n] - \beta[\Delta a]. \tag{3.230}$$

β is the over-relaxation parameter. The value of β is between 0 and 2 (stability limit) and the optimum value is problem-dependent. For $\beta = 1$, the scheme reduces to the classical Gauss–Seidel's method.

The main advantage of using iterative methods lies in the savings of computer memory as a result of eliminating the process of matrix decomposition. The main drawback is the unknown number of necessary iteratives to achieve prescribed accuracy and a related problem of difficulties in estimating the optimal value of the relaxation factor. The method becomes unreliable for asymmetrical matrices.

3.13 APPLICATIONS [3.41; 3.42]

3.13.1 Electromagnet

A two-dimensional magnetic field analysis is conducted for an alternating current electromagnet, shown in Fig. 3.33. The armature is assumed to have a fixed position and the coil is supplied by injecting the following current:

$$i_0 = (1 - e^{-t/T_0})I_0 \tag{3.231}$$

where:

T_0 time constant,
I_0 steady current.

Super-elements have been generated manually (Fig. 3.33), followed by an automatic mesh generator (Fig. 3.34). Owing to symmetry, only half of the system has been modelled.

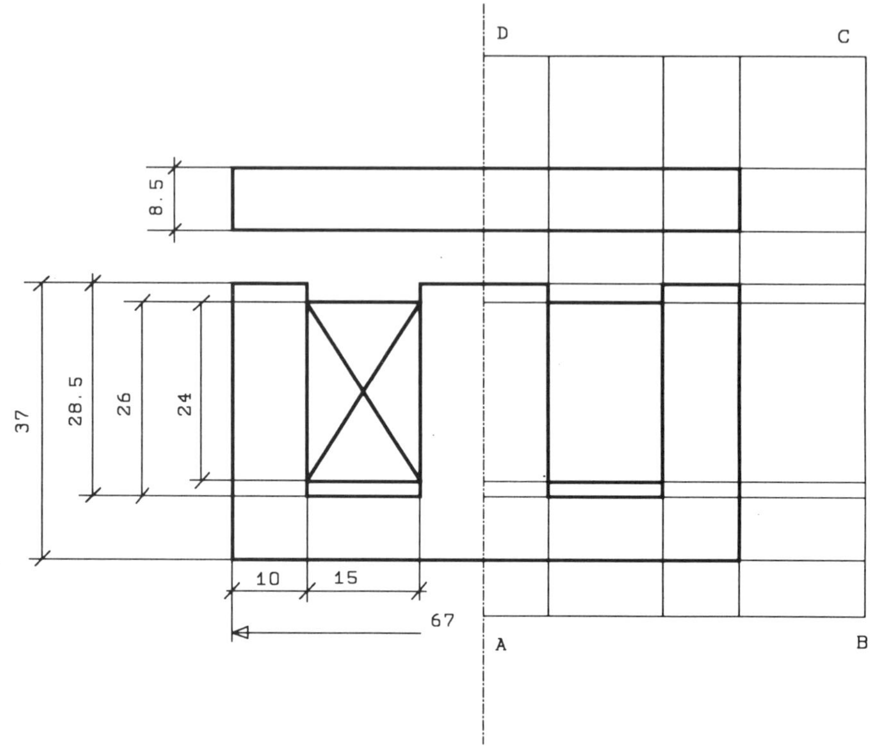

Fig. 3.33 *Electromagnet cross-section and super-elements*

The problem is described in terms of the vector potential $\mathbf{A} = A_z\hat{\mathbf{k}} = A\hat{\mathbf{k}}$. There are four regions with different material properties: 1 - air, 2 - armature, 3 - core, 4 - coil. The following equations hold

$$\frac{\partial^2 A}{\partial x^2} + \frac{\partial^2 A}{\partial y^2} = 0; \qquad \text{in region 1} \tag{3.232}$$

$$\frac{\partial}{\partial x}\left(\nu(\cdot)\frac{\partial A}{\partial x}\right) + \frac{\partial}{\partial y}\left(\nu(\cdot)\frac{\partial A}{\partial y}\right) = \sigma\frac{\partial A}{\partial t}; \qquad \text{in regions 2 and 3} \tag{3.233}$$

$$\frac{\partial^2 A}{\partial x^2} + \frac{\partial^2 A}{\partial y^2} = -\mu_0 i_0, \qquad \text{in region 4} \tag{3.234}$$

where σ is the conductivity, with the boundary conditions

$$A = 0 \tag{3.235}$$

on the symmetry surface Γ_{DA}.

On the external boundary Γ_{ABCD} infinite elements have been applied (see Section 3.9). The following initial condition is assumed.

$$A_{(t=0)} = 0 \tag{3.236}$$

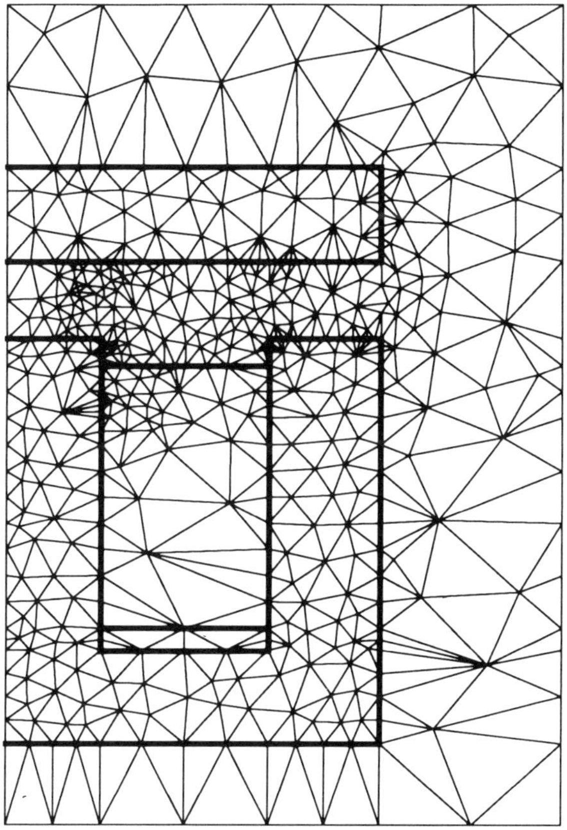

Fig. 3.34 *High density mesh*

In Fig. 3.35 the *B–H* curve for steel E41 used in armature and core construction is given.

Figures 3.36 and 3.37 show field plots under steady state and transient conditions respectively.

3.13.2 Leakage field in a transformer

Taking symmetry into account one quarter of a transformer cross section is considered (Fig. 3.38). We assume that:

- the core is made of nonconducting ideal ferromagnetic material,
- the screen is made of copper,
- screened parts of the tank and core have the same properties,
- the beam and non-screened tank parts are constructed from the same steel.

Let us assume that the magnetic vector potential has one component only

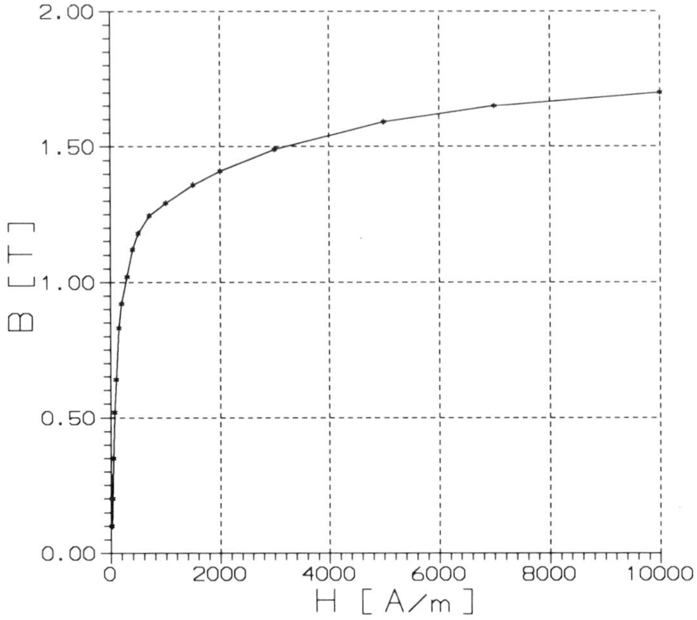

Fig. 3.35 *B–H curve*

$$\underline{\mathbf{A}} = \underline{A}(x, y)\hat{\mathbf{k}} \tag{3.237}$$

In dielectric material potential A fulfils Laplace's equation

$$\nabla^2 \underline{A} = 0 \tag{3.238}$$

and in coils satisfies Poisson's equation

$$\nabla^2 \underline{A} = -\mu_0 \underline{J} \tag{3.239}$$

The influence of the beam on the magnetic field distribution may be modelled by:

- assuming that inside the beam Helmholtz's equation is satisfied:

$$\nabla^2 \underline{A} = \alpha^2 \underline{A} \tag{3.240}$$

- or by a relevant boundary condition.

The boundary condition:

$$\frac{\partial \underline{A}}{\partial n} = 0, \text{ must be satisfied on the surface of the core } (\mu = \infty,\ \sigma = 0) \tag{3.241}$$

and on the plane of symmetry.

Other parts of the boundary, will be considered as:

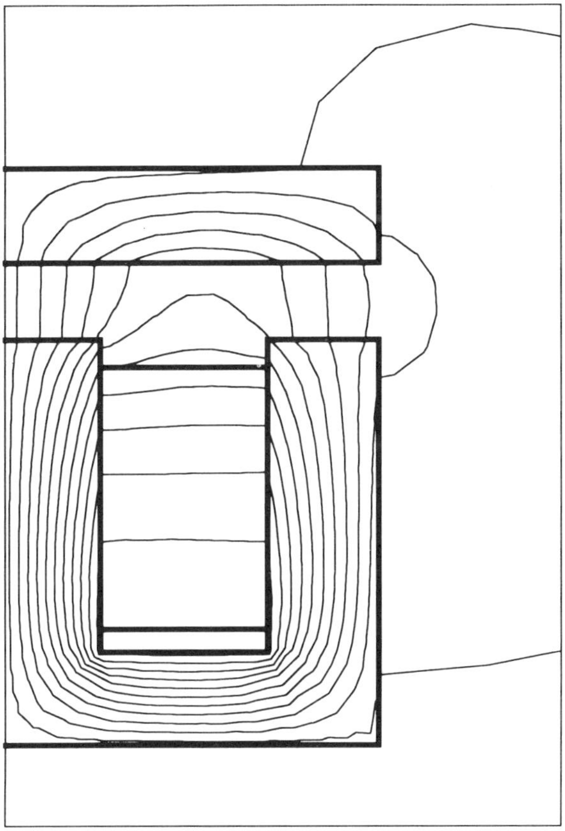

Fig. 3.36 *Magnetic steady state field distribution*

(1) ferromagnetic regions placed in an alternating magnetic field, or
(2) thin conducting layers placed on a ferromagnetic wall.

(1) Assume that a real ferromagnetic material is placed in an alternating magnetic field. A strong skin-effect will occur, and the eddy current will have only one component

$$\underline{\mathbf{J}} = \underline{J}\hat{\mathbf{k}} \tag{3.242}$$

The first Maxwell's equation may be written as

$$curl\,\underline{\mathbf{J}} = -\alpha^2 \underline{\mathbf{H}}_F \tag{3.243}$$

where

$$\alpha^2 = j\omega\mu\sigma,$$

and $\underline{\mathbf{H}}_F$ is the magnetic field strength on the surface of and inside the ferromagnetic. The continuity condition of vector $\underline{\mathbf{H}}$ at the air–ferromagnetic boundary has the form

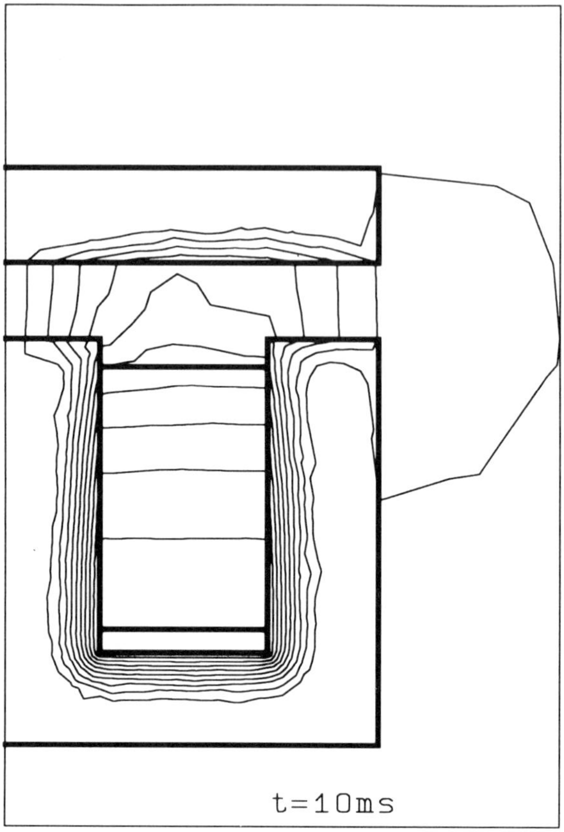

Fig. 3.37 *Transient magnetic field distribution*

$$(\mathbf{n} \times curl\,\underline{\mathbf{J}})_\Gamma = -\alpha^2(\mathbf{n} \times \underline{\mathbf{H}}_F)_\Gamma = -\alpha^2(\mathbf{n} \times \underline{\mathbf{H}}) \tag{3.244}$$

Taking into account that $\mathbf{n}$ and $\hat{\mathbf{k}}$ are perpendicular we obtain

$$grad(\mathbf{n} \cdot \underline{\mathbf{J}}) = (\mathbf{n}\,grad)\,\underline{\mathbf{J}} + \mathbf{n} \times curl\,\underline{\mathbf{J}} = 0 \tag{3.245}$$

After multiplying eqn (3.245) by $\mathbf{n}$ we obtain

$$\begin{aligned}(\mathbf{n} \times grad(\mathbf{n} \cdot \underline{\mathbf{J}}))_\Gamma &= ((\mathbf{n}\,grad)(\mathbf{n} \times \underline{\mathbf{J}}))_\Gamma + (\mathbf{n} \times (\mathbf{n} \times curl\,\underline{\mathbf{J}}))_\Gamma \\ &= \left(\frac{\partial}{\partial n}(\mathbf{n} \times \underline{\mathbf{J}})_\Gamma - \alpha^2(\mathbf{n} \times (\mathbf{n} \times \underline{\mathbf{H}})\right)_\Gamma = 0\end{aligned} \tag{3.246}$$

Taking into account the strong skin effect

$$\underline{\mathbf{J}} = \underline{\mathbf{J}}_\Gamma e^{-\alpha n}, \qquad \underline{\mathbf{H}} = \underline{\mathbf{H}}_\Gamma e^{-\alpha n} \tag{3.247}$$

From eqn (3.246) and eqn (3.247) it follows that

$$(\mathbf{n} \times \underline{\mathbf{J}})_\Gamma = \alpha(\mathbf{n} \times (\mathbf{n} \times \underline{\mathbf{H}}))_\Gamma \tag{3.248}$$

Taking into account that

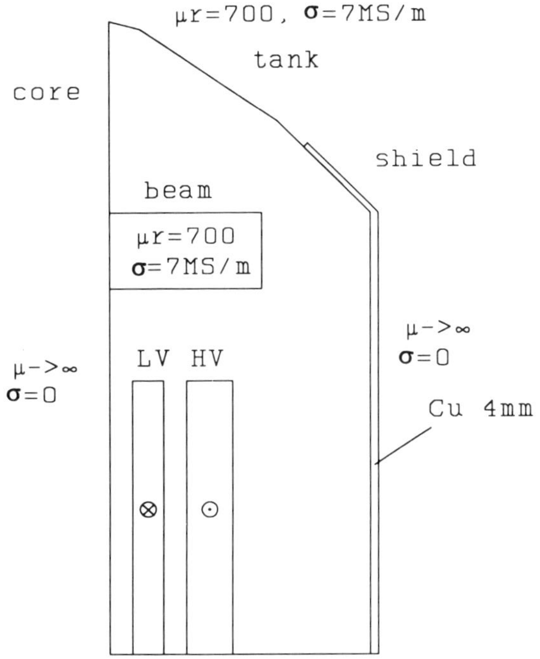

Fig. 3.38 *Cross-section of one quarter of a transformer*

$$\underline{\mathbf{A}} = A\hat{\mathbf{k}} \qquad div\,\underline{\mathbf{A}} = 0, \qquad \underline{\mathbf{H}} = \frac{1}{\mu_0} curl\,\underline{\mathbf{A}}$$

and identity (3.245) we obtain

$$\alpha(\mathbf{n}\times(\mathbf{n}\times\underline{\mathbf{H}}))_\Gamma = \frac{\alpha}{\mu_0}(\mathbf{n}\times(\mathbf{n}\times curl\,\underline{\mathbf{A}}))_\Gamma = -\frac{\alpha}{\mu_0}\{\mathbf{n}\times((\mathbf{n}\cdot\nabla)\underline{\mathbf{A}})\}_\Gamma$$
$$= -\frac{\alpha}{\mu_0}\left((\mathbf{n}\times\hat{\mathbf{k}})\frac{\partial A}{\partial n}\right)_\Gamma \tag{3.249}$$

Inside the ferromagnetic material, the vector potential $\underline{\mathbf{A}}_F$ satisfies Helmholtz's equation:

$$\nabla^2\underline{\mathbf{A}}_F = \alpha^2\underline{\mathbf{A}}_F \tag{3.250}$$

and $\underline{\mathbf{J}}$ is calculated from the formula

$$\underline{\mathbf{J}} = -\mathrm{j}\omega\sigma\underline{\mathbf{A}}_F \tag{3.251}$$

Taking into account continuity of $\underline{\mathbf{A}}$, eqn (3.248) will be transformed into

$$(\mathbf{n}\times\underline{\mathbf{J}})_\Gamma = -\mathrm{j}\omega\sigma((\mathbf{n}\times\hat{\mathbf{k}})\underline{\mathbf{A}}_F)_\Gamma = -\frac{\alpha^2}{\mu}((\mathbf{n}\times\hat{\mathbf{k}})\underline{\mathbf{A}})_\Gamma \tag{3.252}$$

From eqns (3.249) and (3.252) we obtain

$$\left(\frac{\partial \underline{\mathbf{A}}}{\partial n} - \beta \underline{\mathbf{A}}\right)_\Gamma = 0 \tag{3.253}$$

where

$$\beta = \alpha \frac{\mu_0}{\mu}$$

The above relationship expressed the third boundary condition (Hankel's) and is very useful for FEM applications.

(2) Thin layer models simplify calculation of complicated electromagnetic field problems. The structure can be considered as a thin layer if:

- its dimension, in the field penetration direction, is small, compared with other dimensions,
- a weak skin-effect occurs in the domain.

Theoretically, it is assumed that the domain is an infinitely thin conducting region with $\sigma \to \infty$ and it is possible to introduce surface conductivity

$$\sigma_s = \lim_{\substack{d \to 0 \\ \sigma \to \infty}} d\sigma \tag{3.254}$$

which has a limited numerical value.

Let us discuss the model presented in Fig. 3.39.

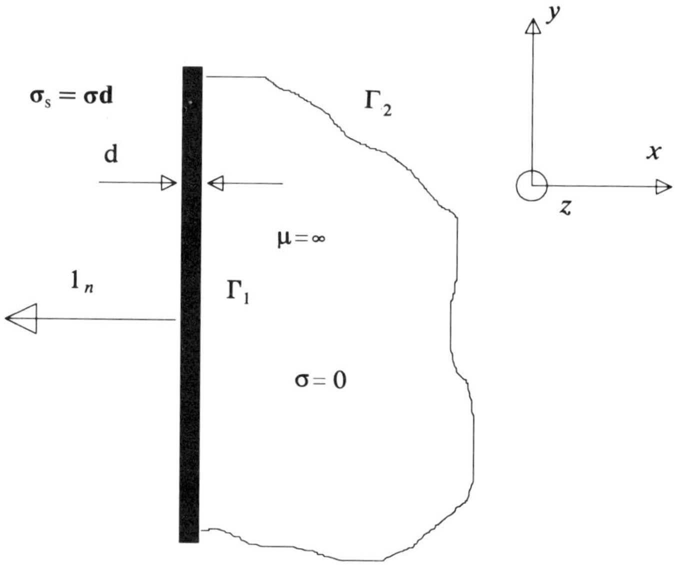

Fig. 3.39 *Thin layer model*

A thin conducting layer is placed on the ferromagnetic material. The induced current flows in the z-direction and on the surface Γ_1 the following equation is satisfied

$$(\mathbf{n} \times (\underline{\mathbf{H}} - \underline{\mathbf{H}}_F)) = (\underline{\mathbf{K}})_\Gamma \tag{3.255}$$

where:

$\underline{\mathbf{H}}$ magnetic field strength on the left side of boundary Γ_1,
$\underline{\mathbf{H}}_F$ magnetic field strength on the right side of boundary Γ_1,
$\underline{\mathbf{K}}$ surface current induced in thin layer.

Because

$$(\mathbf{n} \times \underline{\mathbf{H}})_{\Gamma_1} = 0 \tag{3.256}$$

then

$$(\mathbf{n} \times \underline{\mathbf{H}})_{\Gamma_1} = (\underline{\mathbf{K}})_{\Gamma_1} \tag{3.257}$$

After substitution and transformation we obtain

$$(\mathbf{n} \times \underline{\mathbf{H}})_{\Gamma_1} = \frac{1}{\mu_0}(\mathbf{n} \times curl\,\underline{\mathbf{A}})_{\Gamma_1} = -\frac{1}{\mu_0}((\mathbf{n} \cdot grad)\underline{\mathbf{A}})_{\Gamma_1} = -\frac{1}{\mu_0}\left(\frac{\partial A}{\partial n}\hat{\mathbf{k}}\right)_{\Gamma_1} \tag{3.258}$$

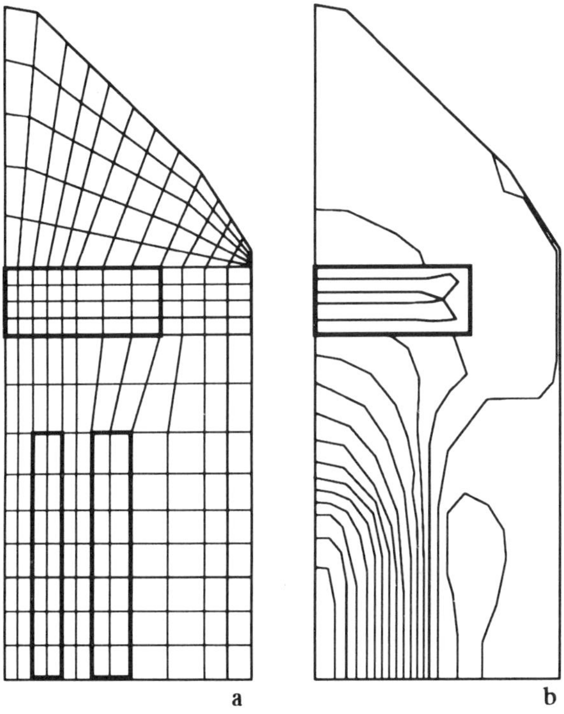

Fig. 3.40 *(a) FEM mesh and (b) magnetic field distribution in a transformer leakage area when the beam is modelled by Helmholtz's equation*

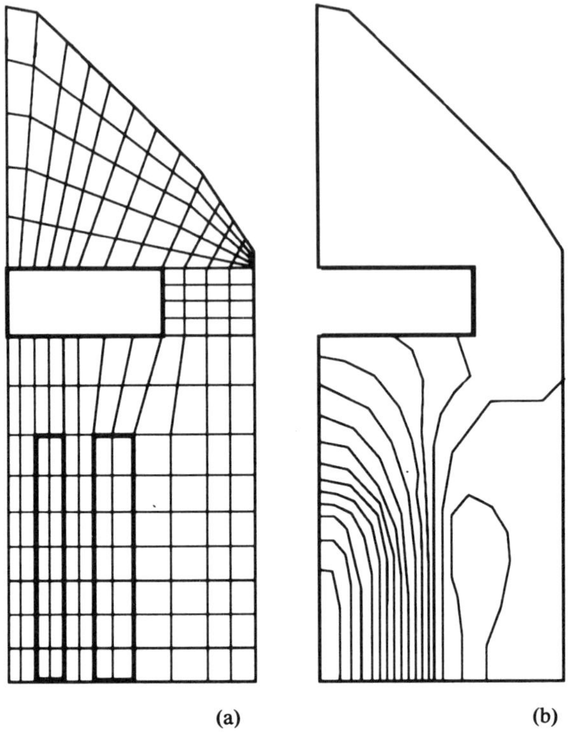

Fig. 3.41 *(a) FEM mesh and (b) magnetic field distribution in a transformer leakage area when the beam is modelled by the boundary condition of the third type*

The surface current density vector is calculated from the formula

$$(\underline{\mathbf{K}})_{\Gamma_1} = (-j\omega\sigma\underline{\mathbf{A}})_{\Gamma_1} \tag{3.259}$$

From eqns (3.257), (3.258) and (3.259) we obtain

$$\left(\frac{\partial\underline{\mathbf{A}}}{\partial n} - \beta_1\underline{\mathbf{A}}\right)_{\Gamma_1} = 0 \tag{3.260}$$

where

$$\beta_1 = j\omega\mu_0\sigma_s = j\omega\mu_0\sigma\,d$$

Equation (3.260) expresses the boundary condition of the third type.

We assume that Helmholtz's equation is satisfied in the beam. The problem formulated above was solved by the 'SONMAP' code. Results of calculations are presented in Figs 3.40 and 3.41.

4
Reluctance Networks

Janusz Turowski

4.1 A SURVEY OF EXISTING SOLUTIONS AND PROGRAMS

The method of *equivalent reluctance networks* (RNM) is based on Ohm's Law for magnetic circuits

$$V_{\mu i} = R_{\mu i}\Phi_i \tag{4.1}$$

and magnetic Kirchhoff's Laws, for nodes

$$\sum_{i=1}^{n} \Phi_i = 0 \tag{4.2}$$

and meshes

$$\sum_{k=1}^{m} V_{\mu k} = 0 \tag{4.3}$$

It is one of the oldest methods of modelling and calculating magnetic circuits in electrical machines and transformers. As such circuits are becoming increasingly complicated, for example in stepping motors or switched reluctance motors, the equivalent networks have been developed into extensive multi-node systems. Moreover, in order to model alternating magnetic fields in the presence of solid metallic bodies, in particular in iron with its nonlinear magnetic characteristics and induced eddy-currents, new complex reluctances have been introduced. The main benefits of using the RNM are evident when calculating three-dimensional fields of complicated geometries, as significant savings in computing time and effort may be achieved owing to the efficiency of the formulation.

The RNM was first used by the author of this Chapter in 1960 [4.22] for two-dimensional (2D) field modelling and calculation of magnetic fluxes in a three-winding power transformer. In 1969 the author introduced a multi-node three-dimensional reluctance network model 'RNM-3D' (Fig. 4.1) for fast approximate calculations of leakage fields in a three-phase power transformer. The RNM-3D was presented and discussed at various international transformer conferences, including CIGRE'81 USA, 3DMAG'89 Japan, ITMA'91 India, Shenyang'91 China, ISST'93 Poland and TRAFOTECH'94 India, and continues to provide the basis for many scientific and technical computations [4.20]; [4.22–4.29].

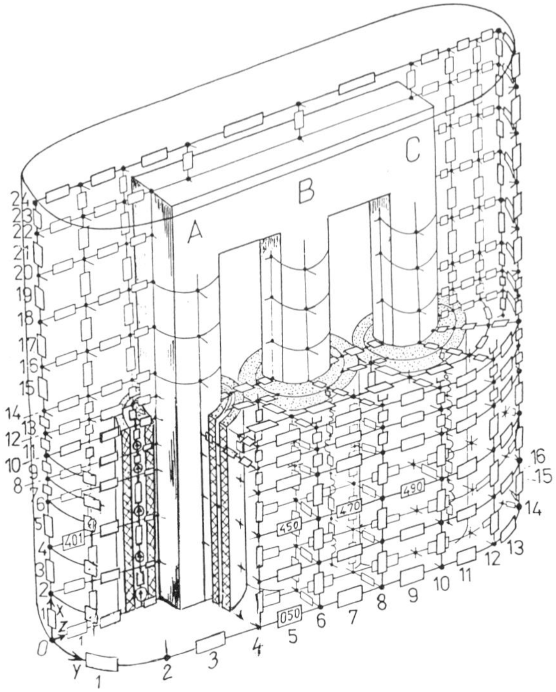

Fig. 4.1 *The basic equivalent reluctance network model RNM-3D for three-dimensional leakage fields in three-phase power transformers (one quarter of a transformer is shown) [4.22; 4.26]*

The equivalent reluctance network method was developed in parallel by many other authors. Since 1966 a series of fundamental papers by Davey and King (e.g. [4.7; 4.15]) and since 1975 works of Carpenter [4.3], Djurovic and Carpenter [4.9], and Djurovic and Monson [4.10], have contributed significantly to the theory of the RNM. Of particular interest is the paper by King [4.15], who demonstrated an equivalence between the reluctance network formulation and the finite difference method, and Carpenter [4.4], who showed similar equivalence with the finite element method. The latter has also been confirmed computationally using an example transformer by Komęza *et al.* [4.16]. In 1979 Kubusch [4.17] effectively applied the equivalent reluctance model to the computation of leakage fields and stray losses in tank walls of power transformers. Since 1980 Anuszczyk [4.2] has been using the two-dimensional version of the method to calculate rotational magnetization of laminated iron cores of electrical machines, especially induction motors. Other successful implementations of the method in recent years include the computation of transient fields, inductances and forces in a linear reversible electromagnetic pump [4.20; 4.29]. Owing to rapid computation the RNM is particularly suitable for analysis of transient processes, which often require a large number of repetitive calculations. The efficiency and good convergence of the method is demonstrated in references [4.26] and [4.29].

The concept of *tubes and slices* put forward by Hammond and Sykulski [4.13], and introduced in Chapter 1, Sections 1.8 to 1.10, is somewhat similar to the idea of reluctance networks, but takes advantage of the inherent duality of electromagnetic fields. At present the RNM uses one solution, although provision of dual bounds for this method could be one of the possible future developments of both techniques.

A series of papers published by the author of this Chapter and his collaborators during the past decade [4.20; 4.25–4.29] have concentrated on developing a convenient engineering computational tool for field modelling and optimization of leakage zones of large power transformers. For such a tool to be acceptable for regular design usage, a number of demanding requirements must be met.

- The program must be easy to use and perform satisfactorily even on low-cost personal computers.
- Computation has to be very fast and interactive, even in three-dimensional fields.
- It must provide useful design parameters.
- Flexibility must be maintained with minimum cost in time and effort.
- The program must allow the incorporation of complicated, three-dimensional, three-phase geometries, magnetic non-linearities, heating effects, deviation of material parameters, and other important practical considerations.

Although, in general, the application of electromagnetic software for design purposes is rapidly increasing (see the discussion in Chapter 7 of this book), for some more complicated devices, such as power transformers, unreasonably high computing times and memory requirements make it too expensive and impractical for standard packages to be used as every-day design tools. Many design and maintenance engineers would argue that the main objective is not to calculate the field with the highest possible accuracy but to quickly, and at an early design stage, localize and eliminate excessive stray losses, crushing forces or overheating hazards. Clearly, compromises have to be made in terms of accuracy or even adequacy of the model. Approximations, simplifications and even rough estimates may have to be introduced. The expertise of the designer becomes invaluable, and experimental verification of the model becomes part of the procedure.

The RNM-3D package offers such a simple, user-friendly, inexpensive engineering tool. The designer is provided with an answer within 15–30 s for each computational variant, which is a clear advantage in the design environment. A dedicated *pre-* and *post-processor* is built in for a selected class of three-phase transformers, and a general purpose *solver* is available (see Fig. 4.12) for solution of multi-node electric networks. In order to model other types or classes of electrical devices, say actuators or rotating machines, the pre- and post-processor has to be modified in accordance with particular requirements. A library of typical geometries, problems or devices could thus be created. Examples of such successful applications include calculation of three-dimensional fields, forces and losses in end-windings of large turbogenerators, by Davey and King [4.7]. Two-dimensional transient fields have been simulated amongst others by Oberretl [4.18] and Demenko [4.8] using equivalent RC networks with mutual capacitances.

4.2 RELUCTANCE NETWORKS AND FINITE DIFFERENCES

The method of equivalent reluctance networks has been extensivly reported, starting with the first paper by King [4.15], in which a theory of the RNM is derived from the well-known method of *finite differences* (see Chapter 2 for a full discussion of this technique). For example, Poisson's equations for the magnetic vector potential **A** in two-dimensions, with the components of current density $\mathbf{J} = J_z$, $J_x = J_y = 0$, and thus $\mathbf{A} = A_z = A$, $A_x = A_y = 0$, at constant permeability μ,

$$\frac{\partial^2 A}{\partial x^2} + \frac{\partial^2 A}{\partial y^2} = -\mu J \tag{4.4}$$

can be transformed into a difference equation by expanding the function $A(x, y)$ into a Taylor's series along both axes x and y (Fig. 4.2), where the grid with a mesh dimensions $a \times h$ is rectangular and can be adjusted to fit a particular shape.

Assuming the coordinates of nodes 1 and 3 to be $x_1 = x + h$ and $x_3 = x - h$ respectively, we find

$$\begin{aligned} A_1 &= A(x+h) = A(x) + \frac{h}{1!}A'(x) + \frac{h^2}{2!}A''(x) + \cdots + \frac{h^n}{n!}A^{(n)}(x) \\ A_3 &= A(x-h) = A(x) - \frac{h}{1!}A'(x) + \frac{h^2}{2!}A''(x) - \cdots + \frac{h^n}{n!}A^{(n)}(x) \end{aligned} \tag{4.5}$$

Adding both sides of eqn (4.5) and assuming $A(x) = A_0$ gives

$$\frac{\partial^2 A(x, y)}{\partial x^2} = A''(x) = \frac{1}{h^2}\left[A_1 + A_3 - 2A_0 - \varepsilon(h^4)\right] \tag{4.6}$$

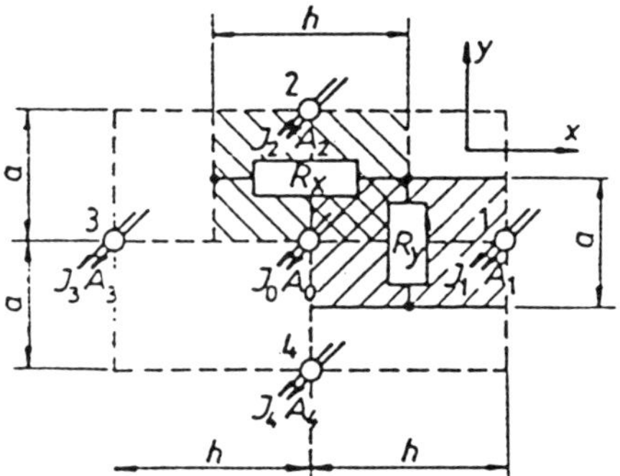

Fig. 4.2 *A rectangular element of the two-dimensional reluctance grid*

By analogy, for nodes 2 and 4 with coordinates $y_2 = y + a$ and $y_4 = y - a$, and taking $A(y) = A_0$, yields

$$\frac{\partial^2 A(x,y)}{\partial y^2} = A''(y) = \frac{1}{a^2}[A_2 + A_4 - 2A_0 - \varepsilon(a^4)] \tag{4.7}$$

The last terms on the right-hand side of eqns (4.6) and (4.7) indicate the order of the error of approximation. Neglecting these errors leads to the following relationships

$$\frac{\partial^2 A}{\partial x^2} \approx \frac{A_1}{h^2} + \frac{A_3}{h^2} - \frac{2A_0}{h^2} \quad \text{and} \quad \frac{\partial^2 A}{\partial y^2} \approx \frac{A_2}{a^2} + \frac{A_4}{a^2} - \frac{2A_0}{a^2} \tag{4.8}$$

Hence eqn (4.4) takes the finite difference form

$$\frac{A_1 - A_0}{h^2} + \frac{A_3 - A_0}{h^2} + \frac{A_2 - A_0}{a^2} + \frac{A_4 - A_0}{a^2} = -\mu J \tag{4.9}$$

Multiplying both sides of eqn (4.9) by ah/μ and introducing values of elementary reluctances per 1 metre of length along the z axis (see Fig. 4.2) in the following form

$$R_x = \frac{h}{\mu a \cdot 1} \quad \text{and} \quad R_y = \frac{a}{\mu h \cdot 1} \tag{4.10}$$

we obtain

$$(A_1 - A_0)R_y + (A_3 - A_0)R_y + (A_2 - A_0)R_x + (A_4 - A_0)R_x = -I_0 \tag{4.11}$$

where $I_0 = Jah$ is the total current in a single mesh.

Considering now the relationships

$$\Phi_i = \iint_{S_i} \mathbf{B}_i \cdot d\mathbf{S}_i = \iint_{S_i} curl\,\mathbf{A}_i \cdot d\mathbf{S}_i = \oint \mathbf{A}_i \cdot d\mathbf{l}_i \tag{4.12}$$

we can write for a single mesh (see Fig. 4.3)

$$\Phi_i = A_0 - A_i, \quad \text{where} \quad i = 1, 2, 3, 4 \tag{4.13}$$

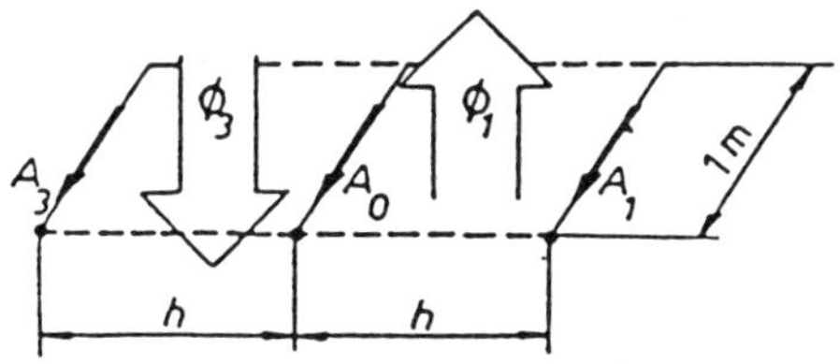

Fig. 4.3 *Determination of mesh fluxes $\Phi_i = \Sigma A_i l_i$ with the help of the magnetic vector potential A_i using eqn (4.12)*

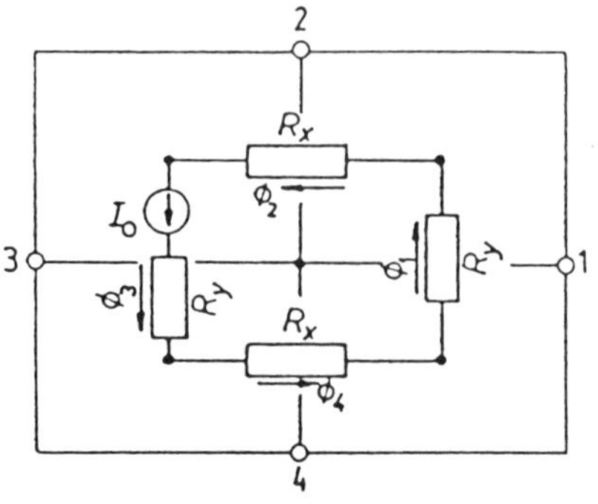

Fig. 4.4 *Equivalent reluctance circuit for a grid of Fig. 4.2*

and then, after substituting eqn (4.13) into eqn (4.11), we obtain Kirchhoff's and Ohm's equations for magnetic circuits

$$\Phi_1 R_y + \Phi_3 R_y + \Phi_2 R_x + \Phi_4 R_x = I_0 \quad \text{or} \quad \sum_{i=1}^{4} \Phi_i R_i = I_0 \tag{4.14}$$

for a corresponding equivalent magnetic circuit (Fig. 4.4).

The problem has now been reduced to setting up an appropriate grid model corresponding to the field in a specified device, and then solving it using any available program for multi-branched electrical networks, such as NAP-2, MIMIC, SPICE-2, or NODAL. There are two particular requirements which must be carefully investigated before the solution is attempted. First, the program must be able to handle a larger number of nodes, typically 500 to 1000 or more, and secondly, the solution times offered by the software must be extremely short, of the order of seconds, in order to make the process a fully interactive session, even on a personal computer.

4.3 LEAKAGE FIELDS IN TRANSFORMERS

In order to illustrate the method, and demonstrate its effectiveness for a typical important design problem, we shall look now at a model of a leakage field in a power transformer (see Fig. 4.5).

Depending on the accuracy required, various grid densities may be used. In the following examples most simulations have been performed on a grid with about 90 nodes (see Fig. 4.5b), but some of the preliminary calculations have also been made on a grid with about 50 nodes (see Fig. 4.5a). The argument here is, as demonstrated in Fig. 4.6a, that the specifics of the transformer structure make the convergence of the solution, with regard to the number of nodes, relatively high, and a coarse grid is often sufficient for preliminary assessment of results. For instance, the calculated distributions of the radial component of flux density in the Low Voltage (B_{LV}) and High Voltage (B_{HV}) windings, as well as the tangential component of field intensity ($H_{S,T}$) on the internal surface of the tank wall (Fig. 4.6a), with 46 nodes

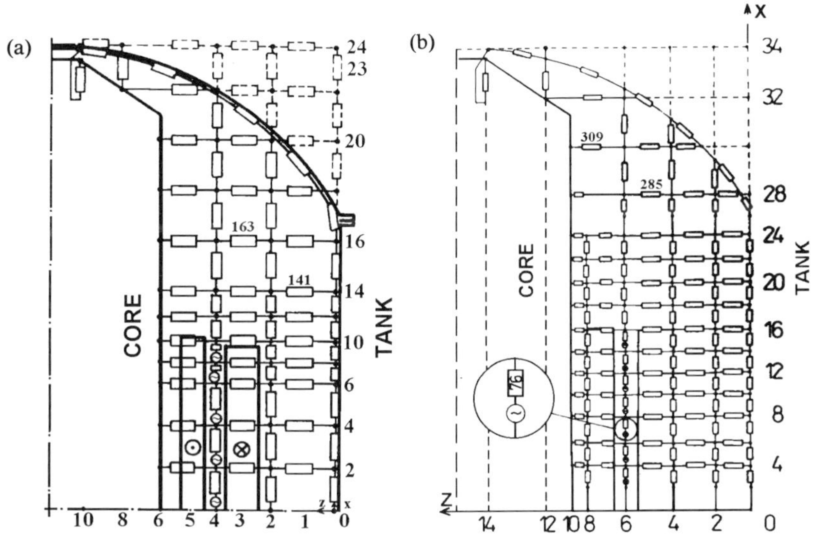

Fig. 4.5 *A reluctance network model of leakage field in a power transformer (a quarter of the cross-section is shown) [4.25]; (a) with 46 or 51 nodes, (b) with 89 or 91 nodes*

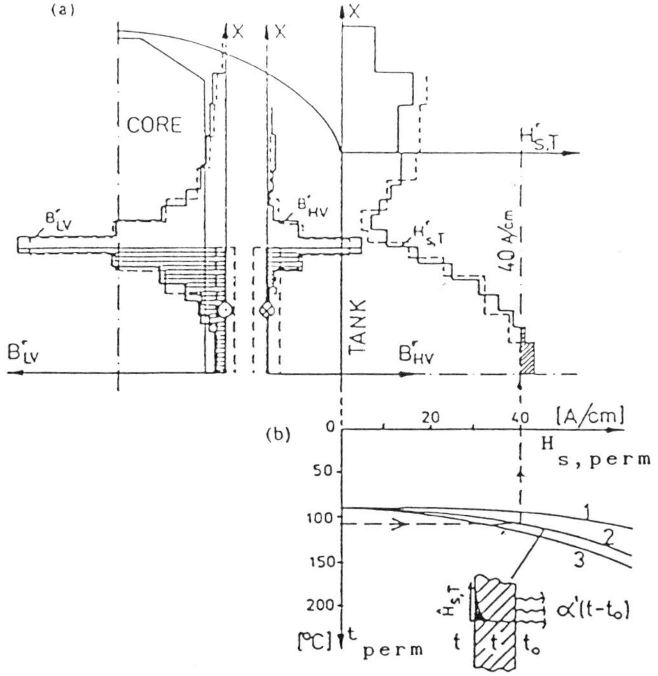

Fig. 4.6 *The RNM-2D analysis of leakage fields [4.25; 4.26]: (a) effects of number of nodes, (b) hot-spot localization at permitted temperature t_{perm} and corresponding permitted field $H_{perm} = 40\ A\ cm^{-1}$*

(dashed lines) and with 89 nodes (continuous lines), are practically the same.

A good accuracy of the network model of the leakage field of transformers, even for a small (50–100) number of nodes (Fig. 4.5), is not surprising if it is remembered that the successful classical method of leakage field calculations is based on a single reluctance of the inter-winding gap, corrected only by the Rogowski's coefficient [4.22; 4.26].

Currently programs with a much higher number of nodes, up to 1000 and more, are available. However, increasing the number of nodes beyond what is reasonably required is not recommended, as a slight improvement in accuracy may not be a sufficient justification for increased computing times, and thus the method's main strength, its fast response time, could be impaired. Therefore, since speed is of the essence, to make the computation faster, whenever possible analytical formulae should be used for computing particular reluctances, as well as for recalculating parameters at the post-processing stage.

In order to illustrate the computation of grid parameters (reluctances) and to facilitate understanding of this process, the example from Fig. 4.5a will be used, with a relatively small number of nodes. The discussion below effectively defines the way in which an analytical pre- and post-processing is created for the interactive program RNM-3D (Fig. 4.20). The example of Figs 4.5 and 4.6 is taken for a 2D plane at $y = 0$.

It follows from Fig. 4.2 that the elementary reluctances (in $1/H$) given by

$$R_{\mu i} = \frac{l_i}{\mu_0 S_i}, \qquad i = x, z \tag{4.15}$$

are calculated from geometric dimensions of meshes (Fig. 4.7), into which the investigated area is subdivided. The number of branches of the network is determined exactly by its discrete coordinates $i = xyz$ of the cartesian coordinate system (Figs 4.1 and 4.5a). For every reluctance in Fig. 4.5 there are corresponding magnetic flux components Φ_x and Φ_z passing through this element. All quantities $V_{xz} = R_{xz}\Phi_{xz}$, R_{xz} and Φ_{xz} are uniquely localized in the network with the aid of indices xz corresponding to discrete coordinates of the system. Complex reluctances $\underline{R}_{x0}$ of the solid iron tank walls and cover contain additional nodes, brought about by the subdivision into real and imaginary components owing to the eddy current reaction.

Determination of magnetomotive forces of sources

Owing to the slenderness of typical windings of large power transformers, in many cases, especially when the field outside the windings is investigated, the *magnetomotive force* (*mmf*) of the windings can be evaluated by assuming that entire current flow of the High Voltage (HV) and Low Voltage (LV) windings

$$F = \sqrt{2}\, I_{HV} N_{HV} = \sqrt{2}\, I_{LV} N_{LV} \tag{4.16}$$

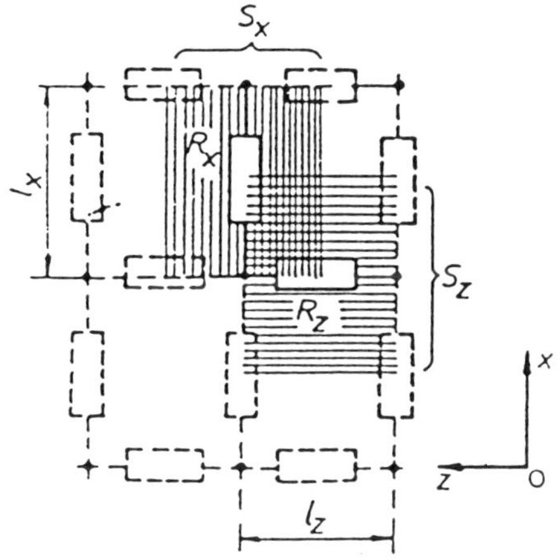

Fig. 4.7 *The principle of calculation of elementary reluctances of a network: R_x is in the x-direction and R_z in the z-direction*

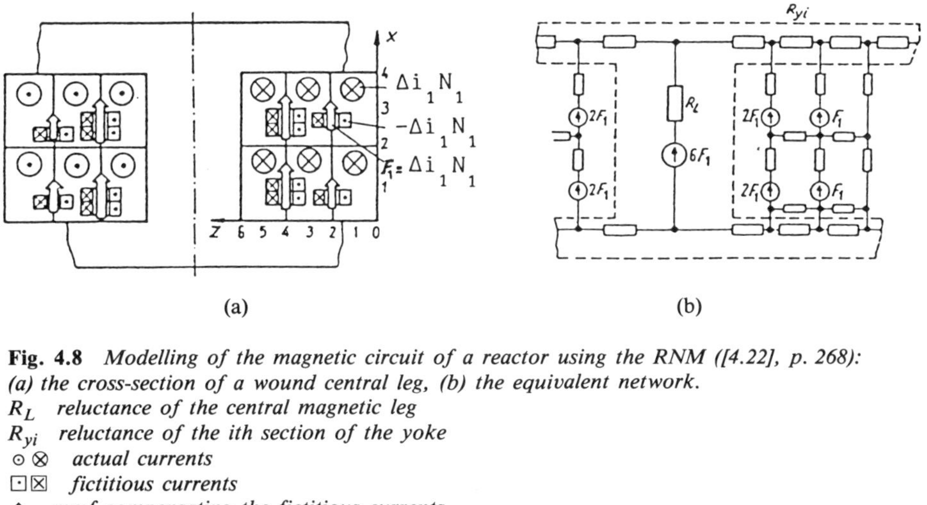

(a) (b)

Fig. 4.8 *Modelling of the magnetic circuit of a reactor using the RNM ([4.22], p. 268): (a) the cross-section of a wound central leg, (b) the equivalent network.*

R_L *reluctance of the central magnetic leg*

R_{yi} *reluctance of the ith section of the yoke*

⊙ ⊗ *actual currents*

⊡ ⊠ *fictitious currents*

⥣ *mmf compensating the fictitious currents*

is concentrated in a filament layer (current sheet) on the axes of the HV or LV windings.

If we wish to account for finite dimensions of winding cross-sections, we can use a well known principle by which the field of a current flow, subdivided into elements $\Delta i_i N_i$ (see Fig. 4.8), is replaced by a network of circuits. This procedure consists of the shifting of the elementary current $\Delta i_i N_i$ from one mesh, e.g. $(x = 3, z = 1)$, across the border between the nodes 22–42, to another mesh, say $(x = 3, z = 3)$. This produces in this border a new fictitious *mmf* source with the value of shifted flow $F = \Delta i_1 N_1$. At the same time the flow in the mesh $(x = 3, z = 3)$ is doubled. As the result the first mesh

($x = 3, z = 1$) becomes empty. This can be explained by postulating that in the axis of the border between adjacent meshes we can insert a fictitious coil (rectangles in Fig. 4.8), with identical but opposite flow ($-\Delta i_1 N_1$) from the first mesh, and simultaneously compensate the action of this coil with the help of a fictitious *mmf* of the same but oppositely directed flow. This implies an additional rule - that every shifting of ampere–turns $\Delta i_i N_i$ across any mesh border creates an *mmf* (flow) $F = \Delta i_i N_i$ in the axis of this border and doubles the flow in the adjacent mesh. The next shifting of ampere–turns follows the same rule, but this time with the entire enlarged mesh current flow - i.e. its own and added.

By repeating this operation we can replace a real coil with a distributed flow (Fig. 4.8a) with a circuit network of reluctances and lumped magnetomotive forces (Fig. 4.8b). In an analogous way one can shift the current flow to a yoke axis.

In the case of height asymmetry of HV and LV windings (Fig. 4.9), an additional transverse flow F_z (Fig. 4.9) should be introduced. This radial flow can be found using a well-known rule of ampere–turn decomposition into axial and radial components ([4.22], page 347). In this way any asymmetry can be considered, with the accuracy depending on the number of nodes introduced.

Because of the slenderness of the windings of large transformers and, usually, the asymmetry of HV and LV windings, we shall limit our attention to the flow F calculated using eqn (4.16). Calculations for higher numbers of nodes involve evaluation of submultiples of the given values [4.25].

Evaluation of network elements

The elementary *mmf* U_p of an inter-winding gap for a bigger mesh (Fig. 4.5a) can be expressed using the formula

$$U_p = \frac{F}{n + \dfrac{n_1}{2}} = \frac{\sqrt{2}\, I_H N_H}{n + \dfrac{n_1}{2}} = 1 \text{ per unit value (basic unit)} \qquad (4.17)$$

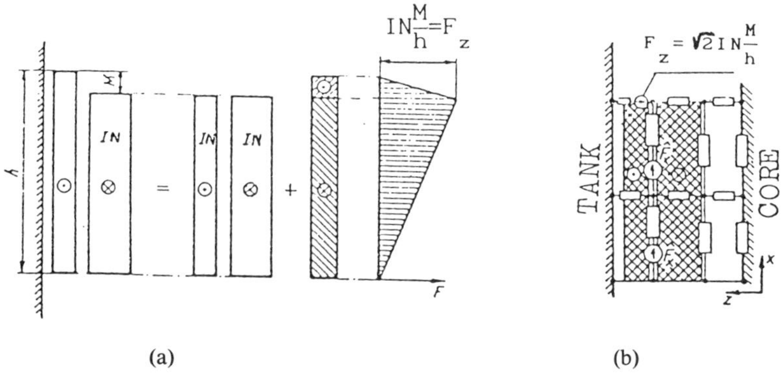

Fig. 4.9 *The influence of winding asymmetry on the RNM model [4.27]: (a) decomposition of ampere–turns into a symmetrical component $F = \sqrt{2}IN$ and a radial component with maximum flow $F = \sqrt{2}IN\,M\,h^{-1}$, (b) equivalent reluctance network with addition source F_z in the radial branch, where F_x is an element of the main flow*

where $n = 6$ is the number of bigger meshes along the entire winding height, $n_1 = 4$ is the number of smaller meshes on both edges of the winding, and $p = 14, 34, 54$ denotes coordinates (numbers) of branches referred to by eqn (4.17).

The elementary *mmf* of smaller meshes at the edges of windings (Fig. 4.5a) are given by

$$U_{p1} = \frac{U_p}{2} \qquad (p1 = 74,\ 94) \tag{4.18}$$

The reluctances of a leakage zone of a transformer are determined from its geometric dimensions (Fig. 4.10). In order to facilitate the calculations, the reluctances have been expressed in relative units with reference to the reluctance of the equivalent inter-winding gap δ', for bigger meshes.

As an example, the reluctances of the inter-winding gap and above it (Fig. 4.5a, 4.10, and 4.11a) are:

(a) for bigger meshes

$$R_p = \frac{1}{n + \dfrac{n_1}{2}} \; \frac{h_R}{\mu_0 \delta' \dfrac{l_{mean}}{s}} = 1 \text{ per unit value (basic unit)} \tag{4.19}$$

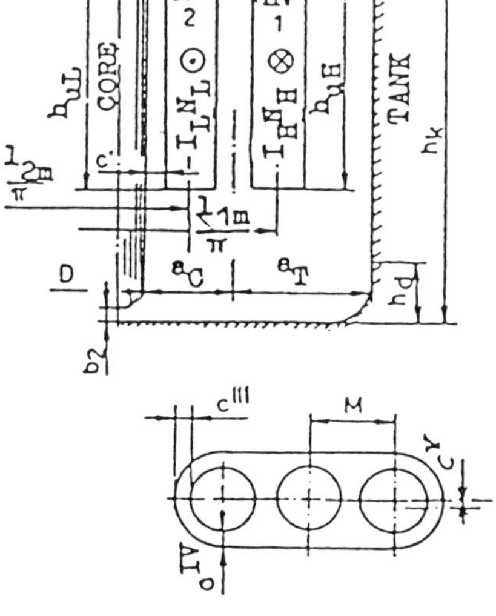

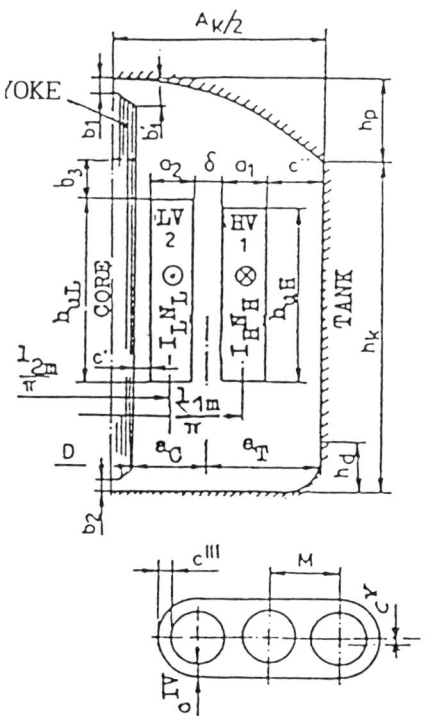

Fig. 4.10 *Geometrical data necessary for calculation of reluctances in the RNM-3D model [4.26]*

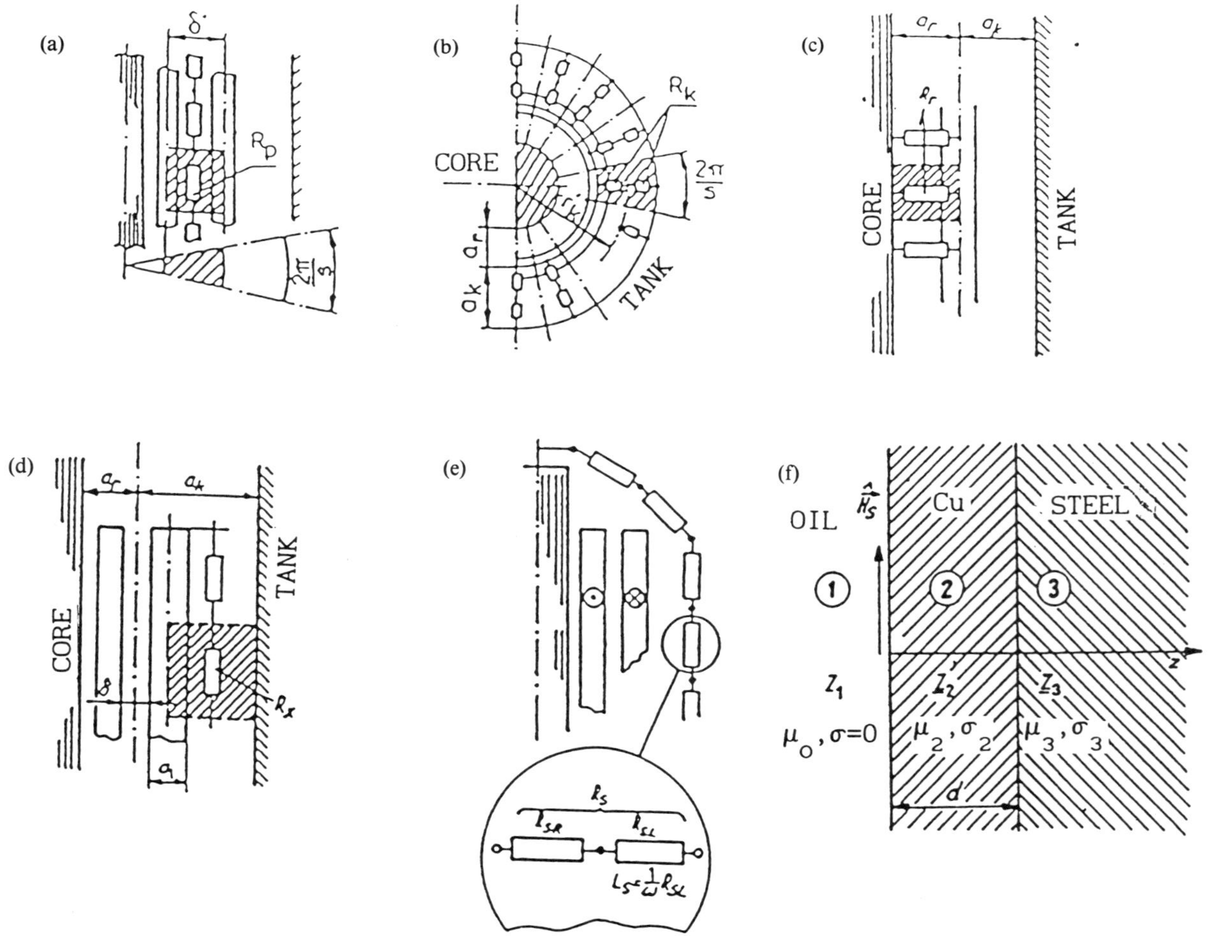

Fig. 4.11 *Computation of reluctances: (a) axial along the gap axis, (b) radial towards the tank, (c) radial towards the core, (d) axial between the tank and the winding, (e) complex, of the tank wall and cover, (f) electromagnetic copper screen*

where $p = 14, 34, 54$ denotes coordinates (numbers) of reluctances, and $s = 16$ is the assumed number of segments on the leg circumference (Fig. 4.11a):

$$R_{p2} = R_p \frac{\delta'}{\delta' + \frac{a_2}{2} + c'} \qquad (p2 = 154,\ 174,\ 194,\ 214) \tag{4.20}$$

$$c'' = \tfrac{1}{2}\,(c^{\mathrm{III}} + c^{\mathrm{IV}})$$

$$h_{mean} = \tfrac{1}{2}\,(h_{uL} + h_{uH})$$

$$\delta' = \delta + \tfrac{1}{2}\,(a_1 + a_2)$$

$$h_R = \frac{1}{K} h_u$$

$$K = 1 - \frac{a_1 + a_2 + \delta}{\pi h}$$

$$l_{mean} = \tfrac{1}{2}\,(l_{1m} + l_{2m})$$

(b) for meshes half as small the reluctances are half as small

$$R_{p1} = \tfrac{1}{2} R_p = 0.5 \quad \text{per unit value} \qquad (p1 = 74,\ 94) \tag{4.20a}$$

$$R_{p3} = \frac{1}{2} R_p \frac{\delta'}{\delta' + \frac{a_2}{2} + c'} \qquad (p2 = 114,\ 134) \tag{4.20b}$$

The radial reluctances towards the tank wall (Figs 4.5a and 4.11b) in $1/H$ are

(a) for bigger meshes

$$R_k = \frac{\frac{a_k}{2}}{\mu_0 \frac{h_u}{n + \frac{n_1}{2}} \frac{2\pi r_k'}{s}} \tag{4.21}$$

(b) for smaller meshes these reluctances are correspondingly increased

$$R_{k1} = \tfrac{4}{3} R_k \quad \text{and} \quad R_{k2} = 2R_k \tag{4.21a}$$

Other reluctances are (Fig. 4.11):

- radial to the core R_r (Fig. 4.11c), with coordinates $r = 25, 45, 165, 185, 205, 225$, $r1 = 65, 145$ and $r_2 = 85, 105, 125$;
- axial along the oil gap between the tank and winding R_x (Fig. 4.11d), with coordinates $x = 12, 32, 52, 152, 172, 192, 212$, $x1 = 72, 92, 112, 132, 238, 231$;
- complex, axial reluctances (or rather magnetic impedances) for conductive areas $\underline{R}_s = R_{sR} + jR_{sL}$ of the tank wall and cover, made of solid

steel (Fig. 4.11e), with coordinates $s = 10, 30, 50, 150, 170$, $s1 = 70, 90, 110, 130$ and $s2 = 191, 213, 235, 239$. These reluctances take into account analytically the skin effect, eddy current reactions with phase shift, nonlinear permeability inside solid metals, equivalent depth of an alternating field penetration into solid metal and the screening effect of a double-layer of the screen and solid metal (Fig. 4.11f).

The unitary reluctances of the solid steel wall have been calculated as complex values with the help of the formulae ([4.22], p. 139)

$$\underline{R}_{\mu 1} = R_{\mu 1} + jR_{\mu 1r} \tag{4.22}$$

where

$$R_{\mu 1} = a_1\sqrt{\left(\frac{\omega\sigma}{\mu_s}\right)} \quad \text{and} \quad R_{\mu 1r} = a_2\sqrt{\left(\frac{\omega\sigma}{\mu_s}\right)} \tag{4.23}$$

and $a_1 \approx 0.37$ and $a_2 \approx 0.61$ are linearization coefficients for solid steel ($a_1 = a_2 = 1$ for non-magnetic metal).

The linearized formulae (4.23) are valid for solid metal walls of thickness d larger than the half wave length λ in the metal, i.e.

$$d > \frac{\lambda}{2}, \qquad \lambda = 2\pi\sqrt{\left(\frac{2}{\omega\mu\sigma}\right)} \tag{4.24}$$

In the case of solid steel the threshold value of $\lambda/2$ is about 4–5 mm, whereas a typical tank wall and cover of a large transformers have a thickness of 10–15 mm or more. Also, $\mu_s = \mu_0\mu_{rs}$, where μ_{rs} is taken from the magnetization characteristics $\mu = \mu(H)$ ([4.22], pp. 68, 71) for the tangential H_{ms} on the internal surface of the steel wall.

Electromagnetic screens of the tank wall

If the tank wall is covered with an electromagnetic screen (e.g. made of copper or aluminium), the complex reluctance of such a conducting wall in per unit values becomes ([4.27], p. 170)

$$\underline{R}_{e1} = -\frac{\alpha_2\alpha_3 \sinh\alpha_2 d\left[1 + \alpha_2 d\sqrt{\left(\frac{\mu_3\sigma_2}{\mu_2\sigma_3}\right)}\right]}{\mu_0\alpha_3\left[1 + \alpha_2 d\sqrt{\left(\frac{\mu_3\sigma_2}{\mu_2\sigma_3}\right)}\right](\cosh\alpha_2 d - 1) + \mu_w\alpha_2\sinh\alpha_2 d}$$

$$\approx \frac{\alpha_2\sinh\alpha_2 d}{\mu_0(\cosh\alpha_2 d - 1)} \approx \frac{2}{\mu_0 d} \tag{4.25}$$

for $|\alpha d|^2 \ll 1$, where d is the screen thickness, $\alpha_i = (1 + j)k_i$, $k_i = (\pi f\mu_i\sigma_i)^{-1/2}$, $i = 2$ for the screen and $i = 3$ for solid steel.

For example, for a screen with $d = 4$ mm and $f = 50$ Hz, $k_2 d = 0.428$ and $\underline{R}_{e1} \approx 0.4 \times 10^9\,\mathrm{H}^{-1}$. This means that with an accuracy sufficient for engineering practice, one can assume, as a first approximation, that in the parts of the tank covered by an electromagnetic screen made of copper or aluminium, the tank reluctance is infinitely large, i.e. $R_s = \infty$.

Magnetic screens (shunts) on the tank wall
If the tank wall is covered with a magnetic screen (shunt), made of laminated iron, the resultant reluctance $\underline{R}_s$ (in H^{-1}) of the solid steel and the screen can be calculated (see reference [4.22], p. 172) as a parallel connection of the reluctance $\underline{R}_s$ of the screen and $\underline{R}_w$ of the solid steel wall:

$$\underline{R}_s = \frac{1}{\dfrac{1}{\underline{R}_e} + \dfrac{1}{\underline{R}_w}} = \frac{1}{\mu_e d + \dfrac{1-j}{2}\sqrt{\left(\dfrac{2\mu_w}{\omega\sigma}\right)}} \cdot \frac{sh_u R_{e1}}{\left(n + n_1 \dfrac{1}{2}\right) \pi (D + 2a_c + 2a_T)} \tag{4.26}$$

where μ_e and d are the magnetic permeability and thickness of the screen, and σ and μ_w are the conductivity and magnetic permeability of the tank wall.

The minimum thickness, with regard to sheets saturation, of the magnetic screen can be evaluated (see reference [4.22], p. 197) from the formula

$$d_{min} \geqslant \frac{a_c}{a_c + a_T} \cdot \frac{\sqrt{2} I N \mu_0 \delta'}{B_{1000} h} \tag{4.27}$$

where B_{1000} is the maximum flux density corresponding to the screen permeability $\mu_{er} = 1000$ (i.e. about 1.7 T for anisotropic and 1.4 T for isotropic transformer steel). Neglecting the reluctance of the solid steel wall provides an additional safety margin for the evaluation of d_{min}.

The reluctance of the steel wall covered with an unsaturated magnetic screen is so small compared with the reluctances of nonmagnetic regions, that one can safely assume, as a first approximation, that this reluctance is zero

$$R_s = R_{s2} \approx 0 \tag{4.28}$$

whereas for a non-screened steel wall and cover, the value of R_{s1} has to be calculated using eqn (4.22).

All the other reluctances of the three-dimensional model (Fig. 4.1), including those in the circumferential direction (y axis), have been calculated as explained above. The detailed formulae are given in reference [4.27], Chapter 3. The appropriate values of these elements, corresponding to the declared transformer parameters (Fig. 4.10) and the frequency of the supply voltage, are placed in the network scheme, along with the voltage sources which model the elementary magnetomotive force in a winding for each phase. All network elements are calculated automatically in a preprocessor on the basis of the analytical formulae. They are usually expressed as per-unit (relative) values referred to the data of the inter-winding gap δ. The superscript 'r' means a relative value referred to the gap value.

The nonlinearity of the permeability of steel with depth has been considered with the help of linearization coefficients $a_p \approx 1.4$ for resistance and active power, and $a_p \approx 0.85$ for reactance and reactive power ([4.22], p. 322). The permeability along the x axis has been taken as constant in the first instant but it is possible to take into account the changes of μ using an iterative method or an analytical approximation.

We shall look now at the post-processor of the program RNM-3D. As a result of the computation, two quantities for each reluctance are of particular interest, the flux density B_{mi} and the magnetic field strength H_{mi}. Hence

$$B_{mi} = \frac{\Phi_{mi}}{A_i} \quad \text{and} \quad H_{mi} = \frac{V_{mi}}{l_i} \tag{4.29}$$

where Φ_{mi} is the maximum value of the flux calculated for the ith reluctance, in Wb, V_{mi} is the maximum value of the magnetic voltage, in A, and A_i and l_i are the cross-section and length of the ith reluctance, in m^2 and m respectively.

The flux density B_{mi} gives rise to forces and additional losses in the windings, whereas the tangential value of H_{msi} on the steel surface can be related to induced eddy-current losses in the tank wall and cover, yoke beams, clamping plates etc., which are often calculated using the following formula ([4.22], pp. 193, 330).

$$P_1 = k(d)\frac{a_p}{2}\sqrt{\left(\frac{\omega\mu_0}{2\sigma}\right)}\iint_s \sqrt{(\mu_{rs})}\,|H_{ms}|^2\,dS \tag{4.30}$$

where $k(d)$ depends on the wall thickness, $k(d) < 1$ at $d < \lambda/2$ and $k(d) = 1$ for $d \geqslant \lambda/2$; $a_p \approx 1.4$ is the linearization coefficient ([4.22] pp. 322–323) for solid steel, $a_p \approx 1$ is a linearization coefficient for a non-magnetic material; $\mu_{rs} = \mu_{rs}(H_{ms})$ is the surface relative permeability, $\mu_s = \mu_0\mu_{rs}$; σ is the conductivity and $\lambda = 2\pi(\omega\mu\sigma/2)^{-1/2}$.

When the calculated quantities are expressed in the per-unit notation as Φ^r_{mi}, B^r_{mi} and H^r_{mi}, the corresponding reference values for the inter-winding gap δ', for a sector of width $2\pi/s$ (see Fig. 4.11b), are given by

$$\Phi_{m\delta} = \frac{\sqrt{2}I_{NH}N_H}{h_R}\mu_0\delta'\frac{l_{\text{mean}}}{s} = 1 \quad \text{per-unit value (Wb)} \tag{4.31}$$

$$B_{m\delta} = \frac{\sqrt{2}I_{NH}N_H}{h_R}\mu_0 = 1 \quad \text{per-unit value (T)} \tag{4.32}$$

$$H_{m\delta} = \frac{\sqrt{2}I_{NH}N_H}{h_R} = \frac{V_{m\delta}}{h_R\left(n + \dfrac{n_1}{2}\right)} = 1 \quad \text{per-unit value (A m}^{-1}\text{)} \tag{4.33}$$

where I_{NH} is the rated current in the HV winding, N_H is the number of turns of the HV winding, and both are phase values. For other symbols refer to Fig. 4.10.

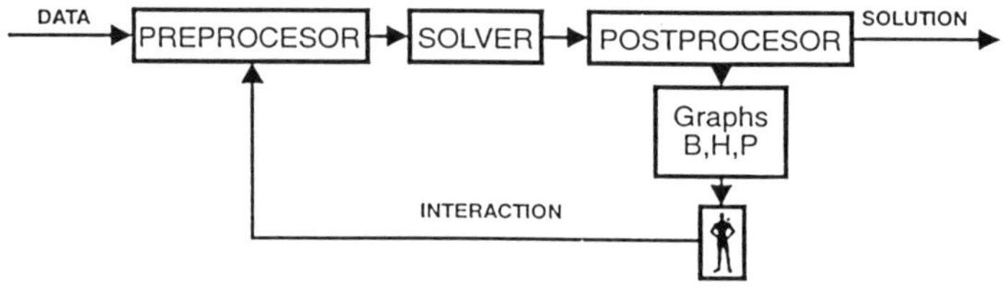

Fig. 4.12 *Interactive design using the RNM-3D package [4.26]*

4.4 COMPUTATIONAL ASPECTS

Having formulated the reluctance model for a given device, or its part, see Fig. 4.1 or 4.5, and calculated all the network parameters using the formulae given in Section 4.3, or in reference [4.27], p. 151, the next step is the solution of the multi-mesh network. This can be done using one of many popular programs for network analysis, such as NAP-2, SPICE-2, MIMIC, NODAL etc. It has been found, however, that it may be advantageous to create a dedicated solving routine, in order to benefit fully from the particular features of the method. Thus a fully integrated system has been developed for fast interactive analysis of the leakage zone of power transformers. The package is called RNM-3D and incorporates the pre- and post-processing as described in Section 4.3 (Fig. 4.12).

The flowchart of the RNM-3D is shown in Fig. 4.13. The basic reluctance model was shown previously as Fig. 4.1. Different variants of the model are stored on a disk. Every variant provides a skeleton framework for calculations and requires actual reluctances and some other data to be specified for each particular case. A node admittance matrix is then created for the network solution, using the modified nodal approach [4.14], and a set of linear equations is solved using a direct sparse matrix method [4.11]. Finally, post-processing is applied and distributions of the magnetic field along selected paths are displayed or printed, as shown in Figs 4.6, 4.14, 4.15, 4.16 etc. For running the simulation and analysis of a three-phase transformer, the user may employ the existing network model (Fig. 4.1), stored on a disk, or define his own model of the same topology, but with a different number of loops and branches. The user's model may also be saved on disk for later use.

The main program (Fig. 4.13) must be provided with the following input data:

(1) a 'PARAM' statement, followed by some constructional parameters of a transformer:
- the maximum current I_{HV} (IHV) in amperes,
- the number of turns Z_{HV} (ZHV),
- the following dimensions in millimetres (see Fig. 4.10): a_1 (A1), a_2 (A2), δ (DEL), h_{LV} (HLV), h_{HV} (HHV), D (D), c^I (C1), $c^{II,III,IV,V}$ (C2, C3, C4, C5), M (M), d_{Al} (DAL), d_{Cu} (DCU), h_k (HK), h_p (HP), h_d (HD), b_1 (B1), b_1' (B1PR), b_2 (B2),
- the relative magnetic permeability of steel μ_{rs} (MRS) – if necessary obtained through iteration from the magnetization curve, Fig. 4.17b, or from the analytical approximation, eqn (4.35);

(2) an 'F' statement, followed by the value of the supply frequency or the range of frequencies and the increment;

(3) a 'MODEL' statement, followed by the model number which is to be retrieved from the disk for analysis:

An example set of data may be found in paper [4.26]. Following the instruction 'RUN' , the calculation of parameters is executed and the results

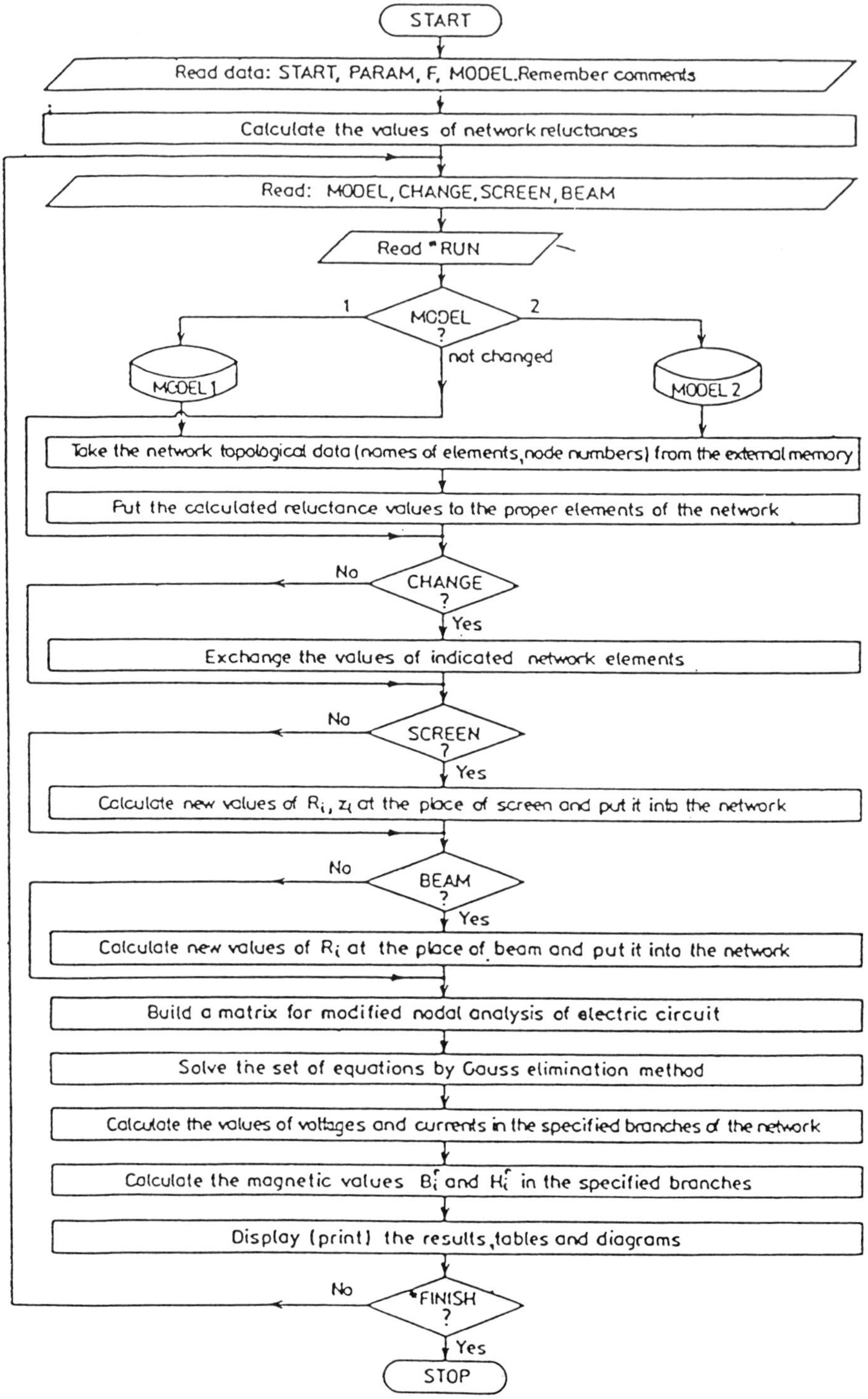

Fig. 4.13 *Flowchart of the RNM-3D algorithm [4.26]*

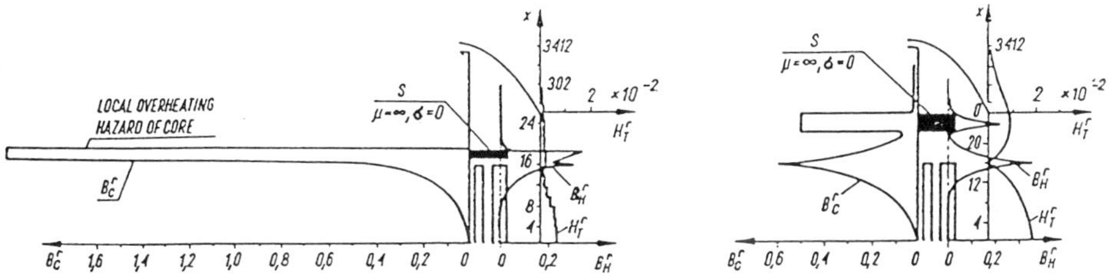

Fig. 4.14 *The RNM-2D analysis of the influence of a laminated magnetic shunt collector on the magnetic field as in Fig. 4.6a [4.25]*

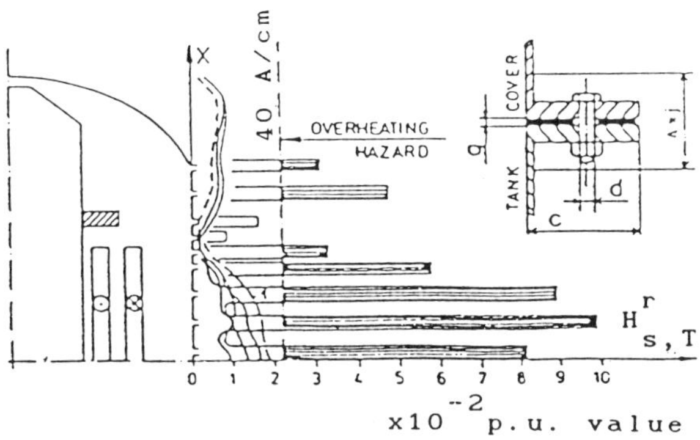

Fig. 4.15 *The RNM-2D analysis of excessive heating hazard of flanged bolt joints with regard to their position on the tank wall ([4.27], pp. 188–194)*

are displayed. The reluctance network thus created can be modified using the following special commands:

- 'CHANGE' – for changing the values of specific elements of the network. The elements can be identified by their position in the *xyz* coordinates of the network (Fig. 4.1),
- 'BEAM' – for the simulation of a horizontal solid-steel yoke beam or a laminated magnetic shunt collector (Fig. 4.14) above the transformer windings,
- 'SCREEN' – for the simulation of copper or aluminium screening plates, of specified thickness, on the tank wall.

Other modifications can also be made. For example, the designer can specify coordinates of the reluctances which are to model the presence of a copper screen or a magnetic shunt (see Fig. 4.16). The five-limb transformer can be represented too, by specifying the reluctances with coordinates ($x = 1, 3, 5, \ldots, 21$; $y = 0$, $z = 2$) and ($x = 20, 22$; $y = 0$; $z = 1, 3$), as well as those on the opposite side of the core, as equal to zero.

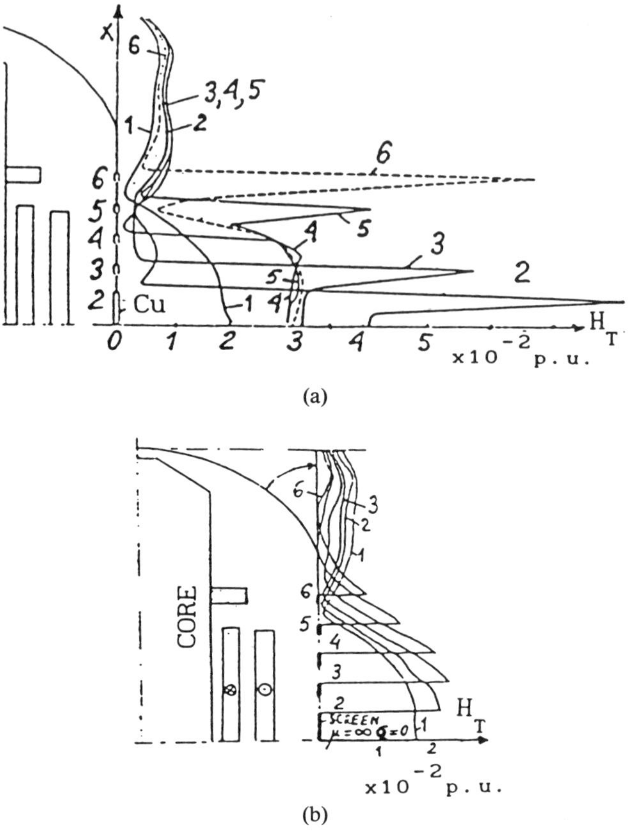

Fig. 4.16 *The RNM-2D analysis of the influence of screens of the tank wall on the field H_{msT} on the tank surface: (a) electromagnetic (copper) screens, (b) magnetic (shunt) screens made of laminated iron*

4.5 INTERACTIVE DESIGN AND DETECTION OF 'HOT SPOTS'

Most of the electromagnetic software used currently to aid the design of electromechanical devices utilizes field analysis, with prescribed boundary conditions and excitation, as the main computational tool. Field synthesis, although fast developing (see Chapter 5) and already capable of dealing with simple geometries and problems, has not yet reached the stage of a fully matured design tool, and very few applications to real engineering problems exist. The most promising, for some time to come, is the idea of an *interactive design*, where the designer is in constant 'conversation' with the computer. Although some elements of the process may be automated, and proper optimization techniques are under development (see Chapter 7), it remains essentially a trial-and-error approach, with the designer making all the important decisions and the computer, with its suitable software, facilitating the computation and providing useful post-viewing and post-processing. Fast computation becomes essential and many techniques, including the most popular finite-element method, may, especially for the more complicated three-dimensional problems, require far too long computing times and thus become impractical for the interactive process. The dedicated approximate computation, with its particularly short solution times as offered by the reluctance network method, becomes an interesting alternative, especially for simplified engineering analysis and design. For example the RNM-3D program provides an approximate answer typically within 15–30 s for one computational variant of a transformer leakage field. This should be compared with possibly even hours of computation using finite elements at this level of complication.

Before we proceed with further discussion of the RNM-3D, it is worthwhile to mention an interesting hybrid approach, proposed by the author (Fig. 4.17), which mixes finite elements and reluctance networks. Thus solid steel elements (tank and cover) are represented in the finite element model by equivalent reluctances of laminated (non-conducting) steel of thickness equal to the skin depth of the element.

Computation of losses in solid metallic elements

Equation (4.30) implies that it is the surface value of the magnetic field strength H_{ms} which governs the value and distribution of local loss density, and thus local heating and 'hot spots', in the tank wall and other solid steel elements of a transformer. Hence the problem of loss computation consists mainly of evaluation of the tangential component of the magnetic field strength on a metal surface.

Although the experimental verification conducted and reported in reference [4.26] confirms a satisfactory accuracy of field computation with the RNM-3D, loss estimation creates new, more difficult problems, and a higher level of non-linearity, which may have to be studied separately.

Dependence of stray losses on rated power and load of a transformer

The active power loss in $\mathrm{W\,m^{-2}}$ generated in a massive steel 'half-space', per unit area on its surface, may be expressed ([4.22], p. 152, 328) using the formula

$$P_1 \approx a_p \sqrt{\left(\frac{\omega\mu_s}{2\sigma}\right)} \frac{|H_{ms}|^2}{2} \approx \frac{\omega}{a_p} \sqrt{\left(\frac{\omega\sigma}{2\mu_s}\right)} \frac{|\Phi_{m1}|^2}{2} \approx \frac{\omega}{2\sqrt{2}} |H_{ms}|\,|\Phi_{m1}| \quad (4.34)$$

where

$$H_{ms} \approx \frac{1}{a_p} \sqrt{\left(\frac{\omega\sigma}{\mu_s}\right)} |\Phi_{m1}|$$

is the field strength on the surface of iron, in $\mathrm{A\,m^{-1}}$, $a_p = 1.3$ to 1.5 (typically 1.4) is a semi-empirical linearization coefficient for solid steel ([4.22], p. 322), and μ_s is the magnetic permeability of the iron surface (Fig. 4.17b).

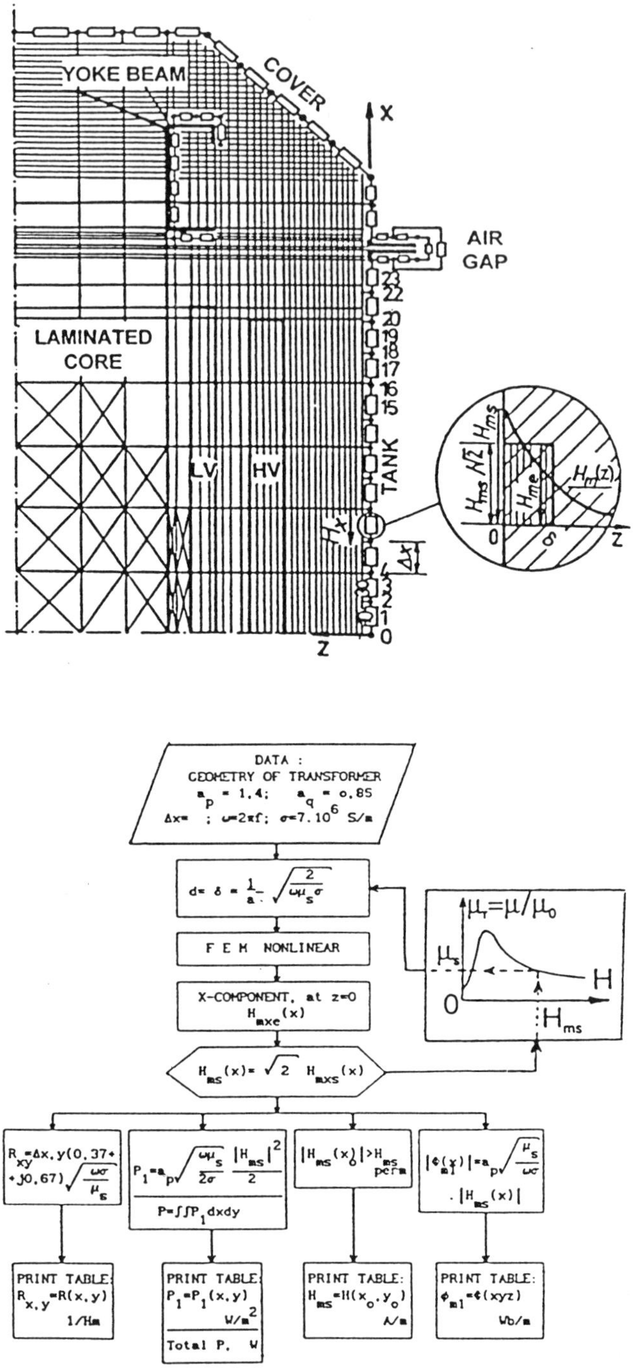

Fig. 4.17 *The hybrid finite-element and reluctance-network approach for modelling magnetic fields in transformers or other devices with solid steel elements: (a) the finite element model, (b) computational flowchart [4.27]*

Although eqn (4.34) is simplified, using a more accurate mathematical description may not necessarily lead to better accuracy of results, as the material data required for computation, in particular conductivity and permeability (or B/H curve), often show significant deviation owing to the varying chemical composition, mechanical and thermal treatment etc., and thus the computational effort may be hindered by the lack of reliable information on material properties. In Fig. 4.18 magnetization curves of various samples of a common steel St3s at 20°C are shown.

In Fig. 4.18 the approximating curve (a) $H = 1.05 \cdot 10^{-5}\ (H\sqrt{\mu_r})^{1.77}$ corresponds to (b) $H^2\sqrt{\mu_r} = 650H^{1.565}$. Introducing into eqn (4.34) the analytical approximation (Fig. 4.18) of the magnetization curve of steel

$$H_{ms} = C(\sqrt{\mu_r}\,|H_{ms}|)^n = C\left[\sqrt{\left(\frac{\omega\sigma}{a_p^2\mu_0}\right)}\,|\Phi_{m1}|\right]^n \tag{4.35}$$

where $C = 1.05 \times 10^{-5}$ and $n = 1.77$, we obtain ([4.22] p. 329]) the relation between power loss in a massive steel element and the field H_{ms} on its surface (in W m^{-2})

$$P_1 = \frac{a_p}{2C^{1/n}}\sqrt{\frac{\omega\mu_0}{2\sigma}}\,|H_{ms}|^{1+1/n} \approx k_1 \times I^{1.6} \times f^{0.5} \times \sigma^{-0.5} \tag{4.36}$$

or

$$P_1 = \frac{C}{2\sqrt{2}}\,\frac{\omega^{1+n/2}\sigma^{n/2}}{a_p^n\mu_0^{n/2}}\,|\Phi_{m1}|^{1+n} \approx k_2 \times I^{2.8} \times f^{1.9} \times \sigma^{0.9} \tag{4.37}$$

The last formulae demonstrate the dependence of losses on physical parameters and current. The crucial factor here is the way in which the field is excited. We can distinguish two typical cases:

(1) when H_{ms} is proportional to I (e.g. the field due to the current in the bushings on the cover plate surface), we should use eqn (4.36), and
(2) when Φ_{m1} is proportional to I (e.g. leakage flux in the inter-winding gap), then eqn (4.37) holds.

Since in the RNM-3D model (Fig. 4.1) the magnetic field H_{ms} is calculated from the leakage flux Φ_{m1} at constant iron reluctance, it follows that value of power in eqn (4.30) should be somewhat higher than 2, say $(H_{ms})^{2x}$, where the semi-empirical correction coefficient x may be selected analytically or experimentally.

The total loss (in W) on the internal surface of the tank wall can be therefore calculated from the following general expression

$$P = \frac{1}{2}\sqrt{\left(\frac{\omega\mu_0}{2\sigma_{st}}\right)}\left[p_e\iint_{A_e}\sqrt{\mu_{rm}}\,|H_{ms}|^2\,\mathrm{d}A_e + p_m\iint_{A_m}\sqrt{\mu_{rm}}\,|H_{ms}|^2\,\mathrm{d}A_m + p_{st}\iint_{A_{st}}\sqrt{\mu_{rs}}\,|H_{ms}|^{2x}\,\mathrm{d}A_{st}\right] \tag{4.38}$$

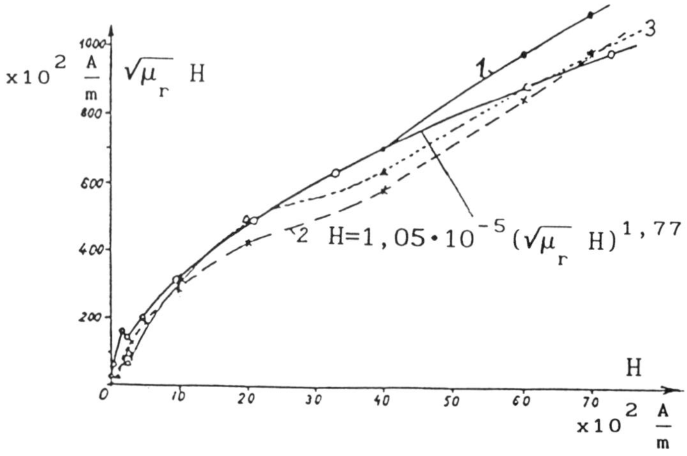

(a)

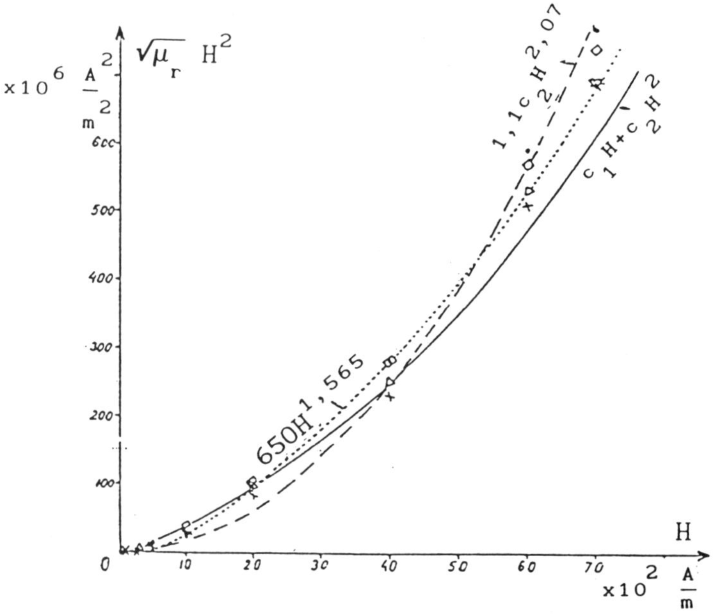

(b)

Fig. 4.18 *Analytical approximations for magnetization curves of steel [4.24]:*
1 - (○○) 0.2C, 0.15Si, 0.03P, 0.5Mn, 0.375, 0.05Cr, 0.03Mo, 0.7W; $\sigma = 7.2 \cdot 10^6$ S m^{-1}
2 - (××) 0.16C, 0.25Si, 0.015P, 0.013S, 0.59Mn; $\sigma = 5.69 \cdot 10^6$ S m^{-1}
3 - (ΔΔ) max 0.22C, (0.1–0.35)Si, max 0.05P, max 0.05S; $\sigma = 6.84 \cdot 10^6$ S m^{-1} *(Polish Standard)*
4 - (□□) 0.3C, hot-rolled

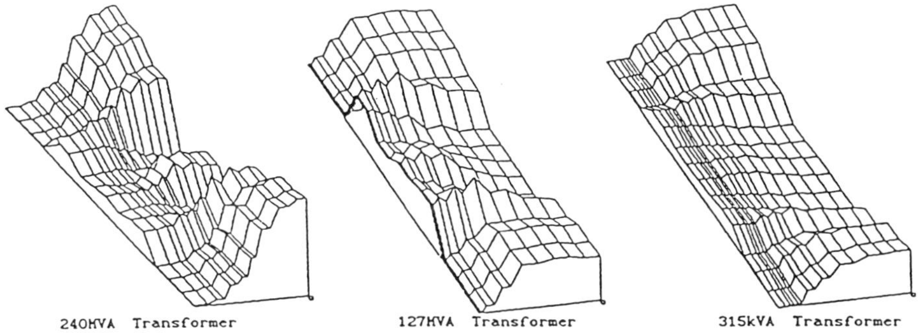

Fig. 4.19 *The 3D loss density distribution in a quarter of a non-screened tank surface (Fig. 4.1) calculated with the RNM-3D package [4.24]*

where $p_e \ll 1$ and $p_m \ll 1$ are the screening coefficients ([4.22], p. 198 and 200) of electromagnetic or laminated magnetic screens, respectively, A_e, A_m and A_{st} are areas covered by a corresponding screen (e or m) or area which is not screened (st). The variation of power for f and σ is less important at typical frequencies and temperatures [4.24]. The value of x varies for different transformers but is typically between 1.1 and 1.14. This has led to the following calculated total losses in the tank wall without a screen: 288 kW for a 240 MV A transformer, 78 kW for a 127 MV A transformer, and 88 W for a 315 kV A unit (Fig. 4.19).

Bearing in mind the inevitable deviation and non-linearity of material parameters, the fact that x usually has to be estimated experimentally should not be considered an unreasonable restriction. The overall accuracy of calculation in any case cannot be better than the natural spread of electric and magnetic properties of materials.

Figure 4.20 shows the results of the interactive design of combined copper (Cu) and magnetic (Fe) screens, including calculation of loss density distribution and optimization of the design with the view to reducing stray losses and local heating.

Detection of 'hot-spots' in the leakage region of a transformer

Peak values of field and loss density distributions in Figs 4.6, 4.14, 4.15, 4.16, 4.19, 4.20 etc. indicate the position of the spots, where excessive local heating is likely to appear. This heating may be linked with the field H_{ms} on the surface of an element. After solving the approximate nonlinear thermal equilibrium equation (see Chapter 5), a graph can be produced (Fig. 4.21) which helps to evaluate a permissible tangential value of $H_{ms,perm}$ for a given permissible temperature t_{perm} ([4.22], p. 381). This temperature is normally recommended by international or national Study Committees of CIGRE (e.g. SC No 12 'Transformers').

Having calculated the distribution of H_{ms} one can easily identify potential 'hot spots' on the graph, where special precautions (temperature detectors,

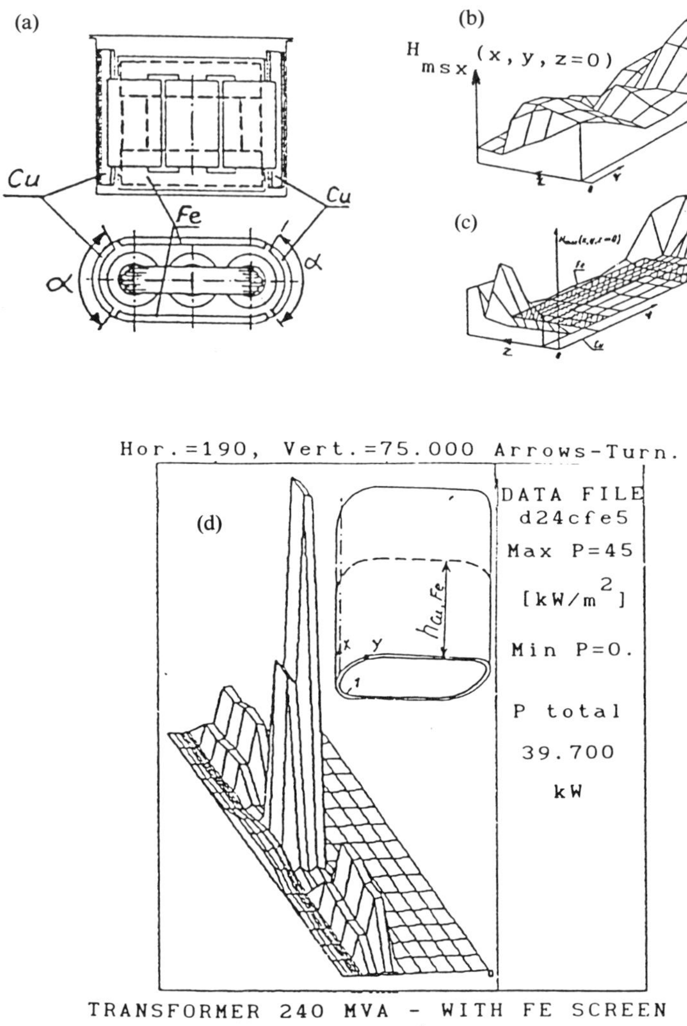

Fig. 4.20 *Interactive design of the tank screens with RNM-3D [4.22, 4.26]: (a) combined copper (Cu) and magnetic (Fe) screen, (b) H_{max} for the case without any screen, (c) the best variant for a frame screen, (d) loss density distribution and total stray losses in a five-limb transformer with continuous Fe screen*

non-magnetic steel, screening, directed cooling etc.), or more accurate calculations should be undertaken. For instance, the three-dimensional thermal analysis (see [4.22], p. 386–389) has shown that strongly non-uniform stray loss distribution can increase the value of $H_{ms,perm}$ even by a factor of two.

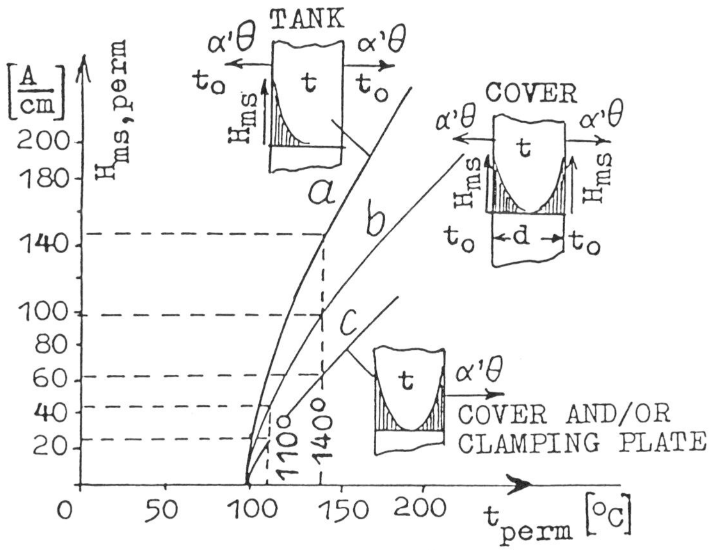

Fig. 4.21 *Permissible values of the tangential component of $H_{ms,perm}$ on the surface of a solid steel plate versus the permissible temperature t_{perm} of the plate [4.22]*

4.6 TESTING RNM ON OTHER ELECTROMAGNETIC DEVICES

In addition to the experimental verification presented in papers [4.20, 4.22, 4.26] etc., further testing of the RNM approach on some other electromagnetic devices has been carried out (Figs 4.22, 4.23). Most internationally recognized benchmark problems for testing various 3D codes are collected and regularly reviewed by the TEAM (Testing Electromagnetic Analysis Methods) Workshop. Nearly all TEAM Workshop Problems have been solved using the RNM at the Technical University of Lodz. One of them is Problem No 7 (Fig. 4.22) solved by RNM [4.28] in reasonable computing time on a PC, with an extremely good agreement with experiments and other computation methods, such as the finite-element nodal A–Φ, nodal T–Ω and edge A–Φ, tested among others by Kamerai, Fujiwara and Olszewski (Okayama University).

An excellent convergence and agreement with experiment (see Fig. 4.23) of a quasi-3D calculation using RNM of a flux density, coil inductance and tractive force of a linear reluctance self-oscillating motor, has been demonstrated in reference [4.29]. The work was done at Sydney University in collaboration with the Technical University of Lodz.

4.7 RELUCTANCE NETWORKS AND FINITE ELEMENTS

The reluctance network formulation may be considered equivalent to some other methods, such as finite differences or finite elements, in a similar way

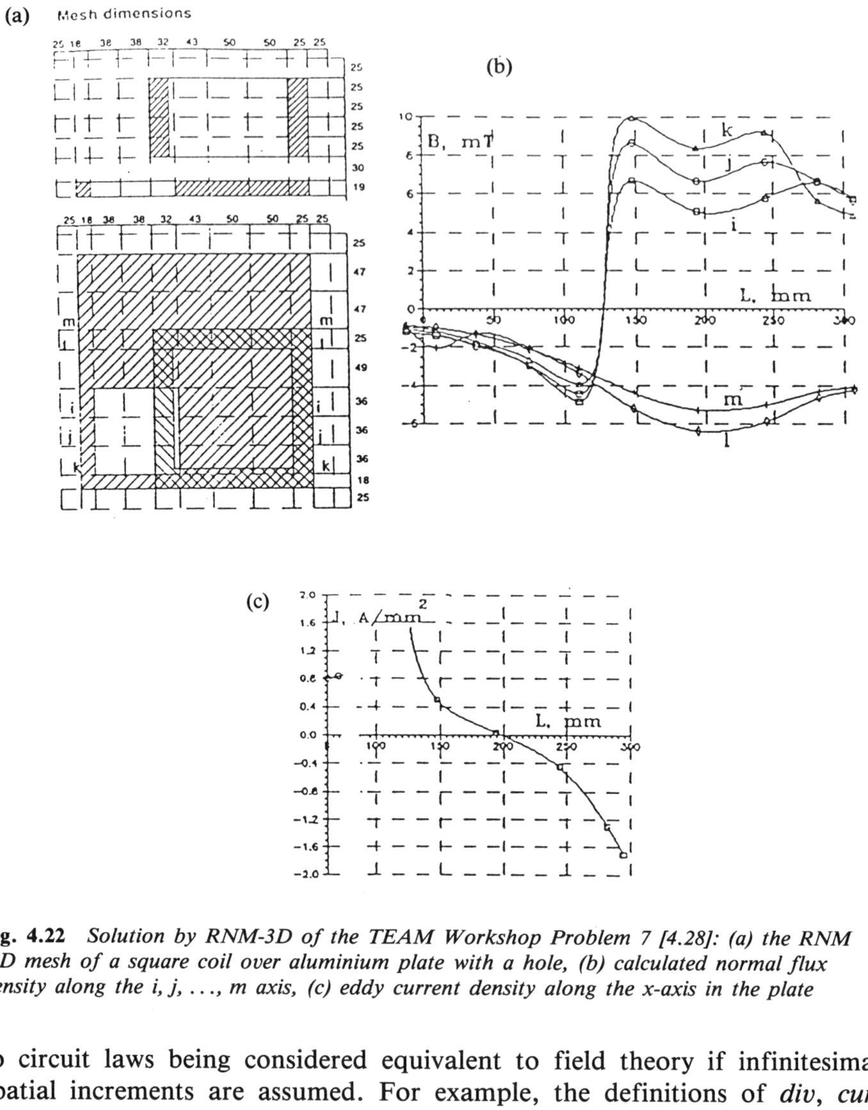

Fig. 4.22 *Solution by RNM-3D of the TEAM Workshop Problem 7 [4.28]: (a) the RNM 3-D mesh of a square coil over aluminium plate with a hole, (b) calculated normal flux density along the i, j, ..., m axis, (c) eddy current density along the x-axis in the plate*

to circuit laws being considered equivalent to field theory if infinitesimal spatial increments are assumed. For example, the definitions of *div*, *curl* and boundary conditions may be obtained in this way by applying Kirchhoff's laws to magnetic circuits.

Consider a scalar Poisson's equation in two dimensions at constant permeability

$$\frac{\partial^2 A}{\partial x^2} + \frac{\partial^2 A}{\partial y^2} = -\mu J \tag{4.39}$$

and a corresponding energy function

$$I = \iint \left\{ \frac{1}{2\mu} \left[\left(\frac{\partial A}{\partial x} \right)^2 + \left(\frac{\partial A}{\partial y} \right)^2 \right] - JA \right\} \mathrm{d}x \, \mathrm{d}y \tag{4.40}$$

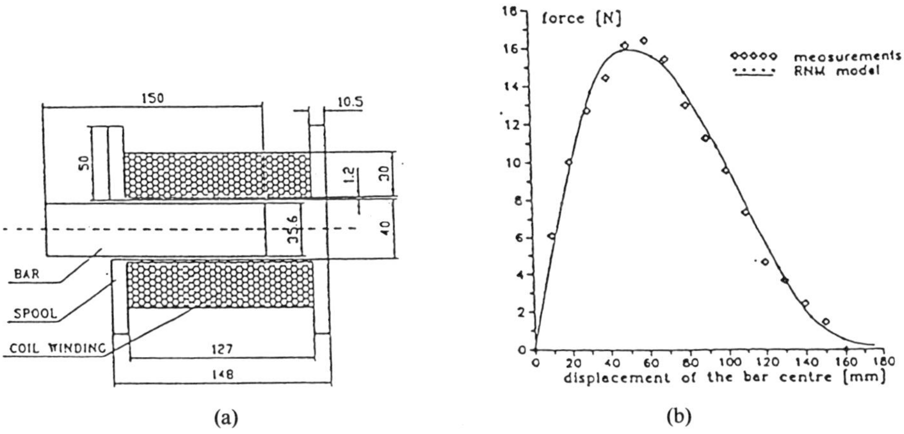

Fig. 4.23 *A quasi 3-D calculation for a linear reluctance motor by RNM [4.29]: (a) cross-section, (b) calculated and measured propulsion force versus bar position*

This functional represents the energy (or the Lagrangian) of a system, which in the finite-element formulation must be a minimum for variation of each of the node potentials, of which a typical one is A_k. If first-order elements are assumed, the potential at an interior point of one is given by linear interpolation between the three vertex values A_i ($i = 1, 2, 3$)

$$A(x, y) = \frac{1}{2\Delta} \sum_{i=1}^{3} (a_i + b_i x + c_i y) A_i \tag{4.41}$$

in which the coefficients depend on the vertex coordinates, and Δ is the area of the triangle (see Chapter 3). As the first-order description demands a linear variation, a similar approximation will also apply to J, i.e.

$$J(x, y) = \frac{1}{2\Delta} \sum_{i=1}^{3} (a_i + b_i x + c_i y) J_i \tag{4.42}$$

The variation in energy when any typical node potential A_k is changed must be zero, which gives, for each value of k, an equation of the form [4.4]

$$\frac{\partial I}{\partial A_k} = \sum_e \sum_i (\alpha_{eki} A_i - \beta_{eki} J_i) = 0 \tag{4.43}$$

where e is the element number, while i denotes the vertices and ranges over three values in each element.

Consider now a linear magnetic network in which a typical component carries a flux Φ and requires an *mmf* of F. The energy input required is $W_m = \Phi F/2 = \Phi^2 R_\mu/2$. If the network is replaced by its dual, Φ becomes a potential quantity, and thus the stored energy in any component, in terms of the potential values Φ_1 and Φ_2 at the two ends, is

$$W_m = \tfrac{1}{2} (\Phi_1 - \Phi_2)^2 R_\mu \tag{4.44}$$

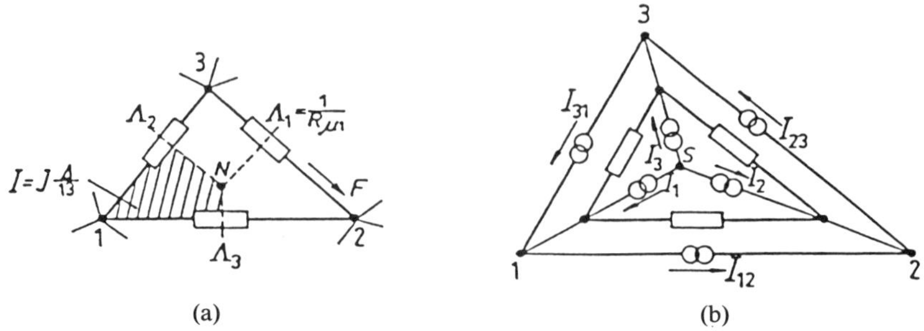

Fig. 4.24 *Elements of a reluctance network equivalent to finite elements [4.4]: (a) linkage of a reluctance network with a flow I, (b) network model with current and reluctance linking*

Varying a single node potential, say Φ_1, therefore changes the stored energy linearly

$$\frac{\Phi W_m}{\partial \Phi_1} = (\Phi_1 - \Phi_2) R_\mu \tag{4.45}$$

Since the magnetic vector potential A is the flux linkage per unit length, comparing the last result with eqn (4.43) shows that the α coefficients correspond to branch reluctances in an equivalent magnetic network representing unit length (see Fig. 4.24). Moreover, this network stores energy in exactly the manner required by the α terms in eqn (4.43).

4.8 TIME-DEPENDENT FIELDS

Ohm's law for magnetic circuits $V_\mu = \Phi R_\mu$ can only be applied to static or quasi-static magnetic fields at a constant frequency f. For such fields magnetic flux becomes a full analogue of electric current. In the case of transient fields, however, it is better to associate the magnetic flux $\Phi = \int \mathbf{B} \cdot \mathrm{d}\mathbf{s}$ with the electric flux $\psi = \int \mathbf{D} \cdot \mathrm{d}\mathbf{s}$ rather than with the electric current $i = \int \mathbf{J} \cdot \mathrm{d}\mathbf{s}$ [4.3; 4.12; 4.19]. The electric flux corresponds to an electric charge $Q = Cu = \psi$. Moreover, $i = \mathrm{d}Q/\mathrm{d}t = C\,\mathrm{d}u/\mathrm{d}t$ and therefore the analogue of electric current i is not the magnetic flux but its rate of change in time. This is consistent with the idea of a *magnetic current* J_μ put forward by Heaviside

$$J_\mu = \frac{\mathrm{d}\Phi}{\mathrm{d}t} = C_\mu \frac{\mathrm{d}V_\mu}{\mathrm{d}t} \quad \text{or} \quad \underline{J}_\mu = \mathrm{j}\omega\Phi = \mathrm{j}\omega C_\mu V_\mu \tag{4.46}$$

where C_μ is the so-called *magnetic capacitance* and V_μ is the magnetic voltage. Hence Kirchhoff's magnetic law for a magnetomotive force F reads

$$F = R_\mu \int J_\mu \mathrm{d}t + L_\mu J_\mu = R_\mu \int \mathrm{d}\Phi + L_\mu \frac{\mathrm{d}\Phi}{\mathrm{d}t} \tag{4.47}$$

where

$$R_\mu = \frac{1}{C_\mu}$$

is the reluctance

$$L_\mu = \frac{\gamma}{J_\mu} = \frac{\gamma}{\frac{d\Phi}{dt}}$$

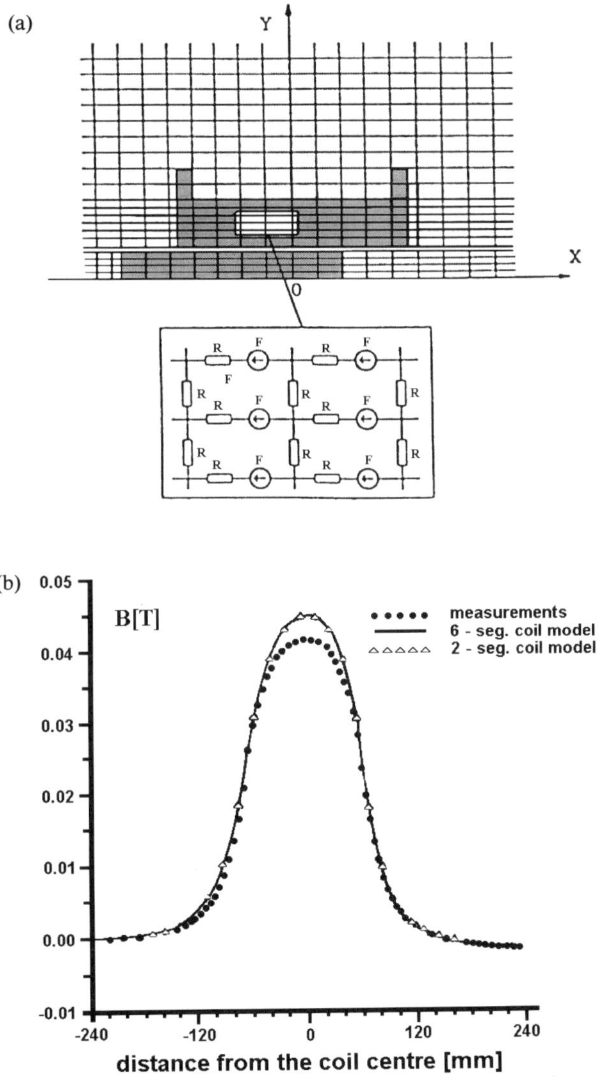

Fig. 4.25 *Experimental verification of flux density calculated using RNM along the axis of a motor of Fig. 4.23 [4.29]: (a) model, (b) distribution in the gap without an iron bar*

is the *transference*, which is a measure of the eddy currents induced by the flux in all conducting paths which are wholly or partially linked with the flux, and

$$\gamma = \sum_{r=1}^{N} i_r$$

is the current linkage. The reluctance is treated as the inverse of the 'magnetic capacitance' rather than a 'magnetic resistance'. Carpenter [4.3] concludes that two electric–magnetic analogies are possible:

electric current:	Φ	$J_\mu = \mathrm{d}\Phi/\mathrm{d}t$
electric voltage:	*mmf F*	*mmf F*
$u = Ri + L\dfrac{\mathrm{d}i}{\mathrm{d}t}$	$F = R_\mu + L_\mu\dfrac{\mathrm{d}\Phi}{\mathrm{d}t}$	–
$u = Ri + \left(\dfrac{1}{C}\right)\int i\,\mathrm{d}t$	–	$F = L_\mu J_\mu + \left(\dfrac{1}{C_\mu}\right)\int J_\mu\,\mathrm{d}t$
resistance R	R_μ	$L_\mu = \dfrac{\gamma}{J_\mu}$

Some interesting practical applications of the reluctance network method for analysis of time-dependent two-dimensional fields are presented in papers by Oberretl [4.18] and Demenko [4.8].

4.9 EXPERIMENTAL VERIFICATION OF THE METHOD

Experimental verifications of the reluctance network method and the RNM-3D package have been reported on several occasions and by many

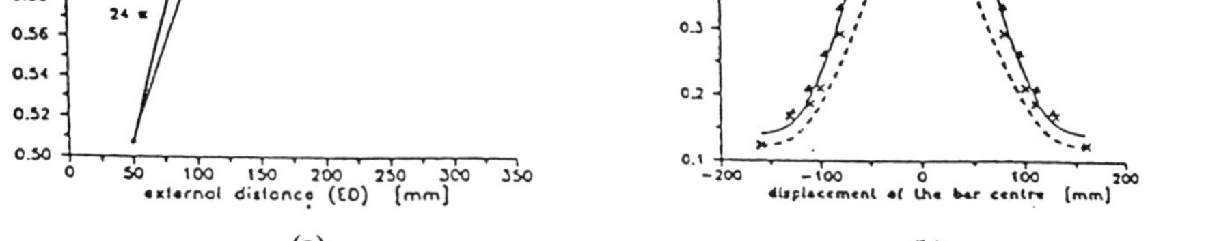

(a) (b)

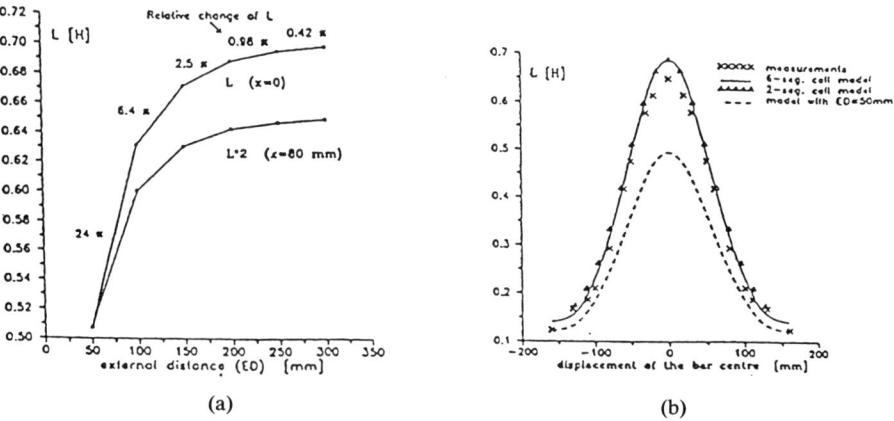

Fig. 4.26 *Experimental verification of calculations by RNM of a coil inductance in a linear motor of Figs 4.23 and 4.25 [4.29]: (a) convergence versus dimensions of the external open boundary region, (b) calculated and measured inductance*

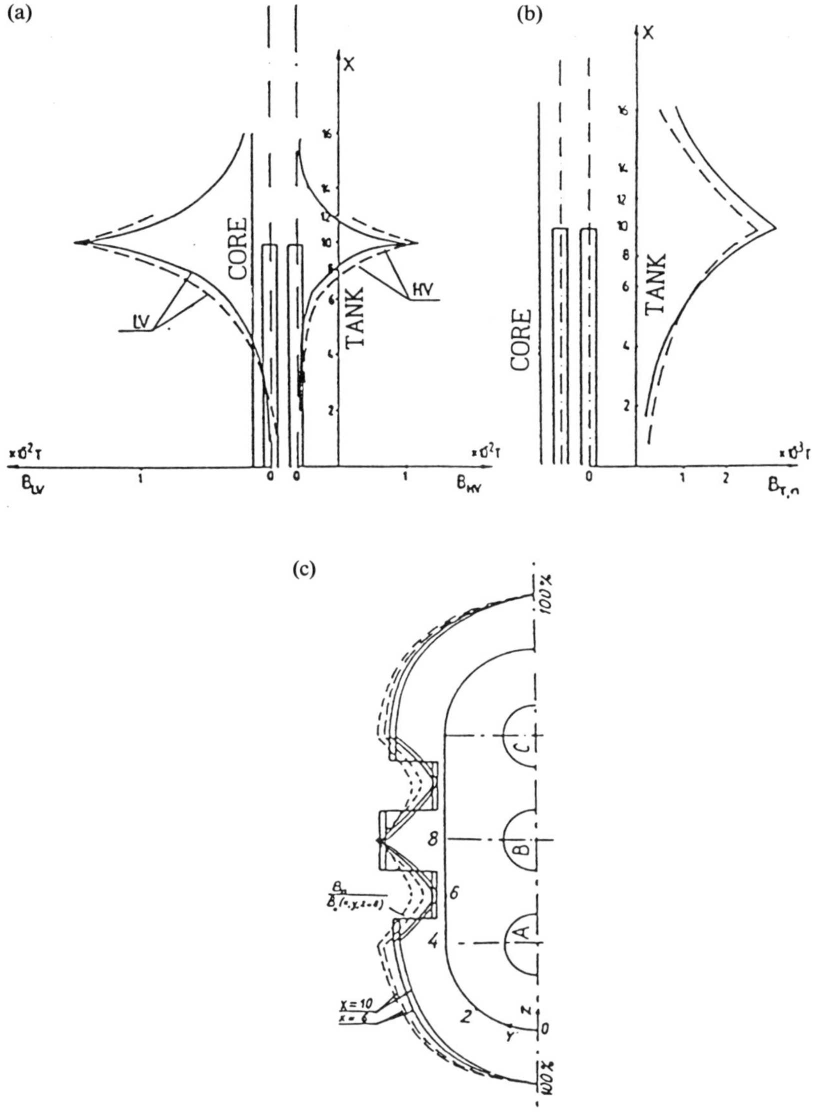

Fig. 4.27 *Measured (dashed) and calculated (bold) fields in a 315 kVA transformer [4.26]: (a) flux density B_n in LV and HV winding, (b) B_n on the tank wall ($y = 0$), (c) B_n along the tank circumference*

authors [4.2; 4.7; 4.20; 4.26; 4.27; 4.28; 4.29]. For example, an axisymmetric problem is relatively easy to investigate (Figs 4.23b, 4.25 and 4.26) and has shown excellent agreement with measurements. Even for an open boundary problem (Fig. 4.25) convergence has been fast (Fig. 4.26b) and computed results compared very favourably with experiment (Figs 4.23b, 4.25b and 4.26b). For a much more complicated TEAM Workshop problem 7

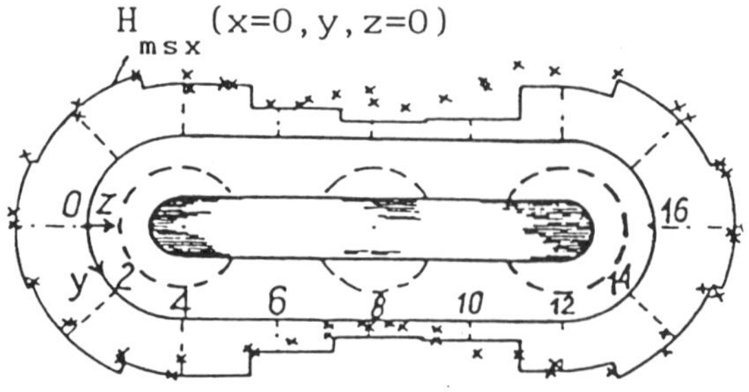

Fig. 4.28 *Axial magnetic field distribution (H_m) along the tank circumference ($x = 0$) of a 150 kVA model transformer [4.26]*
——— *calculation using RNM-3D*
× × × *experiment*

(Fig. 4.22), with 3D solenoidal field and eddy currents, results from the reluctance network computations have been remarkably close to measurements and calculations, using other methods, conducted at Okayama University, Japan [4.28].

The most successful and important application of the method so far, from the engineering point of view, has been to large power transformers, as already demonstrated in this Chapter. Relevant tests on real transformers, especially measurements of power loss associated with leakage fields, are very difficult. Nevertheless many results have been published and may be found for example in references [4.22; 4.24]. Comparisons of calculations performed using RNM-3D with such measurements consistently show good agreement. Some further examples of such comparison are shown in Figs 4.27 and 4.28.

In conclusion it should be emphasized that although the reluctance network method is overshadowed by other techniques, especially the finite element method with its recent significant advances (see Chapter 7), it continues to play an important role in engineering applications. The results produced by this method are approximate, but they help the designer focus his attention on critical issues and not be burdened with those which, from the practical point of view, are less important. Moreover, the expertise of the designer and his knowledge of the device, including existing simplified empirical or analytical partial solutions, may be easily incorporated into the reluctance network formulation. Most significantly, though, the method provides the solution in extremely short computing times which makes it very suitable for interactive design.

5
Field Synthesis

Krystyn Pawluk

5.1 INTRODUCTION

5.1.1 What is field synthesis?

The term *field synthesis* is introduced as the converse of *field analysis*. Field analysis is usually easier and thus preferred, and can be applied when the geometry and material structure of a device are known and when the field sources are given. The analysis is performed by solving a field equation which, generally, is a partial differential equation of the second order with respect to a field quantity of a scalar or vector type. The solution of the analysis problem refers to a region that is geometrically determined, i.e. the shape of the region is given, and moreover, the region has its material characteristics specified. It is assumed that the analysed field is generated by primary field sources, which are given as boundary conditions or internal excitations.

The field synthesis is an inverse problem to that specified by analysis and may be formulated in a variety of ways. The distribution of a field quantity is assumed as given in the whole or part of the region under consideration and we may be looking for:

- outer field sources defined as suitable boundary conditions;
- inner field sources, for instant coils or electric charges;
- material parameters or characteristics;
- the geometric shape of the region.

Thus, the synthesis problem consists of determining unknown effective causes that influence the physical field, the distribution of which is either postulated or previously obtained as the result of a measurement. In electrical engineering we are generally interested in electrostatic, magnetostatic, electromagnetic or current-flow fields, but in addition, the synthesis of a thermal field may also be the object of the design.

The above conclusions are summarized below:

(1) If synthesis is applied to the design of some part of an electrical equipment (electrical machine, apparatus, isolator, antenna etc.) and the desired properties of the equipment can be expressed as a function of the field distribution in part of this equipment, then we shall call this the *design synthesis*.

(2) If synthesis is performed to identify the geometry or material characteristics of the existing equipment and this identification is based either on the measured distribution of some field quantity on part of this equipment, or on the measured integral quantity dependent on the distribution of a field quantity (e.g. a magnetic flux), then we shall call this procedure the *identification synthesis*. There is a very practical aspect to such problems when, for example, we attempt to assess the quality and accuracy of equipment after a long operating time or following a fault.

(3) In the case when synthesis is performed in order to determine an optimal feature of the equipment, we have the *optimal synthesis*.

The design and identification synthesis problems differ significantly in their aims; nevertheless the mathematical formulations are often similar and sometimes identical.

A common property of synthesis problems is that they are not unique. It is generally because the information available is not sufficient to describe the problem mathematically. In some cases the amount of input data may appear to be sufficient, while in fact some equations are either linearly dependent or even contradictory. In the case of deficient data we have an *under-specified problem*, and for an excessive number of independent information we are dealing with an *over-specified problem*.

The design synthesis problems may yield solutions which are technically impossible. This may happen when a designer has insufficient experience in the matter and improper input data has been chosen. The identification synthesis problems, as they refer to existing objects, should always have a solution, but even here the inaccuracies of measurements may lead to inconsistent solutions.

Most tasks concerning the design of electrical equipment are essentially synthesis problems. A traditional approach avoids the proper synthesis formulation altogether. Thus a conventional trial-and-error strategy consists of a series of solutions of analysis problems complemented by empirical or semi-analytical results and often verified by building and testing physical models. An alternative new approach to design uses CAD and optimization, as discussed in Chapter 7. In this Chapter, however, we concentrate on direct methods of synthesis.

The same physical equations describe both the synthesis and the analysis of electromagnetic fields, although the manner in which the problem is formulated is different. Thus, the synthesis problems can be classified with reference to the field problem under consideration:

(1) inverse problems to Laplace's equations, where no internal sources are present (no electrical charges, no currents);

(2) inverse problems to Poisson's equations, where internal field sources exist;

(3) inverse problems to Helmholtz equations, where current is induced in the region;

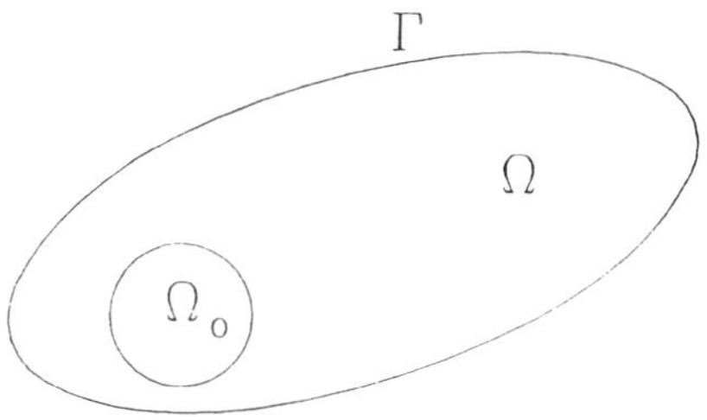

Fig. 5.1 *A field synthesis region*

(4) inverse problems to wave equations, where a full set of Maxwell's equations must be taken into consideration.

The following topological properties of the synthesis problem should be noticed. In the region Ω under consideration, a subregion $\Omega_0 \subset \Omega$ is identified which we shall call the *synthesis controlled subdomain*, see Fig. 5.1. In this subregion the field is subject to verification during the solution. The domain Ω_0 can be of the same or smaller dimension than Ω, and can also be split into a number of discounted parts $\Omega_{01}, \Omega_{02}, \ldots, \Omega_{0n}$.

Synthesis problems can be classified as falling into two categories:

(1) a *group of rigid structure*, when the geometry and environmental characteristics of Ω are fixed and when the field in Ω_0 is to be synthesized by determining the field sources, i.e. the boundary conditions on Γ or/and the internal source distribution in the subregion Ω_S, called the *internal sources subdomain*;
(2) a *group of supple structure*, when the shape and environmental characteristics of Ω are tentatively assumed, and the field in Ω_0 is then synthesized by final determination of its geometry and/or the characteristics. Care has to be taken not to affect the uniqueness of boundary conditions and internal source distributions.

When a problem is formulated in such a manner that it requires specification of a single quantity or condition (e.g. boundary conditions only) then we have a *simple synthesis problem*, whereas if more that one values are needed it leads to a *compound synthesis problem*.

5.1.2 Development of methods and techniques

Synthesis problems were first solved using the least-square method due to Gauss and Legendre. The modern view is that this method is not satisfactory because it often leads to unstable solutions. The development of theory of improperly- and ill-posed problems has been stimulated by progress in numerical techniques. It is generally required that solutions of synthesis problems should be stable and relevant to the physical nature of the objects under consideration. Contemporary synthesis techniques follow the works of

Twomey [5.69], Phillips [5.42] and Tichonov and Arsenin [5.68]. These techniques are known as *regularization methods* and lead to very effective computer codes for problems described by integral equations, as well as those which are derived from the finite element formulation.

It is very likely that methods for solving inverse problems of electromagnetism will continue developing as an independent field. It appears that the majority of necessary theoretical work has already been done and effort is now directed to creating appropriate computer programs and their implementation to the design of electrical equipment. It may be appropriate to point out that even a simple calculation of the magnetizing current in machines and transformers is a synthesis problem, i.e. finding the localized magnetic field source for a prescribed magnetic flux. The same aspect of synthesis may also occur in analysis problems examined by the method of secondary sources (see Pawluk [5.40]). This method uses the principle that the continuity of vector field quantities on the internal and external boundaries is equivalent to a suitable distribution of potentials (single or double layer) on these boundaries.

A sensitivity approach seems to be promising for a large class of design problems – see Henneberger *et al.* [5.16], Il-han Park *et al.* [5.17] and Sikora [5.60]. A general treatment of the synthesis problems may be seen in the papers by Guarnieri *et al.* [5.13], Salon and Istfan [5.55] and Simkin and Trowbridge [5.66].

5.2 A SURVEY OF INTEGRAL EQUATIONS

5.2.1 Types of equations

In this Section we shall discuss such aspects of integral equations which are relevant to the theory of synthesis of electromagnetic fields. Moreover, we shall present only the simplest formulation for one independent variable, although the treatment can be easily broadened to cover 2D and 3D field problems. An *integral equation* is an equation in which the unknown function appears under the integral sign. Thus the integration is performed on the function.

The integral equation in which at least one limit of integration is a variable

$$\int_a^x K(x, \xi) g(\xi)\, \mathrm{d}\xi = \tilde{u}(x) \tag{5.1}$$

is called *Volterra's equation of the first kind*. In the above equation g is the unknown function, and the function $\tilde{u}$ on the right-hand side is prescribed. If $\tilde{u}(x) \equiv 0$, eqn (5.1) becomes *homogeneous*. The function K is called the *kernel*.

The integral equation

$$g(x) + \lambda \int_a^x K(x, \xi) g(\xi)\, \mathrm{d}\xi = \tilde{u}(x) \tag{5.2}$$

is called *Volterra's equation of the second kind*. The unknown function g appears twice in this equation: alone and under the integral sign. In the equation λ is a constant. The particular cases of Volterra's equations are found when limits of integration are fixed. This leads to

Fredholm's equation of the first kind:

$$\int_a^b K(x,\xi)g(\xi)\,d\xi = \tilde{u}(x) \tag{5.3}$$

Fredholm's equation of the second kind:

$$g(x) + \lambda\int_a^b K(x,\xi)g(\xi)\,d\xi = \tilde{u}(x) \tag{5.4}$$

The kernel of an integral equation may be classified as:

(1) symmetric, if

$$K(x,\xi) = K(\xi,x)$$

(2) regularized, if

$$\int_a^b\int_a^b |K(x,\xi)|^2\,d\xi\,dx$$

exists and is not zero,

(3) midcontinuous, if

$$\lim_{\Delta x\to 0}\int_a^b |K(x+\Delta x,\xi) - K(x,\xi)|^2\,d\xi = 0$$

(4) degenerate, if it can be expressed as

$$K(x,\xi) = f(x)\cdot h(\xi)$$

or in the form of a sum of such products,

(5) if the kernel is complex $\underline{K}(x,\xi)$ then it will always have its conjugate counterpart $\underline{K}^*(x,\xi)$,

(6) the transposed kernel to $K(x,\xi)$ is defined as

$$K^T(x,\xi) = K(\xi,x);$$

for the symmetric kernels we have

$$K^T(x,\xi) = K(x,\xi),$$

(7) if the complex kernel $\underline{K}(x,\xi)$ has its conjugate $\underline{K}^*(x,\xi)$ equal to the transposed kernel $\underline{K}(\xi,x)$, then it is called a *Hermitian kernel*. Clearly, real and symmetric kernels are Hermitian,

(8) when the kernel is singular, i.e. when it has the form

$$K(x, \xi) = \frac{L(x, \xi)}{(x - \xi)},$$

then the relevant integral equation is said to be *singular*.

Integral eqns (5.1)–(5.4) are called *linear equations*. If the expression under the integral sign cannot be written as a product of the kernel and the unknown function, for instance

$$\int_a^b K[x, \xi, g(\xi)]\, \mathrm{d}\xi = \tilde{u}(x) \tag{5.5}$$

then such equation is said to be *nonlinear*. In the case of eqn (5.5) it is the nonlinear Fredholm's equation of the first kind.

An integral equation can be written in a compact form by the use of an integral operator:

$$\mathbf{K} = \int_a^b K(x, \xi) \ldots \mathrm{d}\xi \tag{5.6}$$

Thus the Fredholm's equation of the first kind reads

$$\mathbf{K}\, g(x) = \tilde{u}(x) \tag{5.7}$$

The operator **K** attributes a function $f(x)$ (belonging to the functional space $\mathcal{G}$ of some class of the unknown functions) to a function $\tilde{u}(x)$ (which belongs to the space $\mathcal{U}$ of given functions). This attribution or mapping may be symbolically written as $\mathbf{K}: \mathcal{G} \to \mathcal{U}$.

Eigenfunctions and eigenvalues

If for $x \in (a, b)$ a non-negative function $h_i(x)$ satisfies a homogeneous integral equation of the second kind (for example the Fredholm's eqn (5.4)) at any fixed value of the parameter λ_i, so that

$$h_i(x) + \lambda_i \int_a^b K(x, \xi) h_i(\xi)\, \mathrm{d}\xi = 0, \qquad \text{where } i = 1, 2, \ldots, m \tag{5.8}$$

then such functions $h_i(x)$ are called the *eigenfunctions* of the equation and the relevant parameters λ_i are called the *eigenvalues*. The largest absolute value of $1/\lambda_i$ is termed the *spectral radius* of the operator **K** and it is written $\varrho(\mathbf{K})$. Thus:

$$\varrho(\mathbf{K}) = \left|\frac{1}{\lambda_k}\right| = \max\left(\left|\frac{1}{\lambda_1}\right|, \left|\frac{1}{\lambda_2}\right|, \ldots, \left|\frac{1}{\lambda_n}\right|\right) \tag{5.9}$$

On the solution of Volterra's equations

It has been proved that if in the integral equation of the second kind (e.g.

eqn (5.2) or eqn (5.4)), having the midcontinuous kernel $K(x, \xi)$, the following condition holds:

$$\left|\frac{1}{\lambda_k}\right| > \varrho(\mathbf{K}) \tag{5.10}$$

then this equation has a unique and continuous solution $g_1(x)$ for an arbitrary right-hand side function $\tilde{u}(x)$; this solution can be obtained using the method of successive approximations. The above theorem is particularly important for the case of Volterra's equations of the second kind when $\varrho(\mathbf{K}) \to 0$, so that eqn (5.2) has a unique and continuous solution which can be found by successive approximations.

Some Volterra's equations of the first kind (5.1) may be reduced, by differentiation with respect to x, to the equivalent Volterra's equation of the second kind, and hence their solutions may be found. In order to follow this procedure we must have $\tilde{u}(x)_{x=a} = 0$ and $K(x, \xi)_{\xi=x} = K(x, x) \neq 0$. The singular Volterra's equation does not fall into this category. The differentiation of eqn (5.1) may be performed as follows

$$\frac{\mathrm{d}}{\mathrm{d}x}\int_a^x K(x, \xi)g(\xi)\,\mathrm{d}\xi = \frac{\mathrm{d}\tilde{u}(x)}{\mathrm{d}x}$$

$$\int_a^x \frac{\partial K(x, \xi)}{\partial x} g(\xi)\,\mathrm{d}\xi + [K(x, \xi)g(\xi)]_{\xi=x} - [K(x, \xi)g(\xi)]_{\xi=a} = \frac{\mathrm{d}\tilde{u}(x)}{\mathrm{d}x}$$

After appropriate substitutions and division by $K(x, x)$ we have

$$g(x) + \int_a^x K'(x, \xi)g(\xi)\,\mathrm{d}\xi = \tilde{v}(x) \tag{5.11}$$

where the new kernel is

$$K'(x, \xi) = \frac{1}{K(x, x)}\frac{\partial K(x, \xi)}{\partial x}$$

and the right-hand side function becomes

$$\tilde{v}(x) = \frac{1}{K(x, x)}\left[K(x, a)g(a) - \frac{\mathrm{d}\tilde{u}(x)}{\mathrm{d}x}\right]$$

Solution of Fredholm's equations

The following properties of the Fredholm's equations of the second kind (5.4) determine the question of their stability:

(1) if λ is not the eigenvalue of eqn (5.4), then the equation has a unique solution $h(x)$ for an arbitrary function $\tilde{u}(x)$;

(2) if λ is the eigenvalue of eqn (5.4) then the unique solution exists only for certain functions $\tilde{u}(x)$ and only for Hermitian kernels.

The solution of the Fredholm's equation of the second kind may be obtained using many methods including:

(1) the method of successive approximations,
(2) the Hilbert–Schmidt method with the use the *resolvent kernel*,
(3) the Fredholm's method with the use the resolvent kernel,
(4) the kernel iteration.

Readers wishing to study the above methods are referred to the book by Korn and Korn [5.21].

When inverse problems of electromagnetic fields are formulated, they usually involve Fredholm's equations of the first kind. If these equations have Hermitian kernels then they may be solved by expanding them into infinite series in terms of their eigenfunctions $h_i(x)$. Generally, the solutions of Fredholm's integral equations of the first kind occurring in electromagnetic field synthesis are neither unique nor stable. In recent years many numerical algorithms for the solution of such equations have been developed; the method of regularization for example has been efficiently used in some applications, see Section 5.4.

5.2.2 Green's functions

In the theory of integral equations of mathematical physics, *Green's functions* play an important role. Potential theory and Green's functions are closely related. A detailed study of the potential theory is beyond the scope of this book. Nevertheless, in this Chapter we shall introduce some fundamental concepts underlying the Green's functions. In Chapter 3 Green's functions have already been used to describe the boundary element method. In this Chapter we shall give some more consideration to Green's function encountered in inverse electromagnetic problems.

Let us consider Laplace's equation

$$\Delta u = 0 \tag{5.12}$$

In the open n-dimensional space $\Omega^{(n)}$ this equation has the solution

$$\bar{u} = \bar{u}(x_1, x_2, \ldots, x_n), \tag{5.13}$$

which is a harmonic function, i.e. it is twice continuously differentiable, and, if $n \geqslant 3$, it is regular in $\Omega^{(n)}$. This solution is known as a *fundamental solution*, if it is presented in the form

$$\bar{u} = w_n(r) \tag{5.14}$$

where

$$r = \sqrt{[(x_1 - \xi_1)^2 + (x_2 - \xi_2)^2 + \ldots + (x_n - \xi_n)^2]} \tag{5.15}$$

is a modulus of the position vector, i.e. r is the distance between the following two points:

(1) the first point $P = P(x_1, \ldots, x_n)$, or in an abbreviated notation $P = P(P)$, at which the field is to be determined, and
(2) the second point $Q = Q(\xi_1, \ldots, \xi_n)$ or $Q = Q(Q)$, that is unique in $\Omega^{(n)}$, at which the fundamental solution does not obey the Laplace's equation.

Thus, for $n \geqslant 3$, the fundamental solution has an inherent singularity at $P \to Q$, i.e.

$$\lim_{r \to 0} w_n(r) \to \infty \tag{5.16}$$

The fundamental solution (5.14) belongs to a class of functions known as *Green's functions*, which satisfy the Laplace equation. We could say that the fundamental solution is the most elementary of the Green's functions.

We shall demonstrate now how to obtain in a simple way the explicit fundamental solution of Laplace's equation in $\Omega^{(3)}$. Writing Laplace's equation in spherical coordinates r, θ, α we have

$$\frac{1}{r^2}\frac{\partial}{\partial r}\left(r^2\frac{\partial u}{\partial r}\right) + \frac{1}{r^2 \sin\theta}\frac{\partial}{\partial\theta}\left(\sin\theta\frac{\partial u}{\partial\theta}\right) + \frac{1}{r^2\sin^2\theta}\frac{\partial^2 u}{\partial\alpha^2} \tag{5.17}$$

Because the solution has a singularity at the origin of the coordinate system, let us express it as a function of the r coordinate only, so that $\partial u/\partial\theta = \partial u/\partial\alpha = 0$. Thus the following ordinary differential equation in terms of r is found

$$u'' + \frac{2}{r}u' = 0 \tag{5.18}$$

the solution of which is given by $u(r) = C_1 + C_2/r$, where C_1 and C_2 are arbitrary constants. In the fundamental equation we usually put $C_1 = 0$ in order to preserve the regularity of the solution. We also require that the fundamental solution should satisfy Poisson's equation, where the source is described by the unit impulse also called the *Dirac delta function*

$$\Delta u = -\delta(P - Q) \tag{5.19}$$

Thus, if we put $C_2 = 1/4\pi$, we finally obtain

$$w_3(r) = \frac{1}{4\pi r} \tag{5.20}$$

so that the fundamental solution is identical to the Green's function of Laplace's equation in $\Omega^{(3)}$. The integral of $\Delta w_3(r)$ over the whole $\Omega^{(3)}$ space (or any sphere of radius R) is

$$\int_{\Omega^{(3)}} \Delta w_3(r)\, d\xi\, d\eta\, d\zeta = -1 \tag{5.21}$$

which is a consequence of the property of the Dirac's delta function. Instead, let us introduce an additional function

$$w_3(r, \varepsilon) = \frac{1}{4\pi\sqrt{(r^2 + \varepsilon^2)}} \tag{5.22}$$

and find its Laplacian

$$\Delta w_3(r, \varepsilon) = -\frac{3\varepsilon^2}{4\pi\sqrt{(r^2 + \varepsilon^2)^5}} \tag{5.23}$$

Finally, we integrate over the sphere of radius R and put $\varepsilon \to 0$, which yields

$$\lim_{\varepsilon \to 0} \int_{\Omega_R^{(3)}} \Delta w_3(r, \varepsilon)\, d\Omega = \lim_{\varepsilon \to 0} \int_0^{2\pi}\int_0^{\pi}\int_0^{R} -\frac{3\varepsilon^2}{4\pi\sqrt{(r^2 + \varepsilon^2)^5}}\, r^2 \sin\theta\, dr\, d\theta\, d\alpha$$

$$= \lim_{\varepsilon \to 0} \left[-\frac{\varepsilon^2 R^3}{\varepsilon^2 \sqrt{(R^2 + \varepsilon^2)^3}} \right] = -1 \tag{5.24}$$

We conclude that $w_3(r)$, as given by eqn (5.20), is indeed the fundamental solution in $\Omega^{(3)}$ of the Laplace's eqn (5.12), except for the singular point $r = 0$, or of the Poisson's eqn (5.19), too.

In the open 2D space $\Omega^{(2)}$ a similar discussion produces an ordinary differential equation

$$u'' + \frac{1}{r} u' = 0 \tag{5.25}$$

leading to the following 2D fundamental solution:

$$w_2(r) = \frac{1}{2\pi} \ln \frac{1}{r} \tag{5.26}$$

which, like $w_3(r)$, has a singularity at $r \to 0$. The difference in relation to $w_3(r)$ is because $w_2(r)$ is not regular for $r \to 0$.

A fundamental solution for the Helmholtz equation can be found in a similar manner. For example, in the case of a mono-harmonic electromagnetic wave in vacuum, the Helmholtz equation reads

$$\Delta u + k^2 u = 0 \tag{5.27}$$

where $k^2 = \omega^2 \mu_0 \varepsilon_0$. The ordinary differential equation for $u(r)$ in $\Omega^{(3)}$ is found

$$u'' + \frac{2}{r} u' + k^2 u = 0 \tag{5.28}$$

Substitution of $u(r) = \frac{1}{\sqrt{r}} v(kr)$ leads to the Bessel's equation

$$v'' + \frac{1}{kr} v' + \left[1 - \frac{1}{4(kr)^2}\right] v = 0 \tag{5.29}$$

which has a solution

$$v(kr) = \frac{C}{\sqrt{r}} J_{-1/2}(kr)$$

with the presence of the Bessel's function of the minus half order of the first kind. Thus the fundamental solution of wave eqn (5.27) in $\Omega^{(3)}$ may be found as

$$\underline{w}_{3k}(r) = \frac{1}{4\pi r} \exp(\pm \mathrm{j}kr) \tag{5.30}$$

where the signs '+' or '−' refer to the forward or backward travelling waves respectively. Hence the real fundamental solution may be taken as

$$w_{3k}(r) = \frac{\cos(kr)}{2kr} \tag{5.31}$$

leading to the value $w_{3k} = 1/4\pi$ on a sphere with radius equal to the wavelength $r = \lambda = 2\pi/k$.

The solution of eqn (5.27) in $\Omega^{(2)}$ may be found in the form of Hankel's function of zero order

$$\underline{w}_{2k}(r) = \mp \mathrm{j}CY = \mp \mathrm{j}CH_0^{(s)}(kr) \tag{5.32}$$

where for the forward wave we use the minus sign and $s = 1$, and for the backward wave choose the plus sign and $s = 2$. For the real fundamental solution we obtain the Bessel's function of the second kind, following Weber's definition $w_{2k}(r) = CY_0(kr)$. Finally, let us consider the diffusion equation

$$\Delta u - k^2 \frac{\partial u}{\partial t} = 0 \tag{5.33}$$

where $k^2 = \mu\sigma$, describing the penetration of some scalar quantity into a conducting body which is able to resist such penetration, e.g. in a conductor. The fundamental solution in $\Omega^{(3)}$ is found as

$$v_n(r, \tau) = \frac{k^2}{\sqrt{(4\pi\tau)^n}} \exp\left(-\frac{k^2 r^2}{4\tau}\right) \tag{5.34}$$

where $\tau = T - t$ is the time interval between the instant of excitation and the instant of observation.

If the monochromatic wave is subjected to the diffusion, the Helmholtz equation is

$$\Delta u - jk^2 u = 0 \tag{5.35}$$

where $k = \omega\mu\sigma$ and the fundamental equation in $\Omega^{(3)}$ is

$$\underline{v}_3(r) = \frac{C}{\sqrt{r}} [\mathrm{ber}(kr) + \mathrm{j}\,\mathrm{bei}(kr)] \tag{5.36}$$

and in $\Omega^{(2)}$

$$\underline{v}_2(r) = C[\mathrm{ker}(kr) + \mathrm{j}\,\mathrm{kei}(kr)] \tag{5.37}$$

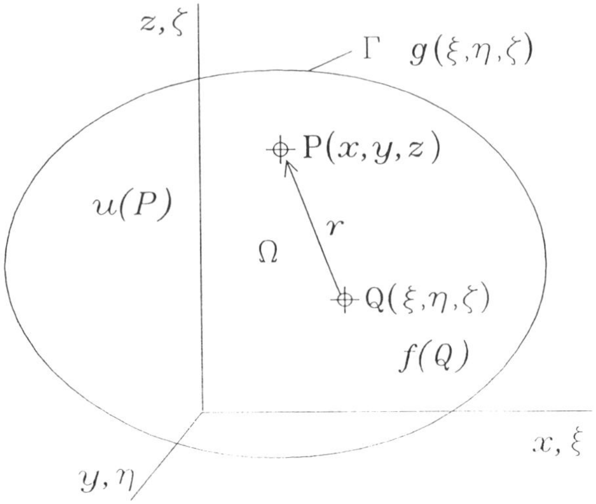

Fig. 5.2 *Region Ω with the boundary Γ*

Green's function in bounded regions
It is well known that the potential at point $P(x, y, x) \in \Omega$, as shown in Fig. 5.2, where Ω is a 3D (or 2D) region bounded by the surface (or line) Γ, may be expressed in terms of three quantities:

(1) the source function $f(\xi, \eta, \zeta)$, in Ω;
(2) the distribution of the normal derivative of the potential on the boundary surface (boundary line) Γ,
(3) the distribution of the potential on Γ.

In relation to (1), the source function may for example be:

- a distribution of volume density of electric charge in electrostatic fields,
- a distribution of heat sources in temperature fields,
- a distribution of d.c. current density for static magnetic field problems.

With reference to cases (2) and (3), the normal derivative of potential and the potential itself could be:

- the normal component $E_n = -\partial V/\partial n$ of the electric field strength $\mathbf{E}$ owing to an electrostatic potential V in a 2D or 3D region,
- the normal component $H_n = -\partial V_m/\partial n$ of the magnetic field strength $\mathbf{H}$ owing to a magnetostatic potential V_m in a 2D or 3D region,
- the tangential component $B_t = \partial A_z/\partial n$ of the magnetic flux density $\mathbf{B}$ owing to a magnetic vector potential $\mathbf{A} = A_z\mathbf{k}$ in a 2D region.

In general the potential $u(x, y, z)$ at $P(x, y, z) \in \Omega$, which normally is written in abbreviated notation as $u(P)$, can be expressed as a superposition of three potentials:

- the volume potential in 3D regions (or surface potential in 2D regions) called the *Newton's potential*, dependent upon the field source distribution in Ω,
- the double layer potential, dependent upon the prescribed distribution of the normal derivative of the resulting potential on Γ,
- the single layer potential, dependent upon the prescribed distribution of the potential on Γ.

The elementary volume potential $\mathrm{d}u(P)$ in the point $P(x, y, z)$ is given by

$$\mathrm{d}u(P) = \frac{f(Q)}{4\pi r(P, Q)}\,\mathrm{d}Q \qquad (5.38)$$

where the following abbreviated notation is used

$$f(Q) = f(\xi, \eta, \zeta)$$

$$\mathrm{d}Q = \mathrm{d}\xi, \mathrm{d}\eta, \mathrm{d}\zeta$$

$$r(P, Q) = \sqrt{[(x-\xi)^2 + (y-\eta)^2 + (z-\zeta)^2]}$$

For example in the case of an electrostatic field

$$\mathrm{d}V(P) = \frac{\varrho(Q)}{4\pi\varepsilon r(P, Q)}\,\mathrm{d}Q$$

where the source function $f(Q)$ is the volume density of electric charge divided by 4π times the electric constant.

The elementary double layer and single layer potentials at $P(x, y, z)$ are given by

$$\mathrm{d}u(P) = \frac{1}{4\pi r(P, Q)}\frac{\partial g(Q)}{\partial n}\,\mathrm{d}Q \qquad (5.39)$$

and

$$\mathrm{d}u(P) = -\frac{\partial}{\partial n}\left[\frac{1}{4\pi r(P, Q)}\right] g(Q)\,\mathrm{d}Q \qquad (5.40)$$

where $g(Q) = g(\xi, \eta, \zeta)$ is the given boundary potential at $Q(\xi, \eta, \zeta) \in \Gamma$. The total potential will be determined by integration over $\Omega \cup \Gamma$.

It can be seen that if in the unbounded $\Omega^{(3)}$ it is assumed that

$$\lim_{Q\to\infty} f(Q) = 0$$

then the volume potential is the only potential and is given by integration over the entire $\Omega^{(3)}$

$$u(P) = \int \frac{f(Q)}{4\pi r(P, Q)}\,\mathrm{d}Q \qquad (5.41)$$

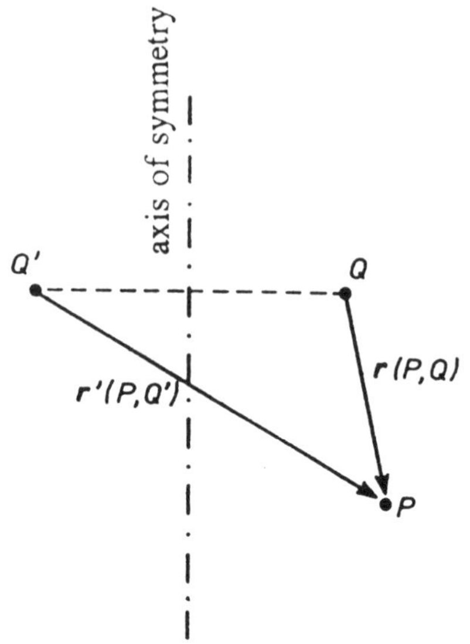

Fig. 5.3 *Image of the field source point*

The function of two points (and thus of six variables) of the form $1/4\pi r(P, Q)$ can be recognized as the kernel in the integral transformation of eqn (5.41). This kernel is singular because

$$\lim_{Q \to P} r(P, Q) = 0.$$

In the case of regions with closed boundaries, or boundaries limited by, say, lines of symmetry, the kernel will be influenced by the appropriate boundary conditions or types of field symmetry. Let us study a simple case of a symmetrical field in $\Omega^{(3)}$. The field symmetry axis is shown in Fig. 5.3, forming the half-space $\Omega_h^{(3)}$. As a consequence of field symmetry we can determine a point Q', which forms an image of a point Q relative to the symmetry axis, where we have $f(Q') = f(Q)$. The integration can be performed either over the entire space $\Omega^{(3)}$ with the use of the kernel $1/4\pi r(P, Q)$, or can be limited to $\Omega_h^{(3)}$, with the kernel

$$\frac{1}{2\pi}\left[\frac{1}{r(P, Q)} + \frac{1}{r(P, Q')}\right]$$

Both the above kernels can be interpreted as Green's functions.

Similar considerations will apply to the choice of Green's functions in the electrostatic or magnetostatic fields in regions limited by planes separating two media of finite and infinite (idealized) electric or magnetic constants. The Green's functions are also affected by the boundary conditions (Dirichlet, Neumann, mixed).

Generally, we notice that the total potential at $P(x, y, z)$ can be expressed as the sum of three components: the volume potential, the double layer potential and the single layer potential. We have, for each component part, a kernel composed of the fundamental solution $1/4\pi r(P, Q)$ and the functional supplement $F(P, Q)$, which depends upon the shape, material characteristics and boundary conditions inherent to the problem under consideration. Such a kernel can be interpreted as a Green's function, and for 3D problems it is usually written as

$$G(P, Q) = \frac{1}{4\pi r(P, Q)} + F(P, Q) \tag{5.42}$$

In order to determine the main property of Green's functions, we notice that the Green's function expresses the potential at P under the assumption that the source function tends to unity at a unique point Q_1 and disappears at all other points Q

$$f(Q) = \begin{cases} 1, & \text{for } Q = Q_1 \\ 0, & \text{for } Q \neq Q_1 \end{cases} \tag{5.43}$$

This can be described with the use of the Dirac delta function $\delta(P, Q) = \delta(x - \xi)\delta(y - \eta)\delta(z - \zeta)$ as

$$\int_\Omega \delta(P - Q)\, dQ = \begin{cases} 1, & \text{for } Q \in \Omega \\ 0, & \text{for } Q \notin \Omega \end{cases} \tag{5.44}$$

so that

$$\Delta G(P, Q) = -\delta(P - Q) \tag{5.45}$$

where the Laplacian is expressed in terms of variables ξ, η, ζ. Expression (5.45) simply states that the Green's function satisfies Laplace's equation almost everywhere, except the singular point at which it obeys Poisson's equation with unity source function.

The second important property of Green's function is that

$$G(P, Q) > 0 \quad \text{in} \quad \Omega \tag{5.46}$$

and the third property is its symmetry:

$$G(P, Q) = G(Q, P) \tag{5.47}$$

It can be verified by inspection that the fundamental equation $1/4\pi r(P, Q)$ has all three properties mentioned above. As a consequence, the function $F(P, Q)$ occurring in eqn (5.42), will also possess these properties.

General equation for potential

Following the considerations of the previous section we can write the potential at $P(x, y, z) \in \Omega$ with the boundary Γ (see Fig. 5.2) in the following form

$$u(P) = \int_\Omega G(P, Q) f(Q)\, dQ + \int_\Gamma \left[G(P, Q) \frac{\partial g(Q)}{\partial n} - g(Q) \frac{\partial G(P, Q)}{\partial n} \right] ds \tag{5.48}$$

where:

$G(P, Q)$ Green's function,
$g(Q)$ boundary potential,

n, s denote the variable normal to Γ, and the surface (or line) boundary element, respectively.

Making use of the fact that there is a certain degree of arbitrariness in the choice of the Green's function, it is possible to introduce some special conditions for $G(P, Q)$ leading to such a form of the potential formula which will contain the dependence upon just one type of boundary condition – the one that actually occurrs in the problem.

For Dirichlet boundary conditions (i.e. for the given potential $g(Q)$ on Γ), Green's function could be selected to satisfy:

$$G(P, Q) = 0, \qquad P, Q \in \Gamma \tag{5.49}$$

and then the first term under the sign of integration will disappear.

For a Neumann boundary condition (i.e. for the given $\partial g/\partial n$ on Γ) a similar requirement may be introduced

$$\frac{\partial G(P, Q)}{\partial n} = 0, \qquad P, Q \in \Gamma; \qquad P \neq Q, \tag{5.50}$$

However, in the case where Γ is a closed surface (or a closed line), the non-uniqueness of the Neumann boundary problem necessitates supplementary specifications of the average potential on Γ.

For the Dirichlet boundary problem the *surface Green's function* may be defined as

$$G_S(P, Q) = -\left.\frac{\partial G(P, Q)}{\partial n}\right|_{Q \in \Gamma} \tag{5.51}$$

where n is an external normal to Γ.

Let us express $G_S(P, Q)$ for two types of bounded regions: a half-space $\Omega_h^{(3)}$ and a half plane $\Omega_h^{(2)}$. According to eqn (5.20) we have the following $G(P, Q)$ as the fundamental solution in $\Omega^{(3)}$

$$G(P, Q) = \frac{1}{4\pi\sqrt{[(x-\xi)^2 + (y-\eta)^2 + (z-\zeta)^2]}} \tag{5.52}$$

In the case of half-space $z > 0$, i.e. when Γ is a plane $z(x, y) = 0$, the external normal to Γ is $-z$. Thus, following the idea of an image we have

$$G(P, Q) = \frac{1}{4\pi}\left\{\frac{1}{\sqrt{[(x-\xi)^2 + (y-\eta)^2 + (z-\zeta)^2]}} - \frac{1}{\sqrt{[(x-\xi)^2 + (y-\eta)^2 + (z+\zeta)^2]}}\right\} \tag{5.53}$$

It can be easily verified that for $z \to 0$ we have $G(P, Q) \to 0$, so that the condition of eqn (5.43) holds. It can also be shown, by differentiating twice with respect to the variables ξ, η, ζ, that the function $G(P, Q)$ satisfies Laplace's equation.

Following the definition of eqn (5.51) let us now formulate the surface Green's function for half-space $z > 0$

$$G_S(P,Q) = -\frac{\partial G(P,Q)}{\partial(-\zeta)} = \left.\frac{\partial G(P,Q)}{\partial z}\right|_{\zeta=0}$$

$$= \frac{z}{2\pi[(x-\xi)^2 + (y-\eta)^2 + z^2]^{3/2}} \tag{5.54}$$

Similarly we can find Green's function for a plane

$$G(P,Q) = \frac{1}{2\pi}\ln\frac{1}{\sqrt{[(x-\xi)^2 + (y-\eta)^2]}} \tag{5.55}$$

In the case of a half-plane $y > 0$, where Γ is a line $y(x) = 0$, we have

$$G(P,Q) = \frac{1}{2\pi}\left\{\ln\frac{1}{\sqrt{[(x-\xi)^2 + (y-\eta)^2]}} - \ln\frac{1}{\sqrt{[(x-\xi)^2 + (y+\eta)^2]}}\right\} \tag{5.56}$$

or

$$G(P,Q) = \frac{1}{2\pi}\ln\frac{(x-\xi)^2 + (y+\eta)^2}{(x-\xi)^2 + (y-\eta)^2} \tag{5.56a}$$

The surface Green's function becomes

$$G_S(P,Q) = -\frac{\partial G(P,Q)}{\partial(-\eta)} = \left.\frac{\partial G(P,Q)}{\partial \eta}\right|_{\eta=0} = \frac{y}{\pi[(x-\xi)^2 + y^2]} \tag{5.57}$$

Finally let us examine the Green's function for an inside of a cylinder of radius R. Starting from the Green's function in $\Omega^{(3)}$, we may express eqn (5.52) in cylindrical coordinates, then for the points $P(r, \alpha, z)$ and $Q(\varrho, \beta, \zeta)$ we obtain, after relevant substitutions,

$$G(P,Q) = \frac{1}{4\pi}\frac{1}{\sqrt{[r^2 + \varrho^2 - 2r\varrho\cos(\alpha-\beta) + (z-\zeta)^2]}} \tag{5.58}$$

The Green's function for the region limited by $r < R$ can be found by taking into consideration the images on the inner surface of the cylinder

$$G(P,Q) = \frac{1}{4\pi}\left\{\frac{1}{\sqrt{[r^2 + \varrho^2 - 2r\varrho\cos(\alpha-\beta) + (z-\zeta)^2]}} - \frac{1}{\sqrt{[r^2 + (2R-\varrho)^2 - 2r(2R-\varrho)\cos(\alpha-\beta) + (z-\zeta)^2]}}\right\} \tag{5.59}$$

The surface Green's function for the inner cylinder surface may be obtained by differentiating eqn (5.58) with respect to the external normal ϱ and then substituting $\varrho = R$

$$G_s(P,Q) = -\left.\frac{\partial G(P,Q)}{\partial \varrho}\right|_{\varrho=R}$$

$$= -\frac{1}{4\pi}\frac{R - r\cos(\alpha-\beta)}{\sqrt{[r^2+R^2+2rR\cos(\alpha-\beta)+(z-\zeta)^2]^3}} \qquad (5.60)$$

When the point P is situated on the axis of the cylinder (i.e. the z axis) then $r = 0$ and the function $G_S(P, Q)$ has the form

$$G_S(P,Q) = \frac{1}{4\pi}\frac{R}{\sqrt{[R^2+(z-\zeta)^2]^3}} \qquad (5.61)$$

5.3 TYPES OF PROBLEMS

We use the notion of *field synthesis*, as the opposite to *field analysis* which is commonly considered as a mathematically modelled physical problem that consists of solving the field equation, which, generally, is a partial differential equation of the second order with respect to some scalar or vector field quantity. In analysis this solution is accomplished in the environmentally and geometrically specified region which may be finite, bounded infinite, or unbounded infinite. Field sources in the analysis are known and given in the form of boundary conditions or/and internal sources.

The synthesis of the field can be formulated in many ways and we must define precisely what kind of problem is to be investigated. The field distribution, or some integral of it, is given *a priori* in any synthesis problems. Some or all of the following quantities synthesizing the field may therefore be investigated:

(1) boundary conditions (i.e. outer field sources),
(2) inner excitations (i.e. inner field sources),
(3) geometric shape of a region including the localization of inner sources,
(4) material parameters.

5.3.1 Inverse problem to Laplace's equation

In this section the basic formulation of a synthesis problem is presented as the inverse to the problem described by Laplace's differential equation with Dirichlet boundary conditions. A closed 3D (or 2D) region Ω with a boundary surface (or boundary line) Γ is shown in Fig. 5.4. Let Γ be the surface of a fixed shape, hence the considered problem belongs to the *group of a rigid structure*. Suppose that in Ω a scalar potential field $u(P)$ may exist satisfying Laplace's equation

$$\Delta u(P) = 0 \qquad \text{for } P \in \Omega \qquad (5.62)$$

Let the following considerations be limited to the functions $u \in \mathfrak{U}$, where $\mathfrak{U}$ is a metric space with a distance function $\varrho_{\mathfrak{U}}[\bar{u}(P), \tilde{u}(P)]$ defined as a *metric*. Suppose that in the subregion $\Omega_0 \subset \Omega$, called the *synthesis controlled*

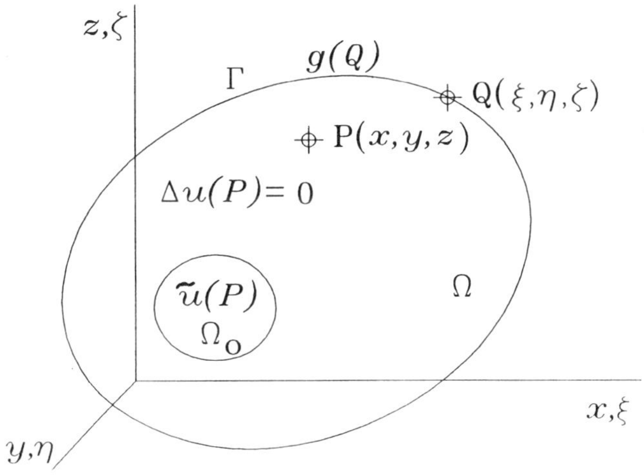

Fig. 5.4 *Region in which Laplace's equation holds*

subdomain, the function $\tilde{u}(P)$ is prescribed. The synthesis problem, which we wish to consider here, consists of searching for such a boundary function $g(P) = u(P)_{P\in\Gamma}$, that the solution $\bar{u}(P)$ satisfying Laplace's equation (5.62) and uniquely dependent on $g(P)$, would be in Ω_0 as close to $\tilde{u}(P)$ as possible. This condition implies the metric $\varrho_{\mathfrak{u}} = \min$ for $P \in \Omega_0$. It is evident, that the solution of the problem is not unique because it depends on the choice of $\varrho_{\mathfrak{u}}$. If the identity $\bar{u}(P) \equiv \tilde{u}(P)$ was required for the solution, then the choice of $\tilde{u}(P)$ would be limited to functions satisfying accurately the Laplace's equation with the given Dirichlet boundary conditions, i.e. to the unique harmonic functions only. In practice, various functions $\tilde{u}(P)$ are taken into consideration and accurate solutions are rarely sought.

Suppose that $u(P)$ represents a scalar potential function. Then the formula (5.48) can be applied with the first term becoming zero. The second term can be reduced to zero too, if $G(P, Q)$ is chosen in such a way that $G(P, Q)_{Q\in\Gamma} = 0$, see eqn (5.49). In consequence, after the appropriate choice of Green's function, $\tilde{u}(P)$ will be linked with $g(P)$ via a simple integral formula. For the inverse problem, i.e. when $\tilde{u}(P)$ is a given function and the Dirichlet boundary conditions $g(P)$ is unknown, this integral formula becomes the integral equation, or more specifically the Fredholm's equation of the first kind similar to eqn (5.3), but formulated for a 3D (or 2D) region. Thus, the unknown function $g(P)$ satisfies the Fredholm's equation of the first kind

$$\int_{\Gamma} K(P, Q)\, g(Q)\, \mathrm{d}Q = \tilde{u}(P), \qquad P \in \Omega_0 \tag{5.63}$$

where the kernel $K(P, Q)$ is a function of two points: the field point $P \in \Omega_0$ and the source point $Q \in \Gamma$. It is necessary to differentiate between the coordinate symbols of these points, namely $P(x, y, z)$ and $Q(\xi, \eta, \zeta)$. Hence, eqn (5.63), valid in a 3D subregion Ω_0, is

$$\int_{\Gamma} K(x, y, z, \xi, \eta, \zeta)\, g(\xi, \eta, \zeta)\, \mathrm{d}\xi\, \mathrm{d}\eta\, \mathrm{d}\zeta = \tilde{u}(x, y, z) \tag{5.64}$$

where x, y, z are the coordinate symbols in Ω as well as in Ω_0, and ξ, η, ζ are the coordinates of the same absolute coordinate system describing the point on Γ. When a local coordinate system s, t is introduced on Γ, the integral equation takes the form

$$\int_{\Gamma} K(x, y, z, s, t)\, g(s, t)\, \mathrm{d}s\, \mathrm{d}t = \tilde{u}(x, y, z) \tag{5.65}$$

and is valid together with the transformation equations of the coordinates. Similarly, we have for 2D regions

$$\int_{\Gamma} K(x, y, \xi, \eta)\, g(\xi, \eta)\, \mathrm{d}\xi\, \mathrm{d}\eta = \tilde{u}(x, y) \tag{5.66}$$

and with the use of a local coordinate s on Γ

$$\int_{\Gamma} K(x, y, s)\, g(s)\, \mathrm{d}s = \tilde{u}(x, y) \tag{5.67}$$

A kernel in the Fredholm's equation must be a known function. If both the geometrical structure and physical features of Ω were given, one would be able, in principle, to determine the kernel. In practice, it is possible only for simpler shapes of Ω, as the explicit form of kernel for a general case is difficult to obtain. The kernel occurring in eqn (5.63) is the Green's function of the second kind $K(P, Q) = G_S(P, Q)$ (see Section 5.2.2) called the *surface Green's function*, which satisfies the Laplace's equation

$$\Delta G_S(P, Q) = 0, \qquad P \in \Omega, \qquad Q \in \Gamma \tag{5.68}$$

with the homogeneous Dirichlet boundary condition

$$G_S(P, Q) = 0, \qquad P, Q \in \Gamma, \quad P \neq Q \tag{5.69}$$

and the following unique kernel integral equation:

$$\int_{\Gamma} G_S(P, Q)\, \mathrm{d}Q = -1, \qquad P, Q \in \Gamma \tag{5.70}$$

A Fredholm's equation is said to be *linear* (see Chapter 5.1.1) if the kernel is independent on the unknown function g. In consequence, the synthesis problem under consideration can be formulated by the use of the linear Fredholm's equation when:

(1) the problem belongs to the group of rigid structure, and
(2) material parameters of the region Ω are independent of both the field function u and its gradient.

Other problems

We will now specify some variants of the synthesis problems, all being inverse to the Laplace's equation, and all having a mathematical formulation not much different from that already described.

(1) When the dimension of Ω_0 is less then that of Ω, the reduced number of coordinates occurs in the kernel. This is the case in numerous engineering problems. The simplest problem is when Ω and Ω_0 are assumed as 2D and 1D regions respectively and a single boundary coordinate s is determined on Γ. Then the synthesis problem is simply formulated by the one-dimensional Fredholm's equation of the first kind, like eqn (5.3), namely:

$$\int_{\Gamma} K(x, s)\, g(s)\, ds = \tilde{u}(x), \qquad x \in \Omega_0. \tag{5.71}$$

(2) There are in Ω some number of split synthesis controlled subdomains Ω_{0i}, where $i = 1, 2, \ldots, n$.
(3) The unknown function of Γ is of a different type from the Dirichlet boundary condition $g(P) = u(P)_{P\in\Gamma}$. In the case of the Neumann boundary condition the kernel would be a Green's function, i.e. $K(P, Q) = G(P, Q)$.
(4) The function prescribed in Ω_0 is not a potential u but its gradient, for instance $V(P) = -grad\,(P)$.
(5) The vector field $U(P)$, satisfying the vector Laplace's equation $\Delta U(P) = 0$, occurs in Ω and the distribution $U(P)$ is prescribed in Ω_0.
(6) A case like the one mentioned above but with $\tilde{W}(P) = curl\, U(P)$ prescribed in Ω_0.
(7) Ω is not a closed region but it is a limited space (for instant the half-space or half-plane). Γ is in that case the limiting surface (or line).

5.3.2 *Inverse problem to Poisson's equation*

In this Section we shall consider a basic formulation of the synthesis problem that is inverse to Poisson's equation with homogeneous Dirichlet boundary conditions.

Suppose that in a closed region Ω, with boundary Γ shown in Fig. 5.5, a scalar potential field may exist satisfying Poisson's differential equation

$$\Delta u(P) = f(P), \qquad P \in \Omega \tag{5.72}$$

In a synthesis controlled subdomain $\Omega_0 \subset \Omega$ let the function $\tilde{u}(P)$ be prescribed for $P \in \Omega_0$. The synthesis problem is to determine such a source function $f(P)$ in Ω that the function $\bar{u}(P) = u(P)_{P\in\Omega}$, which is a unique solution of Poisson's equation (5.72) subject to the homogeneous boundary conditions $g(P) = u(P)_{P\in\Gamma} = 0$, would be in Ω_0 as close to $\tilde{u}(P)$ as possible, in the sense of the metric chosen for the problem

$$\varrho_u[\bar{u}(P), \tilde{u}(P)] = \min, \qquad P \in \Omega_0 \tag{5.73}$$

We note that the integral formulation considered above is generally covered

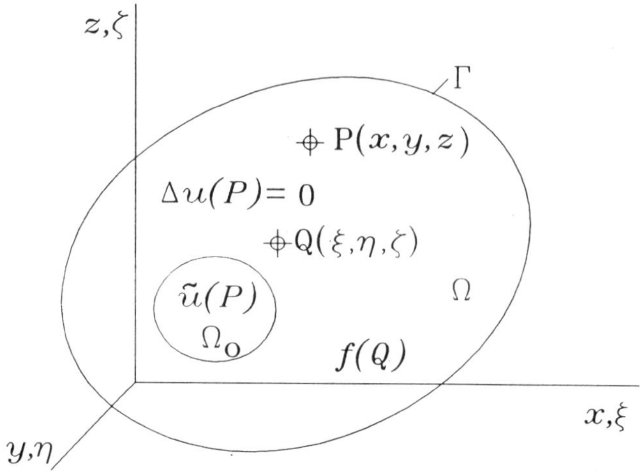

Fig. 5.5 *Region in which Poisson's equation holds*

by eqn (5.48), but only the first term needs to be applied. Thus the unknown source function $f(P)$ satisfies Fredholm's equation of the first kind

$$\int_\Omega K(P,Q)\,f(Q)\,\mathrm{d}Q = \tilde{u}(P) \qquad P \in \Omega_0 \tag{5.74}$$

where the kernel $K(P, Q)$ is a function of two points: the field point $P \in \Omega_0$ and the source point $Q \in \Omega$, and the kernel here is simply the Green's function $K(P, Q) = G(P, Q)$, which satisfies the Laplace's equation

$$\Delta G(P,Q) = 0, \qquad P, Q \in \Omega; \quad P \neq Q \tag{5.75}$$

with the homogeneous Dirichlet boundary conditions

$$G(P,Q) = 0, \qquad P, Q \in \Gamma; \quad P \neq Q \tag{5.76}$$

and the following unique kernel integral equation

$$\int_\Omega G(P,Q)\,\mathrm{d}Q = -1 \tag{5.77}$$

Fredholm's equation (5.74) is linear when the material parameters of Ω are independent on both $u(P)$ and grad (P).

Other problems

(1) A field function can be controlled in the whole Ω, i.e. $\Omega_0 = \Omega$.
(2) Two subregions are given in Ω: a synthesis controlled subdomain Ω_0 and a source subdomain Ω_S, in which the unknown source function is to be determined (see Fig. 5.6). Then the source function in $\Omega - \Omega_S$ is given and, in many practical cases, it is simply zero.
(3) There are in Ω a number of subdomains Ω_0 and Ω_S, which may or may not be separate from each other.

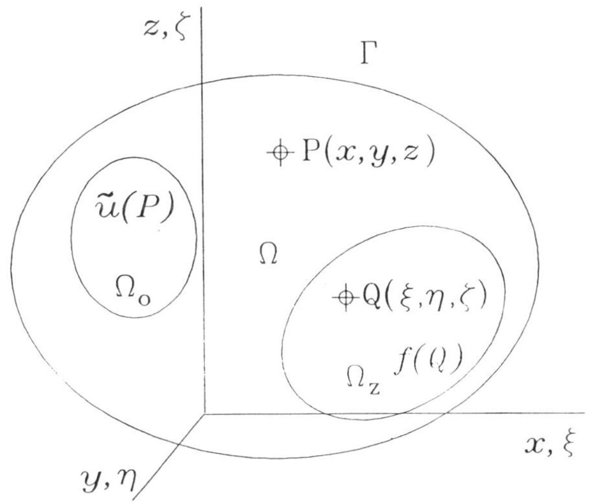

Fig. 5.6 *Region in which the field is synthesized by the internal sources*

(4) Boundary conditions on Γ are of different type than the Dirichlet boundary conditions.
(5) $V(P) = -grad(P)$ is prescribed in Ω_0.
(6) A vector field $U(P)$ exists in Ω and $U(P)$ or $W(P) = curl\, U(P)$ are prescribed in Ω_0.

5.3.3 *Variable boundary problem*

In this Section we shall discuss a basic synthesis problem applied to the *group of supple structure* that is inverse to the Laplace's boundary problem with Dirichlet boundary conditions.

Let Ω shown in Fig. 5.7 be a closed region with a boundary that is separated into two parts $\Gamma = \Gamma_1 \cup \Gamma_2$. Suppose that in Ω may occur the scalar field $u(P) \subset \mathcal{U}$ satisfying a Laplace's equation and suppose also, that in the synthesis controlled subdomain $\Omega_0 \subset \Omega$ the distribution of the field function $\tilde{u}(P)$ is prescribed. The following Dirichlet boundary conditions are given on Γ

$$u(P) = g_1(P), \qquad P \in \Gamma_1 \tag{5.78}$$
$$u(P) = g_2(P) = g_2 = \text{const}, \qquad P \in \Gamma_2$$

Γ_1 is assumed to be fixed, while Γ_2 is variable and the equation $\gamma(P) = 0$ describing this surface is to be determined in the synthesis problem. A constant function $g(P)$ on Γ_2 with the Dirichlet boundary conditions (5.78) has been chosen in order to conserve this boundary condition independently from γ. A set of the available γ functions should be determined. Let us impose two limits Γ_2' and Γ_2'' on the shape of γ. In practical cases these limits will be set as evident requirements by a designer. Figure 5.7 shows the region under consideration for which Ω_0 is independent of the variable part of the boundary.

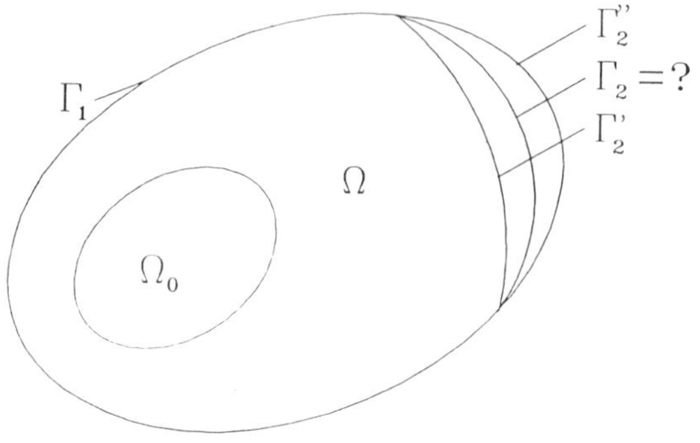

Fig. 5.7 *Forming the boundary shape*

The synthesis problem is to determine such a boundary shape Γ_2 conforming to the imposed limits Γ_2' and Γ_2'', or to put it in a different way, to determine such a function γ comprised between the given limits, that the solution $\bar{u}(P)$, satisfying Laplace's eqn (5.62) subject to boundary conditions (5.78), would be in Γ_0 as close to $\tilde{u}(P)$ as possible in sense of the metric.

The Fredholm's equation valid for the above synthesis problem is

$$g_2 \int_{\Gamma_2} K_2[P, Q, \gamma_2(Q)]\, \mathrm{d}Q = \tilde{u}(P) - u_1(P), \qquad P \in \Omega_0 \tag{5.79}$$

where

$$u_1(P) = \int_{\Gamma_1} K_1(P, Q)\, g_1(Q)\, \mathrm{d}Q$$

can be obtained by integration of two known functions K_1 and g_1. The characteristic property of Fredholm's eqn (5.79) is that the unknown function γ_1 occurs under the kernel. Thus the variable boundary synthesis problem leads to the nonlinear Fredholm's equation of the first kind.

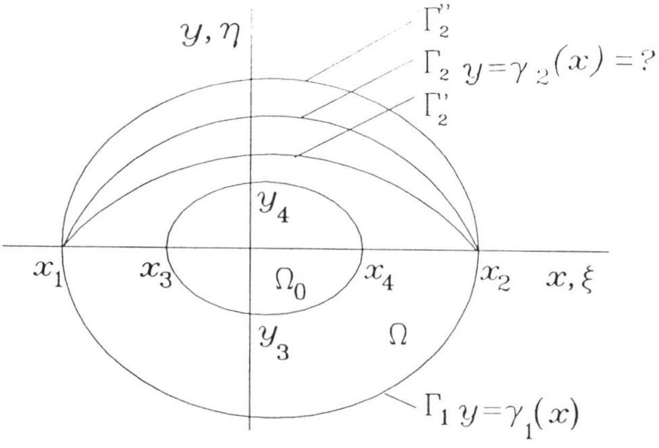

Fig. 5.8 *Forming the boundary shape described by $y = \gamma(x)$*

Let us study the above problem, i.e. Laplace's equation (5.62) and boundary conditions (5.78), in a particular case for the 2D region shown in Fig. 5.8. We suppose that the equation of the boundary line Γ_1 can be written in the explicit form $y = \gamma_1(x)$ and, similarly, the equation of γ_2 can be of the form $y = \gamma_2(x)$, with the supposition that $\gamma_2 \subset \mathfrak{U}_E$, the elements of which are half-ellipses of the major axis fixed to a and of the minor axis comprised between b' and b'', hence $\gamma_2 = b[1-(x/a)^2]^{1/2}$. Equation (5.79) becomes

$$g_2 \int_{a}^{-a} K_2[x, y, \xi, \gamma(\xi)]\, d\xi = \tilde{u}(x, y) - u_1(x, y); \qquad x, y \in \Omega_0 \qquad (5.80)$$

where the second term on the right-hand side is

$$u_1(x, y) = \int_{-a}^{a} K_1(x, y, \xi)\, g_1(\xi)\, d\xi$$

5.3.4 *Synthesis of material parameters*

In this Section a synthesis problem of an another kind will be formulated. Let us consider once again the inverse formulation of the Laplace's boundary problem with Dirichlet boundary conditions, but let the inversion be different from that discussed in Section 5.3.1. Thus, assume an electrostatic field in the 3D region Ω with boundary Γ, as shown in Fig. 5.9. Suppose the electric charge density is absent in Ω and the unique field source is the given boundary potential $\varphi_\Gamma = \varphi(P)_{P \in \Gamma}$. We wish to obtain the prescribed potential $\varphi_\Omega = \varphi(P)_{P \in \Omega}$ inside the region Ω by forming a suitable distribution of the permittivity $\varepsilon = \varepsilon(P)_{P \in \Omega}$.

Starting from eqn (1.41) for the field that is deprived of electric charges we have the following equation

$$div\,(\varepsilon\, grad\, \phi) = 0 \qquad (5.81)$$

Substituting the common distribution of the given $\bar{\varphi}_\Gamma$ and the prescribed $\tilde{\varphi}_\Omega$ for φ we obtained a known function $\varphi = \varphi(P)_{P \in \Gamma \cup \Omega}$ and similarly *grad* ϕ is a known function. Equation (5.81) can be transformed into the following form

$$-grad\, \varphi \cdot grad\, w = \nabla^2 \varphi \qquad (5.82)$$

giving the partial differential equation of the first order for the unknown function of three variables $w(x, y, z) = \ln \varphi(x, y, z)$. Equation (5.82) is linear with the functional coefficients $-grad_x\, \varphi = E_x(x, y, z)$, $-grad_y\, \varphi = E_y(x, y, z)$, $-grad_z\, \varphi = E_z(x, y, z)$, occurring at the differential terms of $w(x, y, z)$, i.e. at $\partial w/\partial x$, $\partial w/\partial y$, $\partial w/\partial z$ respectively. The right-hand function is the Laplacian of φ.

Therefore, the problem of synthesis of environmental parameters leads to a differential formulation. A practical use of this equation may concern the one dimensional variability of φ rather than the case of the whole 3D variability of it that seems to lead to difficult numerical algorithms.

5.4 SOLUTION OF SYNTHESIS PROBLEMS

5.4.1 *General remarks*

Many design issues pertaining to the electrical technology are, in fact, problems of synthesis. The way in which satisfactory solutions are obtained can be classified as:

- methods based on the theory of optimization,
- methods related to the inverse formulation of appropriate differential equations describing the electromagnetic phenomena.

Optimization techniques will not be discussed here (but see Section 7.7 later in this book). A fundamental approach to the solution of synthesis problems formulated with the use of integral equations, can be associated with the solution methods of Fredholm's and Volterra's equations. Some numerical problems concerning these methods are presented in this Section. The majority of the inverse problems of electromagnetism are ill-posed, see Section 5.4.4, which leads to unstable numerical solutions. Many problems have no unique solutions. Because of this, particular procedures must be applied in order to obtain the numerical solutions of such problems. The aim of these procedures is to stabilize the results of solutions and, moreover, to attribute a physical relevance to the numerical solutions, which otherwise could be formulated in a different way. One of the most promising approaches is the *technique of regularization* which is the subject of Sections 5.4.6 and 5.4.7. To apply this technique, suitable computers are necessary. The simplest method is to use various methods of approximate solution of algebraic equations. Some of them are presented in Sections 5.4.2 and 5.4.3.

5.4.2 *Quadrature*

Let us consider eqn (5.67) for a 2D region, presented in the form

$$\int_{\Gamma} K(P,s)\, g(s)\, \mathrm{d}s = \tilde{u}(P), \qquad P \in \Omega_0 \tag{5.83}$$

It is the synthesis equation for the potential $\tilde{u}(P)$ in Ω_0 created by the boundary potential $g(s)$ on Γ. Suppose g is continuous and limited. It follows that eqn (5.83) can be discretized. This means that this equation can be replaced by an operation over the finite set of the values of $g(s)$ related to the finite set of the chosen $\tilde{u}(P)$ values. This discretization is a *quadrature.*

We designate, for simplicity, the expression under the integral sign of eqn (5.83) as $h(s)$ and divide the whole Γ into m parts (arcs, straight-line sections etc.). The division points are s_ν, where $\nu = 1, 2, \ldots, n$. Thus, the integral in eqn (5.83) can be approximated by the sum

$$\int_{\Gamma} h(s)\, \mathrm{d}s = \sum_{\nu=1}^{n} a_\nu h(s_\nu) + R \tag{5.84}$$

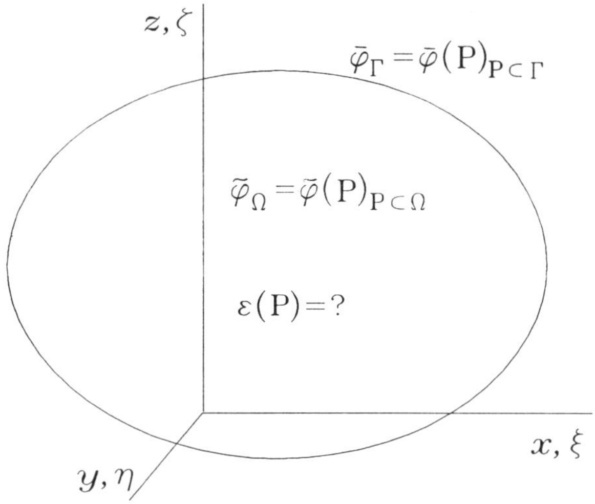

Fig. 5.9 *Forming the distribution of the permittivity*

where $h(s_\nu)$ denotes the values of $h(s)$ at s_ν, a_ν are the numerical coefficients termed *weights of a quadrature*, and R is the residue expressing an inaccuracy due to replacing the integral by a sum; when a 'dense' discretization is made then $R \approx 0$ is expected.

The synthesis controlled subdomain Ω_0 can also be discretized by the arbitrarily chosen points P_μ, where $\mu = 1, 2, \ldots, m$. In consequence, the integral equation (5.83) can be substituted by the following system of algebraic equations

$$\sum_{\nu=1}^{n} A_{\mu\nu} g_\nu = \tilde{u}_\mu, \qquad \mu = 1, 2, \ldots, m \tag{5.85}$$

where $A_{\mu\nu} = a_\nu\ K(P_\mu, s_\nu)$ are coefficients of a set of equations being the products of the weights of quadrature and the values of kernels, $g_\nu = g(s_\nu)$ are unknown values of the boundary function $g(s)$ on Γ, and $\tilde{u}_\mu = \tilde{u}(P_\mu)$ are given values of $\tilde{u}(P)$ at the points P_μ in Ω_0. A set of equations (5.85) can be expressed in a matrix form

$$[A_{\mu\nu}]\,[g_\nu] = [\tilde{u}_\mu] \tag{5.85a}$$

When $m = n$ is chosen, then $[A_{\mu\nu}]$ becomes an $n \times n$ square matrix, but this does not automatically mean that the solution of eqn (5.85a) will be reached. In practice, if Γ is a boundary or source line, two types of quadrature can be effectively applied: the *trapezoidal* and *Simpson's rules*.

For the curved line Γ_1 limited by two points s_a and s_b, where s is a curved line or coordinate on Γ_1, the quadrature can be done by introducing the division of Γ_1 into $n - 1$ curved elements of the same length Δs. When the trapezoidal rule is applied n may be any real number but using the Simpson's rule an odd number is recommended. When the division points s_ν are denoted together with the limits point as: $s_1 = s_a$ and $s_n = s_b$ then the weights of trapezoidal rule are given by

$$a_\nu = \Delta s \begin{cases} 1, & \text{if} \quad \nu \neq 1 \text{ or } \nu \neq n \\ \frac{1}{2}, & \text{if} \quad \nu = 1 \text{ or } \nu = n \end{cases} \tag{5.86}$$

and for Simpson's rule by

$$a_\nu = \Delta s \begin{cases} \frac{4}{3}, & \text{if} \quad \nu = 2, 4, \ldots, n-1 \\ \frac{2}{3}, & \text{if} \quad \nu = 3, 5, \ldots, n-2 \\ \frac{1}{3}, & \text{if} \quad \nu = 1 \text{ or } \nu = n \end{cases} \tag{5.87}$$

When Γ is a closed line then any point s_0 can qualify as the first point as well as the last $s_0 = s_1 = s_n$. In this case we have for the trapezoidal rule

$$a_\nu = \Delta s \tag{5.88}$$

and for Simpson's rule

$$a_\nu = \Delta s \begin{cases} \frac{4}{3}, & \text{if} \quad \nu = 2, 4, \ldots, n-1 \\ \frac{2}{3}, & \text{if} \quad \nu = 1, 3, \ldots, n \end{cases} \tag{5.89}$$

The discretization by quadrature can also be effectively applied to functions of two variables. Consider eqn (5.65) and suppose the boundary conditions are to be determined on the boundary surface $\Gamma_1 \subset \Gamma$. If Γ_1 is a rectangle with corner points (s_a, t_a), (s_b, t_a), (s_a, t_b), (s_b, t_b) and if the coordinate system on Γ_1 is rectangular, the following Fredholm's equation holds

$$\int_{s_a}^{s_b}\int_{t_a}^{t_b} K(P, s, t)\, g(s, t)\, \mathrm{d}s\, \mathrm{d}t = \tilde{u}(P) \tag{5.90}$$

The 2D quadrature can be introduced on the above rectangle Γ_1. The simplest way is to use a uniform division with the step h for both coordinates. Denoting the coordinates of division points by

$$S_k \quad \text{for} \quad k = 1, 2, \ldots, M; \quad t_l \text{ for } \quad l = 1, 2, \ldots, N$$

where M and N are odd numbers and $s_1 = s_a$, $s_b = s_M$, $t_1 = t_a$, $t_b = t_N$ respectively, we have for the trapezoidal rule

$$a_{kl} = \frac{2h^2}{3} \begin{cases} 0, & \text{if } k, l = \text{odd numbers} \\ 1, & \text{if } k (\text{or } l) = \text{odd number and } l (\text{or } k) = 1 \text{ or } N (\text{or } M) \\ 2, & \text{for any other } k, l \end{cases} \tag{5.91}$$

and for Simpson's rule

$$a_{kl} = \frac{h^2}{9\omega} \begin{cases} 4, & \text{if } kl = \text{odd number} \\ 8, & \text{if } kl = \text{even number and } k + l = \text{odd number} \\ 16, & \text{if } k, l = \text{even numbers} \end{cases} \tag{5.92}$$

where

$$\omega = \begin{cases} 4, & \text{for corners} \\ 2, & \text{for sides} \\ 1, & \text{inside} \end{cases}$$

The integral over the function of two variables $h(s, t)$ can be substituted by the double sum formula

$$\int_{s_a}^{s_b} \int_{t_a}^{t_b} h(s, t)\,\mathrm{d}s\,\mathrm{d}t = \sum_{k=1}^{M} \sum_{l=1}^{N} a_{kl} h(s_k, t_l) + R \tag{5.93}$$

When the controlled points $P_\mu, \mu = 1, 2, \ldots, m$ are chosen in Ω_0 then eqn (5.93) can be written as the following system of algebraic equations

$$A_{\mu kl} g_{kl} = \tilde{u}_\mu \qquad \mu = 1, 2, \ldots, m; \quad k = 1, 2, \ldots, M; \quad l = 1, 2, \ldots, N \tag{5.94}$$

where

$$A_{\mu kl} = a_{kl} K(P_\mu, s_k, t_l)$$

This set of equations can easily be transformed into a matrix equation if the indices k, l are replaced by one common index $\nu = 1, 2, \ldots, n = M \times N$.

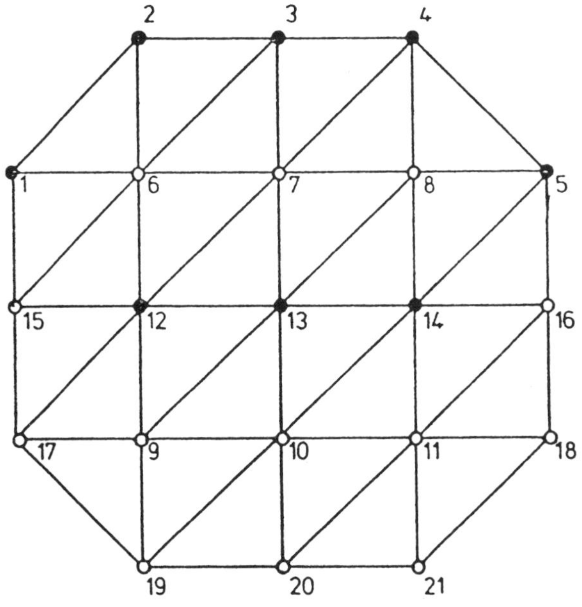

Fig. 5.10 *The synthesis of scalar potential by searching the boundary conditions*

5.4.3 Direct discretization

In many practical problems the explicit form of the kernel is not easy to find. In such cases the region under consideration may be discretized directly. This means that a *finite element method* may be used, but it has to be modified to suit the synthesis formulation. Let us give an example of such a procedure which transforms a matrix equation from a form used commonly in analysis into a form suitable for the inverse problems.

A region Ω is shown in Fig. 5.10, in which Laplace's equation holds. The region is discretized by using 28 triangles that form 21 nodes. Let the synthesis problem be to obtain three prescribed values of potential $\tilde{u}_{12}, \ldots, \tilde{u}_{14}$ at points 12, ..., 14 by determining five values of boundary potentials $g_1, \ldots, g_5$ at boundary points 1, ..., 5. Suppose seven values of boundary potential $g_{15}, \ldots, g_{21}$ are given at nodes 15, ..., 21. It is clear that six values of potential $u_6, \ldots, u_{11}$ at inner points in Ω may assume any values.

Let us start from the general matrix equation in the form that we would have had by formulating the analysis problem using finite elements. We will, however, partition the matrices in accordance with the formulation of the synthesis problem

$$\left[\begin{array}{c|c} [A1]_{6\times 6} & [A2]_{6\times 3} \\ \hline [A2]^{\mathrm{T}}_{3\times 6} & [A3]_{3\times 3} \end{array}\right] \left[\begin{array}{c} u_6 \\ \vdots \\ u_{11} \\ \hline \tilde{u}_{12} \\ \vdots \\ \tilde{u}_{14} \end{array}\right] = \left[\begin{array}{c|c} [B1]_{6\times 5} & [B2]_{6\times 7} \\ \hline [B3]_{3\times 5} & [B4]_{3\times 7} \end{array}\right] \left[\begin{array}{c} g_1 \\ \vdots \\ g_5 \\ \hline \bar{g}_{12} \\ \vdots \\ \bar{g}_{21} \end{array}\right] \tag{5.95}$$

The next step is to rewrite the above equation in a decomposed form as a system of two matrix equations with respect to the potentials $u_6, \ldots, u_{11}$ and $\tilde{u}_{12}, \ldots, \tilde{u}_{14}$

$$[A1]_{6\times 6}\begin{bmatrix} u_6 \\ \vdots \\ u_{11} \end{bmatrix} + [A2]_{6\times 3}\begin{bmatrix} \tilde{u}_{12} \\ \vdots \\ \tilde{u}_{14} \end{bmatrix} = [B1]_{6\times 5}\begin{bmatrix} g_1 \\ \vdots \\ g_5 \end{bmatrix} + [B2]_{6\times 7}\begin{bmatrix} \bar{g}_{12} \\ \vdots \\ \bar{g}_{21} \end{bmatrix}$$

$$[A2]^{\mathrm{T}}_{3\times 6}\begin{bmatrix} u_6 \\ \vdots \\ u_{11} \end{bmatrix} + [A3]_{3\times 3}\begin{bmatrix} \tilde{u}_{12} \\ \vdots \\ \tilde{u}_{14} \end{bmatrix} = [B3]_{3\times 5}\begin{bmatrix} g_1 \\ \vdots \\ g_5 \end{bmatrix} + [B4]_{3\times 7}\begin{bmatrix} \bar{g}_{12} \\ \vdots \\ \bar{g}_{21} \end{bmatrix} \tag{5.95a}$$

The potentials $u_6, \ldots, u_{11}$ are not the subject of synthesis in our problem and they can be eliminated from the matrix equation. After multiplying the first of the above equations by $[A1]^{-1}_{6\times 6}$, it is then solved with respect to $u_6, \ldots, u_{11}$. These potentials are now substituted into the second matrix equation and we obtain the following equation governing the synthesis problem

$$[B5]_{3\times5}\begin{bmatrix} g_1 \\ \vdots \\ g_5 \end{bmatrix} = [B6]_{3\times7}\begin{bmatrix} \bar{g}_{15} \\ \vdots \\ \bar{g}_{21} \end{bmatrix} + [A4]_{3\times3}\begin{bmatrix} \tilde{u}_{12} \\ \vdots \\ \tilde{u}_{14} \end{bmatrix} \qquad (5.96)$$

where

$$[B5]_{3\times5} = [A2]^{\mathrm{T}}_{3\times6}\cdot[A1]^{-1}_{6\times6}\cdot[B1]_{6\times5} - [B3]_{3\times5},$$

$$[B6]_{3\times7} = [B4]_{3\times7} - [A2]^{\mathrm{T}}_{3\times6}\cdot[A1]^{-1}_{6\times6}\cdot[B2]_{6\times7},$$

$$[A4]_{3\times3} = [A2]^{\mathrm{T}}_{3\times6}\cdot[A1]^{-1}_{6\times6}\cdot[A2]_{6\times3} - [A3]_{3\times3}.$$

Thus in our example the problem of finding the unknown boundary potentials $g_1, \ldots, g_5$ is reduced to a system of three equations with five unknowns, as expressed by the matrix $[B5]_{3\times5}$ with three rows and five columns. An infinite number of solutions is possible and each is represented by a set of five potentials $g_1, \ldots, g_5$ with a linear dependence between them. There are no mathematical reasons to choose any particular solution and the choice has to be made by considering the physical properties of the system under consideration.

Suppose $g_1, \ldots, g_5$ are the chosen potentials on the boundary. Then the whole set of 12 boundary potentials $g_1, \ldots, g_5$, $g_{15}, \ldots, g_{21}$ forms an acceptable solution for all inner potentials. The potentials at points 12, ..., 14 will have the values $\bar{u}_{12}, \ldots, \bar{u}_{14}$, that are the best approximations for the required values $\bar{u}_{12}, \ldots, \bar{u}_{14}$ from the engineering point of view. The values of the potentials $u_6 \ldots u_{11}$ are not crucial, but may nevertheless be determined in

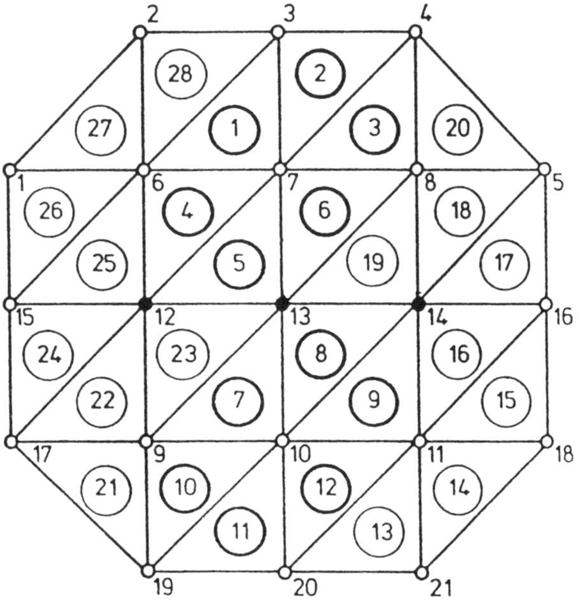

Fig. 5.11 *The synthesis of vector potential by inner current densities*

order to have a complete picture of potential distribution for the structure given in Fig. 5.10.

Application of such algorithms faces two practical difficulties. The first is due to the necessity of matrix inversion (e.g. matrix $[A1]_{6\times6}$). In problems in which Ω_0 is significantly smaller then Ω, such matrices tend to be large and lead to difficulties of a numerical nature. Secondly, a special node numbering technique must be employed which is usually different from the numbering provided by most finite element codes used for analysis, see Fig. 5.10.

A similar approach may also be adopted to the synthesis of internal field sources in a 2D region Ω shown in Fig. 5.11. In the region under consideration Poisson's equation is valid for a z-component A_z of the magnetic vector potential A. A system of potentials $\tilde{A}_{12}, \ldots, \tilde{A}_{14}$ is to be generated by determining appropriate current densities $S_1, \ldots, S_{12}$. Suppose all boundary current densities at points $1, \ldots, 5$ and $12, \ldots, 21$ are zero and, in addition let, $\bar{S}_{13}, \ldots, \bar{S}_{28}$ be known. The basic matrix equation is

$$\left[\begin{array}{c|c} [A1]_{6\times6} & [A2]_{6\times3} \\ \hline [A2]^{\mathrm{T}}_{3\times6} & [A3]_{3\times3} \end{array}\right] \cdot \left[\begin{array}{c} A_6 \\ \vdots \\ A_{11} \\ \hline \tilde{A}_{12} \\ \vdots \\ \tilde{A}_{14} \end{array}\right] = \left[\begin{array}{c|c} [C1]_{6\times12} & [C2]_{6\times16} \\ \hline [C3]_{3\times12} & [C4]_{3\times16} \end{array}\right] \cdot \left[\begin{array}{c} S_1 \\ \vdots \\ S_{12} \\ \hline \bar{S}_{13} \\ \vdots \\ \bar{S}_{28} \end{array}\right].$$

It is clear that a similar operation to that applied in the previous example will lead to a matrix equation of a form analogous to eqn (5.95a), if we replace $[B1]_{6\times5}$ by $[C1]_{6\times12}$, $[B2]_{6\times7}$ by $[C2]_{6\times16}$, $[B3]_{3\times5}$ by $[C3]_{3\times12}$ and $[B4]_{3\times7}$ by $[C4]_{3\times16}$, and

$$\begin{bmatrix} g_1 \\ \vdots \\ g_5 \end{bmatrix} \text{ by } \begin{bmatrix} S_1 \\ \vdots \\ S_{12} \end{bmatrix} \text{ and } \begin{bmatrix} \bar{g}_{12} \\ \vdots \\ \bar{g}_{21} \end{bmatrix} \text{ by } \begin{bmatrix} \bar{S}_{16} \\ \vdots \\ \bar{S}_{28} \end{bmatrix}$$

The synthesis equation will have the form of eqn (5.96), but there are now 12 unknown variables and only three equations. The problem is thus strongly under-specified and ill-posed and may be solved using various special techniques discussed in Sections 5.4.4 to 5.4.6.

5.4.4 Conditions for solutions

The synthesis problem can be expressed in a generalized form as

$$\mathbf{A}g = \tilde{u} \tag{5.97}$$

where $\mathbf{A}$ may be either an integral midcontinuous operator

$$\mathbf{A} \equiv \int K(P, Q) \ldots \mathrm{d}Q$$

or a matrix

$$\mathbf{A} \equiv [A_{\mu\nu}]$$

The operator transforms the space of solutions $\mathcal{G}$ into the space of given functions $\mathcal{U}$

$$\mathbf{A}: \mathcal{G} \rightarrow \mathcal{U}$$

where $g \in \mathcal{G}$ and $\bar{u} \in \mathcal{U}$.

The problem formulated by eqn (5.97) is correctly posed in the Hadamard sense [5.15] if three conditions are satisfied:

(1) for each $\tilde{u} \in \mathcal{U}$ the solution $g \in \mathcal{G}$ exists,
(2) the existing solution is unique,
(3) the problem is stable on the pair of spaces $(\mathcal{U}, \mathcal{G})$, i.e. for any $\varepsilon > 0$ there is such $\delta(\varepsilon)$ that for any two solutions $g_1, g_2 \in \mathcal{G}$ and $\tilde{u}_1, \tilde{u}_2 \in \mathcal{U}$ we have

$$\varrho_{\mathcal{U}}(\tilde{u}_1, \tilde{u}_2) \leqslant \varepsilon \Rightarrow \varrho_{\mathcal{G}}(g_1, g_2) \leqslant \delta(\varepsilon)$$

In design synthesis the second condition may not be necessary; in identification synthesis, on the other hand, the uniqueness is strictly required. The majority of synthesis problems do not satisfy the third condition of the solution stability. As a consequence, in numerical treatment of the problems, even very small disturbances in the value of the $\tilde{u}$ vector, or small inaccuracies in numerical calculations of the matrix **A**, may lead to large changes in the solution vector g.

Most synthesis problems are in fact ill-posed in the Hadamard sense. The reasons are of two-fold origin: they are either due to the physical nature of the problem or are caused by inaccuracies introduced by the numerical treatment of the solution. We shall classify such cases for a discretized equation of the form

$$\sum_{\nu=1}^{n} A_{\mu\nu} g_\nu = \tilde{u}_\mu \qquad \mu = 1, 2, \ldots, m \tag{5.98}$$

(1) $m > n$ – there is more information than unknowns; the problem is over-specified,
(2) $m = n$, $\det[A_{\mu\nu}] = 0$ or $\det[A_{\mu\nu}] \approx 0$ – the problem is singular or quasi-singular, respectively,
(3) $m < n$ – not enough information for the given number of unknowns; the problem is under-specified,
(4) values of $\tilde{u}$ are approximate, e.g. they are taken from a measurement,
(5) calculated coefficients $A_{\mu\nu}$ are not accurate, e.g. due to discretization error.

Many physical and mathematical problems are ill-posed in the Hadamard sense. The well known example is Laplace's equation with Cauchy's boundary condition.

When a quasi-singular matrix $[A_{\mu\nu}]_{n\times n}$ is of rank n, its solution may be found. The probability of reaching the satisfactory solution can be estimated using a quantity that we will term the *index of conditioning*, defined as

$$\operatorname{cond}(\mathbf{A}) = \|\mathbf{A}\| \cdot \|\mathbf{A}^{-1}\| \tag{5.99}$$

The smaller the value of cond(**A**), the better the conditioning of the problem under consideration. The index of conditioning defined by eqn (5.99) can be directly applied only when the operator **A** is inversible. Hence $[\mathbf{A}_{\mu\nu}]$ must be square and non-singular. Definition (5.99) describes a general way of addressing the problem of quality of conditioning irrespective of the norm applied, although the numerical value of cond(**A**) depends on the choice of the norm.

The following norms are used in order to evaluate the ill-conditioned systems of equations reduced to a square $n \times n$ matrix of the rank n, i.e. $\mathbf{A} = [A_{\mu\nu}]$, where $\mu, \nu = 1, 2, \ldots, n$ and $\det \mathbf{A} \neq 0$.

The *first norm*:

$$\|A\|_1 = \max_{1\leqslant\nu\leqslant n} \sum_{\mu=1}^{n} |A_{\mu\nu}| \tag{5.100}$$

is equal to the maximum sum of the absolute values of elements in one column of the matrix.

The *norm of maximum*:

$$\|A\|_\infty = \max_{1\leqslant\mu\leqslant n} \sum_{\nu=1}^{n} |A_{\mu\nu}| \tag{5.101}$$

is equal to the maximum sum of the absolute values of elements in one row of the matrix.

The *Euclidean norm*:

$$\|A\|_E = \sqrt{\sum_{\mu=1}^{n} \sum_{\nu=1}^{n} |A_{\mu\nu}|^2} \tag{5.102}$$

is equal to the square root of sum of squares of the absolute values of all matrix elements.

The *second norm* also called the *spectral norm*:

$$\|A\|_2 = \max_{1\leqslant\mu\leqslant n} (\sqrt{\lambda'_\mu}) \tag{5.103}$$

is equal to the square root of the greatest eigenvalue λ'_μ of the matrix $\underline{\mathbf{B}} = \underline{\mathbf{A}}^{\mathrm{H}} \cdot \underline{\mathbf{A}}$, where $\underline{\mathbf{A}}^{\mathrm{H}}$ is the conjugate (in the Hermitian sense) to the matrix $\underline{\mathbf{A}}$, i.e. for a real matrix **A**, $\mathbf{B} = \mathbf{A}^{\mathrm{T}} \cdot \mathbf{A}$.

If a square matrix **A** can be diagonalized and, in particular, if it is symmetric, then

$$\max_{1 \leqslant \mu \leqslant n} (\sqrt{\lambda'_\mu}) = \max_{1 \leqslant \mu \leqslant n} (\lambda_\mu) \tag{5.104}$$

where λ_μ are the eigenvalues of **A**. Therefore, the norm $\|\mathbf{A}\|_2$ is, in this case, the greatest eigenvalue of **A**. In consequence, the eigenvalues of diagonalized matrices $\mathbf{A}^{-1}$ are the reciprocals of the eigenvalues of **A** and the definition for the index of conditioning (5.99) can be simplified to

$$\text{cond}_2(A) = \frac{\lambda_n}{\lambda_1} \tag{5.105}$$

where λ_n and λ_1 are the maximum and minimum eigenvalues of **A**, respectively.

Example
Let us consider two matrices

$$[A1] = \begin{bmatrix} 1 & 0{,}5 \\ 2 & 1{,}1 \end{bmatrix}; \qquad [A2] = \begin{bmatrix} 1 & 0{,}5 \\ 2 & 1{,}01 \end{bmatrix}$$

where $\det[A1] = 0.1$ and $\det[A2] = 0.01$. The inverse matrices are

$$[A1]^{-1} = \begin{vmatrix} 11 & -5 \\ -20 & 10 \end{vmatrix}; \qquad [A2]^{-1} = \begin{vmatrix} 101 & -50 \\ -200 & 100 \end{vmatrix}$$

Although the matrices $[A1]$ and $[A2]$ are not symmetrical, nevertheless they can be diagonalized because both of them have two real eigenvalues: $\lambda_1 = 0.04875$ and $\lambda_2 = 2.0512$ for $[A1]$ and $\lambda_1 = 0.004987$ and $\lambda_2 = 2.0050$ for $[A2]$. It is easily shown that the matrices $[A1]^{-1}$ and $[A2]^{-1}$ have the eigenvalues equal to the reciprocals of the above given eigenvalues and the matrices $[B1] = [A1]^{\mathrm{T}}[A1]$ and $[B2] = [A2]^{\mathrm{T}}[A2]$ have the eigenvalues equal to the squares of these eigenvalues. The indices of conditioning determined using the second norm can be calculated following the definition (5.105)

$$\text{cond}_2[A1] = \frac{2.0512}{0.04875} = 42.08$$

and

$$\text{cond}_2[A2] = \frac{2.0050}{0.004987} = 402.05$$

The indices of conditioning calculated using the first norm, see eqn (5.100), are:

$$\text{cond}_1[A1] = (1 + 2)(11 + 20) = 93$$

$$\text{cond}_1[A2] = (1 + 2)(101 + 200) = 903$$

Calculation of $\mathrm{cond}_\infty[A1]$ and $\mathrm{cond}_\infty[A2]$ is left to the reader. We notice, finally, that the index of conditioning for the Euclidean norm (5.102) is

$$\mathrm{cond}_E[A1] = \sqrt{(1^2 + 0.5^2 + 2^2 + 1.1^2)} \cdot \sqrt{(11^2 + 5^2 + 20^2 + 10^2)} = 64.6$$

and

$$\mathrm{cond}_E[A2] = \sqrt{(1^2 + 0.5^2 + 2^2 + 1.01^2)} \cdot \sqrt{(101^2 + 50^2 + 200^2 + 100^2)} = 627$$

However, the use of eqn. (5.99) for examining *a priori* the solvability of the system of ill-posed equations is in practice limited. Application of the formula requires that matrix $[A_{\mu\nu}]$ is inverted but, whether it can be inverted is precisely the question we seek to answer.

For the rectangular matrices that are singular or quasi-singular, the basic equation

$$[A_{\mu\nu}]\,[g_\nu] = [\tilde{u}_\mu], \qquad \mu = 1, 2, \ldots, m; \quad \nu = 1, 2, \ldots, n \tag{5.106}$$

can be transformed by pre-multiplying by $[A_{\mu\nu}]^{\mathrm{T}}$ into

$$[A_{\omega\nu}]\,[g_\nu] = [A_{\omega\mu}]^{\mathrm{T}}\,[\tilde{u}_\mu], \qquad \omega = 1, 2, \ldots, n \tag{5.107}$$

where $[B_{\omega\nu}] = [A_{\omega\nu}]^{\mathrm{T}} \cdot [A_{\mu\nu}]$ is a square $n \times n$ matrix and the definition of eqn (5.99) can then be directly applied as

$$\mathrm{cond}([B_{\omega\nu}]) = \|[B_{\omega\nu}]_+\| \cdot \|[B_{\omega\nu}]\| \tag{5.108}$$

where $[B_{\omega\nu}]_+$ is the inverse matrix of $[B_{\omega\nu}]$, i.e. $[B_{\omega\nu}]_+ = [B_{\omega\nu}]^{-1}$. The matrix $[B_{\omega\nu}]_+$ is termed the *pseudo-inverse matrix* to the primary matrix $[A_{\mu\nu}]$.

5.4.5 Least-squares methods

Basic approach

Systems of algebraic equations where $m > n$, as for example eqn (5.85), have been solved in the past using Gauss' method known as the *least-squares method* (*LSM*). This method provides the foundation for new approaches developed recently and generally called the *regularization techniques*.

If the classical Gaussian form of the method is applied to an ill-posed generalized eqn (5.97), it must be supposed that the *solution space* $\mathcal{G}$ and the *space of given function* $\mathcal{U}$ have metrics, and therefore a 'distance' between two functions is determined. For example the distance between two solutions g_1, and g_2, say $\varrho_G(g_1 - g_2)$, may be expressed by a real number called the *norm* of the function g in the space $\mathcal{G}$ and it is designated $\|g_1 - g_2\|_G$, i.e.

$$\varrho_G(g_1, g_2) = \|g_1 - g_2\|_G \tag{5.109}$$

If $\bar{g}$ is the exact solution of eqn (5.97) then the norm vanishes

$$\|\mathbf{A}\bar{g} - \tilde{u}\|_{\mathcal{U}} = 0 \tag{5.110}$$

When an exact solution $\bar{g}$ of eqn (5.97) cannot be obtained, e.g. because the equation is ill-posed in the Hadamard sense, we instead seek an approximate solution $\tilde{g}$ such that the square of the norm is minimum

$$\|\mathbf{A}g - \tilde{u}\|_{\mathcal{U}}^2 = \min_{g \in \mathcal{G}} \tag{5.111}$$

as a consequence the square of the norm can be recognized as a *functional*

$$\mathfrak{M}[\tilde{u}, g] = \|\mathbf{A}g - \tilde{u}\|_{\mathcal{U}}^2, \tag{5.112}$$

and a suitable procedure for finding the solution $\tilde{g}$ could be applied by minimizing this functional in the space $\mathcal{U}$.

When eqn (5.97) is discretized to the form (5.98), the spaces $\mathcal{U}$ and $\mathcal{G}$ are recognized as the Euclidean spaces: $\mathcal{G} \equiv \mathcal{R}^n$ (n-dimensional Euclidean space) and $\mathcal{U} \equiv \mathcal{R}^m$ (m-dimensional Euclidean space). In this case the functional (5.112) becomes

$$\mathfrak{M}[\tilde{u}, g] = \|A_{\mu\nu} g_\nu - \tilde{u}_\mu\|_{\mathcal{R}^m}^2 = \sum_{\mu=1}^{m} \left[\sum_{\nu=1}^{n} A_{\mu\nu} g_\nu - \tilde{u}_\mu \right]^2 \tag{5.113}$$

The Euler–Lagrange equation minimizing the functional (5.112) can be directly obtained from eqn (5.97) by multiplying both sides by the operator $\mathbf{A}^*: \mathcal{U} \to \mathcal{G}$, conjugate with respect to $\mathbf{A}$.

$$\mathbf{A}^* \mathbf{A} g = \mathbf{A}^* \tilde{u} \tag{5.114}$$

In the case of discretized eqns (5.98) the Euler–Lagrange system of equations has n unknown variables

$$\sum_{\mu=1}^{m} \sum_{\nu=1}^{n} A_{\omega\mu} A_{\mu\nu} g_\nu = \sum_{\mu=1}^{m} A_{\omega\mu} \tilde{u}_\mu, \qquad \omega = 1, 2, \ldots, n \tag{5.115}$$

Using a matrix notation we have

$$[A]^{\mathrm{T}}_{n\times m} \cdot [A]_{m\times n} \cdot [g]_n = [A]^{\mathrm{T}}_{n\times m} \cdot [\tilde{u}]_m \tag{5.116}$$

The system of eqns (5.115) is called the *normal system of Gauss' equations*. If a function $\tilde{g}$ is the solution of Euler–Lagrange's equation (5.114) then this function is seen as an approximate solution of the ill-posed eqn (5.97) obtained by the least-squares method. Similarly, if the system of eqns (5.116) is solvable or, what is equivalent, the matrix

$$[B]_{n\times n} = [A]^{\mathrm{T}}_{n\times m} \cdot [A]_{m\times n}$$

is inversible, then the set of values $\tilde{g}_\nu$ for $\nu = 1, 2, \ldots, n$ is the approximate solution of the ill-posed system of eqns (5.98).

The matrix $[A]_{n\times n}$ is inversible when the rank of the primary matrix $[A]_{m\times n}$ is equal to the lower of two numbers m, n i.e. when the following relation holds

$$\operatorname{rank}([A]_{m\times n}) = \min(m, n) \tag{5.117}$$

It holds in the case when no linear dependence exists between any two rows or any two columns of the matrix.

Example

Consider an over-specified system of three equations with two unknown variables written in a matrix form

$$\begin{bmatrix} 1 & 1 \\ 1 & 2 \\ 1 & 3 \end{bmatrix} \begin{bmatrix} x \\ y \end{bmatrix} = \begin{bmatrix} 3 \\ 5 \\ 5 \end{bmatrix}$$

The reader is encouraged to make some guesses as to the possible solutions by ignoring one equation at a time. If the third line of the matrix is ignored the solution reads: $x = 1$, $y = 2$; without the second line we have $x = 2$, $y = 1$; and ignoring the first line gives: $x = 5$, $y = 0$. If we use the least-squares method we obtain

$$\begin{bmatrix} 1 & 1 & 1 \\ 1 & 2 & 3 \end{bmatrix} \begin{bmatrix} 1 & 1 \\ 1 & 2 \\ 1 & 3 \end{bmatrix} = \begin{bmatrix} 3 & 6 \\ 6 & 14 \end{bmatrix} \text{ and } \begin{bmatrix} 1 & 1 & 1 \\ 1 & 2 & 3 \end{bmatrix} \begin{bmatrix} 3 \\ 5 \\ 5 \end{bmatrix} = \begin{bmatrix} 13 \\ 28 \end{bmatrix}$$

i.e. the normal Gauss's equation has the form

$$\begin{bmatrix} 3 & 6 \\ 6 & 14 \end{bmatrix} \begin{bmatrix} x \\ y \end{bmatrix} = \begin{bmatrix} 13 \\ 28 \end{bmatrix}$$

which gives the solution $x = 2\frac{1}{3}$, $y = 1$. We can verify the square of the Euclidean norm (5.102) applied to the above normal Gauss' equation, i.e. we express the sum of the squares of differences between the left-hand and right-hand sides of this equation.

$$(2\tfrac{1}{3} + 1 - 3)^2 + (2\tfrac{1}{3} + 2 - 5)^2 + (2\tfrac{1}{3} + 3 - 5)^2 = \tfrac{2}{3}$$

The method guarantees that the value 2/3 is the smallest of all values given by the combination of pairs x, y satisfying the basic over-specified equation. To verify this we calculate the squares of the norm for each solution

for $x = 1$ and $y = 2$: $(1 + 2 - 3)^2 + (1 + 4 - 5)^2 + (1 + 6 - 5)^2 = 4 > 2/3$

for $x = 2$ and $y = 1$: $(2 + 1 - 3)^2 + (2 + 2 - 5)^2 + (5 + 0 - 5)^2 = 1 > 2/3$

for $x = 5$ and $y = 0$: $(5 + 0 - 3)^2 + (5 + 0 - 5)^2 + (5 + 0 - 5)^2 = 4 > 2/3$

In many design synthesis such solutions are sought with additional conditions imposed. They may concern, for instance, the desired structural symmetry of the object under consideration. The solution by the least-squares method, while minimizing the Euclidean norm, does not of course promise to fulfil any such extra conditions. In order to include these conditions, a suitable enlarged system of equations has to be solved. The following example shows a suitable procedure.

Example
An over-specified system of four algebraic equations with three unknown quantities is given

$$\begin{bmatrix} 1 & 1 & 1 \\ 1 & 2 & 3 \\ 1 & 3 & 4 \\ 1 & 4 & 6 \end{bmatrix} \begin{bmatrix} x \\ y \\ z \end{bmatrix} = \begin{bmatrix} 3.1 \\ 6.2 \\ 8.3 \\ 11.2 \end{bmatrix}$$

and an additional condition $y = z$ is imposed. First, we enquire into the solution without the extra constraint. The normal Gauss's equation is

$$\begin{bmatrix} 4 & 10 & 14 \\ 10 & 30 & 43 \\ 14 & 43 & 62 \end{bmatrix} \begin{bmatrix} x \\ y \\ z \end{bmatrix} = \begin{bmatrix} 28.8 \\ 85.2 \\ 122.1 \end{bmatrix}$$

with the solution $x = 1.55$; $y = 1.0$; $z = 0.9$. The square of the Euclidean norm is equal to 0.01. We shall introduce the condition $y = z$ and we notice that this condition does not strongly contradict the approximate solution for y and z. After adding the third and the second columns of the matrix we obtain a new system of four equations with two unknown quantities

$$\begin{bmatrix} 1 & 2 \\ 1 & 5 \\ 1 & 7 \\ 1 & 10 \end{bmatrix} \begin{bmatrix} x \\ y \end{bmatrix} = \begin{bmatrix} 3.1 \\ 6.2 \\ 8.3 \\ 11.2 \end{bmatrix}$$

the normal Gauss's equation of which is

$$\begin{bmatrix} 4 & 24 \\ 24 & 178 \end{bmatrix} \begin{bmatrix} x \\ y \end{bmatrix} = \begin{bmatrix} 28.8 \\ 207.3 \end{bmatrix}$$

and leads to the solution $x = 1.1118$; $y = 1.0147$. The square of the Euclidean norm is 0.01265 which is not considerably higher than the 0.01 determined for the system without the additional constraint.

Singular value decomposition
Recently, the least-squares method has been enhanced and improved by the use of a special decomposition of the rectangular $m \times n$ matrix $[A_{\mu\nu}]$, see Sikora *et al.* [5.61]. We consider once again, see Section 5.4.4, the square $n \times n$ matrix $[B_{\omega\nu}]$ obtained from $[A_{\mu\nu}]$:

$$[B_{\omega\nu}] = [A_{\omega\nu}]^{\mathrm{T}} \cdot [A_{\mu\nu}] \tag{5.118}$$

The single value decomposition consists of replacing the matrix $[A_{\mu\nu}]$ by the product of three matrices

$$[A_{\mu\nu}] = [U_{\mu\mu'}]\cdot[S_{\mu'\nu'}]\cdot[V_{\nu'\nu}]^{\mathrm{T}} \tag{5.119}$$

The matrix $[U_{\mu\mu'}]$ is an orthogonal $m \times m$ matrix, the columns of which are the eigenvectors of the transposed matrix $[B_{\nu\omega}]^{\mathrm{T}}$; the matrix $[V_{\nu\nu'}]$ is an orthogonal $n \times n$ matrix, the columns of which are the orthonormal eigenvectors of $[B_{\omega\nu}]$; the matrix $[S_{\mu'\nu'}]$ is the diagonal matrix containing not more than n (if $n < m$), and in practice $k < n$, non-zero diagonal coefficients, or containing not more than m (if $n > m$), and in practice $l < m$ non-zero diagonal coefficients:

$$[S_{\mu'\nu'}] = \left[\begin{array}{cccc} \sigma_1 & & & \\ & \sigma_2 & & \\ & & \ddots & \\ & & & \sigma_n \\ \hline & & [0] & \end{array}\right] \quad \text{or} \quad [S_{\mu'\nu'}] = \left[\begin{array}{cccc|c} \sigma_1 & & & & \\ & \sigma_2 & & & \\ & & \ddots & & [0] \\ & & & \sigma_m & \end{array}\right] \tag{5.120}$$

where

$$\sigma_1 \geqslant \sigma_2 \geqslant \cdots \geqslant \sigma_k > \sigma_{k+1} = \cdots = \sigma_n = 0$$

or

$$\sigma_1 \geqslant \sigma_2 \geqslant \cdots \geqslant \sigma_1 > \sigma_{l+1} = \cdots = \sigma_m = 0$$

The coefficients σ_i of $[S_{\mu'\nu'}]$ are called the *singular values* of $[A_{\mu\nu}]$. They can be obtained as the square roots of the eigenvalues of the matrix $[B_{\omega\nu}]$ given by eqn (5.118).

The solution of the matrix equation by the use of the least-squares method with the singular value decomposition leads to the following result

$$[g_\nu] = [A_{\nu\mu}]^{+}\cdot[\tilde{u}_\mu] \tag{5.121}$$

where

$$[A_{\nu\mu}]^{+} = [V_{\nu\nu'}]\cdot[S_{\nu'\mu'}]^{+}\cdot[U_{\mu'\mu}]^{\mathrm{T}},$$

and

$$[S_{\nu'\mu'}]^{+} = \left[\begin{array}{ccccccc} \frac{1}{\sigma_1} & & & & & & \\ & \frac{1}{\sigma_2} & & & & & \\ & & \ddots & & & & \\ & & & \frac{1}{\sigma_k} & & & \\ & & & & \ddots & & \\ & & & & & 0 & \\ & & & & & & \ddots \\ & & & & & & 0 \\ \hline & & & [0] & & & \end{array}\right]$$

or

$$[S_{\nu'\mu'}]^{+} = \left[\begin{array}{cccccc:c} \frac{1}{\sigma_1} & & & & & & \\ & \frac{1}{\sigma_2} & & & & & \\ & & \ddots & & & & \\ & & & \frac{1}{\sigma_1} & & & [0] \\ & & & & \ddots & & \\ & & & & & 0 & \\ & & & & & \ddots & \\ & & & & & & 0 \end{array}\right]$$

In the above formulation the given elements σ_k (or σ_1) denote the smallest arbitrarily chosen non-zero singular values. If $[S_{\nu'\mu'}]^{+}$ has large dimension, the practical choice of σ_k(or σ_l) is difficult and thus it becomes the crucial stage in the procedure.

5.4.6 Regularization

Many authors have tried to reduce the sensitivity of the least-squares method to the ill-conditioning of the normal Gauss's equation. The principal idea was to add some stabilizing term $\mathcal{O}(g)$ to the functional $\mathfrak{M}(\tilde{u}, g)$, see eqn (5.112). The first attempt was due to Twomey [5.69], who numerically solved the Fredholm's equation of the first order by minimizing the sum of the square of the Euclidean norm together with the second derivative of the solution. The idea was followed by Philips [5.42] and Tihonov [5.68] who developed the fundamentals of this approach, which nowadays is called the *method of regularization*.

Following the method of regularization, a special regularized solution is sought, instead of the solution that cannot be reached directly by the least-square method. Let us form the functional $\mathfrak{M}^{\alpha}[\tilde{u}, g]$ that consists of two terms. The first term is the square of the Euclidean norm as expressed by eqn (5.112); it tends to zero when a good solution is found. The second term is the square of the norm of a special function, which is related to derivatives of g. Let the second term be $\mathbf{L}g$; it has the form

$$\mathbf{L} = w_0 + w_1 \frac{\mathrm{d}}{\mathrm{d}Q} + w_2 \frac{\mathrm{d}}{\mathrm{d}Q^2} + \cdots \tag{5.122}$$

where Q denotes the suitable coordinate, for instance $Q = \xi$. The function $\mathbf{L}g$ belongs to the space $\mathcal{W}$ of the *Sobolev's smooth functions*.

Hence the functional $\mathfrak{M}^{\alpha}[\tilde{u}, g]$ has the following form

$$\mathfrak{M}^{\alpha}[\tilde{u}, g] = \|\mathbf{A}g - \tilde{u}\|^2 + \alpha\|\mathbf{L}g\|^2 \tag{5.123}$$

where α is a chosen real and small number called the *parameter of regularization*. The second term in the functional (5.123) is the *functional of regularization*

$$\Theta[g] = \alpha\|\mathbf{L}g\|^2 \tag{5.124}$$

that can be written as

$$\Theta[g] = \alpha \sum_{i=1}^{M} \int_{\Gamma} w_i [g^{(i)}]^2 \,\mathrm{d}Q \tag{5.125}$$

where: w_i are the weight coefficients and $g^{(i)}$ denotes the ith derivative with respect to the chosen coordinate Q of the solution $g(Q)$. The number M determines the order of regularization. Numerous synthesis problems have been solved by restricting the formulation to the first order, i.e. with the assumption that $w_1 = 1$ and $w_0 = 0$ or $w_0 = 1$. The 0-order regularization has also been taken under consideration.

For the Fredholm's equation of the first kind (for instance for eqn (5.63)) the functional $\mathfrak{M}^\alpha$ can be written as

$$\mathfrak{M}^\alpha[\tilde{u}(P), g(Q)] = \int_{\Omega_0} \left[\int_{\Gamma} K(P,Q)\, g(Q)\,\mathrm{d}Q - \tilde{u}(P) \right]^2 \mathrm{d}P + \\ + \alpha \int_{\Gamma} \{ w_0 [g(Q)]^2 + w_1 [g'(Q)]^2 \} \,\mathrm{d}Q \tag{5.126}$$

If the discretized form of the Fredholm's equation is given (for instance eqn (5.85)) then $\mathfrak{M}^\alpha$ becomes

$$\mathfrak{M}^\alpha[\tilde{u}_\mu, g_\nu] = \sum_{\mu=1}^{m} \left[\sum_{\nu=1}^{n} A_{\mu\nu} g_\nu - \tilde{u}_\mu \right]^2 + \alpha \sum_{\nu=1}^{n} (w_0 g_\nu^2 + w_1 g_\nu'^2) \tag{5.127}$$

where g_ν' are the values of the derivative of $g(Q)$.

The solution of eqn (5.97) by the use of the regularization method is found by minimizing the functional (5.123) upon a set of functions $g \in \mathcal{W}$. The Euler–Lagrange's equation associated with this functional is

$$\mathbf{A}^*\mathbf{A}g + \alpha(w_0 g - w_1 g'') = \mathbf{A}^*\tilde{u} \tag{5.128}$$

where

$$g'' = \frac{\mathrm{d}^2 g}{\mathrm{d}g^2}$$

and $\mathbf{A}^*$ is the integral operator that is conjugate to $\mathbf{A}$.

Let us discuss the problem described by eqn (5.63). If the operator $\mathbf{A}$ is expressed in the form

$$\mathbf{A} = \int_{\Gamma} K(P,Q) \ldots \mathrm{d}Q \qquad \text{for } P \in \Omega_0$$

then $\mathbf{A}^*$ must be

$$\mathbf{A}^* = \int_{\Omega_0} K(Q,P) \ldots \mathrm{d}P \qquad \text{for } Q \in \Gamma$$

The combined operator $\mathbf{A}^* \cdot \mathbf{A}$ occurring in Euler–Lagrange's functional (5.128) has the following kernel

$$\bar{K}(Q,R) = \int_{\Omega_0} K(P,Q)\, K(P,R)\, \mathrm{d}P \qquad \text{for } R,\ Q \in \Gamma \tag{5.129}$$

and, as a consequence, the Euler-Lagrange's equation, which represents the synthesis problem given by eqn (5.63) and formulated with the use of the regularization approach, becomes

$$\int_{\Gamma} \bar{K}(Q,R)\, g(R)\, \mathrm{d}R + \alpha[w_0 g(Q) - w_1 g''(Q)] = \tilde{v}(Q) \tag{5.130}$$

where

$$\tilde{v}(Q) = \mathbf{A}^* \tilde{u}(P) = \int_{\Omega_0} K(Q,P)\, \tilde{u}(P)\, \mathrm{d}P \qquad \text{for } Q \in \Gamma$$

Equation (5.130) is an integral–differential equation. To solve it the boundary conditions for $g''(Q)$ should be chosen. They could be formulated in an arbitrary manner, but suitable in the physical sense for the problem under consideration. In this way the solution can be influenced by designers themselves, providing they can predict suitable types of solutions of the problem.

The extremum of the functional $\mathfrak{M}^\alpha$ depends on the chosen value of the parameter of regularization α. It is clear that for $\alpha = 0$ we deal with the problem in its natural, i.e. not regularized, form which leads in many cases, to an unstable solution. The larger α is the more stable solutions may be achieved. Moreover, the solutions are also smooth like the corresponding distributions of the field quantities occurring in physical objects. For very large values of α the solutions are very stable, but unfortunately larger errors will be introduced.

If we prefer to use the discretized form of the synthesis equation, for instance eqn (5.85), Euler–Lagrange's equation for the functional (5.127) is

$$\sum_{\mu=1}^{m} \sum_{\nu=1}^{n} A_{\omega\mu} A_{\mu\nu} g_\nu + \alpha(w_0 g_\omega - w_1 g''_\omega) = \sum_{\mu=1}^{m} A_{\omega\mu} \tilde{u}_\mu \tag{5.131}$$

where $\omega = 1, 2, \ldots, n$ and the second derivative g''_ω occurs in this equation as the central finite difference of the second order

$$g''_\omega = \frac{g_{\omega+1} - 2g_\omega + g_{\omega-1}}{h^2} \tag{5.132}$$

where h denotes the step of discretization.

The matrix form of eqn (5.131) reads

$$[A_{\omega\mu}]^{\mathrm{T}} \cdot [A_{\mu\nu}] \cdot [g_{\nu}] + \alpha(w_0[g_{\omega}] - w_1[g''_{\omega}]) = [A_{\omega\mu}]^{\mathrm{T}}[\tilde{u}_{\mu}] \quad (5.131a)$$

where $[g''_{\omega}]$ is a band matrix, with the band width consisting of three terms

$$\frac{g_{\omega-1}}{h^2}, \quad \frac{-2g_{\omega}}{h^2}, \quad \frac{g_{\omega+1}}{h^2},$$

where $[g_{\omega}]$ is a diagonal matrix.

All discussion in this section refers to problems formulated by the Fredholm's equation of the first kind that describes the inverse problem to Laplace's eqn (5.74). For problems inverse to Poisson's equation, expressed for instance by Fredholm's eqn (5.74), for determining the unknown source function $f(P)$ in Ω at the given homogeneous Dirichlet's conditions, we have the following equations

the functional:

$$\mathfrak{M}^{\alpha}[\tilde{u}(P), f(Q)] = \int_{\Omega_0} \left[\int_{\Omega} K(P,Q) f(Q)\, \mathrm{d}Q - \tilde{u}(P) \right]^2 \mathrm{d}P + $$

$$+ \alpha \int_{\Omega} \{ w_0[f(Q)]^2 + w_1[f'(Q)]^2 \}\, \mathrm{d}Q \quad (5.133)$$

and Euler–Lagrange's equation

$$\int_{\Omega} \bar{K}(Q,R) f(R)\, \mathrm{d}R + \alpha[w_0 f(Q) - w_1 f''(Q)] = \tilde{v}(Q) \quad (5.134)$$

where

$$\tilde{v}(Q) = \int_{\Omega_0} K(Q,P)\, \tilde{u}(P)\, \mathrm{d}P \qquad \text{for } Q \in \Omega$$

5.4.7 Parameter of regularization

Solutions obtained from eqns (5.130) or (5.134) are not unique because they depend on the parameter of regularization α. The choice of α is a particular procedure that is associated with the techniques of synthesis. In many problems, the examples of which we will present in Section 5.5, the value of α has been chosen by numerical experiment. The idea is to establish an error function $\varepsilon = f(\alpha)$ which has a clear minimum. The value of α that corresponds to that minimum is considered as optimum, and the solution of the regularized problem at $\alpha = \alpha_{opt}$ is accepted as the best one.

The error function ε, also known as the *index of quality*, can be defined as the norm

$$\|u^{\alpha} - \tilde{u}\| = \|\mathbf{A}g^{\alpha} - \tilde{u}\| \quad (5.135)$$

where g^{α} denotes the approximate solution obtained from the Euler–Lagrange's equation minimizing the functional $\mathfrak{M}^{\alpha}[\tilde{u}, g]$. When the problem is discretized the Euclidean norm can be effectively accepted as the error function

$$\varepsilon = \varrho_{\mathcal{U}}([A_{\mu\nu}]\cdot[g_\nu^\alpha],\ [\tilde{u}_\mu]) = \sqrt{\left(\sum_{\mu=1}^{m}([A_{\mu\nu}]\cdot[g_\nu^\alpha] - [\tilde{u}_\mu])^2\right)} \quad (5.136)$$

where $[g_\nu^\alpha]$ is the solution matrix of eqn (5.131a) obtained for arbitrarily chosen values of α.

Other possibilities of defining the error function are to use either a gradient of the scalar potential or a curl of the vector potential. When, for instance, $-grad_x u = V_x$ is determined in Ω_0, then the error function may take the form

$$\varepsilon = \varrho_{\mathcal{U}}(grad_x \mathbf{A}_g^\alpha,\ \tilde{V}_{\mathbf{x}}) \quad (5.137)$$

A generalized approach to the choice of α_{opt}, which does not require a significant number of solutions to determine the curve $\varepsilon(\alpha)$, has also been proposed, see Rudnicki [5.50]. The procedure is called the *generalized discrepancy method*. In this method a *generalized discrepancy function* $\varrho(\alpha)$ is introduced into the equation

$$\mathbf{A}g = \tilde{u} \quad (5.138)$$

In order to determine $\varepsilon(\alpha)$, two positive numbers are chosen. The first expresses the maximum error of the discretized operator, i.e. the *discrepancy* of the operator

$$\|A_{\mu\nu} - \mathbf{A}\| \leqslant h \quad (5.139)$$

and, in a similar manner, the second number describes the maximum discrepancy of the vector on the right-hand side of eqn (5.138):

$$\|\tilde{u}_\mu - \tilde{u}\| \leqslant \delta \quad (5.140)$$

The value of h depends on the discretization that has been applied. If, for instance, the quadrature is used, see Section 5.4.2, the residue R occurring in eqn (5.84) can be accepted as h. The value of δ has to be evaluated. For identification synthesis problems the value of δ is dependent on the accuracy of the measuring instruments which are used to determine the values of $\tilde{u}$. For the design synthesis problems we determine δ as a value dependent on the truncation errors of the computer. The generalized discrepancy function can be described as

$$\varrho(\alpha) = \|A_{\mu\nu}g_\nu^\alpha - \tilde{u}_\mu\|^2 - (\delta + h\|g_\nu^\alpha\|^2) \quad (5.141)$$

It has been shown that $\varrho(\alpha)$ is monotonous and it has one real root only, which can be interpreted as α_{opt}. The solution of a synthesis problem with the use of the generalized discrepancy method consists of determining this root in any approximate manner, e.g. using the bisection method. We start by choosing a rather large value of $\alpha = \alpha_0$ and then solve the regularized equation with α_0 as the 0-order approximation that leads to the calculated value $\varrho(\alpha_0)$. Next we put $\alpha_1 = \alpha_0/2$ and compute the solution of the 1-order approximation and calculate $\varrho(\alpha_1)$. We follow this procedure until $\varrho(\alpha)$ takes a negative value. Finally, we increase α slightly so that $\varrho(\alpha) = 0$.

5.5 EXAMPLES OF SOLVED PROBLEMS

5.5.1 Synthesis by searching for boundary conditions

Let us briefly present a problem which was investigated by Sikora and Palka [5.63], in order to examine the effectiveness of the regularization techniques applied to a potential field problem. Let the region Ω be a half-plane $y > 0$ on which Laplace's equation holds. The boundary Γ is the x-axis. The synthesis controlled subdomain Ω_0 is a line that is limited by points $(x_2, y_1) = (0.2; 1)$ and $(x_4, y_1) = (0.8; 1)$, see Fig. 5.12. The electric potential $\phi = V = 1$ V is to be generated by the boundary potential $\phi(x) = g(x)$ on Γ, i.e. on the x-axis (the Dirichlet boundary conditions). The following condition is imposed on the unknown function $g(x)$

$$g(x) \begin{cases} = 0, & \text{if } x < x_1 \quad \text{or} \quad x > x_5 \\ \neq 0, & \text{if } x \in (x_1, x_5) \end{cases}$$

at $x_1 = 0$ and $x_5 = 1$.

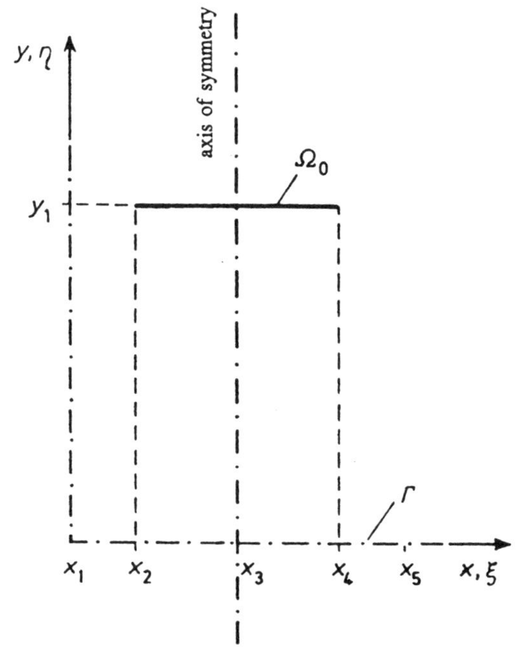

Fig. 5.12 *Synthesis of the potential on* Ω_0

To use the Fredholm's equation, a surface Green's function for a half-space (see eqn (5.57)) should be taken as the kernel, so that

$$\int_{x_1}^{x_5} K(x,\xi)\, g(\xi)\, \mathrm{d}\xi = \tilde{V}, \qquad x \in (x_2, x_4) \tag{5.142}$$

where:

$$K(x,\xi) = \frac{y_1}{\pi[(x-\xi)^2 + y_1^2]} \tag{5.143}$$

Equation (5.142) can be handled by the 1-order regularization at $w_0 = 0$ and $w_1 = 1$, i.e. the following integral–differential equation is to be solved

$$\int_{x_1}^{x_5} \bar{K}(\xi,\xi')\, g(\xi')\, \mathrm{d}\xi' - \alpha g''(\xi) = \tilde{v}(\xi) \tag{5.144}$$

with the composed kernel, see formula (5.129)

$$\bar{K}(\xi,\xi') = \int_{x_2}^{x_4} \frac{y_1^2\, \mathrm{d}x}{\pi^2\{[(x-\xi)(x-\xi')]^2 + y_1^2[(x-\xi)^2 + (x-\xi')^2] + y_1^4\}}$$

$$\tilde{v}(\xi) = \int_{x_2}^{x_4} \frac{y_1}{\pi[(\xi - x)^2 + y_1^2]}\, \tilde{V}\, \mathrm{d}x$$

The following boundary condition is chosen for eqn (5.144): $g'(x_1) = g'(x_5) = 0$. The error function is defined as

$$\varepsilon = \int_{x_2}^{x_4} \left[\int_{x_1}^{x_5} K(x,\xi')\, g^{\alpha}(\xi')\, \mathrm{d}\xi' - \tilde{V} \right]^2 \mathrm{d}x \tag{5.145}$$

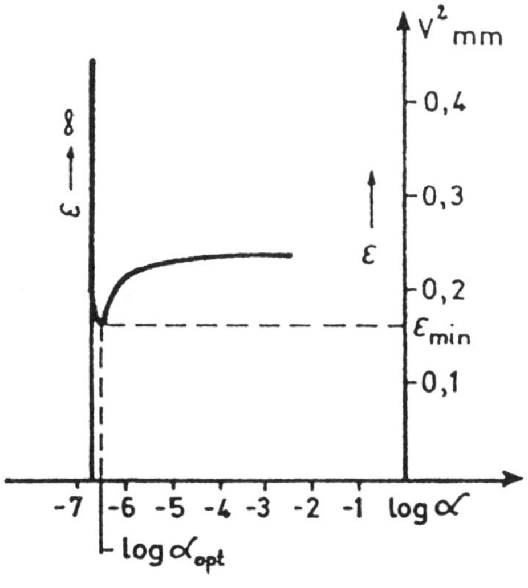

Fig. 5.13 *The error as a function of the regularization parameter*

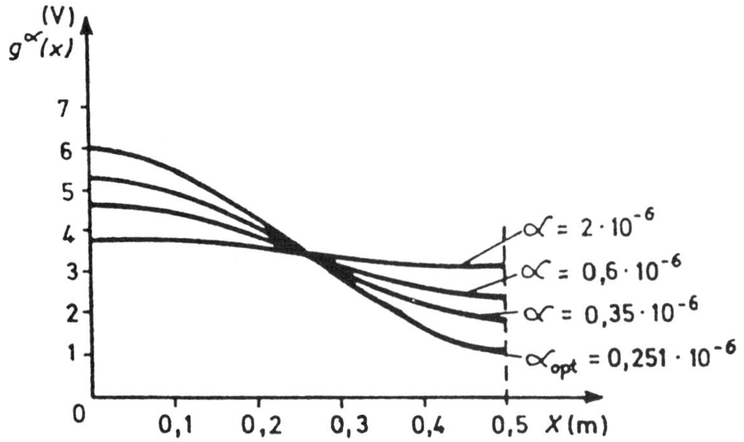

Fig. 5.14 *Boundary potential – the solution of the synthesis problem*

The region under consideration was discretized by using a grid with a mesh size 0.02×0.02 m then eqn (5.144) was numerically solved several times for different values of the parameter of regularization α. The distribution of the error function $\varepsilon(\alpha)$ is shown in Fig. 5.13. The minimum of this function is $\varepsilon_{min} = 0.17\ \mathrm{V}^2\,\mathrm{mm}$ corresponding to $\alpha_{opt} = 0.251 \times 10^{-6}$.

Figure 5.14 demonstrates the computed boundary potential on the x-axis; due to symmetry with respect to $x_3 = 0.5$ m only half of the interval (x_1, x_5) is shown. The boundary potential function corresponding to α_{opt} is characterized by the largest variation. It yields the synthesized potential shown in Fig. 5.15 of minimum deviation from the prescribed value 1 V, that reaches

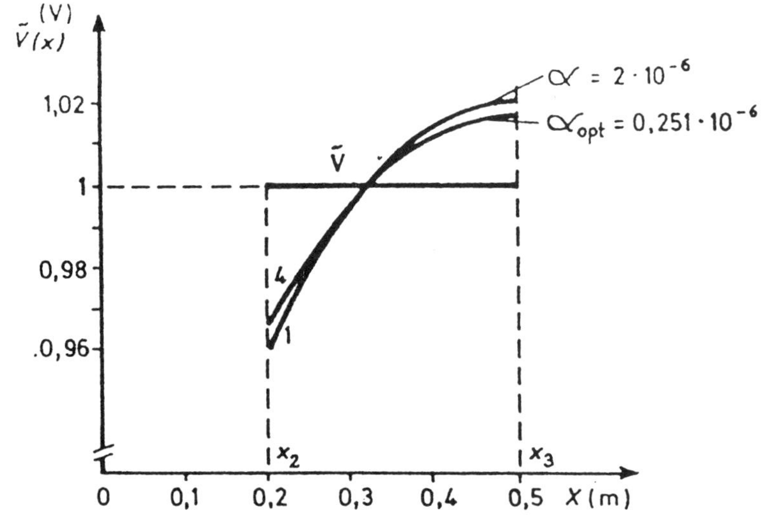

Fig. 5.15 *The synthesized electric potential*

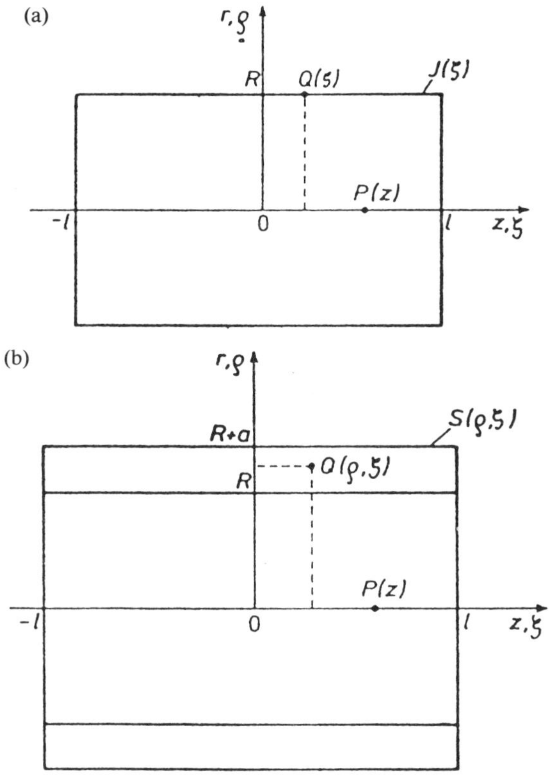

Fig. 5.16 *Air-cored coil: (a) with an infinitely thin winding tape, (b) with a winding tape of finite thickness*

3.4% at the end of the Ω_0 line. At $\alpha = 0.2 \times 10^{-6}$ the solutions become unstable.

5.5.2 *An air-cored coil*

Let us present the problem of synthesis of the magnetic field strength in an air-cored coil. The problem shown in Fig. 5.16 has been investigated by Adamiak [5.2] and Sikora *et al.* [5.64]. It has be formulated in two versions, both postulating the magnetic field strength **H** to have a unique z-component $H_z(z) = constant$ on the coil axis along the whole length $L = 2l$ of the coil. This field is to be achieved by:

(1) the distribution of a d.c. current-sheet $J(z)$ localized on the outer cylindrical surface of radius R, or
(2) the distribution of a d.c. current density $S(z)$ in the winding tape of thickness to a.

Both formulations represent an open synthesis problem in which Ω is a 3D space, Ω_0 is the synthesis controlled subdomain confined to a part of the

z-axis (from $-l$ to l) and Ω_s is the field source subdomain which is a cylinder of radius r, length $2l$ and wall thickness which is either infinitesimal or equal to a.

To obtain Fredholm's equation we could use the surface Green's function for the inside of the cylinder, see eqn (5.61). Another approach is to apply Biot–Savart's rule expressing the magnetic field strength $\mathrm{d}H_z(z)$ at the point P(z) on the z-axis, due to electric current $\mathrm{d}I = J(\zeta)\,\mathrm{d}\zeta$ of the current ring at the point $Q(\zeta)$:

$$\mathrm{d}H_z(z) = \frac{R^2}{2\sqrt{[R^2 + (z-\zeta)^2]^3}} \tag{5.146}$$

Thus, we have for variant (1)

$$\int_{-1}^{1} \frac{R^2}{2\sqrt{[R^2 + (z-\zeta)^2]^3}} J(\zeta)\,\mathrm{d}\zeta = \tilde{H}_z(z), \qquad z \in (-1, 1) \tag{5.147}$$

and for variant (2)

$$\int_{-l}^{l}\int_{R}^{R+a} \frac{r^2}{2\sqrt{[r^2 + (z-\zeta)^2]^3}} S(\zeta)\,\mathrm{d}r\,\mathrm{d}\zeta = \tilde{H}_z(z), \qquad z \in (-1, 1) \tag{5.148}$$

When the first order regularization is applied with $w_0 = 1$ and $w_1 = 1$, the following equations with the complex kernel are obtained:

for case (1)

$$\int_{-l}^{l} \bar{K}(\zeta, \zeta')\, J(\zeta')\,\mathrm{d}\zeta' + \alpha\,[J(\zeta) - J''(\zeta)] = \int_{-l}^{l} K(z, \zeta)\,\tilde{H}_z(\zeta)\,\mathrm{d}z \tag{5.149}$$

$$\bar{K}(\zeta, \zeta') = \int_{-l}^{l} K(z, \zeta)\,K(z, \zeta')\,\mathrm{d}z \tag{5.150}$$

for case (2)

$$\int_{l}^{l} \bar{K}_a(\zeta, \zeta')\, S(\zeta')\,\mathrm{d}\zeta' + \alpha[S(\zeta) - S''(\zeta)] = \int_{-l}^{l} K_a(z, \zeta)\,\tilde{H}_z(\zeta)\,\mathrm{d}z \tag{5.151}$$

$$\bar{K}_a(\zeta, \zeta') = \int_{-1}^{l} K_a(z, \zeta)\,K(z, \zeta')\,\mathrm{d}z \tag{5.152}$$

The boundary conditions that are either $J'(-l) = J'(l) = 0$ or $S'(-l) = S'(l) = 0$ lead to the solutions presented in Figs 5.17 and 5.18. It can be seen from Fig. 5.17a and 5.17b that the maximum magnetic field uniformity of about 99% is achieved at $\alpha = 10^{-9}$. Similar results apply to the coil of finite thickness tape shown in Fig. 5.18a and 5.18b. In both cases the necessary distribution of current must be precisely followed. In practice, the coil has to

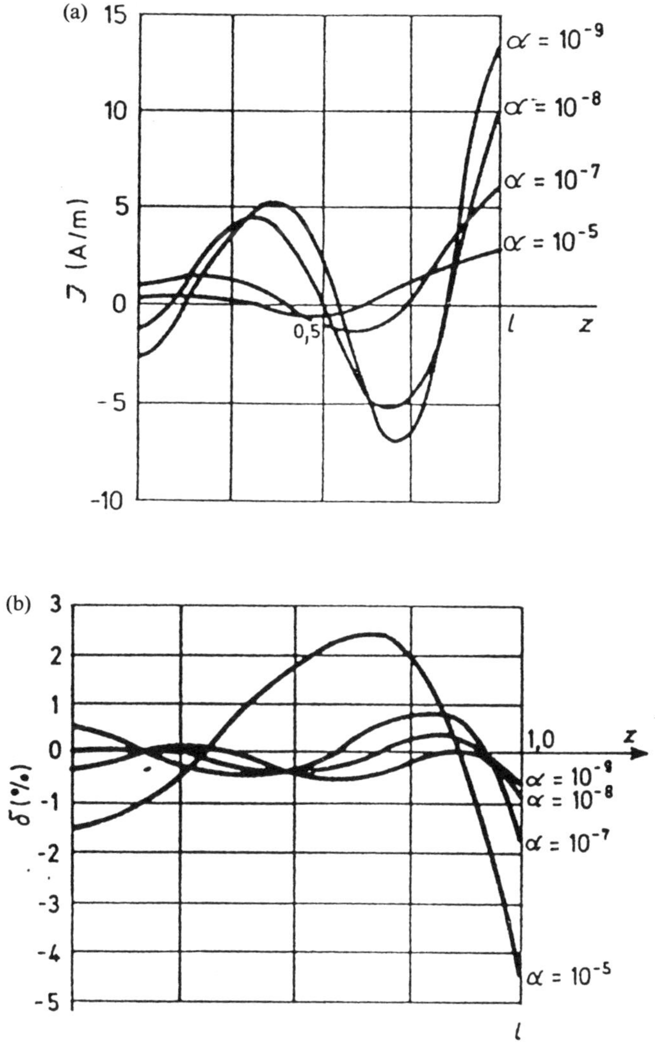

Fig. 5.17 *(a) Current-sheet distribution of the coil with an infinitely thin winding tape, (b) relative error of the magnetic field strength*

be partitioned into a number of sections, each of them supplied by a current of some average value which is determined from the distributions shown in Figs 5.17 or 5.18.

5.5.3 Boundary conditions for a transmission line

In this Section the synthesis problem inverse to the diffusion equation will be presented following the paper by Sikora and Palka [5.62]. Let us consider the transmission line equation for the voltage u:

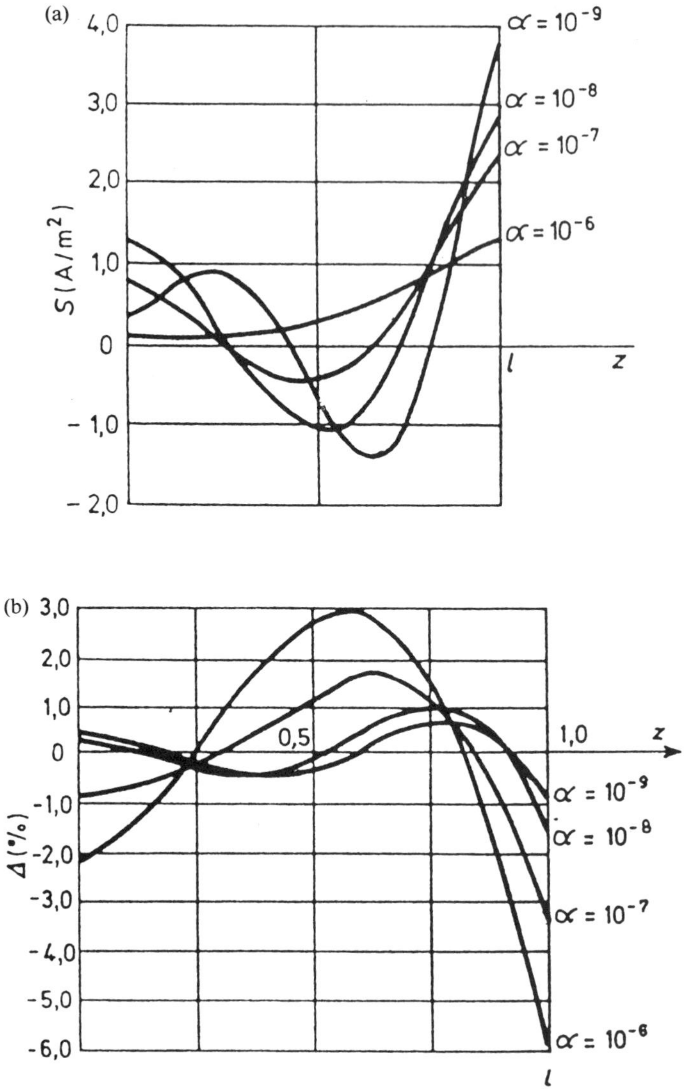

Fig. 5.18 *(a) Current-sheet distribution of the coil with a finite thin winding tape, (b) relative error of the magnetic field strength*

$$\frac{\partial^2 u}{\partial x^2} = \frac{1}{\beta^2}\frac{\partial u}{\partial t} \tag{5.153}$$

where x determines the distance from the beginning of the line, t denotes the time and $B = 1/\sqrt{(RC)}$ where R and C are the resistivity and capacity of the line, respectively, both expressed per unit length of the line.

The following identification synthesis problem is considered. Suppose at the distance x_1 the time-dependent voltage $u(x_1, t) = \tilde{u}(t)$ has been measured.

The problem is to determine the Dirichlet boundary conditions $u(0, t) = g(t)$ at the beginning of the line, with the assumption that the initial conditions for the whole line are zero, i.e. $u(x, 0) = 0$. The complete formulation of the problem is therefore as follows:

(1) the given voltage function at the point x_1 is $u(x_1, t) = \tilde{u}(t)$ for $t \geqslant 0$,
(2) the initial conditions are $u(x, 0) = 0$ for $x \geqslant 0$,
(3) the boundary conditions: $u(0, t) = g(t)$ for $t \geqslant 0$ should be determined.

The problem is described by Volterra's equation of the first kind

$$\int_0^t \tilde{K}(t, \tau)\, g(\tau)\, d\tau = \tilde{u}(t) \tag{5.154}$$

with the kernel

$$\tilde{K}(t, \tau) = \frac{x_1}{2\beta\sqrt{[\pi(t-\tau)^3]}} \exp\left[\frac{-x_1^2}{4\beta(t-\tau)}\right].$$

having the singularity at $t = \tau$. Equation (5.154) can be reduced to the equivalent Fredholm's equation of the first kind

$$\int_0^T K(t, \tau)\, g(\tau)\, d\tau = \tilde{u}(t) \tag{5.155}$$

with the kernel

$$K(t, \tau) = \begin{cases} \tilde{K}(t, \tau) & \text{for } 0 \leqslant \tau < t \\ 0 & \text{for } t \leqslant \tau \leqslant T \end{cases}$$

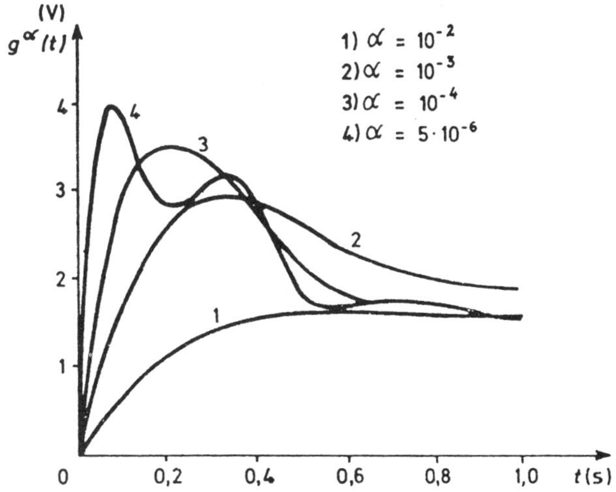

Fig. 5.19 *Time-dependent function of the voltage at the beginning of the line*

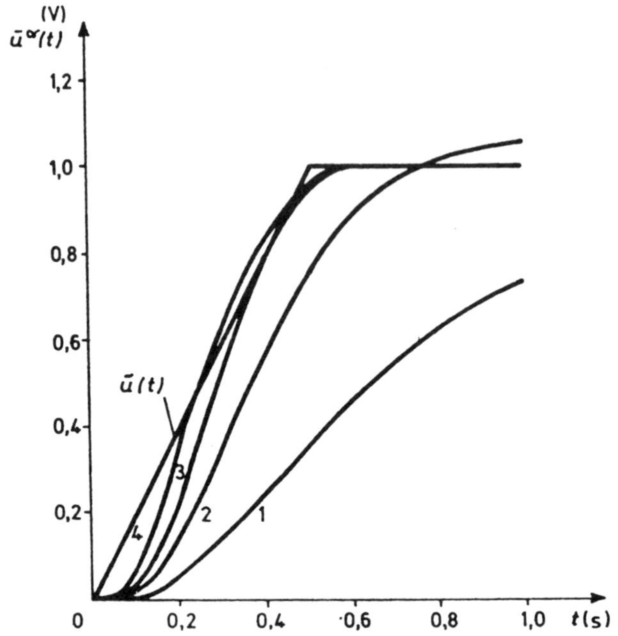

Fig. 5.20 *Time-dependent function of the voltage at x_l, $\tilde{u}(t)$ is the measured function, 1–4 are recalculated functions*

where T is the time duration of the whole process. The solutions of the problem obtained by the use of the regularization are presented in Figs 5.19 and 5.20 for the following measured voltage, in volts

$$\tilde{u}(t) = \begin{cases} 2t/T, & \text{for } 0 \leqslant t < T/2 \\ 1, & \text{for } T/2 \leqslant t \leqslant T \end{cases}$$

Equation (5.155) has been discretized and then numerically solved for the following data: $R = 1\,\Omega\,\mathrm{m}^{-1}$, $c = 1\,\mathrm{F\,m}^{-1}$, $x_l = 1$ m. Figure 5.19 shows the voltage $u(0, t) = g^{\alpha}(t)$ traced at four chosen values of the regularization parameter, and Fig. 5.20 presents the recalculated voltage function $\tilde{u}^{\alpha}(t)$ that is expected to be close to $\tilde{u}(t)$. It can be seen that at $\alpha = 5 \times 10^{-6}$ the function $\tilde{u}^{\alpha}(t)$ (curve no. 4) approaches closely the $\tilde{u}(t)$ and, in consequence, curve 4 on Fig. 5.19 is the best solution for the voltage function at the beginning of the line. For smaller values of α the function $g^{\alpha}(t)$ becomes unstable and this is the main feature of an inverse problem that is ill-posed in the Hadamard sense.

5.5.4 Other problems

For an increased treatment of solved synthesis problems pertaining to electrical technology the reader may wish to refer to the papers published in majority in the *IEEE Transactions on Magnetics* and in other journals.

An identification synthesis is reported by Iwamura and Miya [5.18] and Lord *et al.* [5.24]. A shape design of an electromagnet is presented by Armstrong *et al.* [5.4], Arumugam *et al.* [5.5], Di Barba *et al.* [5.8], Marrocco and Pironneau [5.25], Nakata and Takahashi [5.30] and Siebold *et al.* [5.58]. A design of various electric devices is reported by Preis *et al.* [5.43], Ratnajeevan and Hoole [5.48], Rudnicki *et al.* [5.50], Russenschuck [5.52] and Sikora [5.59]. The use of synthesis techniques for a design of NMR tomography is presented by Gottvald [5.11], Miyata *et al.* [5.28] and Mustarelli *et al.* [5.29]. For an optimization strategy pertaining to the synthesis problems the reader may wish to refer to the papers by Bellina *et al.* [5.6], Grago *et al.* [5.6]. Gitosusastro *et al.* [5.10], Gottvald *et al.* [5.12], Kasper [5.20], Michalski and Sikora [5.26], Nakata *et al.* [5.31], Osama [5.32] and Saldanha *et al.* [5.54].

6
Coupled Fields

Janusz Turowski

6.1 TYPES OF COUPLED FIELDS. STRONG AND WEAK COUPLING

By the term *coupled fields* we understand a system of directly interacting physical fields: electric, magnetic, mechanical forces and stresses, thermal, gravitational fields and flow field of currents, liquids and gaseous conducting media, plasma and similar. Coupled fields also encompass the interaction of electromagnetic reaction of secondary circuits to primary fields. The latter is especially evident when changes of forces and torques take place at the start, reverse, braking, automatic restart and similar transient conditions of electrical machines. Such transients are usually investigated with the help of Euler–Lagrange's equation (6.23) using equivalent lumped electromagnetic parameters - flux linkages λ_k, inductances L_{ki} and forces f_{ek}, calculated from field analysis. Typical examples of coupled fields include magnetohydrodynamic, magnetogasdynamic and electromechanical levitation fields, radial traction and axial fields of rotating and linear electrical machines, electromechanical fields of transient electrodynamic short circuit forces, transient electromagnetic/thermal fields in skin hardening of steel parts, electromagnetic pumps etc. Such fields can be considered as having a *strong coupling*. They are often associated with the strong eddy-current effect.

In static, quasi-static and slow-varying fields the diferences in time constants between electromagnetic, thermal and mechanical processes are such that the coupling is practically in one direction, i.e. the thermal or mechanical stress sources are of electromagnetic origin, but there is little or no feed-back effect on the primary magnetic field. Such fields, which cover a wide spectrum of technical problems, are said to have a *weak coupling*. These fields can often be investigated independently or regarded as a limiting case for fully coupled fields, after transients have decayed.

The degree of coupling between physical fields, as well as the distinction between strong and weak coupling, can be evaluated with the help of Reynolds number R_M (6.74). For instance a definite electromechanical coupling in the form of levitation of linear motors occurs with high-speed motors at $R_M = 10, \ldots, 25$, whereas a noticeable reaction of induced eddy-currents already exists at $R_M > 0.5$ [6.22].

An electromagnetic field is in its nature the coupling of electric and magnetic fields, which is observed as the propagation of electromagnetic waves. This occurs at higher frequencies and is beyond the scope of this book. The diffusion process in conducting media, on the other hand, is very much the subject of our discussion.

The emphasis of this chapter is on a physical and mathematical description of coupled fields. Once the problem is formulated its solution may be found using one of the methods discussed in other chapters of this book. Numerical solutions are usually very difficult owing to the complexity of the problem formulation. Thus analytical, semi-analytical and empirical solutions are often sought, after suitable approximations and simplifications to the model have been made.

6.2 ELECTRO-MAGNETO-MECHANICAL FIELDS

6.2.1 Hamilton's principle

In the analysis and synthesis of coupled fields we concentrate on investigating the state, motion and energy conversion. In Lagrangian mechanics of charged particles moving in electromagnetic fields, the state of a dynamic system at a given instant of time can be described as a point in a $2n$-dimensional space of n generalized coordinates $q_k = q_k(t)$ and n generalized momenta $p_k = p_k(t)$ or corresponding generalized velocities $\dot{q}_k = \dot{q}_k(t)$ [6.34]. This approach is usually applied to systems with lumped parameters [6.31] or if field effects can be expressed in terms of discrete parameters Ψ_{ij}, M_{ij}, R_{ij}.

The Lagrangian description of interactions of electromagnetic fields with external field sources of charge density ρ or current density J, runs in strong analogy to the description of particles ([6.8], p. 571). The finite number of coordinates $q_k(t)$ and $\dot{q}_k(t)$, where $k = 1, 2, \ldots, n$ is substituted here by an infinite number of degrees of freedom, and the generalized coordinate q_k is substituted by continuous fields $\Phi_i(x)$. Therefore both Hamilton's principle, which according to Hammond ([6.7], p. 31) '... is more general than the law of the conservation of energy in its usual form ...' (see also reference [6.34], p. 31) and following on from it, as the necessary condition for an extremum, Euler-Lagrange's equation, refer to electromagnetic fields as well as to discrete systems with lumped parameters. Both these relationships follow from energy interactions, whereas the system of generalized coordinates depends on the method of calculation of these energies (see Section 6.2.4). We shall start our discussion with the system of lumped parameters.

The conservative (loss-less) kinetic energy T of the system is a quadratic form of the generalized velocities

$$T(q, \dot{q}) = \sum_{i=1}^{n} \sum_{k=1}^{n} p_{ik}(q_1, q_2, \ldots, q_n; t)\dot{q}_i \dot{q}_k \tag{6.1}$$

In the case of nonlinear media, using the more convenient coordinate system (x_k, i_k), it is better [6.34] to employ the so-called kinetic *coenergy* T', evaluated for example from magnetization curves for flux linkage $\psi' = \psi'(i)$ (Fig. 6.1) as

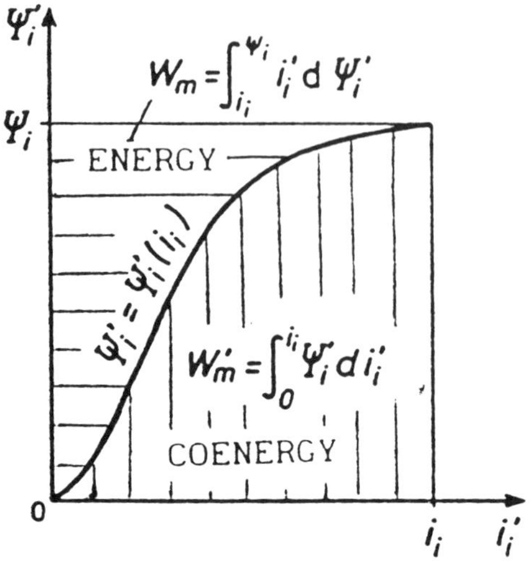

Fig. 6.1 *Magnetic coenergy $W_m' = i_i \psi_i - W_m$ of the ith element of an electromechanical system [6.34]*

$$T'(q, \dot{q}) = \int_{0,\ldots,0}^{\dot{q}_1,\ldots,\dot{q}_n} \sum_{k=1}^{n} p_k'(q_1, \ldots, q_n; \dot{q}_1', \ldots, \dot{q}_n'; t) \, \mathrm{d}\dot{q}_k' \tag{6.2}$$

where primes denote variables of integration.

With the help of the *Legendre transformation* of coordinates [6.34] we obtain a simple relationship between the magnetic energy W_m and magnetic coenergy W_m'. For the ith element

$$W_{mi} + W_{mi}' = \psi_i i_i \tag{6.3}$$

from which it follows that for linear systems $W_m' = W_m$ and $T' = T$. The potential conservative energy V is linear form of general coordinates $q_i(t)$

$$V(q) = \sum_{i=1}^{n} \sum_{k=1}^{n} V_{ik}(q_1, \ldots, q_n; t) \tag{6.4}$$

The most general mathematical description of the transformations and processes mentioned above is *Hamilton's principle* [6.34], which can be derived for a mechanical system from *d'Alambert's principle* and the *principle of virtual work*, using the *calculus of variations*. It comprises not only all kinds of fields but also electric and magnetic circuits with lumped parameters, the movement of discrete charges and masses as well as mechanical bodies and particles.

The Hamilton's variational principle states that the variation δI of the time integral of the Lagrange state function, which is also called the *conservative Lagrangian*

$$L(q, \dot{q}, t) = T - V \tag{6.5}$$

must be zero during motion of a system between fixed end states $q_i(t_1)$ and $q_i(t_2)$, i.e.

$$\delta I = \delta \int_{t_1}^{t_2} L(q_i,\ \dot{q}_i,\ t)\,\mathrm{d}t = 0 \tag{6.6}$$

This means that motion of the system in time from t_1 to t_2 between the two end points $q_i(t_1)$ and $q_i(t_2)$ is accomplished in such a way that the functions $q(t)$ forming a functional

$$I = \int_{t_1}^{t_2} L(q,\ \dot{q},\ t)\,\mathrm{d}t \tag{6.7}$$

make this functional stationary (usually a minimum) [6.34]. Under static conditions and only then, this condition reduces to the principle of minimum potential energy $\partial V/\partial q_i = 0$.

Since

$$\frac{1}{t_1 - t_2} \int_{t_1}^{t_2} (T - V)\,\mathrm{d}t$$

expresses the average value of energy difference $(T - V)$ in time interval $t_2 - t_1$, hence $\delta I = (t_2 - t_1)\,\delta(T - V)_{mean} = 0$ and we can state that nature tends to the equalization of mean kinetic energy with potential energy of a system during its motion.

In the case of nonlinear *dissipative* systems (with power losses and energy sources) we replace the Lagrange function (6.5), also called the *conservative Lagrangian* or kinetic potential, with the *nonconservative Lagrangian*

$$L_{FG} = (T' + T_F') - (V + V_G) = L + (T_F - V_G), \qquad L = T' - V \tag{6.8}$$

which also satisfies the condition of *Hamilton's principle* [6.34], where

$$T_F'(\dot{q}) = \int_0^t F\,\mathrm{d}t$$

is the non-conservative kinetic coenergy of the system (energy of power losses),

$$F(\dot{q}) = \sum_{i=1}^{n} \frac{1}{2} r_i (\dot{q}_i)^2$$

is the *Rayleigh dissipation function*, equal to half the Joule's losses (r_i - resistance, $\dot{q}_i$ - current) or half the frictional losses (r_i - viscous frictional resistance, $\dot{q}_i$ - velocity of the body),

$$V_G(q) = -\int_0^{q_k} G_k(t)\,\mathrm{d}q_k'$$

is the potential energy of non-conservative forces of constraints imposed by external energy sources, and

$$G_k = G_k(t)$$

is the non-conservative force of constraint applied from the sources, e.g. input voltages applied at the terminals of an electric network or externally applied mechanical torques or forces.

From Hamilton's principle many physical laws and principles of mechanics, electrodynamics, electrothermics, thermodynamics, electrochemistry etc. can be derived.

6.2.2 *Euler's equation*

It follows from the previous Section that coupled fields and systems can be investigated by solving a variational problem of the form

$$I = \iint_{\Omega} F(u, u_x, u_y, x, y)\, \mathrm{d}x\, \mathrm{d}y = extremum \tag{6.9}$$

where the integrand F represents the energy or quantity proportional to it, contained in the system with n degrees of freedom. Every variable x and y has a vector in the space $\mathbf{R}^m$ [6.15], and $u = u(x, y) \in \mathbf{R}^n$. Coordinates of the vector $\mathbf{R}^n$ are *general coordinates* of the system. It is assumed that general coordinates and *general velocities* of the system are known at the instant x_0, y_0 and

$$u \in C^1[x_0, y_0; x, y], \qquad u_x = \frac{\partial u}{\partial x}, \qquad u_y = \frac{\partial u}{\partial y}$$

The functional (6.9) has a more general significance than (6.6) and comprises both Hamilton's principle

$$\delta \int_{t_1}^{t_2} \mathrm{d}t = 0$$

and the *principle of least action* [6.23]

$$\delta \int_{x_1}^{x_2} \mathbf{P}\, \mathrm{d}\mathbf{l} = 0$$

The two principles have a different physical interpretation [6.23]. In the case of Hamilton's principle the subject of variation is the path of integration between two points, which is not changing in space and time, and on this path the equations of motion and conservation laws are not observed. In the principle of least action, on the other hand, the end points are fixed in space - but not in time, and on the path being subject of variation the energy conservation law is satisfied.

The task of finding the extremum of the functional (6.9) in region Ω can be resolved with the help of calculus of variation. We are looking for the surface $u(x, y)$ which satisfies the condition (6.9) and boundary conditions $u(C)$. Let us assume that

$$u(x, y) + \alpha\eta(x, y) \tag{6.10}$$

represents an adjacent, sufficiently near surface on which $\eta(x, y)$ is an arbitrary function satisfying the boundary condition $\eta(C) = 0$ and α is a small constant, of positive or negative value, which is the parameter of this surface.

For the surface (6.10) we have

$$I + \delta I = \iint_{\Omega} F(u + \alpha\eta, u_x + \alpha\eta_x, u_y + \alpha\eta_y, x, y)\,\mathrm{d}x\,\mathrm{d}y \tag{6.11}$$

where

$$\eta_x = \frac{\partial\eta}{\partial x}, \qquad \eta_y = \frac{\partial\eta}{\partial y}$$

After expanding (6.11) in powers of the parameter α with the help of Taylor's formula

$$I(\alpha) = I(0) + \alpha I'(0) + \frac{\alpha^2}{2!}I''(0) + \alpha^3 k(a) \tag{6.12}$$

we see, that the necessary condition of existence of the extremum of the functional (6.9) is that the first variation δI must be zero, i.e.

$$\delta I = a\left(\frac{\partial I}{\partial \alpha}\right)_{\alpha=0} = 0 \tag{6.13}$$

The analysis of the second variation $\delta^2 I = \alpha^2 I''(0)$, with the help of Legendre's, Jacobi's and Weierstrass's conditions [6.15], reveals that usually $\delta^2 I \geqslant 0$, which means that the functional attains a minimum.

From eqn (6.13) it follows that

$$\delta I = \alpha \iint_{\Omega} \left(\frac{\partial F}{\partial u}\eta + \frac{\partial F}{\partial u_x}\eta_x + \frac{\partial F}{\partial u_y}\eta_y\right)\mathrm{d}x\,\mathrm{d}y = 0 \tag{6.14}$$

The last two components of eqn (6.14) are transformed using Riemann's formula [6.23]

$$\iint_{\Omega(C)} \left(\frac{\partial Q}{\partial x} - \frac{\partial P}{\partial y}\right)\mathrm{d}x\,\mathrm{d}y = \int_{C(\Omega)} (P\,\mathrm{d}x + Q\,\mathrm{d}y) \tag{6.15}$$

by substituting

$$Q = \frac{\partial F}{\partial u_x}\eta \quad \text{and} \quad -P = \frac{\partial F}{\partial u_y}\eta$$

so that

$$\iint_\Omega \left[\frac{\partial}{\partial x}\left(\frac{\partial F}{\partial u_x}\eta\right) + \frac{\partial}{\partial y}\left(\frac{\partial F}{\partial u_y}\eta\right)\right] \mathrm{d}x\,\mathrm{d}y = \iint_\Omega \left(\eta\frac{\partial}{\partial x}\frac{\partial F}{\partial u_x} + \eta_x\frac{\partial F}{\partial u_x} + \eta\frac{\partial}{\partial y}\frac{\partial F}{\partial u_y} + \eta_y\frac{\partial F}{\partial u_y}\right)\mathrm{d}x\,\mathrm{d}y$$

and thus from eqn (6.15) we find

$$\int_c \eta\left(-\frac{\partial F}{\partial u_y}\mathrm{d}x + \frac{\partial F}{\partial u_x}\mathrm{d}y\right) = \iint_\Omega \eta\left(\frac{\partial}{\partial x}\frac{\partial F}{\partial u_x} + \frac{\partial}{\partial y}\frac{\partial F}{\partial u_y}\right)\mathrm{d}x\,\mathrm{d}y + \iint_\Omega\left(\frac{\partial F}{\partial u_x}\eta_x + \frac{\partial F}{\partial u_y}\eta_y\right)\mathrm{d}x\,\mathrm{d}y \tag{6.16}$$

From the boundary condition $\eta(C) = 0$, the line integral in eqn (6.16) vanishes and after substituting eqn (6.16) into eqn (6.14) we obtain

$$\delta I = \alpha \iint_\Omega \eta\left(\frac{\partial F}{\partial u} - \frac{\partial}{\partial x}\frac{\partial F}{\partial u_x} - \frac{\partial}{\partial y}\frac{\partial F}{\partial u_y}\right)\mathrm{d}x\,\mathrm{d}y = 0 \tag{6.17}$$

Since $\eta = \eta(x, y)$ is an arbitrary function, eqn (6.17) can be satisfied only if the term within the brackets is identically equal to zero. In this way we have obtained the *necessary condition* for the function $u(x, y)$ to minimize the functional (6.9). This condition is known as *Euler's differential equation*

$$\frac{\partial F}{\partial u} - \frac{\partial}{\partial x}\left(\frac{\partial F}{\partial u_x}\right) - \frac{\partial}{\partial y}\left(\frac{\partial F}{\partial u_y}\right) = 0 \tag{6.18}$$

6.2.3 Euler–Lagrange equation

If we express the integrand in eqn (6.9) in the form of Lagrange's state function as

$$F(u, u_x, u_y, x, y) = L(q_k, \dot{q}_k, t) = T' - V \tag{6.19}$$

then eqn (6.9) will represent Hamilton's principle (6.7)

$$\int_{t_1}^{t_2} L(q_k, \dot{q}_k, t)\,\mathrm{d}t = min. \tag{6.20}$$

After the substitution of $x = t$, $y = 0$, $u(x, y) = u(x) = q_k(t)$ and $u_y = 0$, we obtain from eqn (6.18) the well-known *Euler–Lagrange equation for conservative* electromechanical systems

$$\frac{\partial L}{\partial q_k} - \frac{\mathrm{d}}{\mathrm{d}t}\left(\frac{\partial L}{\partial \dot{q}_k}\right) = 0, \qquad k = 1, 2, \ldots, n \tag{6.21}$$

For an electromagnetic field eqn (6.21) takes the form [6.8]

$$\frac{\partial \mathcal{L}}{\partial \phi_1} - \partial^\beta \frac{\partial \mathcal{L}}{\partial (\partial^\beta \phi_i)} = 0 \tag{6.21a}$$

where $\mathcal{L} = \mathcal{L}(\phi_i, \partial^\alpha \phi_i)$ is the density of the Lagrangian, whereas 'coordinates' and 'velocities' are I^α and $\partial^\beta I^\alpha$ respectively; the finite number of k coordinates q_k, $\dot{q}_k$ in eqn (6.21) corresponds to an infinite number of degrees of freedom and a point x^α in the space-time coordinate system. The continuous fields $\phi_i(x)$ contain a discrete indicator $i = 1, 2, \ldots, n$ and a continuous indicator x^α; the generalized velocity $\dot{q}_k$ is replaced by a four-vector $\partial^\beta \phi_i$, which has the form of a gradient; the action integral (6.7) takes the following form for fields

$$I = \iint \mathcal{L}\, \mathrm{d}^3x\, \mathrm{d}t = \int \mathcal{L}\, \mathrm{d}^4x \tag{6.7a}$$

In practical calculations of system dynamics, eqn (6.21) is preferred to eqn (6.21a) and the emphasis is shifted to finding appropriate equivalent lumped parameters M_{ij}, R_{ij} (6.62) of the system.

On the other hand, taking into account the nonconservative Lagrangian (6.8)

$$L_{FG} = (T + T_F) - (V + V_G) \tag{6.22}$$

we obtain the Euler–Lagrange equation for *dissipative* electromechanical systems

$$\frac{\partial L}{\partial q_k} - \frac{\mathrm{d}}{\mathrm{d}t}\left(\frac{\partial L}{\partial \dot{q}_k}\right) - \frac{\partial F}{\partial \dot{q}_k} = -G_k \tag{6.23}$$

where $k = 1, 2, 3, \ldots, n$ is the number of degrees of freedom of the electrical and mechanical systems and other symbols have already been explained. Other equations of electrodynamics can be obtained in a similar way (see references [6.31; 6.32; 6.33]).

6.2.4 Static electric and magnetic fields in isotropic media

Many technical problems are resolved using static equations

$$curl\,\mathbf{H} = \mathbf{J},\; curl\,\mathbf{E} = 0,\; \mathbf{J} = \sigma\mathbf{E},\; \mathbf{D} = \varepsilon\mathbf{E},\; \mathbf{B} = \mu\mathbf{H},\; div\,\mathbf{D} = \rho,\; div\,\mathbf{B} = 0 \tag{6.24}$$

In current-free regions ($curl\,\mathbf{H} = 0$) we have a potential (irrotational or conservative) field

$$\mathbf{E} = -grad\, V_e, \qquad \mathbf{H} = -grad\, V_m \tag{6.24a}$$

The energy of an electric field per unit volume ($\mathrm{J\,m^{-3}}$) is

$$w_e = \frac{ED}{2} = \frac{\varepsilon E^2}{2} = \frac{D^2}{2\varepsilon} = \frac{1}{2}\varepsilon\, grad^2\, V_e = \varepsilon\left(\frac{\partial V_e}{\partial x}\right)^2 + \varepsilon\left(\frac{\partial V_e}{\partial y}\right)^2 + \varepsilon\left(\frac{\partial V_e}{\partial z}\right)^2 \tag{6.25}$$

In a region V containing a charge with the volume density ρ, the energy of the system is

$$W_e = \frac{1}{2}\sum_{i=1}^{n} \iiint_V V_e \rho \, dV = \frac{1}{2}\sum_{i=1}^{n} Q_i V_{ei} \tag{6.26}$$

The energy of a magnetic field per unit volume ($J\,m^{-3}$) is

$$w_m = \frac{HB}{2} = \frac{\mu H^2}{2} = \frac{B^2}{2\mu} = \frac{1}{2}\mu\, grad^2\, V_m = \mu\left(\frac{\partial V_m}{\partial x}\right)^2 + \mu\left(\frac{\partial V_m}{\partial y}\right)^2 + \mu\left(\frac{\partial V_m}{\partial z}\right)^2 \tag{6.27}$$

In terms of the magnetic vector potential $\mathbf{A}$ defined as $\mathbf{B} = curl\,\mathbf{A}$ the energy of the magnetic field can be expressed as

$$w_m = \frac{B^2}{2\mu} = \frac{1}{2\mu} curl^2\,\mathbf{A} = \frac{1}{2\mu}\left[\left(\frac{\partial A_z}{\partial y} - \frac{\partial A_y}{\partial z}\right)^2 + \left(\frac{\partial A_x}{\partial z} - \frac{\partial A_z}{\partial x}\right)^2 + \left(\frac{\partial A_y}{\partial x} - \frac{\partial A_x}{\partial y}\right)^2\right]$$

$$= \frac{1}{2}\mathbf{H}\, curl\,\mathbf{A} = \frac{1}{2}\left[\mathbf{A}\cdot curl\,\mathbf{H} - div\,(\mathbf{H}\times\mathbf{A})\right] \tag{6.27a}$$

which follows from the identity $div\,(\mathbf{H}\times\mathbf{A}) = \mathbf{A}\cdot curl\,\mathbf{H} - \mathbf{H}\cdot curl\,\mathbf{A}$.

In a region containing currents with density $\mathbf{J} = curl\,\mathbf{H}$, the energy of the power loss per unit volume ($W\,m^{-3}$) is

$$w_j = \rho J^2 = EJ = \frac{1}{\rho}E^2 = \sigma E^2 \tag{6.28}$$

where resistivity $\rho = 1/\sigma$.

The energy in the entire region of volume V is

$$W_m = \frac{1}{2}Li^2 = \frac{1}{2}\int_V \mathbf{H}\cdot\mathbf{B}\,dV = \frac{1}{2}\int_V \mathbf{A}\cdot\mathbf{J}\,dV \tag{6.29}$$

Equation (6.29) follows from eqn (6.27a), because

$$\int_{V\to\infty} div\,(\mathbf{H}\times\mathbf{A})\,dV = \int_{S\to\infty} (\mathbf{H}\times\mathbf{A})\,dS = 0$$

since at infinity $H = k_1/r^2$, $A = k_2/r$, $dS = k_3 r^2$, and thus $HA\,dS = k_4/r \to 0$.

The potential energy V of a charge Q in an electric field with a scalar potential V_e, in a region of volume Ω, is given by

$$V = \iiint_\Omega \rho\, V_e\, d\Omega = QV_e \tag{6.30}$$

The kinetic energy of a charge Q in a magnetic field described by a magnetic vector potential $\mathbf{A}$, in a region of volume Ω, is

$$T = \iiint_\Omega \rho(\mathbf{A}\cdot\mathbf{v})\,d\Omega = Q(\mathbf{A}\cdot\mathbf{v}) \tag{6.31}$$

where $\mathbf{v}$ is the velocity of charge Q in the magnetic field of flux density $\mathbf{B} = curl\,\mathbf{A}$. The energies given by eqns (6.30) and (6.31) are included in the Lagrangian $L = T - V$ (eqn (6.5)), expressed in terms of values of electric and magnetic fields.

6.2.5 *Maxwell's equations as a consequence of Hamilton's principle*

In the case of motion of a particle of mass m and charge Q with velocity $\mathbf{v}$ in eqn (6.6), which represents the Hamilton principle, the conservative Lagrangian (6.5) consists of kinetic mechanical and electrical energy [6.7]

$$L = T - V = (\tfrac{1}{2}mv^2 + Q\mathbf{A}\cdot\mathbf{v}) - QV_e \tag{6.32}$$

Substituting (6.32) in the Euler–Lagrange equation (6.21), where $q_k = x_1$, $\dot{q} = v$, $i = 1, 2, 3$, we have

$$\frac{\partial}{\partial x_1}\left(\frac{1}{2}mv_i^2 + Q\mathbf{A}\cdot\mathbf{v} - QV_e\right) - \frac{\mathrm{d}}{\mathrm{d}t}\frac{\partial}{\partial v_i}\left(\frac{1}{2}mv_i^2 + Q\mathbf{A}\cdot\mathbf{v} - QV_e\right) = 0$$

$$Q\nabla(\mathbf{A}\cdot\mathbf{v}) - Q\nabla V_e = \frac{\mathrm{d}}{\mathrm{d}t}(m\mathbf{v} + Q\mathbf{A}) \tag{6.33}$$

From the vector identity ([6.15], p. 90)

$$\begin{aligned}\nabla(\mathbf{A}\cdot\mathbf{v}) &= (\mathbf{A}\nabla)\mathbf{v} + (\mathbf{v}\nabla)\mathbf{A} + \mathbf{A}\times(\nabla\times\mathbf{v}) + \mathbf{v}\times(\nabla\times\mathbf{A})\\ &= (\mathbf{v}\nabla)\mathbf{A} + \mathbf{v}\times(\nabla\times\mathbf{A})\end{aligned}$$

since the left-hand side of (6.33) is a partial derivative with respect to x_i, and the operation ∇ on the velocity v_i equals zero. Hence eqn (6.33) takes the form

$$Q(\mathbf{v}\nabla)\mathbf{A} + Q\mathbf{v}\times(\nabla\times\mathbf{A}) - Q\nabla V_e = \frac{\mathrm{d}}{\mathrm{d}t}(mv) + Q\frac{\mathrm{d}\mathbf{A}}{\mathrm{d}t} + Q\sum_{i=1}^{3}\frac{\partial\mathbf{A}}{\partial x_i}v_i$$

from which Lawrynowicz [6.15] finds

$$\frac{\mathrm{d}}{\mathrm{d}t}\mathbf{p} = Q(\mathbf{E} + \mathbf{v}\times\mathbf{B}) \tag{6.34}$$

where:

$\mathbf{p} = m\mathbf{v}$	mass momentum	(6.35)
$E = -\dfrac{\partial\mathbf{A}}{\partial t} - \nabla V_e$	electric fields strength,	(6.36)
$\mathbf{B} = \nabla\times\mathbf{A}$	vector of flux density.	(6.37)

The right-hand side of eqn (6.34) represents the Lorentz force, which is thus given by the time derivative of the particle momentum $\mathrm{d}\mathbf{p}/\mathrm{d}t$. After calculating *curl* of both sides of eqn (6.36)

$$curl\,\mathbf{E} = -\frac{\partial}{\partial t}curl\,\mathbf{A} - curl\,grad\,V_e$$

and considering eqn (6.37) and vector identity *curl grad* $V_e = 0$, we obtain from (6.36) the *second Maxwell* equation

$$curl\,\mathbf{E} = -\frac{\partial \mathbf{B}}{\partial t} \tag{6.38}$$

and then, from eqn (6.37) and the vector identity *div curl* $\mathbf{A} = 0$, we have

$$div\,\mathbf{B} = 0 \tag{6.39}$$

In a similar way, from Hamilton's principle, the first Maxwell equation and $div\,\mathbf{D} = \rho$ can be derived [6.15].

6.2.6 Generalized vector of power density of coupled fields [6.31]

Extending *Poynting's theorm* using Hamilton's principle and the Lagrangian state function ([6.34], p. 38) to other kinds of energy, one can, by analogy with energy conservation law, formulate an integral principle applying to a *generalized vector* **N** of power density of the set of physical fields

$$P_s = \oint_{S(V)} \mathbf{N} \cdot \mathrm{d}\mathbf{S} \tag{6.40}$$

where P_s is the power of physical fields flowing in or out from a space of volume V and surface $S(V)$, **N** is the generalized power density vector of physical fields, where

$$\underline{\mathbf{N}} = \underline{\mathbf{S}} + \mathbf{U} + \mathbf{q}_F + \mathbf{q}_N + \mathbf{q}_{SB} + \mathbf{M} + \mathbf{E}_e + \mathbf{I} + \mathbf{S}_{ch} + \cdots \tag{6.41}$$

where

$$\underline{\mathbf{S}} = \underline{\mathbf{E}} \times \underline{\mathbf{H}} \tag{6.42}$$

is the complex Poynting's vector, which for sinusoidal fields has complex components

$$\begin{aligned} \underline{S}_x &= \tfrac{1}{2}(\underline{E}_{my}\underline{H}^*_{mz} - \underline{E}_{mz}\underline{H}^*_{my}) \\ \underline{S}_y &= \tfrac{1}{2}(\underline{E}_{mz}\underline{H}^*_{mx} - \underline{E}_{mx}\underline{H}^*_{mz}) \\ \underline{S}_z &= \tfrac{1}{2}(\underline{E}_{mx}\underline{H}^*_{my} - \underline{E}_{my}\underline{H}^*_{mx}) \end{aligned} \tag{6.43}$$

and

$$-div\,\mathbf{S} = \frac{\partial}{\partial t}\left(\frac{ED}{2} + \frac{HB}{2}\right) + \sigma E^2 + \rho\mathbf{E}\cdot\mathbf{v}_\rho + \sigma\mathbf{E}(\mathbf{E}_{\mathrm{ext}} + \mathbf{v}\times\mathbf{B}) \tag{6.42a}$$

$\mathbf{U} = \mathbf{v}\dfrac{\mathrm{d}W_{mech}}{\mathrm{d}V}$ Umov's vector of flux density of mechanical power in elastic media;

$\mathbf{q}_E = -\lambda\,grad\,\theta$ vector of thermal power flux density dissipated by conduction (Fourier's law);

$\mathbf{q}_N = \mathbf{1}_v a(\theta_p - \theta_{fluid})$ vector of thermal power flux density dissipated by convection (Newton's law);

$\mathbf{q}_{SB} = \mathbf{1}_v K\nu(\theta^4 - \theta^4_{amb.})$	vector of thermal power flux density dissipated by radiation (Stefan–Boltzman law);
$\mathbf{M}$	vector of light radiation power, emitted by a surface (emittance power);
$\mathbf{E}_e$	vector of radiant power flux density incident to a surface (irradiance);
$\mathbf{I}$	vector of acoustic power density (sound intensity)
$\mathbf{S}_{ch}$	vector of electrochemical power flux density etc.

Vectors $\mathbf{q}_{SB}$, $\mathbf{M}$, $\mathbf{E}_e$ can be also expressed by Poynting's vector $\mathbf{S}$, but here they are isolated to emphasize their role and means of calculation. Equation (6.40) therefore expresses the first rule of thermodynamics ([6.21], p. 13)

$$\frac{\mathrm{d}}{\mathrm{d}t}(W_k + W_i) = P_{ex} + P_{em}$$

where W_k and W_i are kinetic and internal energy, P_{ex} is the power of external forces and P_{em} is the flow of electromagnetic energy through the surface S surrounding the volume V of the body, which according to the energy conservation principle means that the increment in time of the kinetic and internal energy is equal to the power of external forces and flow of electromagnetic energy.

6.2.7 *Field of forces*

An elementary force acting on a conductor with current $I = \mathbf{J}\cdot\mathrm{d}\mathbf{S}$ placed in a field of flux density $\mathbf{B}$ can be calculated using one of three fundamental methods: Ampere's and Biot–Savart's law, the Lorentz force and the principle of virtual work. Maxwell's stress tensor (6.66) is also very useful when calculating local forces from electromagnetic field parameters.

Ampere's and Biot–Savart's laws

Ampere's law is also called the law of flow [6.8] $\oint \mathbf{H}\cdot\mathrm{d}\mathbf{l} = i$ and the formula for force $\mathbf{F}_2$ which is the action of the current loop 1 on current loop 2 (Fig. 6.2)

$$\mathbf{F}_2 = \frac{\mu_0 J_1 J_2}{4\pi}\oint_{l_2}\oint_{l_1}\frac{\mathrm{d}\mathbf{l}_2\times(\mathrm{d}\mathbf{l}_1\times\mathbf{r}_{12})}{r_{12}^3} = -\frac{\mu_0 J_1 J_2}{4\pi}\oint_{l_1}\oint_{l_2}\frac{\mathrm{d}\mathbf{l}_1\cdot\mathrm{d}\mathbf{l}_2}{r_{12}^3}\mathbf{r}_{12} \tag{6.44}$$

Using expressions for loop and flux linkage ψ we have

$$\mathbf{A} = \frac{\mu_0}{4\pi}\oint_{l_1} I_1\frac{\mathrm{d}\mathbf{l}_1}{r_{12}} \quad \text{and} \quad \psi = \frac{\mu_0}{4\pi}\oint_{l_1}\oint_{l_2} I_1\frac{\mathrm{d}\mathbf{l}_1\cdot\mathrm{d}\mathbf{l}_2}{r_{12}} \tag{6.45}$$

A different form of eqn (6.44), expressed by the flux linkage $\psi(x)$ and current I, can be obtained by writing

$$\mathbf{F}_2 = \frac{\mu_0 I_1 I_2}{4\pi}\,grad\oint_{l_1}\oint_{l_2}\frac{\mathrm{d}\mathbf{l}_1\cdot\mathrm{d}\mathbf{l}_2}{r_{12}} = I_2\,grad\,\psi \tag{6.46}$$

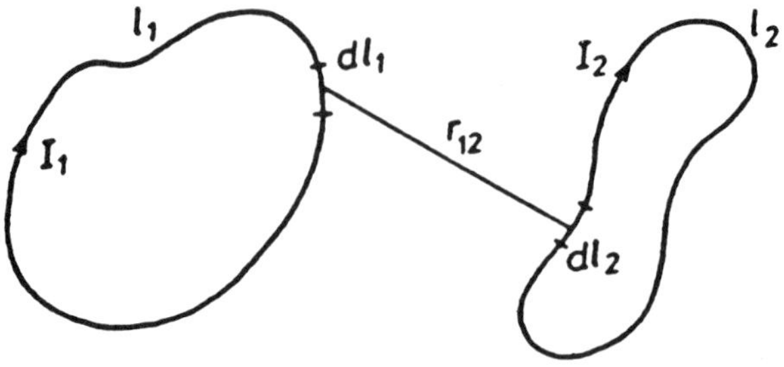

Fig. 6.2 *Interaction of current loops*

Then, considering eqn (6.45), we have the general expression for components of force as in the method of virtual work

$$F_x = I\frac{\partial\psi(x)}{\partial x} = I_1 I_2 \frac{\partial M(x)}{\partial x}$$

where

$$M = \frac{\psi}{I} = \frac{\mu_0}{4\pi}\oint_{l_1}\oint_{l_2}\frac{\mathrm{d}\mathbf{l}_1\cdot\mathrm{d}\mathbf{l}_2}{r_{12}} \tag{6.47}$$

is the mutual inductance of conductors ($M_{12} = M_{21}$) without iron. The presence of iron provides additional sources for $\mathbf{A}$ ($\mathbf{B} = curl\,\mathbf{A}$) and the integral in eqn (6.47) may then no longer be symmetrical. The statement $M_{12} = M_{21}$ is not in general true when there is iron in the magnetic field.

If conductors cannot be treated as filaments with negligible thickness it is better to calculate inductance from the field vectors in volume V

$$L = \frac{1}{I^2}\iiint_V \mathbf{H}\cdot\mathbf{B}\,\mathrm{d}V = \frac{1}{I^2}\iiint_V \mathbf{A}\cdot\mathbf{J}\,\mathrm{d}V \tag{6.48}$$

The force acting on element $\mathrm{d}\mathbf{l}$ of a conductor with current I in a magnetic field $\mathbf{B}$ can be calculated from Biot–Savart's [6.11] law

$$\mathrm{d}\mathbf{F} = I(\mathrm{d}\mathbf{l}\times\mathbf{B}) = \mathbf{n}\,BI\,\mathrm{d}l\sin\angle(\mathrm{d}\mathbf{l},\mathbf{B}) \tag{6.49}$$

The force per unit volume ($\mathrm{N\,m^{-3}}$) is

$$\mathbf{F}_1 = \mathbf{J}\times\mathbf{B} \tag{6.50}$$

and the force acting on a conductor of length l with current I is

$$\mathbf{F} = I\int_l \mathrm{d}\mathbf{l}\times\mathbf{B} \tag{6.51}$$

Lorentz force

Lorentz's force density in $\mathrm{N\,m^{-3}}$, according to eqn (6.50) is

$$\mathbf{f}_L = \mathbf{J}\times\mathbf{B} = \mathbf{J}\times curl\,\mathbf{A} = \rho(\mathbf{E} + \mathbf{v}\times\mathbf{B}) \tag{6.52}$$

where ρ is the volume density of a charge moving with velocity $\mathbf{v}$ in field $\mathbf{B}$.

Lorentz's force is therefore

$$\mathbf{F}_L = Q(\mathbf{E} + \mathbf{v} \times \mathbf{B}) \tag{6.52a}$$

The total force acting on an arbitrary part of a conductor or winding of volume V is

$$\mathbf{F} = \iiint_V (\mathbf{J} \times \mathbf{B})\, \mathrm{d}V = \iiint_V (\mathbf{J} \times curl\,\mathbf{A})\, \mathrm{d}V \tag{6.53}$$

For sinusoidal time variation and complex amplitudes $\underline{\mathbf{J}}_m, \underline{\mathbf{B}}_m, \underline{\mathbf{A}}_m$ the force can be expressed in the form

$$F_{mean} = \frac{1}{2}\mathrm{Re}(\underline{\mathbf{J}}_m \times \underline{\mathbf{B}}_m^*) = -\frac{\omega\sigma}{2}\mathrm{Re}(\mathrm{j}\underline{\mathbf{A}}_m \times curl\,\underline{\mathbf{A}}_m^*) \tag{6.54}$$

with the components

$$\begin{aligned} F_{xmean} &= \tfrac{1}{2}\mathrm{Re}(\underline{J}_{my}\underline{B}_{mz}^* - \underline{J}_{mz}\underline{B}_{my}^*) \\ F_{ymean} &= \tfrac{1}{2}\mathrm{Re}(\underline{J}_{mz}\underline{B}_{mx}^* - \underline{J}_{mx}\underline{B}_{mz}^*) \\ F_{zmean} &= \tfrac{1}{2}\mathrm{Re}(\underline{J}_{mx}\underline{B}_{my}^* - \underline{J}_{my}\underline{B}_{mx}^*) \end{aligned} \tag{6.55}$$

Principle of virtual work

The principle of virtual work states that arbitrary displacements $\mathrm{d}x_j$ of elements of a winding in the j-direction under the influence of an electromagnetic force F_{ej} (Fig. 6.3) are so small, that they cause no changes of current i_k. At the same time the work of the force is equal to an increment of energy of the system

$$F_{ej}\,\mathrm{d}x_j = \mathrm{d}(W)_{i_k=constant}$$

The instantaneous force acting on the kth element of a conductor or winding in direction x_j is equal to the partial derivative of energy W_m of the magnetic field in that direction where $i_k = constant$. From the *Legendre transformation* it follows that for nonlinear systems it should be the coenergy W_m'. Therefore the force

$$F_{ej} = \left(\frac{\partial W_m'}{\partial x_j}\right)_{i_k=constant} = -\left(\frac{\partial W_m}{\partial x_j}\right)_{\psi_k=constant} \tag{6.56}$$

The magnetic energy

$$W_m = \int_V \int_0^B \mathbf{H} \cdot \mathrm{d}\mathbf{B}\, \mathrm{d}V$$

and coenergy (Fig. 6.1)

$$W_m' = \int_V \int_0^H \mathbf{B} \cdot \mathrm{d}\mathbf{H}\, \mathrm{d}V$$

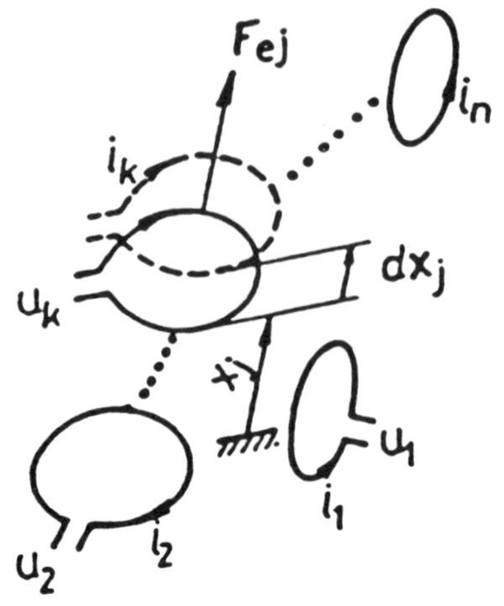

Fig. 6.3 *Virtual work in a system of currents*

can be calculated directly from the fields **H**, **B**, **A** and **J**. Alternatively, self- or mutual inductances according to eqns (6.1), (6.2) and (6.48) can be used

$$W_m = \frac{1}{2}\sum_{i=1}^{n}\sum_{k=1}^{n} M_{ik}(x) i_i i_k = \frac{1}{2}\sum_{i=1}^{n} i_i \phi_i = \frac{1}{2}\sum_{i=1}^{n} i_i \oint \mathbf{A}\cdot \mathrm{d}\mathbf{l} = \frac{1}{2}\iiint_V \mathbf{B}\cdot\mathbf{H}\,\mathrm{d}V$$
$$= \frac{1}{2}\iiint_V \mathbf{A}\cdot\mathbf{J}\,\mathrm{d}V = \frac{\mu}{2}\sum_{i=1}^{n}\sum_{k=1}^{n}\int_{V_i}\int_{V_k}\frac{J_i J_k\,\mathrm{d}V_i\,\mathrm{d}V_k}{r} \tag{6.57}$$

Having a field of magnetic vector potential

$$\mathbf{A} = \mu\sum_{k=1}^{n}\int_{V_k}\frac{J_k\,\mathrm{d}V_k}{r}$$

the integration can be limited only to the conducting regions, where the current density $J \neq 0$.

Examples

(1) Lifting force of an electromagnet (Fig. 6.4)
For simplicity we assume an analytical approximation of the magnetizing characteristics of the iron core [6.23] in the form

$$B = cH^{1/n}$$

For isotropic transformer laminations the constants are typically $n = 10$, $c = 0.9$.

The magnetic coenergy W'_m in a magnetic circuit is therefore

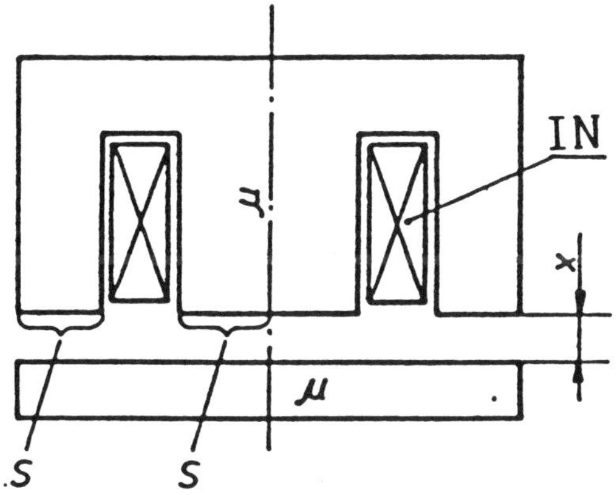

Fig. 6.4 *Lifting electromagnet*

$$W'_m = \int_V \int_0^H \mathbf{B} \cdot \mathbf{H} \, \mathrm{d}V = \int_{V_{Fe}} c \int_0^H H_{Fe}^{1/n} \, \mathrm{d}H \, \mathrm{d}V_{Fe} + \int_{V_{air}} \int_0^H \mu_0 H_{air} \, \mathrm{d}H \, \mathrm{d}V_{air}$$

$$= c \frac{1}{1/n + 1} H^{(n+1)/n} V_{Fe} + \frac{1}{2} \mu_0 H_p^2 \, \mathrm{d}V_{air} = 0.91 \, c H_{Fe}^{1.1} V_{Fe} + \frac{\mu_0}{2} H_{air}^2 V_{air}$$

$$\simeq \frac{\mu_0}{2} H_{air}^2 V_{air} = \frac{\mu_0}{2} \left(\frac{IN}{2x} \right)^2 4Sx = \mu_0 \frac{(IN)^2}{2x} S$$

According to eqn (6.56) the lifting force is

$$F = \left(\frac{\partial W_m}{\partial x} \right)_{I = constant} = -\frac{\mu_0 (IN)^2 S}{2x^2} = -\frac{\mu_0}{2} H^2 S = 4S \frac{B^2}{2\mu_0}$$

In the where the excitation current is sinusoidal, and the flux $\phi = \phi \sin \omega t$, the instantaneous force is

$$F(t) = 4 \frac{\phi_m^2}{4\mu_0 S} (1 - \cos 2\omega t)$$

Conclusion

The magnetic pull in $\mathrm{N\,m^{-2}}$ per unit surface of the magnet (cf. Fig. 6.7b) is

$$p = \frac{B^2}{2\mu_0} = \frac{HB}{2} \tag{6.58}$$

(2) Magnetic pole pull

The magnetic pole pull (in N) of one pole of an electrical machine within one pole pitch $\tau = \pi D/2p$ and armature length l_i is

$$F_N = l_i \int \frac{B^2(x)}{2\mu_0} \mathrm{d}x \tag{6.59}$$

(3) Magnetic attractive force
The magnetic attractive force of field B_t parallel to the steel surface ($B_n = 0$) is normal (cf. Fig. 6.7c) to the surface ([6.30], p. 368), but much weaker

$$p \simeq \frac{B_t^2}{2\mu_r\mu_0} \tag{6.60}$$

6.2.8 *Vibration of busbars and conductors*

Instantaneous electrodynamic forces, together with local excessive heating brought about by strong leakage fields, are a significant factor in decreasing the reliability of large electrical machines, power transformers and other electric power equipment in power systems, power stations, substations, electric furnaces etc. Particularly important is the calculation of these forces and tensions during transient short-circuit conditions and other transient faults, such as self-operated switching on, impulse overloading, etc.

Nowadays there is plentiful literature on the calculation of electrodynamic forces in transformers in quasi-stationary states which neglect changes of self- and mutual inductances M_{ij} during winding vibration. Less researched is the influence of changes of M_{ij} in the course of vibration of busbars and windings. This effect was widely investigated [6.34] in the case of rotating machines, and it was found that transient torques and linear forces could exceed the nominal torque of the motor, in some cases by even ten times (for instance for fast self-operated repeated switching on and off of induction motors). This is the effect of strong coupling of electromechanical and electrodynamic fields. There is much less information on this subject in the area of windings of large transformers, end-windings of large electrical machines, busbars etc.

In reference [6.31] a method was proposed for the generalized solution of vibration and transient displacement of conductors taking into account changes of inductances. The task consisted of solving the investigated system of transformer windings or end-windings of electrical machines (Fig. 6.5) by subdividing the winding into elementary bars, evaluating the field of elementary transient forces from the principle of virtual work (6.56)

$$F_x = \left(\frac{\partial W'_m}{\partial x}\right)_{i_i=constant}, \quad F_y = \left(\frac{\partial W'_m}{\partial y}\right)_{i_i=constant}, \quad F_z = \left(\frac{\partial W'_m}{\partial z}\right)_{i_i=constant} \tag{6.61}$$

and adding the elementary forces in such a way as to obtain the field of components F_x, F_y, F_z of electrodynamic forces in the form

$$F_i = \frac{1}{2}\sum_{j=1}^{n}\sum_{k=1}^{n} i_j i_k \frac{\partial M_{jk}(x_i)}{\partial x_i} \tag{6.62}$$

where $i = x$, y, z and M_{ij} is self- or mutual inductance of particular elementary conductors.

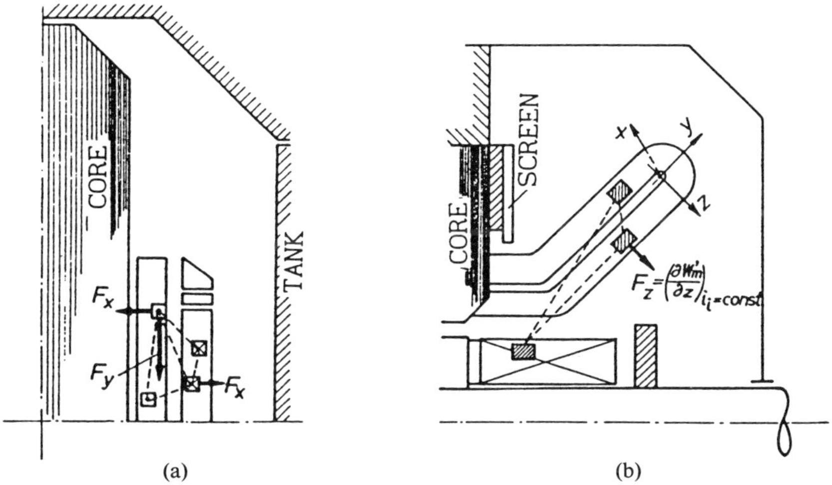

Fig. 6.5 *Calculation of components F_x, F_y, F_z of the field of forces by the method of virtual work: (a) transformer windings, (b) end-winding of electrical machines*

For the elementary conductors we attribute corresponding masses m_1, m_2 per unit length of conductor, damping coefficients D_{x1}, D_{x2}, D_{y1}, D_{y2} due to external and internal friction, the mechanical work of forces etc. and the elasticity constants K_{x1}, K_{x2}, K_{y1}, K_{y2} of the material, construction stiffness, spring action etc.

The process of vibration and its amplitudes for conductor pairs or parts of windings can be calculated using the Euler–Lagrange equation (6.23), taking into account the mutual interaction of electromagnetic and mechanical coupled fields during the impulse of short circuit.

In the case of two bars (Fig. 6.6) we obtain the system of six electromechanical equations for electrical $q_k = Q_1$, Q_2 and mechanical $q_k = x_1$, x_2, y_1, y_2 coordinates and generalized velocities $\dot{q}_k$, i.e. $\dot{Q}_1 = i_1$, $\dot{Q}_2 = i_2$, $\dot{x}_1 = v_{x1}$, $\dot{x}_2 = v_{x2}$, $\dot{y}_1 = v_{y1}$, $\dot{y}_2 = v_{y2}$. The inductances $M_{ij}(xy)$ are calculated from the classical formulae [6.11; 6.32], and then a system of partial differential equations is formulated [6.33] which is solved with the help of a computer library program, for example using an interactive program GOSPEL [6.33]. This package solves a system of nonlinear partial differential equations with variables $x_1, x_2, \ldots, x_n$. The program enables not only interactive analysis of the system, but also selection of the most efficient algorithm from several numerical methods available. In addition these methods can be changed during the analysis by the computer, to select the most suitable method at any one time.

In reference [6.32] the principle of calculation of elementary electrodynamic forces on the basis of equations (6.61) and (6.62) is given. In this approach the method of 'mean geometrical distances' for the calculation of self- and mutual inductances M_{ij} of the conductors has been applied.

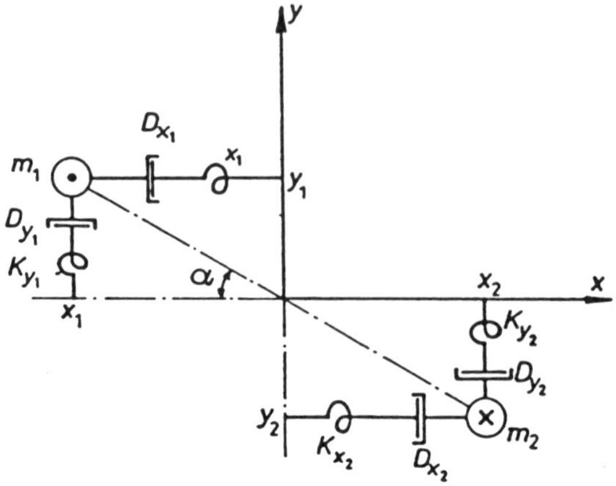

Fig. 6.6 *Elementary pair of busbars for calculating the field of forces and winding vibration*

6.2.9 Maxwell's stress tensor

Maxwell's stress tensor **T** is used for the expression of volume forces (6.50) and (6.52) inside a body with the help of field and forces on the surface of the body. Examples are the formulae (6.58) and (6.60).

For linear media with charge $\rho = div\,\mathbf{D}$ and current density $\mathbf{J} = \rho\mathbf{v} = curl\,\mathbf{H} - \varepsilon\,\partial\mathbf{E}/\partial t$, the resultant surface force is

$$\mathbf{F} = \int_V \mathbf{f}_L\,dV = \oint_S \mathbf{T}^n\,dS \tag{6.63}$$

where the *Lorentz force density* [6.8] is

$$\mathbf{f}_L = \rho\mathbf{E} + \mathbf{J}\times\mathbf{B} = \varepsilon\left(\mathbf{E}\,div\,\mathbf{E} + \mathbf{B}\times\frac{\partial\mathbf{E}}{\partial t}\right) - \mu\mathbf{H}\times curl\,\mathbf{H} \tag{6.64}$$

From *Gauss's theorem* and eqn (6.63) it follows at the same time that

$$\mathbf{f}_L = div\,\mathbf{T}^n \tag{6.65}$$

where $\mathbf{T}^n$ is the vector of surface force, which is the projection of Maxwell's tensor on a normal to the body surface and **n** is the normal vector.

Generally, Maxwell's stress tensor at $\varepsilon = constant$ and $\mu = constant$ is

$$\mathbf{T} = \varepsilon(\mathbf{n}\cdot\mathbf{E}) + \mu(\mathbf{n}\cdot\mathbf{H})\mathbf{H} - \tfrac{1}{2}(\varepsilon E^2 + \mu H^2)\mathbf{n} \tag{6.66}$$

Example

The force **F** acting on a surface S of the stator or rotor of an electrical machine is

$$\mathbf{F} = \iint_S \mathbf{T}\,\mathrm{d}S = \iint_S \frac{1}{\mu_0}\left[(\mathbf{B}\cdot\mathbf{n})\mathbf{B} - \frac{1}{2}|\mathbf{B}|^2\mathbf{n}\right]\mathrm{d}S$$
$$= \left(\mu_0\iint_S \frac{H_n^2 - H_t^2}{2}\,\mathrm{d}S\right)\mathbf{n} + \left(\mu_0\iint_S H_nH_t\right)\mathrm{d}S\,\mathbf{t} = F_n\mathbf{n} + F_t\mathbf{t} \tag{6.66a}$$

It can be seen that two forces F_n and F_t act on each metal element. Normally motor cores are made of laminated iron and are not highly saturated. Therefore one can take $H_t = 0$, hence force $F_t = 0$ and on the iron core there only acts the normal force $F_n = 0.5\iint_S BH\,\mathrm{d}S$.

If however the core is made of solid iron, a permanent magnet of low permeability, or if it is covered with a copper screen and the field is alternating, equivalent permeability μ is low and the values H_n and H_t are comparable. In this case both forces F_n and F_t exist and should be considered, especially at impulse excitation of switchgears or fast relays.

From eqn (6.66a) it can easily be seen that the modulus of the surface force density of the magnetic field satisfies eqn (6.60) and vector **H** divides the angle between **n** and **T** in half (Fig. 6.7a). In this way the electromagnetic volume forces can be transferred to the surface.

These forces in general do not coincide with the direction of the field **H** or the vector **n** normal to a surface (Fig. 6.7). Their behaviour can be compared to an elastic belt pull [6.11].

Depending on the direction of **H** the force may be stretching the conductor in the direction of the field (Fig. 6.7b), may be compressive perpendicular to the field (Fig. 6.7c), or may even act to move the conductor as in Fig. 6.7d.

Depth of penetration of magnetic pressure

In the case of the action of the magnetic field impulse $H = H_y(0, t)$ along a metal surface (Fig. 6.7c) we introduce to eqn (6.50) the current density [6.11]

$$J_x = -\frac{\partial H_y}{\partial z} \tag{6.67}$$

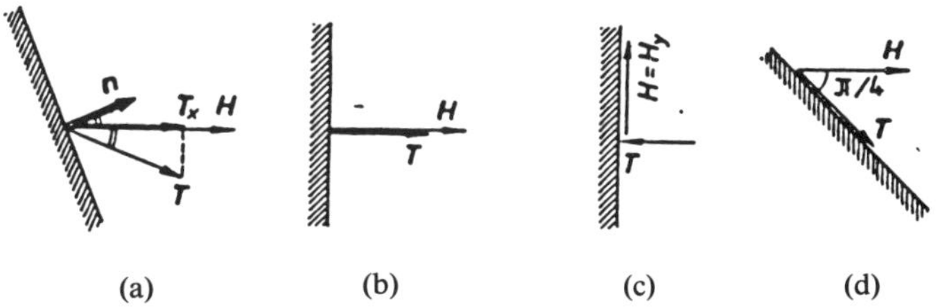

Fig. 6.7 *Maxwell's stress tensor* $|T| = 0.5\,\mu H^2$ *according to the position of field* **H** *on the body surface*

Integrating this force with respect to the z-axis we find the *hydrodynamic pressure* on the depth z brought about by the coupling of the magnetic field with the incompressible body

$$p(z,t) = -\int_0^t \frac{\partial H_y}{\partial z} B_y \, dz = p_H - \frac{1}{2}\mu H^2(z,t) \tag{6.68}$$

where

$$p_H = \tfrac{1}{2}\mu H^2 \tag{6.69}$$

is the pressure of the magnetic field on the surface. From eqn (6.68) it follows that a pressure drop in time and space inside the metal is

$$\frac{\partial p}{\partial t} = \mu H_0 \frac{\partial H_0}{\partial t} - \mu H \frac{\partial H}{\partial t} \quad \text{and} \quad \frac{\partial p}{\partial z} = -\mu H \frac{\partial H}{\partial z}$$

This means that the full magnetic pressure decays faster in time than the external field and acts on a smaller depth than that of the field penetration.

In the case when the conducting body is compressible an impulse of hydrodynamic pressure penetrates the metal with finite speed. According to Knoepfel ([6.11], p. 125) if the magnetic field strength is of the order of megaoersteds (several hundreds $\mathrm{A\,m^{-1}}$), and the pulse rise in time is of the order of microseconds, a shock wave develops (one or several). In this case it is necessary to consider that a diffusion effect of the field is linked with a pressure distribution in the conductor. This relatively complex dynamic problem is resolved with the help of a number of hydrodynamic equations, including Maxwell's equations.

6.3 MHD

Magnetic hydrodynamics (*magnetohydronamics*), MHD, has a broad range of applications: from astrophysics and thermonuclear reactions to typical industrial equipment such as in transport, melting and mixing of liquid metals and electrolytes, plasma motion in the electric arc of electrical apparatus etc. Here, a complicated interaction between a conducting liquid media and an electromagnetic field occurs.

These phenomena are generally described by the system of equations created from Maxwell's and Navier–Stokes' equations, which extend the Navier–Stokes equation with the additional component

$$\mathbf{f}_L = \mathbf{J} \times \mathbf{B} \tag{6.70}$$

This represents the volume density of the Lorentz force (6.64), driving the conducting media to motion with a velocity $\mathbf{v}$.

The *Navier–Stokes* equation which describes the heat exchange by convection between the fluid and solid body surface, as well as the vector field of fluid velocity $\mathbf{v}$ is usually presented in the form

$$\frac{\partial \mathbf{v}}{\partial t} + (\mathbf{v}\, grad)\mathbf{v} = -\frac{1}{\rho_m} grad\, p + \nu\Delta\mathbf{v} + \frac{\nu}{3} grad(div\, \mathbf{v}) + \mathbf{P} \tag{6.71}$$

where:

$\rho_m = \rho_m(x, t)$	fluid density,
$p = p(x, t)$	hydrodynamic pressure,
ν	coefficient of kinematic viscosity of incompressible fluid,
$\nu\Delta\mathbf{v}$	density of force of internal friction,
$\mathbf{P}$	force acting on mass unit.

Here the vector field of forces **P** can be not only the field of Lorentz forces (6.70), but also the gravitational field etc.

The *Navier–Stokes* equation (6.71), together with Maxwell's equation, create among other things a basis for the analysis of induction pumps and liquid metal mixers [6.29]. It is difficult to obtain the solution of the system of coupled equations of magnetohydrodynamics, therefore various simplifications are used, especially in analytical solutions. There are a number of analytical solutions in this field, for example Sajdak [6.28; 6.29] and others. The solution of fields equations (**J**, **B**) is first carried out at the assumed value $v = 0$ or $v = constant$, then by using the Navier–Stokes equations the electrodynamic force densities are resolved. If the speed obtained is much different from that assumed, the calculation is repeated.

In reference [6.28] the calculation of volume force density (6.70), where

$$\mathbf{J} = \sigma\left(-\frac{\partial \mathbf{A}}{\partial t} + \mathbf{v} \times curl\, \mathbf{A}\right) \tag{6.72}$$

is carried out from the equation for a medium moving with velocity v

$$\Delta\mathbf{A} - \mu\sigma(\mathbf{v} \times curl\, \mathbf{A}) = \mu\sigma\frac{\partial \mathbf{A}}{\partial t} \tag{6.73}$$

In induction pumps and mixers a cylindrical model is normally used, which allows a significant simplification of calculations of current density, forces, power and efficiency of the system. The accuracy of the solution depends mainly on the degree of simplification of the geometry of the system and on its Reynolds number (6.74). At small Reynolds numbers $R_M \ll 1$ (e.g., after [6.28], at $v = 0.5\,\mathrm{m\,s^{-1}}$ and $R = 0.05\,\mathrm{m}$, for liquid Al $R_M = 0.16$ and for Zn $R_M = 0.085$) one can assume a model with $v = 0$. For a large capacity and high metal speed and low frequencies (R_m large) ignoring the motion of the metal can lead to significant errors [6.28]. Among the several simplified solutions of electric heating problems which have been published, using Navier–Stokes equation, Wieczorek (1st Conference on Mathematical Methods in Electrothermics, MMET; Wisla, Poland, 1986) proposed an analysis of induction heating processes of metals, which comprises interdependent electromagnetic, thermal and thermoelastic phenomena while neglecting the dynamic effects and the effect of coupling of the fields of stresses and temperature. In this case the task reduces to the consecutive, separated solution (Fig. 6.8)

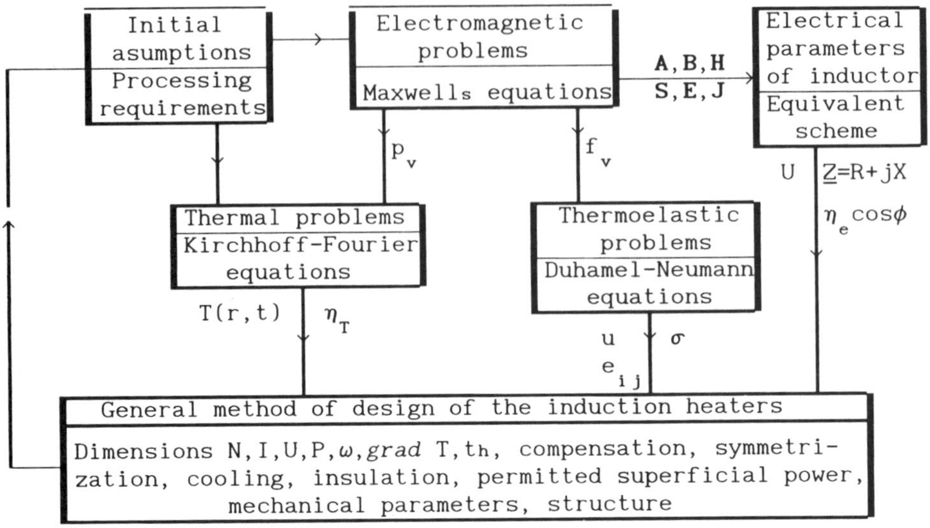

Fig. 6.8 *Flow chart of the complex analysis and design of induction heaters of metals (Wieczorek, MMET, Wisla 1986, Poland): where N, I, U, P, ω, t_h are the number of turns, current, active power, angular frequency and heating time respectively, p_v is the volume active power density in metal charge, f_v is the volume density of electrodynamic forces in the charge; Z is the system impedance; η_e, η_T are the electrical and thermal efficiency of the system respectively; T(r, t) is the non-stationary, source temperature field in the charge; u, σ, e are the vectors of dispacements and stress, and deformation tensors in the charge; and* **A, B, H, S, E, J** *are the vectors of the fields explained in text*

of equations of electrodynamics (Maxwell), thermal conductivity (Kirchhoff–Fourier) and quasi-static thermoelastic (Duhamel–Neumann).

In electroheating problems and in the heating of electrical machines and devices, the structures and processes are complex, and so coupled fields are still often resolved using an analytical-numerical method rather than methods based on volume discretization. In such cases when coupling of both fields is weaker and computational 'decoupling' of the electromagnetic field from the thermal one is possible, the finite-element method is often successfully applied, e.g. in the works of Krsteva *et al.*, Schultze and Andree, Napieralska-Juszczak and Dems and others (see papers II-1, II-3 and II-6, respectively, of the MMET Conference, Wisla, Poland, 1993).

According to Voldek ([6.35], p. 29) it is impossible to obtain full solution of the magnetohydrodynamic equations in MHD induction a.c. machines, and simplifications in this case are absolutely necessary. Since liquid metals and electrolytes can be considered as incompressible, and as their temperature along the MHD machine is practically invariable, it is sufficient to add to eqn (6.71) the *equation of continuity* of incompressible liquids

$$div\,\mathbf{v} = 0$$

and consider, after Voldek, that the liquid secondary circuit acts as a solid body, i.e. as a secondary part of the linear induction motor.

Magnetic Reynolds number
As a measure of the degree of coupling of electromechanical and electromagnetic primary and secondary fields the magnetic *Reynolds number*

$$R_m = v\mu\sigma L = \frac{L}{L_e} = \frac{v}{v_e} \tag{6.74}$$

can be used. It is a dimensionless parameter, in which v is the velocity of liquid flow or motion of the conducting body, μ is the magnetic permeability, σ is the electric conductivity, L is the linear dimension of the field of flow (e.g. radius of channel with the liquid metal [6.30]) and

$$L_e = \frac{1}{v\mu\sigma} \quad \text{or} \quad v_e = \frac{1}{L\mu\sigma} \tag{6.75}$$

L_e is the characteristic length, over which the conducting medium is subjected to the magnetic field and v_e is the characteristic velocity with which the magnetic field is propagated ([6.30], p. 135) in a conductor

$$v = \lambda f = \sqrt{\left(\frac{2\omega}{\mu\sigma}\right)}$$

If $R_M \gg 1$ (i.e. $L \gg L_e$ or $v \gg v_e$) the magnetic field moves jointly with the conductor and the entrainment of the field or its extension by the conductor will prevail over its diffusion (freezing). Such effects takes place very strongly in geophysical and astrophysical phenomena. On the other hand, when $R_m \ll 1$ (i.e. $L \ll L_e$ or $v \ll v_e$) the motion of the conductor does not observably influence the magnetic field. This occurs, for example, in laboratory experiments with the flow of mercury, sodium or salt water.

In electrical induction machines with a relative speed of the field with respect to the rotor

$$v_2 = \frac{s\omega\tau}{\pi} = \frac{s\omega}{\pi}\frac{\pi D}{2p} = \frac{s\omega D}{2p}$$

the magnetic Reynolds number (6.74) can be expressed as

$$R_M = v\mu\sigma L = s\omega\mu\sigma\frac{\tau}{\pi}L = 2k^2\frac{\tau}{\pi}L = \frac{DL}{p\delta^2} \tag{6.76}$$

where

$$\delta = \sqrt{\left(\frac{2}{s\omega\mu\sigma}\right)}, \qquad \left(s = \frac{n_1 - n}{n_1}, \; \omega = 2\pi f\right) \tag{6.77}$$

which is equivalent to the depth of penetration of the field into the solid rotor.

In electrical machines, the Reynolds number is large only if the slip s is sufficiently large. This is due to high copper conductivity ($\sigma_{\text{Cu}, 20°\text{C}} = 58 \times 10^6\,\text{S m}^{-1}$) and high iron permeability ($\mu_r = 2000$–4000).

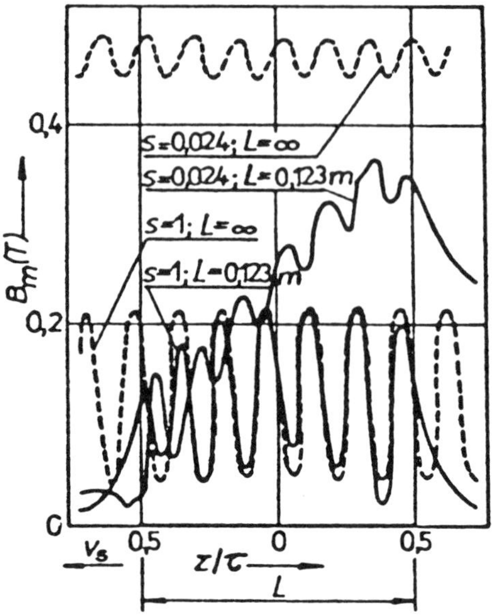

Fig. 6.9 *Normal flux density distribution in a gap of the induction motor at different slips s [6.6]:*
——— *LIM with finite armature length;*
--- *rotating and LIM with armature of infinite length*

At a synchronous speed of induction machines ($s = 0$) $R_m = 0$ and the field penetrates the rotor without any obstruction. With the rotor at standstill ($s = 1$), on the other hand, R_M reaches its highest value (at the smallest depth of penetration δ of the field) and then the stator field is strongly deformed by the reaction of the rotor currents. In the intermediate states ($0 < s < 1$) the rotating field entrains the rotor and is deformed at the same time (Fig. 6.9). Smaller oscillations in Fig. 6.9 are caused by the influence of slots.

In rotating and linear induction motors of infinite length ($L = \infty$; dashed lines in Fig. 6.9) the change of slip from $s \cong 0$ to $s = 1$ brings about over twofold suppression of flux density in the air-gap. In linear motors with a short armature ($L = 0.123$ m; continuous lines in Fig. 6.9), diffusion damping of the wave front under the moving armature and an electromagnetic tail on the abandoned armature edge occurs. This is due to edge armature effects, even at $s \cong 0$.

Similarly a twist of the linear motor axis by an angle α with respect to the direction of motion of the secondary brings about significant field deformation (Fig. 6.10), which effects a decrease in the drawing force, causing additional power losses and a force realignmenting of the twisted armature.

The field deformations can be considered as a decayed imperfect *'freezing'* of the armature field on the conducting surface of the secondary, which according to *Lenzs' rule* acts towards maintaining the initial state. In this case in the Reynolds number (6.76), instead of $s = (v_1 - v)/v_1$, as for motor

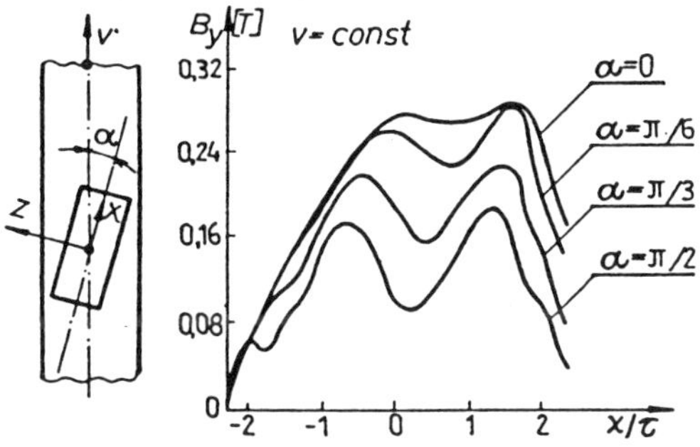

Fig. 6.10 *Deformation of flux density distribution in the air-gap of LIM with the twist of the armature axis with respect to a forced secondary speed $v = 0.9v_1$ [6.17]*

with $L = \infty$, we put $s' = v/v_1 \cong 1$ to 0.5, which gives the large value of R_M.

In linear induction motors the Reynolds number is expressed [6.22] in the form

$$R_M = v_1 \mu_0 \sigma d \tag{6.78}$$

where d is the thickness of the conducting surface of the secondary.

In these motors the Reynolds number changes from $R_M = 0.265$ for small motors with $v_1 = 3\,\mathrm{m\,s^{-1}}$, $\sigma_{\mathrm{Al}} = 35 \times 10^6\,\mathrm{S\,m^{-1}}$, $d = 2\,\mathrm{mm}$, through $R_M = 8$ for average motors with $v_1 = 100\,\mathrm{km\,h^{-1}}$, $\sigma_{\mathrm{Cu}} = 57 \times 10^6\,\mathrm{S\,m^{-1}}$, $d = 4\,\mathrm{mm}$, up to $R_M = 200$ for high speed trains with $v_1 = 500\,\mathrm{km\,h^{-1}}$, $\sigma_{\mathrm{Cu}} = 57 \times 10^6\,\mathrm{S\,m^{-1}}$ and $d = 20\,\mathrm{mm}$.

The magnetic Reynolds number R_M, as a discriminant of coupled electromechanical fields, can be considered as a quantity factor of Lenz's rule. This rule reveals not only the inertial field deformation, but also the repulsive action of the eddy current reaction and in the levitation of linear motors during fast motion.

According to reference [6.22] the normal force F_n (attraction and repulsion) and drawing force F_i of a single-sided linear motor (Fig. 6.11) moving with a slip s can be expressed

$$F_n = \frac{\tau}{2\mu_0 b}\left(\frac{E_1}{v_1 w_1}\right)^2 \frac{1 - R_M s}{\left(\mathrm{ch}\,\frac{\pi}{\tau}\delta_i\right)^2 + \left(R_M s\,\mathrm{sh}\,\frac{\pi}{\tau}\delta_i\right)^2} \tag{6.79}$$

$$F_i = \frac{\tau}{\mu_0 b}\left(\frac{E_1}{v_1 w_1}\right)^2 \frac{R_M s}{\left(\mathrm{ch}\,\frac{\pi}{\tau}\delta_i\right)^2 + \left(R_M s\,\mathrm{sh}\,\frac{\pi}{\tau}\delta_i\right)^2} \tag{6.80}$$

where:

$$\mu_0 = 4\pi \times 10^{-7}\,\mathrm{H\,m^{-1}}$$

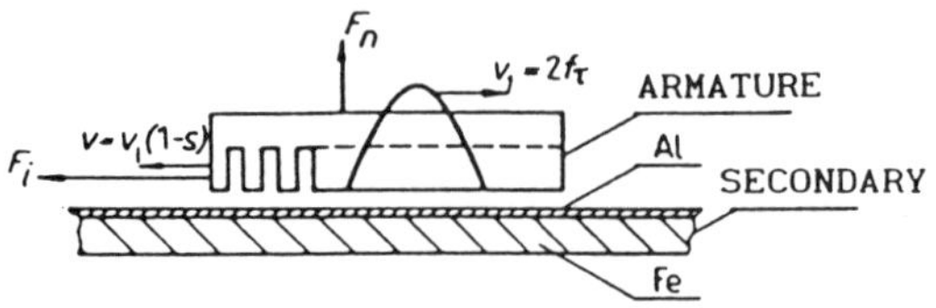

Fig. 6.11 *Single-sided linear induction motor*

E_1	emf in primary winding,
w_1	quantity proportional to number of turns per pole pair,
$v_1 = 2f\tau$	synchronous speed,
τ	pole pitch
b	magnetically active width of motor,
$\delta_i = \delta_{gap} k_C$	ideal gap,
k_C	*Carter's coefficient.*

From eqns (6.79) and (6.80) it follows that at the synchronous speed ($s = 0$) only the pulling force F_n exists and there is no drawing force F_i. At $s > 1/R_M$ the levitation (repulsive) force exceeds the attractive force. Since $0 < s < 1$, the levitation effect will not occur for small low-speed motors having $R_M < 1$. Levitation only appears during braking conditions ($s > 1$) of small motors, or in large motors with large values of R_M, of the order of 10 to 25 and more.

6.4 NON-HOMOGENOUS MEDIA

In the electrodynamics of superconductors ([6.30], p. 117) the *London's equations* are

$$curl\, \Lambda \mathbf{J}_s = -\mathbf{B} \quad \text{and} \quad \frac{\partial(\Lambda \mathbf{J}_s)}{\partial t} = \mathbf{E} \tag{6.81}$$

where $\mathbf{J}_s$ is the density of the superconducting current, $\Lambda = m/(e^2 n_s)$, the constant parameter of the superconductor; m, e, n_s are the mass, charge and density respectively of the superconductive electrons and become a necessary addendum to Maxwell's equations. London's equations (6.81) take into account the structural changes of a superconductor under the influence of the external effects of field and temperature and help to explain the *Maissner* and other effects.

Another quite independent structural correction derives from other assumptions by Rawa [6.25] with respect to all kinds of non-homogeneous media in which physical parameters of the medium are continuous functions of the coordinate vector $\mathbf{x}$, i.e.

$$\sigma = \sigma(\mathbf{x}) = \overline{\sigma}, \qquad \varepsilon = \varepsilon(\mathbf{x}) = \overline{\varepsilon}, \qquad \mu = \mu(\mathbf{x}) = \overline{\mu} \tag{6.82}$$

The work of Rawa has led him to the conclusion that the classical formulations of Maxwell's equations

$$curl\,\mathbf{H} = \mathbf{J} + \frac{\partial \mathbf{D}}{\partial t} \quad \text{and} \quad curl\,\mathbf{E} = -\frac{\partial \mathbf{B}}{\partial t} \tag{6.83}$$

with the condition $div\,\mathbf{D} = 0$ do not describe sufficiently the electromagnetic field in non-homogeneous media. From reference [6.25] these equations are only satisfied if we assume $div\,\mathbf{D} = \rho$, where ρ is not defined solely as the density of free charge introduced externally to the investigated region. It is instead a quantity which characterizes the variable concentration of free charge (e.g. electrons), created as a result of medium non-homogenity. In order to eliminate this complication Rawa has proposed complementing Maxwell's equations (6.83) with the additional equation

$$\frac{\partial \rho}{\partial t} + \frac{\bar{\sigma}}{\bar{\varepsilon}}\rho = \bar{\sigma}\mathbf{E} \cdot grad\ln\frac{\bar{\varepsilon}}{\bar{\sigma}} \tag{6.84}$$

Only in the special case, when $\bar{\varepsilon}/\bar{\sigma} \cong constant$, are eqns (6.83) satisfied without correction, this occurs in homogeneous media or in special cases of non-homogenity. This hypothesis, in order to be confirmed, needs further investigations connected with the electron conductivity (σ), dielectric irregularities of solid bodies, (ε), and the dynamics of charge motion ($-\partial\rho/\partial t = \mathbf{J}$).

6.5 MAGNETOSTRICTION

Magnetostriction is the name for the complex phenomena connected with the change of crystals and body dimensions under the influence of magnetization (Fig. 6.12). This is caused by the change in separation between atoms of the crystal lattice. Magnetostriction, which is characterized by unit elongation $\varepsilon = \Delta l/l$ can be positive or negative. It is one of the causes of noise in electrical machines and power transformers, and is used in ultrasonic generators.

6.6 COUPLED MAGNETIC AND THERMAL FIELDS

6.6.1 Magnetoelasticity

A typical example of coupled fields is the complex phenomena called magnetoelasticity. This is a relatively new branch of science, which is devoted to investigating the creation of additional electromagnetic fields as a result of mechanical and thermal disturbances in a body placed in a strong magnetic field. These additional field effects depend on the value of stresses, and the speed of propagation of elastic and electromagnetic waves ([6.21], p. 7).

The essence of the problem is the fact that if a body is placed in a strong external magnetic field $\mathbf{H}^0$, the mechanical impact or thermal shock generate within it a field of deformation coupled with an electromagnetic field ([6.21], p. 118), which is characterized by small fluctuations $\mathbf{h}(x, t)$ and $\mathbf{e}(x, t)$ which are excited in the body by mechanical or thermal effects. From the moment of activation ($t = 0$) the resultant field at the moment $t > 0$ has the form

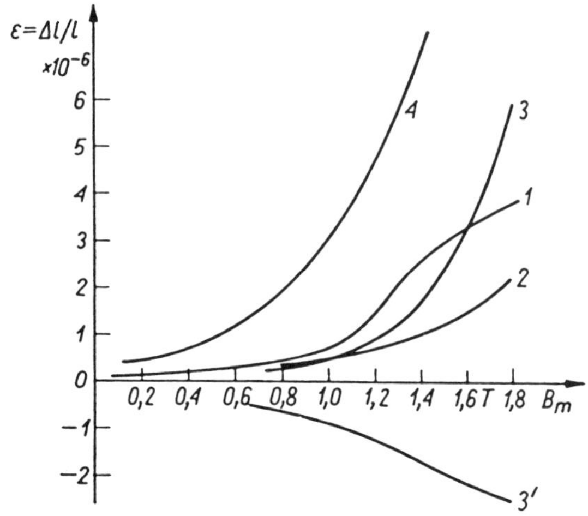

Fig. 6.12 *Magnetostriction curves of transformer steel:*
1. *hot rolled after Knowlton,*
2. *cold rolled Armco,*
3. *cold rolled not annealed after BBC,*
3′. *as 3 but annealed,*
4. *cold rolled not annealed*

$$\mathbf{H}(\mathbf{x}, t) = \mathbf{H}^0 + \mathbf{h}(\mathbf{x}, t), \qquad \mathbf{E} = \mathbf{e}(\mathbf{x}, t) \tag{6.85}$$

where **x** is a point of body moving with a velocity **v** in an external electric or magnetic field.

At small velocities **v** of displacement of deformed body the *constitutive* equations can be expressed ([6.21], p. 120) in the form

$$\begin{aligned} &\mathbf{D} = \varepsilon\mathbf{E} + \alpha\mathbf{v} \times \mathbf{H} \\ &\mathbf{B} = \mu\mathbf{H} - \alpha_v\mathbf{v} \times \mathbf{H} \qquad \alpha_v = \varepsilon\mu - \varepsilon_0\mu_0 \\ &\mathbf{J} = \sigma(\mathbf{E} + \mathbf{v} \times \mathbf{B}) + \rho\mathbf{v} \end{aligned} \tag{6.86}$$

In the theory of magnetoelasticity it is assumed that the only mechanical effect in the electromagnetic field is Lorentz's force ([6.21], p. 120) with density (6.64) which is developed from the equations of motion (6.87).

Using the principle of conservation of momentum and after several mathematical transformations, Nowacki [6.21] has developed a system of displacemental equations of motion of an isotropic body without initial stresses

$$\mu_L \Delta\mathbf{u} + (\lambda_L + \mu_L)\, grad\, div\, \mathbf{u} + \mathbf{X} + f_L = \rho_m \frac{\partial^2 u}{\partial t^2} \tag{6.87}$$

where μ_L, λ_L are the Lamé constants concerning an adiabatic state, **u** is the vector of deformations (displacements u_i) with tensor of deformations ε_{ij} and tensor of rotation ω_{ij} ([6.21], p. 15) which are described with the help of equations

$$\varepsilon_{ij} = \tfrac{1}{2}(u_{ij} + u_{ji}), \quad \omega_{ij} = \tfrac{1}{2}(u_{ij} - u_{ji}), \quad u_{ij} = \varepsilon_{ij} + \omega_{ij} \tag{6.88}$$

X is the vector of mass forces in an external load causing body motion; mass forces are a function of position **x** and time t; the power of these forces is equal to $\int_v \mathbf{X}_i \dot{u}_i \, \mathrm{d}V$; f_L is the density of the Lorentz force (6.64) and ρ_m is the density of the medium.

The complete set of differential equations of magnetoelasticity, apart from eqn (6.87), encompasses further equations of electrodynamics (6.83), divergence equations (6.24) and constitutive equations (6.86).

The boundary conditions for these equations ([6.21], p. 123) are

$$[\mathbf{E} + \mathbf{v} \times \mathbf{B}]_t = 0, \qquad [\mathbf{H} - \mathbf{v} \times \mathbf{D}]_t = J_{ms} - \rho_s v_m \tag{6.89a, b}$$

$$[\sigma_{ji} + T_{ji} + g_i v_j] n_j = 0 \tag{6.90}$$

where the left-hand side of eqn (6.89b) represents the difference between the tangential components on the surface both inside and outside the body; J_{ms}, ρ_s are surface values and σ_{ji} is the symmetric stress tensor, connected with external surface load $\mathbf{p}(\mathbf{x}, t)$ by the equation [6.21]

$$p_i = \sigma_{ji} n_j \tag{6.91}$$

where n_j is the component of **n** normal to the surface S of region V

$$T_{ji} = E_i D_j + H_i B_j - \tfrac{1}{2}(E_k D_k + H_k B_k)\delta_{ji} \tag{6.92}$$

is the Maxwell's stress tensor (6.66); δ_{ji} is the Kronecker's delta; and

$$g_i = (\mathbf{D} \times \mathbf{B})_i \tag{6.93}$$

Electromagnetic-pulse forming

The immediate application of magnetoelasticity in electrical engineering is in electrodynamic forces in transient states, and especially in the magnetic-pulse forming of metal parts. These are mostly cylindrical or flat elements [6.3]. Maxwell stresses (6.66) acting on a conducting surface are produced by discharging a battery capacitor through an excitation coil. The coil produces a damped out, oscillating impulse with a magnetic field amplitude of the order of 20–50 tesla. Stress impulses produced in this way according to reference [6.3] reach $10^9\,\mathrm{N\,m^{-2}}$ in about a dozen microseconds.

Krajewski [6.13] has resolved the system of magnetoelasticity equations (6.87) for infinitely long cylinders in which the field is excited with the help of

- a cylindrical coil or
- a rod inductor.

Considering an axisymmetric system, with assumptions of linearity, isotropy and homogeneity of the medium, Krajewski has obtained the following system of equations in cylindrical coordinates

$$\frac{\partial^2 H}{\partial r^2} + \frac{1}{r}(1 - \sigma\mu v r)\frac{\partial H}{\partial r} - \sigma\mu\frac{1}{r}H\frac{\partial}{\partial r}(rv) - \sigma\mu\frac{\partial H}{\partial t} = 0 \tag{6.94a}$$

$$\frac{\partial^2 H}{\partial r^2} + \frac{1}{r}(1 - \sigma\mu v r)\frac{\partial H}{\partial r} - \frac{1}{r^2}\left(1 - \sigma\mu r^2 \frac{\partial v}{\partial r}\right) H - \sigma\mu \frac{\partial H}{\partial t} = 0 \quad (6.94b)$$

$$\frac{\partial^2 u}{\partial r^2} + \frac{1}{r}\frac{\partial u}{\partial r} - \frac{u}{r^2} + \frac{1}{c^2 \rho_m} F_L - \frac{1}{c^2}\frac{\partial^2 u}{\partial t^2} = 0 \quad (6.95)$$

$$F_L = -\mu H \frac{\partial H}{\partial r} \quad (6.96a)$$

$$F_L = -\mu H \left(\frac{\partial H}{\partial r} + \frac{1}{r} H\right) \quad (6.96b)$$

$$v = \frac{\partial u}{\partial t} \quad (6.97)$$

Equations (6.94a) and (6.96a) describe the circumferential excitation currents whereas eqns (6.94b) and (6.96b) describe the axial current, where $H = H_z$ for the circumferential excitation (a) and $H = H_\phi$ for the axial excitation (b)

$$F_L = F_{LF}, \quad v = v_r, \quad u = u_r \quad \text{and} \quad c^2 = \frac{E}{\rho_m(1 - \nu^2)}$$

where E is the modulus of elasticity, and ν is Poisson's coefficient. Equations (6.94) to (6.97) have been resolved by Krajewski [6.13], using the following iteration method.

The solution of eqns (6.94a, b) for a stationary medium ($v = 0$) provides the first step of approximation $H_1(r, t)$, $J_1(r, t)$ for the electromagnetic field distribution. The Lorentz force, calculated from these distributions, was used to evaluate the first approximation of the displacement field $u_1(r, t)$ and the field of velocity $v_1(r, t)$ with the help of equation of motion (6.95). Using the values u_1 and v_1, by applying the generalized Ohm's law (6.86), a second approximation of current density

$$J_2 = J_1 + \sigma\mu v_1 H_1 \quad (6.98)$$

is obtained.

A second approximation of the magnetic field intensity H can be found using Ampère's law:

(1) for a system with circumferential currents, when the cylinder is placed inside a coil

$$H_2(r, t) = \int_r^{r_2} J_2(\rho, t)\, d\rho + A_e(t) \quad (6.99)$$

where $A_e(t)$ is the specific electric loading of excitation current in A m^{-1}

(2) for a system with an axial current

$$H_2(r,t) = \frac{1}{r}\int_{r_2}^{r} \rho J_2(\rho, t)\,\mathrm{d}\rho \tag{6.100}$$

where r_1 and r_2 are the internal and external radii of the cylinder.

Proceeding in this way Krajewski obtained consecutive approximations of the electromagnetic field distribution. Since functions $J_n(r, t)$ are obtained in the form of discrete integrals (6.99) and (6.100) they are replaced by the corresponding quadratures. The proposed procedure, using relatively simple analytical solutions [6.13], obviates the need to find repeated solutions of the relatively complicated equations (6.94a) and (6.94b). Next in reference [6.13] a method was proposed to solve the hyperbolic equation of motion (6.95) with the help of the Boundary Elementary Method (BEM). The method proposed in reference [6.13] is useful for the analysis of electromechanical problems in which the velocity of the medium is considerable, i.e. where the magnetic Reynolds number (6.74) is much higher than 1.

6.6.2 Magnetothermoelasticity

In electrical machines and electrical power equipment, especially of high power, and in electrothermic equipment, the thermal effects resulting from electromagnetic fields play an exceptionally important role. They very often determine the reliability of these devices. Of lesser technical significance from this point of view is magnetothermoelasticity. Nevertheless it will be mentioned now for completeness of the description.

Equations of magnetothermoelasticity

Nowacki [6.21] and Parkus [6.24], in a similar way to eqns (6.40) and (6.41), as a starting point to the analysis of magnetothermoelasticity, have taken balance equation of mechanical, electrodynamic and thermal energy, which in space notation for a volume V with surface S reads

$$\frac{\mathrm{d}}{\mathrm{d}t}\int_V \left(\frac{1}{2}\rho_m v_i v_i + \rho_m W + W_e\right) \mathrm{d}V = \int_V (X_i v_i + \rho_m Q_1)\,\mathrm{d}V + \int_S [p_i v_i - q_i n_i - (\mathbf{E}\times\mathbf{H})_i + W_e v_i n_i]\,\mathrm{d}S \tag{6.101}$$

where:

$\frac{1}{2}\frac{\mathrm{d}}{dt}\int_V \rho_m v_i v_i \,\mathrm{d}V$ increment in time of kinetic energy;

$\frac{\mathrm{d}}{\mathrm{d}t}\int_V \rho_m W \mathrm{d}V$ increment in time of mechanical energy;

$\frac{\mathrm{d}}{\mathrm{d}t}\int_V W_e \mathrm{d}V$ increment in time of internal electrodynamic energy;

ρ_m and v	density and velocity of body;
W	mechanical and electromagnetic internal energy density per unit mass;
$W_e = \frac{1}{2}(\varepsilon E^2 + \mu_e H^2)$	electrodynamic energy density per unit volume;
$\int_V X_i v_i \, \mathrm{d}V$	power of external mass forces X (see eqn 6.87), which cause motion of the body and are the functions of position $\mathbf{x}$ and time t ([6.21], p. 13);
$\int_V \rho_m Q_1 \, \mathrm{d}V$	thermal power; Q_1 is heat quantity generated per unit mass and time;
$\int_S p_i v_i \, \mathrm{d}S$	power of external, surface (contact) forces which cause motion of the body and are functions of position $\mathbf{x}$ and time t;
$\int_S q_i n_i \, \mathrm{d}S$	flow of thermal flux q_i through the surface S, where n_i is the vector normal to the surface $\mathrm{d}S$;
$\int_S (\mathbf{E} \times \mathbf{H})_i \, \mathrm{d}S$	flow of energy of electromagnetic field;
$\int_S W_e v_i n_i \, \mathrm{d}S$	electromagnetic energy flow caused by body motion in an external magnetic field;
$\mathbf{q}$	density of thermal flux vector;
$(\mathbf{E} \times \mathbf{H})_i$	ith component of Poynting's vector.

For a body moving, or deforming, slowly with velocity $\mathbf{v}$, Ohm's law may be applied. It is written here in a modified form to account for body deformation $(\mathbf{v} \times \mathbf{B})$ and heat flow under the influence of temperature $\theta = T - T_0$

$$\mathbf{J} = \sigma(\mathbf{E} + \mathbf{v} \times \mathbf{B} - \pi_0 \, grad \, \theta) \tag{6.102}$$

Considering now equations of mass ρ_m continuity and entropy η balance, Nowacki [6.21] derived a set of three equations of magnetothermoelasticity

$$\left(\Delta - \frac{1}{\kappa}\partial_t\right)\theta - \eta \, div \, \dot{\mathbf{u}} = -\frac{\rho_m Q_1}{k_0} \tag{6.103}$$

where:

$\kappa = k_0/C_\varepsilon, \; k_0 > 1, \; C_\varepsilon$	constants; $\partial_t = \partial/\partial t$;
$\eta = \hat{\eta}/k_0$	entropy referred to mass unit;
$\hat{\eta} = \sigma' T_0, \; \sigma' = (3\lambda + 2\mu')\alpha_t, \; \mu', \; \lambda$	Lamé constants in generalized Hook's equations referred to the adiabatic state;
α_t	coefficient of linear thermal expansion; $k_0 > 0$ a constant;
$\mathbf{u}$	vector of displacements at body deformation;

ρ_m and Q_1	density, heat generated per unit mass per second.

$$(\Delta - \beta\partial_t)\mathbf{H} - \beta_0\partial_t^2\mathbf{H} = -\beta\, curl\,(\mathbf{v} \times \mathbf{H}) \tag{6.104}$$

where $\beta = \mu\sigma$, $\beta_0 = \mu\varepsilon$;

$$\mu'\Delta\mathbf{u} + (\lambda + \mu')\, grad\, div\, \mathbf{u} + \mathbf{x} + \mu[(curl\,\mathbf{H} - \varepsilon\dot{\mathbf{E}}) \times \mathbf{H}] = \rho_m\ddot{\mathbf{u}} + \sigma'\, grad\,\theta \tag{6.105}$$

The equations of magnetothermoelasticity (6.103) to (6.105) illustrate the complexity of coupled fields which have been discussed, in spite of the fact that they still contain limitations due to the assumption of slow moving media and simplifications in the form of $\mu = constant$, $\varepsilon = constant$ and $\rho = 0$.

Thermoelasticity, as a part of elasticity theory plays an important role in the investigation of stresses and strength of structural elements under the influence of changes and non-uniform distributions of temperature. It concerns particularly the end regions of rotors, rods of squirrel cages and short-circuiting rings of large induction motors, winding barstocks of turbo-generators etc. Uncontrolled dilation (linear $L_T = L_0(1 + \lambda\Delta T)$ or volume $V_T = V_0(1 + \alpha\Delta T)$ thermal expansion), can cause serious failures, especially with the assembly of parts of different coefficients of thermal expansion, as for instance $\lambda_{Cu} = 16.5 \times 10^{-6}\,K^{-1}$, $\lambda_{Al} = 24.1 \times 10^{-6}\,K^{-1}$, $\lambda_{bronze} = 17.5 \times 10^{-6}\,K^{-1}$ and $\lambda_{steel} = 11.10 \times 10^{-6}\,K^{-1}$. For isotropic bodies $\alpha = 3\lambda$.

6.6.3 Thermokinetics

In electrical power equipment it is not usually necessary to apply a full set of coupled equations (6.71) and (6.101). Indeed, the difference in time constants between electromagnetic, thermal and mechanical processes is such that they may be studied independently, i.e they are weakly coupled. However, strong coupling may be found in certain cases, for example in induction surface hardening [6.30].

The fundamental equation of thermokinetics dealing with heat exchange through contact is contained in the Fourier–Kirchhoff's law [6.10].

$$\frac{\partial\theta}{\partial t} + \mathbf{v}\, grad\,\theta = \frac{P_1}{c\rho_m} + \frac{1}{c\rho_m}\left[div(\lambda\, grad\,\theta) + \frac{\partial p}{\partial t} + \mathbf{v}\, grad\, p\right] \tag{6.106}$$

where:

$\theta = \theta(x, y, z, t)$	temperature in K;
t	time in s;
$\mathbf{v} = \mathbf{v}(x, y, z, t)$	velocity of the medium in $m\,s^{-1}$;
P_1	power of the heat source in $W\,m^{-3}$;
$p = p(x, y, z, t)$	pressure in $N\,m^{-2}$;
c	specific heat in $J\,kg^{-1}\,K^{-1}$;
ρ_m	material density in $kg\,m^{-3}$;
$\lambda = \lambda(x, y, z, t)$	thermal conductivity in W

in anisotropic media λ is a symmetric tensor.

In the case of solid bodies ($v = 0$, $p = constant$) eqn (6.106) takes the form of the heat conduction equation

$$c\rho_m = \frac{\partial\theta}{\partial t} = div\,\lambda\, grad\,\theta + P_1(\mathbf{x}, t) \tag{6.107}$$

The electrical conductivity σ and magnetic permeability μ of the body are often temperature dependent. Thus a feed-back loop is created, where the temperature field reacts, via σ and μ, on the original electromagnetic field and power loss which provided heat sources.

For thermally isotropic media ($\lambda_x = \lambda_y = \lambda_z = \lambda$), using the identity $div\, grad\,\theta = \nabla^2\theta$, we have

$$\nabla^2\theta - \frac{1}{\kappa}\frac{\partial\theta}{\partial t} = -\frac{1}{\lambda}P_1(\mathbf{x}, t) \tag{6.108}$$

where $\kappa = \lambda/(c\rho_m)$ is the coefficient of equalization of temperatures, called *thermal diffusivity*.

6.6.4 Thermometric method of power loss measurement

Equation (6.108) provides the basis for the thermometric method of measurement of local power losses in W m^{-3} and magnetic field strength H_{ms} on the surface of the body. The power P_1 developed in a unit volume inside the body can be expressed in terms of the temperature θ as

$$P_1 = c\rho_m\frac{\partial\theta}{\partial t} - \left(\lambda_x\frac{\partial^2\theta}{\partial x^2} + \lambda_y\frac{\partial^2\theta}{\partial y^2} + \lambda_z\frac{\partial^2\theta}{\partial z^2}\right) \tag{6.109}$$

In the case where the sample is in the form of a thin sheet (Fig. 6.13a) in which, in its whole thickness, a uniform power P_1 is developed, the thermal balance equation (6.108) should include a term to account for heat dissipation to the environment of ambient temperature θ_0

$$P_1 = c\rho_m\frac{\partial\theta}{\partial t} - \left(\lambda_x\frac{\partial^2\theta}{\partial x^2} + \lambda_y\frac{\partial^2\theta}{\partial y^2} + \lambda_z\frac{\partial^2\theta}{\partial z^2}\right) + \alpha'(\theta - \theta_0) \tag{6.110}$$

where α' is the resultant coefficient of heat dissipation to the surroundings through convection and radiation in W m^{-2} K^{-1}. If the body temperature is measured sufficiently rapidly at the instant of field excitation ($t = 0$), so that the temperature of the body may be considered uniform, i.e.

$$\nabla^2\theta = \lambda_x\frac{\partial^2\theta}{\partial x^2} + \lambda_y\frac{\partial^2\theta}{\partial y^2} + \lambda_z\frac{\partial^2\theta}{\partial z^2} = 0 \quad \text{and} \quad \theta = \theta_0 \tag{6.111}$$

then from eqns (6.109) and (6.110) we have

$$P_1 = c\rho_m\left(\frac{\partial\theta}{\partial t}\right)_{t=0} = c\rho_m \tan\alpha_0 \tag{6.112}$$

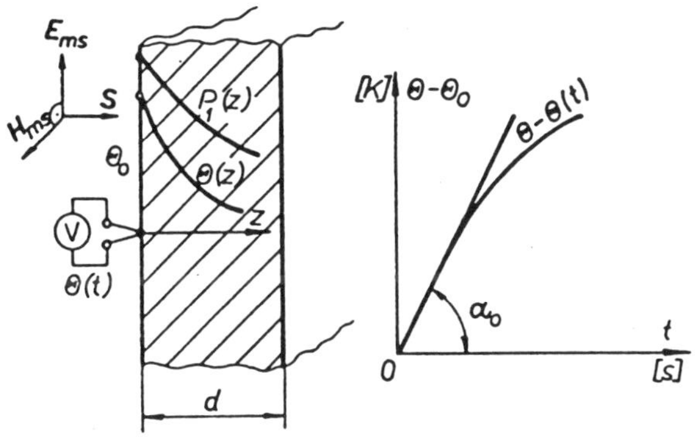

(a) (b)

Fig. 6.13 *Measurement of unit power losses* W m^{-3} *with thermometric method* $P_1 = c\rho_m \tan \alpha_0$ *([6.30], p. 406–416): (a) method of measurement, (b) reading of* tan α_0 *from the curve*

where α_0 is the gradient of the initial rectilinear part of the heating curve (Fig. 6.13b). Performing as fast and accurate as possible a measurement of the temperature with the help of built-in thermoelements (thermocouples) we can evaluate from eqn (6.112) a power loss density distribution in structural parts of electrical machines and devices [6.12] without endangering the equipment by excessive heating during the test. A disadvantage of this method is that it requires a sudden switching-on of full power. This can be difficult for large objects (e.g. power transformers). Moreover the device needs to be cooled down before subsequent measurements. This cooling can be avoided if simultaneous automatic measurements are taken with many built-in thermoelements in a short space of time, as described in reference [6.12]. Applications and errors of the method are discussed in references [6.20] and [6.30].

The less heat outflow during the measurement, that is the more closely the condition (6.111) is observed, the better is the accuracy of the thermometric method. Best results will therefore be obtained in regions which have uniform loss distribution, e.g. transformer cores. Largest errors occur when power loss and temperature distribution are varying rapidly, which takes place if the skin-effect is strong.

A typical example of such rapid changes in power and temperature distributions is in solid steel walls (Fig. 6.13a). In such solid plates the gradient of the resulting power loss $P_1(z)$ is so large [6.30] that even during the short time of measurement of the heating curve $\theta = \theta(t)$ there is considerable heat outflow from the point of measurement to colder parts of the metal. This significantly decreases the temperature on the surface and, at the same time, the value of $\tan \alpha_0$ and of the measured losses. This error has been investigated in reference [6.20]. This is a typical example of coupled fields described by eqn (6.108). If the analysis is limited to a plane wave, we have

$$\frac{\partial^2\theta}{\partial z^2} - \frac{1}{\kappa}\frac{\partial\theta}{\partial t} = -\frac{1}{\lambda}P_1(\mathbf{x}, t) \tag{6.113}$$

The right-hand side of eqn (6.113) represents eddy-current power loss calculated from distribution of the alternating electromagnetic field ([6.30], pp. 180, 189). The distribution of volume power loss density $P_1(z)$ can be calculated analytically only with the assumption of constant iron permeability μ, as

$$P_1(z) = \frac{1}{2\sigma}|J_m(z)|^2 = 2k\frac{\mathrm{ch}\,2k(d-z) + \cos 2k(d-z)}{\mathrm{ch}\,2kd - \cos 2kd}\sqrt{\frac{\omega\mu}{2\sigma}}\frac{|H_{ms}|^2}{2} \tag{6.114}$$

For strong skin effect we can assume $d \to \infty$ and eqn (6.114) reduces to [6.30]

$$P_1(z) = \frac{\omega\mu}{2}|H_{ms}|^2\,\mathrm{e}^{-2kz} \tag{6.114a}$$

In the case of ferromagnetic materials with strong nonlinearity $\mu = \mu(H)$ it is necessary [6.20] to apply a numerical method of solution of Maxwell's equations (6.83) with nonlinear magnetic permeability

$$\mu_n = \frac{\partial B}{\partial H} = \mu(H) + \frac{\partial\mu(H)}{\partial H}H \tag{6.115}$$

These equations have been solved numerically using a finite-difference method by Zakrzewski [6.36] for plane polarized waves. The magnetization curve $B = f(H)$ has been automatically interpolated for every instantaneous value $H(z)$ at a depth z from the steel surface. The program has been applied in reference [6.20] to the right-hand side of eqn (6.113). After harmonic analysis of the non-sinusoidal dependence $E_m = E_m(z, t)$ the resultant rms value for each nth elementary layer has been defined by

$$E_{rms,n} = \frac{1}{\sqrt{2}}\sum_{\nu=1}^{m} E_{\nu mn}^2 \tag{6.116}$$

where $E_{\nu mn}$ are amplitudes of corresponding time harmonics $E_n(t)$ in the nth layer. Only harmonics with amplitudes $E_{\nu n} \geqslant 0.01\,E_{\nu=1,n}$, where $\nu = 1, 2, 3, \ldots, m$ have been considered. After introducing eqn (6.116) into (6.113) the distribution of volume power loss density $P_1(z)$, necessary for further numerical solution of eqn (6.113) has been obtained (Fig. 6.14). The solution for θ of the full equation (6.113) has been compared with an idealized case of θ_{id}, for any adiabatic ($\partial\theta/\partial z = 0$) process. For such a case

$$\frac{\partial\theta_{id}}{\partial t} = \frac{1}{c\rho_m}P_1 \quad \text{and} \quad \theta_{id} = \frac{P_1}{c\rho_m}t \tag{6.117}$$

From the comparison of both solutions at the same value of Δt and H_{ms} the correction coefficient k_t has been determined to account for heat outflow

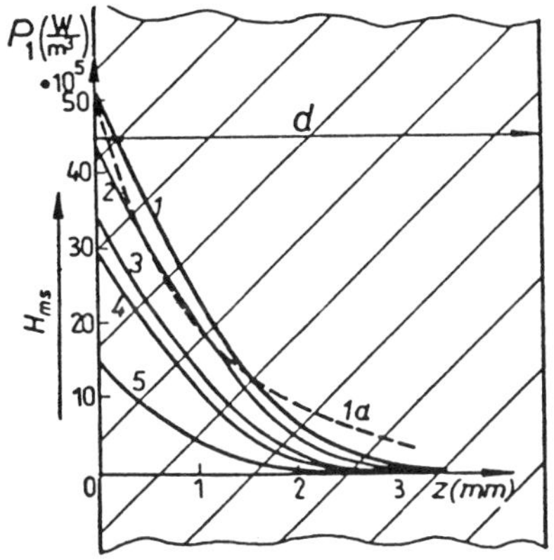

Fig. 6.14 *Volume power density $P_1(z)$ distribution at $\mu = \mu(H)$ inside solid steel wall for different H_{ms} values on its surface:*
1–5: 10.0, 8.82, 7.0, 6.17, 0.4 kA m^{-1}
1a: linear model $\exp(-2kz)$ at $H_{ms} = 10$ kA m^{-1}, $\mu_r = 150 = \text{constant}$, $k = 450\ m^{-1}$ [6.20]

from the point where the thermoelement is attached away inside the steel as

$$k_t = \left|\frac{P_{1real}}{P_{1meas}}\right|_{t=\Delta t} = \left|\frac{\theta_{id}}{\theta}\right|_{t=\Delta t} \tag{6.118}$$

In this way the actual (real) loss P_{1real} is k_t times greater than the measured P_{1meas} loss after time Δt. As $\Delta t \to 0$, $P_{1meas} \to P_{1real}$ and $k_t \to 1$ (Fig. 6.15). The stronger the excitation field the more uniform is the power distribution at a depth d inside the plate and the smaller is the correction k_t. The curves in Fig. 6.15 refer to plates placed in still air.

Using a similar method, reference [6.20] presented a method and curves which allow the evaluation of H_{ms} on the steel plate surface, on the basis of

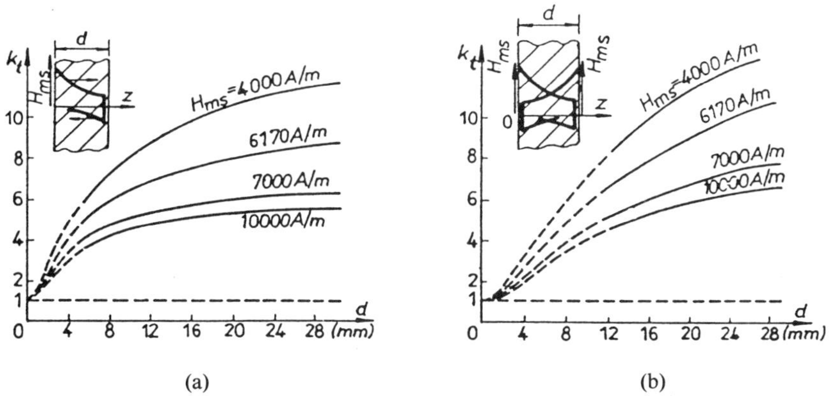

Fig. 6.15 *Coefficient $k_t = k_t(d)$ by which measured, at time $\Delta t = 1$ s, unit power loss $P_s = c\rho_m \tan \alpha_0$ should be multiplied: for (a) one-sided and (b) two-sided excitation [6.20]*

body temperature, measured after 30 seconds. The method also allows a direct measurement of the power loss density P_1 both from the side of the excitation field, and from the opposite side of the plate, e.g. the outside of a transformer tank.

6.6.5 Local overheating criteria

Uncontrolled heating of structural parts of electrical machines, power transformers, parts of switch-gear etc. affect their reliability and are among the most hazardous.

In 1960 [6.30] the method of electromagnetic criterion $H_{ms,perm}$ was proposed and has proved very useful for fast estimation and location of hot-spots, as well as places endangered by excessive heating caused by eddy currents and strong leakage fields. This method is especially convenient when combined with the reluctance network numerical modelling of leakage fields (Figs 4.6, 4.15, 4.16, 4.21) in large power transformers.

As a criterion of local overheating, the author selected the permissible tangential value of magnetic field strength $H_{ms,perm}$ on the surface of parts under investigation, at which the body temperature t and its rise $\theta = t - t_0$ do not exceed the permitted values (as recommended, for example, by CIGRE) t_{perm} and θ_{perm}. Values of $H_{ms,perm}$ can vary for different materials, excitation methods and heat dissipation ([6.30], pp. 375–392).

The unit power loss density in steel walls of thickness $d > \lambda$ ($\lambda = 2\pi\delta$, $\delta = \sqrt{[2/(\omega\mu\sigma)]}$), with one-sided excitation by an alternating uniform magnetic field of amplitude H_{ms} (Fig. 6.16), in W m^{-2} can be expressed ([6.30], p. 152) in the form

$$P_1 = a_p \sqrt{\left(\frac{\omega\mu_s}{2\sigma}\right)} \frac{|H_{ms}|^2}{2} \tag{6.119}$$

Introducing this value to the equation of thermal balance (Fig. 6.16)

$$P_1 = \alpha'\theta \tag{6.120}$$

where $\alpha' = \alpha'_0(\theta/\theta_0)^{0.25}$ is the coefficient of heat dissipation by convection and radiation, and using the analytical approximation (Fig. 4.18) of the recalculated [6.30] magnetization curve in the form

$$\sqrt{\mu_r} H^2 = c_1 H + c_2 H^2 \tag{6.121}$$

where $c_1 = 310 \times 10^2$ A m^{-1} and $c_2 = 7.9$, we finally obtain ([6.30], pp. 377–381) the permitted field value (Fig. 4.21)

$$|H_{ms,perm}| \leqslant 1962[\sqrt{(1 + 3.288 \times 10^{-8} c')} - 1] \tag{6.122}$$

where for steel at the temperature t

$$c' = \alpha'_0 \frac{\theta_{perm}^{1.25}}{\theta_0^{0.25}} \frac{2}{a_p} \sqrt{\left(\frac{2\sigma_0}{\omega\mu_0} \cdot \frac{192}{172 + t_{perm}}\right)}, \quad \sigma = \sigma_0 \frac{172 + t}{172 + t_{perm}}$$

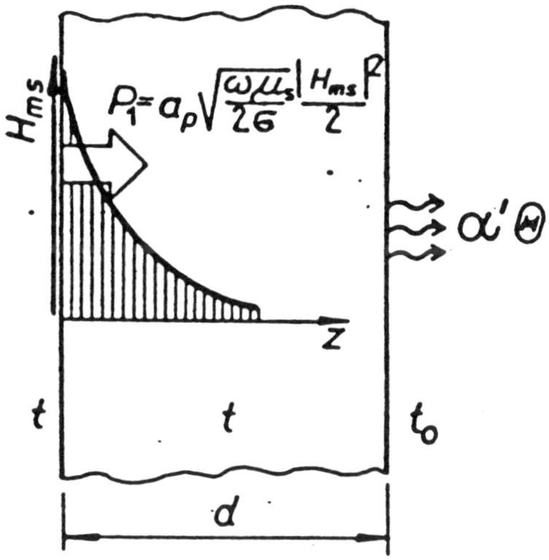

Fig. 6.16 *Thermal balance in steel wall, one-side excited and cooled*

If necessary, the values $H_{ms,perm}$ can be calculated by the numerical solution of nonlinear eqn (6.108).

Examples of the application of these criteria are given in Chapter 4. If the permitted values are found to be exceeded other materials can be used to increase the limits [6.30], or changes of geometry, screening and/or better cooling can be applied.

Effects of non-uniform field distribution

If the magnitude of the magnetic field strength H_{ms} on the surface of a metal wall exceeds the permitted value given by eqn (6.122) we can expect excessive heating. In this case a more accurate calculation should be carried out and additional precautions mentioned above may be needed.

In large electrical machines and power transformers the permissible value $H_{ms,perm}$ could easily be exceeded locally. However, the non-uniform character of the power loss density distribution can significantly reduce this hazard. Typical example are the cover plate and tank wall of a transformer, and clamping plates and end-plates of electrical machines, where the heat is carried along the plate from warmer to colder parts (Fig. 6.17).

The non-uniform power loss distribution can almost always be approximated analytically with the help of the exponential function in two-dimensions

$$P_1 = A\,\mathrm{e}^{-B|y|} \tag{6.123}$$

Since the plate thickness d is usually much smaller than others dimensions, and the thermal conductivity λ is large, the changes of temperature in the plate thickness can be neglected, and only the changes along the plate need to be considered. The heat exchange can be expressed by the equation

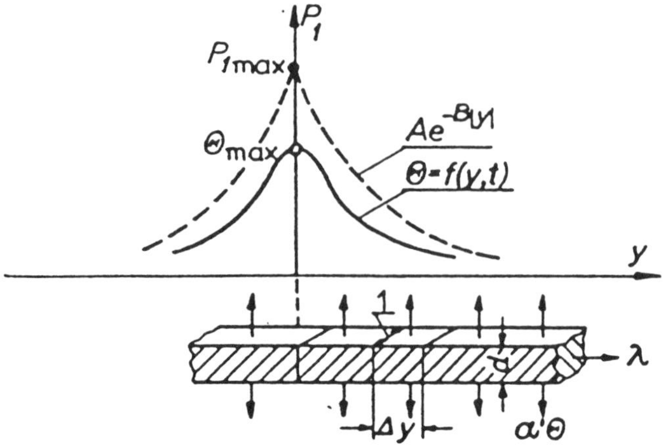

Fig. 6.17 *Non-uniform loss density and temperature distribution in a steel plate [6.30]*

$$c\rho_m = \frac{\partial\theta(y,t)}{\partial t} = \lambda\frac{\partial^2\theta(y,t)}{\partial y^2} + q \tag{6.124}$$

where the symbols are the same as those in eqn (6.106) and q is heat quantity per unit volume, which comprises power losses $2P_1/d$ as a positive source, and heat carried away $\alpha'\theta$ as a negative heat source, i.e.

$$q = \frac{2P_1\Delta y}{d\Delta y} - \frac{2\alpha'\theta\Delta y}{d\Delta y} = \frac{2A\mathrm{e}^{-By}}{d} - \frac{2\alpha'\theta}{d} \tag{6.125}$$

Neglecting transients, eqn (6.125) was resolved [6.30] for the thermal quasi-stationary state and the following maximum stationary temperature rise θ_{max} was obtained (Fig. 6.17)

$$\theta_{max} = \frac{2A}{\lambda dB^2 - 2\alpha'}\left[\frac{B}{\sqrt{\left(\frac{2\alpha'}{\lambda d}\right)}} - 1\right] = \frac{A}{\alpha'}K \tag{6.126}$$

where

$$K = \frac{1}{\frac{B}{\sqrt{\left(\frac{2\alpha'}{\lambda d}\right)}} + 1} \leqslant 1 \tag{6.127}$$

The coefficient K (6.127) accounts for the heat carried away from the hot-spot to a colder place. At $B = 0$ or $\lambda = 0$ the factor $K = 1$. In practice ([6.30]) K is often 0.5 to 0.7 or even less.

The values $H_{ms,perm}$ given in Fig. 4.21 can be considered as a criterion of overheating due to eddy currents, and the coefficient K (6.127) as the necessary safety factor in such cases. By dividing by K we can make the criteria mentioned above more accurate (Fig. 4.21)

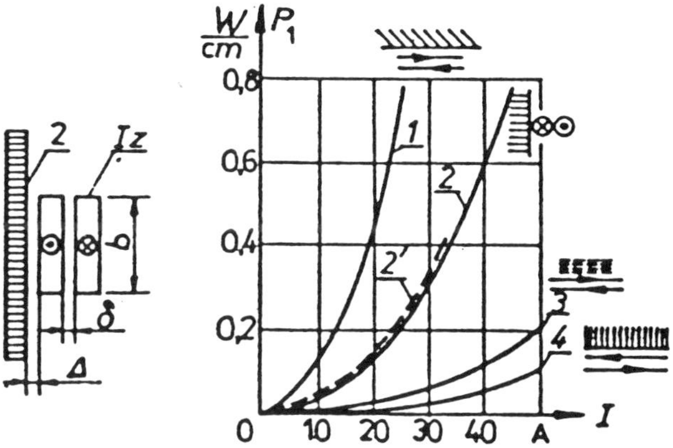

Fig. 6.18 *Influence of iron core lamination on decrease of eddy current losses in a steel wall placed in a 50 Hz field of two busbars [6.30]*

$$H_{ms,perm,K} = \frac{H_{ms,perm}}{K} \tag{6.128}$$

By decreasing the value of K through changing the parameters present in eqn (6.127), we can increase the value of the permitted field $H_{ms,perm}$ and so reduce the thermal hazard of the part under investigation. In a more accurate analysis of eqn (6.124) the matter of steel nonlinearity should be considered, because this nonlinearity may have an additional influence on design. In reference [6.30], pp. 385–392 criteria for other materials, methods of field excitations and different cooling conditions have been collected.

Effects of the lamination of steel walls

Power loss in laminated iron cores is much smaller than in solid steel walls (Fig. 6.18). This results from both the reduction of eddy currents owing to the laminated core, and the higher resistivity of silicon iron. This has been confirmed by measurements on models (Fig. 6.18) with dimensions $b = 198$ mm, $\delta = 50$ mm, $\Delta = 10$ mm and number of turns $N = 288$ in one 'bar' (winding). Power losses in a solid steel wall of thickness 12 mm (curve 1), after recalculation [6.30] in the ratio $P_2' = P_1(\sigma_2/\sigma_1)^{0.9}$ have given curve 2′, which almost exactly coincides with the curve 2 of losses measured in a stack of laminates of *Epstein's apparatus*, when placed at right-angles with respect to the flux lines. The same laminations placed approximately along the field lines (curve 3) give significantly smaller power losses, and when placed completely along the field lines (curve 4) produce negligible losses.

For the development of the criterion $H_{ms,perm}$ for the local overheating on iron core surface (transformers, machines) the models should first be based on curve 3 (Fig. 6.18). Such investigations, initiated by Mankin and Morozov [6.16] have been extended significantly by Kazmierski, Kozlowski and others [6.9], Fig. 6.19. Full solution of this problem requires three-dimensional (3D)

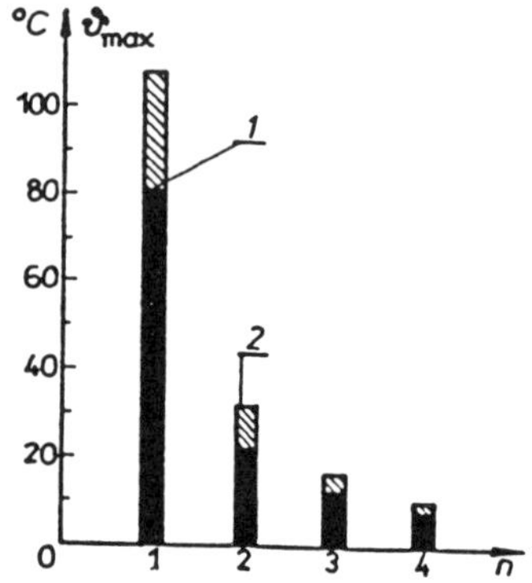

Fig. 6.19 *Influence of the subdivision of external packet into n narrower packets on the maximum packet temperature:*
1. with a magnetic clamping plate,
2. with non-magnetic plates [6.9]

analysis, which poses a considerable challenge for contemporary numerical techniques.

The leakage flux of density B_C (Fig. 4.14) which penetrates perpendicularly the surface of sheets and clamping plates of the core, may cause local overheating, especially at the top and bottom edges of windings. Such effects were investigated using full scale models and are reported in [6.9]. It has been found that the normal component of flux density on the surface of the core distributes itself inside the external packet of laminations according to the relationship

$$B_C(y) = B_C(0)\,(1 + ay^3) \tag{6.129}$$

where y is the distance from the symmetry plane of the packet and $B_C(0)$ is the normal flux density at the centre.

The local temperature rise is approximately proportional to $(f\phi_c)^2$, where $\phi_c = \int B_C\,dy$ (Wb m)$^{-1}$. It follows that the best way of reducing the temperature is subdividing the external core packet into n narrower packets. This was confirmed by computations (Fig. 6.19) made for a 250 MV A, 220 kV autotransformer.

6.6.6 Induction heating

Coupled electromagnetic and thermal fields occur in many kinds of induction heating, e.g. heating in forging and annealing, melting, transportation and batching of liquid metals. This coupling is usually stronger than in other electrical machines and devices. Typical examples of interaction of an electromagnetic field with a thermal field include the penetration of energy into metal for induction heating or skin hardening of solid steel (Fig. 6.20).

The electromagnetic field distribution in metals in the initial cold state (σ_3, μ_3) and the final hot state (σ_2, μ_2) may be described by the same equations

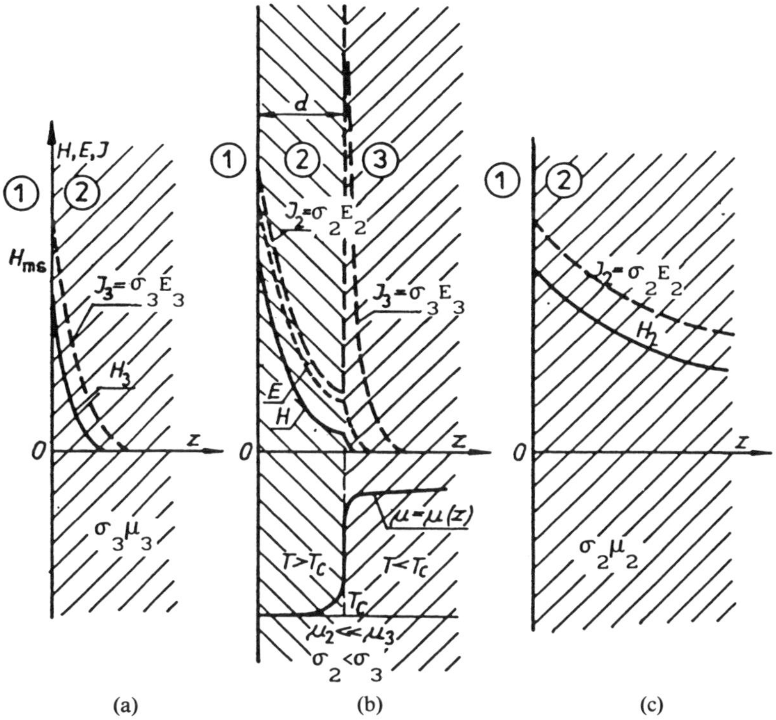

Fig. 6.20 *The penetration of electromagnetic fields* **H, E, J** *into steel for induction heating: (a) initial, cold state; (b) transition state with increasing depth d of layer with Curie temperature $T = T_C$ and (c) final, hot state ([6.30], p. 300)*

$$H_m = H_{ms}\,e^{-\alpha z}; \quad J_m = \sigma E_m = \frac{\alpha}{\sigma} H_{ms}\,e^{-\alpha z}; \quad P_1 = \sqrt{\left(\frac{\omega\mu}{2\sigma}\right)}\,\frac{|H_{ms}|^2}{2} \tag{6.130}$$

only if the thickness d (diameter) of the body is larger than the wave-length λ in metal [6.30], where

$$\lambda = 2\pi\sqrt{\frac{2}{\omega\mu\sigma}} = 2\pi\delta, \quad \delta = \sqrt{\frac{2}{\omega\mu\sigma}} \tag{6.131}$$

and δ is the skin depth. If the thicknesses $d < \lambda$ the internal reflected waves should also be considered ([6.30], pp. 178–194).

The transition from the initial (cold) state to the stationary (hot) state is accompanied in steel by an almost 90-times increase of the skin depth δ owing to the 1000–2000-times decrease of μ and the 6–10-times decrease of σ. As $E_3 = E_2$ on the interface between layers 2 and 3 (Fig. 6.20b), a sudden jump of current density $J_3 = \sigma_3 E_3$ takes place on the surface of region 3. This helps maintain uniform heating of the external layer of steel 2 for surface hardening, without a risk of burning the surface of the hardened element. This is

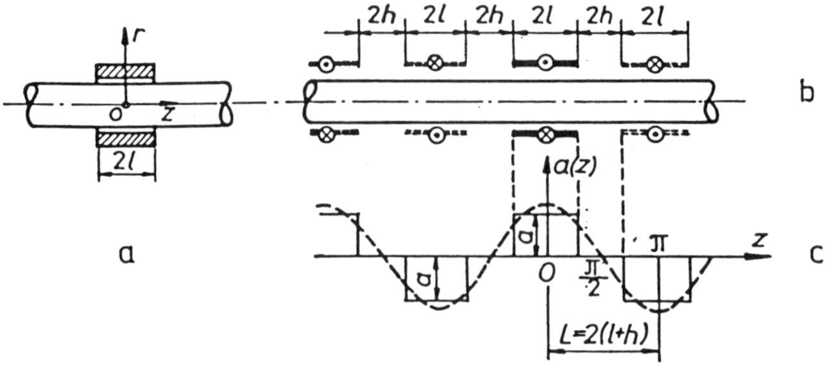

Fig. 6.21 *Application of the method of analytical prolongation: (a) core with inductor, (b) infinite sequence of fictitious inductors, (c) periodic flow distribution ([6.32], p. 289)*

obviously only a qualitative explanation. As a first approximation such calculations can be made in a similar way as for electromagnetic screens ([6.30], pp. 176–194). If necessary, the nonlinear permeability μ of the cold solid steel layer can be considered using the corresponding linearization coefficients a_p, a_q ([6.30], pp. 139, 318–328).

Calculations of inductors using the method of analytical prolongation
Induction heating is often applied to cylindrical metal charges by an inductor of finite length (Fig. 6.21). Such a system is often solved with the help of the so-called method of analytical prolongation, which uses a Fourier series ([6.30], pp. 207–211, 333–345). This method gives in general good results however the problem now arises in selecting a suitable distance $2h$ between the fictitious inductors of an infinite sequence of mirror reflections. The greater the distance $2h$, the smaller the interference from the nearest fictitious inductors. At the same time however, account needs to be taken of the higher harmonics on summation of the Fourier series. For this, a small distance $2h$ should be selected, but not too small to avoid interference between adjacent inductors.

The optimal distance $2h$ can be evaluated as follows. The field in the air gap of the inductor for the magnetic scalar potential V_m can be calculated from Laplace's equation $\nabla^2 V_m = 0$. After solving Laplace's equation we can find the components of the magnetic field strength $\mathbf{H} = -grad\, V_m$, for a given specific electric loading $a = a(z)$, which in the case of cylindrical inductors (Fig. 6.21) can be expanded into the Fourier series as

$$a(z) = \frac{4a}{\pi} \sum_{n=1}^{\infty} \frac{1}{n} \sin \frac{n\pi l}{2(l+h)} \cos \frac{n\pi z}{2(l+h)} \tag{6.132}$$

When studying the edge effects of the inductor we can use the approximate formula ([6.32], p. 131) for radial flux density B_{mr} (Fig. 6.22)

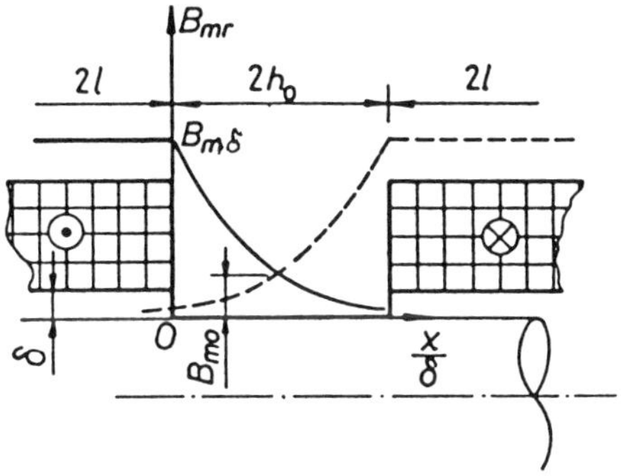

Fig. 6.22 *Evaluation of the optimal distance $2h_0$ of the fictitious inductors ([6.30], p. 259)*

$$\frac{B_{mr}}{B_{m\delta}} \cong \frac{1}{1 + \frac{\pi}{2}\frac{x}{\delta}} \tag{6.133}$$

where $B_{m\delta}$ is the maximum flux density in the gap δ on the edge of the inductor.

If we assume the smallest permissible value of the relative flux density $b_0 = B_{m0}/B_{m\delta}$ in the centre of the armature, we can evaluate the corresponding optimum distance

$$2h_0 = \frac{4}{\pi}\delta\left(\frac{1}{b_0} - 1\right) \tag{6.134}$$

which for $b_0 = 0.05$ gives $2h_0 = 24\,\delta$. More generally the condition may be defined as $2h_0 \geqslant \delta$.

Equation (6.134) refers to stationary inductors. If the inductor moves along the z-axis (as is the case in linear motors or induction heaters using movement), the reaction of eddy currents in the secondary circuit causes a weakening of the field on the incoming edge of the core, and increase on the edge moving away. The latter decays slowly in air and creates a so-called 'magnetic tail' (Fig. 6.9). In this case the optimum distance $2h_0$ significantly changes and depends on the speed of the core (Fig. 6.23a). At constant speed it is a function of the coefficient $b_s = B_m(x = 2h_0)/B_{m\delta}$. The values can be evaluated from graphs shown in Fig. 6.23. In order to avoid errors, the smallest value of b_s should always be used. In practice satisfactory accuracy is obtained by assuming $b_s \leqslant 0.02$.

Numerical methods for induction heating

The most popular methods nowadays for field computation are the finite-element, finite-difference and boundary-element methods. However, in the electrothermic industry these methods are used rather reluctantly [6.14], because simple models often give quite satisfactory results.

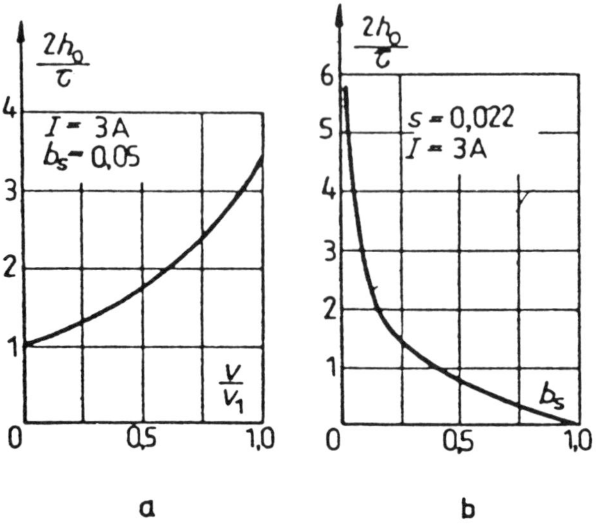

Fig. 6.23 *Dependence of the optimum distance $2h_0/\tau$ of fictitious armatures in a linear induction motor SL-5-270: (a) on speed, (b) on assumed flux density b_s*

One of the oldest finite-difference formulations, still in use, for a heating problem of a steel cylinder, uses two closely coupled differential equations

$$\frac{1}{r}\frac{d}{\mathrm{d}r}\left(r\frac{\mathrm{d}H_m}{\mathrm{d}r}\right) - \alpha^2 H_m = 0 \tag{6.135}$$

$$\rho_m c\frac{\partial T}{\partial t} = \frac{1}{r}\frac{\partial}{\partial r}\left(r\lambda\frac{\partial T}{\partial t}\right) + q \tag{6.136}$$

with a boundary condition on the surface

$$-\lambda\frac{\partial T}{\partial n} = \alpha_c(T_s - T_a) + C_s(T_s^4 - T_a^4) + Q_s \tag{6.137}$$

where:

$$\alpha^2 = \mathrm{j}\omega\mu\sigma, \qquad \mu = \mu(r, H_m, \theta), \qquad \sigma = \sigma(r, \theta)$$

T	temperature (K);
Q	power loss density (W m^{-3});
$\rho_m = \rho_m(T)$	mass density;
$c = c(T)$	specific heat;
$\lambda = \lambda(T)$	thermal conductivity;
s and a	refer to surface and ambient temperature indices respectively;
α_c	coefficient of heat dissipation by convection;
c_s	heat dissipated by radiation;
Q_s	constant surface losses (W m^{-2}).

Using simple transformations, the above solution may be extended to cross-sections other than circular [6.14].

In references [6.26; 6.27] the induction mixing of liquid metal using finite differences has been solved. In this case the coupling between electromagnetic and flow equations through the field $\mathbf{v} \times \mathbf{B}$ is weak and equations can be solved independently.

For electroheat eddy-current or liquid metal mixing problems, reference [6.2] reports a successful application of the volume integral equation. This method is a very convenient approach to the solution of coupled axisymmetrical, mechano-thermo-electromagnetic problems [6.14]. For this application, the boundary element method, on the other hand, is seen as less useful.

Simulation of induction heating using finite elements, in contrast to the problems of induction mixing, is shown to be very appropriate according to reference [6.14]. However the final matrix equations for finding the value H_m in internal nodes

$$\{[S(\rho)] + j\omega[T][\mu]\}[H] = \{b\} \tag{6.138}$$

where $\rho = \rho(T)$ and $\mu = \mu(H_0, T)$ can be uncoupled from the thermal equations

$$\rho_m c \frac{\partial T}{\partial t} = \frac{\partial}{\partial x}\left(\lambda \frac{\partial T}{\partial x}\right) + \frac{\partial}{\partial y}\left(\lambda \frac{\partial T}{\partial y}\right) + w \tag{6.139}$$

if the ferromagnetic conductor is in a non-magnetic state, i.e. above the Curie temperature (Fig. 6.20). In eqn (6.139) w is the internal heat source in $\mathrm{W\,m^{-3}}$.

A comprehensive analysis of the state of the art of induction heaters and their methods of computation has been made by Skoczkowski, cited in reference [6.33]. From his study it follows that most of the models applied so far are one-dimensional and linear, and use averaged values of material parameters. The most popular methods used are separation of variables (Fourier), Green functions, the Laplace and Hankel integral transformations, Fourier series and integrals, integral equations and others. Skoczkowski has also elaborated the simplified numerical solution of the system of coupled Maxwell's and Fourier–Kirchhoff's equations for induction heating of steel charges, e.g. steel tubes. Applications of finite elements to similar problems of coupled fields for thermal anisotropy of a body are also presented in reference [6.18].

6.7 LEVITATION

Levitation is a technical term used to describe all phenomena connected with lifting or suspension of ferromagnetic bodies under the influence of magnetic fields of permanent magnets or electromagnets, or under the influence of induced currents in solid conducting bodies placed in alternating fields or as a result of rapid movement with respect to the field. Currently this effect is

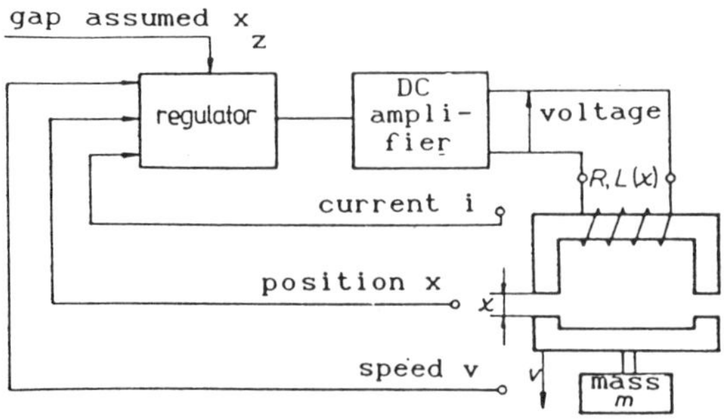

Fig. 6.24 *Levitation of an iron cramp of an electromagnet [6.5]*

used on an industrial scale both for suspension of trains on magnetic cushions [6.19] and in technical processes (melting of metals, welding technology, separation of steel sheets etc.), in machine design (magnetic bearings), diagnostics (e.g. free suspension of bodies in aerodynamic tunnels) etc. Gieras [6.5] has investigated linear induction motors with levitating secondary parts, driven and stabilized electrodynamically, using electromagnets with automatically controlled air gaps, magnetic bearings etc. In experimental systems of electromagnetic levitation (Fig. 6.24) the stability of air-gap x is achieved with the help of negative current feed-back, controlled by sensing position, linear speed v and coil current i. The operation of the magnet is described by the system of equations

$$\frac{\partial x}{\partial t} = v, \quad m\frac{\partial v}{\partial t} = mg + \frac{1}{2}i^2\frac{\mathrm{d}L(x)}{\mathrm{d}x}, \quad u = \frac{d}{\mathrm{d}t}[L(x)i] + Ri \tag{6.140}$$

where

$$L(x) = L_\infty + \frac{L_0 x_0}{x}, \quad L_\infty = L(x \to \infty) \tag{6.141}$$

x_0 is the nominal air-gap.

When applied in the feedback state variables x, m and i affect the supply voltage u of the electromagnet according to the relationship

$$u = K(x - x_z) + K_v v - K_i i + C(x_z, i) \tag{6.142}$$

where K, K_v, K_i are the feedback coefficients and $C(x_z, i)$ is evaluated for the stationary state ($x = x_z$, $v = 0$) and equals

$$C(x_z, i) = u + K_i i \tag{6.143}$$

Equations (6.140) to (6.143) are used among others for the evaluation of the stability zone of the system.

The force of attraction of the electromagnet with an air-gap x (Fig. 6.24) can be calculated from the formula

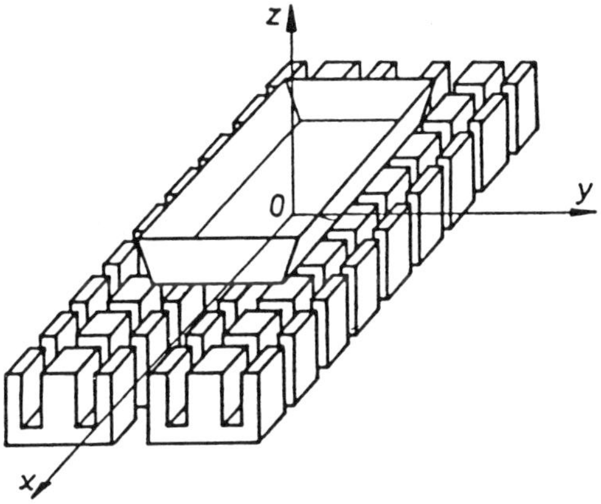

Fig. 6.25 *Levitation of secondary element of linear induction with transverse field*

$$F = \frac{\mu_0 S_\delta (iN)^2}{4\left(\dfrac{l_{Fe}}{2\mu_r \dfrac{S_{Fe}}{S_\delta}} + x\right)^2} \cong \frac{\mu_0 S_\delta (iN)^2}{4x^2} \tag{6.144}$$

where:

- S_δ cross-section of the air gap;
- S_{Fe} cross-section of the iron core;
- l_{Fe} length of path of magnetic flux in the core and magnet keeper;
- μ_r relative magnetic permeability of the core depending on coil current i.

In the model of the linear motor with levitating secondary element (Fig. 6.25) the inductor (armature) of the motor is made up from two rows of cores with mk cores in one row, where m is the number of phases, $k = 1, 2, 3, \ldots$. The aluminum block has inclined sides angled at 60°, which ensures the highest levitaion force with side stabilization.

The boundary condition for Laplace's equation in air and Helmholtz equation inside the walls of the secondary is the distribution of the normal flux density component on the inductor surface, which with a three-phase supply with a sequence $R, -S, T, -R, S, -T, \ldots$ has the form (Fig. 6.26):

$$B_z(x, y) = 0.5 B_0 \sum_{\nu=1}^{\infty} \sum_{n=1}^{\infty} b_\nu b_n [e^{j(\omega_\nu^+ t - \beta_\nu x)} (1 + a^{\nu+2} + a^{2\nu+1}) + e^{j(\omega_\nu^- t + \beta_\nu x)} (1 + a^{2\nu+2} + a^{\nu+1}) \cos \eta_n y] \tag{6.145}$$

where:

- B_0 flux density at the centre of symmetry of the main column;
- $\nu = 1, 3, 5, \ldots$ space harmonics in the x-direction;
- $n = 1, 3, 5, \ldots$ space harmonics in the y-direction;

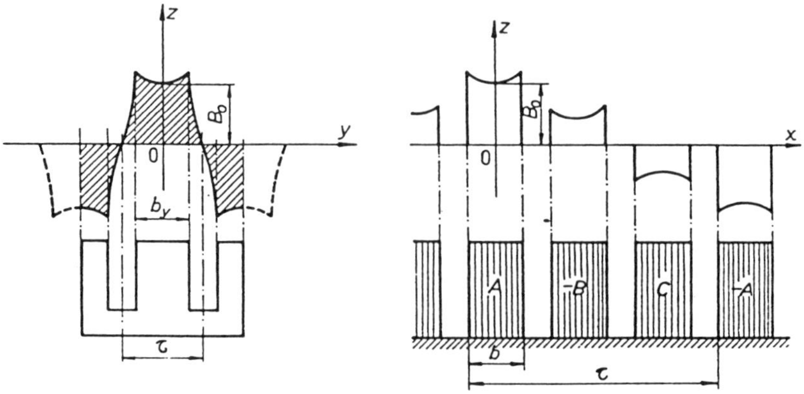

Fig. 6.26 *Magnetic flux density distribution on the surface of the inductor*

$\omega_\nu^+ = s_\nu^+ \omega$; $\omega_\nu^- = s_\nu^- \omega$; $\omega = 2\pi f$; s_ν slip of νth harmonics;
$\beta_\nu = \nu\pi/\tau$; $\eta_n = n\pi/\tau_y$; τ; τ_y (Fig. 6.26);
$a = \exp\left(j\dfrac{2\pi}{3}\right)$; b_ν; b_n coefficients of Fourier's series.

Indices '+' and '−' indicate direct and reverse components of the rotating field respectively.

Gieras [6.5] has also given a solution of the equations for the field distribution. He recommends the application of Maxwell's stress tensor for the calculation of forces acting on the unit surface of a secondary

$$f_{z\nu n} = -0.5\,\mu_0\,\mathrm{Re}\left[H_{z\nu n} H^*_{x\nu n}\right]_{z=\delta} \tag{6.146}$$

$$f_{z\nu n} = 0.5\,\mu_0\,\mathrm{Re}\left[0.5\,H_{z\nu n} H^*_{z\nu n} - 0.5\,H_{x\nu n} H^*_{x\nu n} - 0.5\,H_{y\nu n} H^*_{y\nu n}\right]_{z=\delta'} \tag{6.147}$$

where δ is the air-gap dimension.

The total forces acting on the secondary part, taking into account that the field is excited by two rows of electromagnets, are

$$F_x = 2L \sum_{\nu=1}^{\infty} \sum_{n=1}^{\infty} \int_0^{\tau_y} f_{x\nu n}\,\mathrm{d}y \tag{6.148}$$

$$F_z = 2L \sum_{\nu=1}^{\infty} \sum_{n=1}^{\infty} \int_0^{\tau_y} f_{z\nu n}\,\mathrm{d}y \tag{6.149}$$

where L is the length of secondary part of motor in x-direction.

7
Computer Aided Design in Magnetics

Jan K. Sykulski

7.1 THE CAD ENVIRONMENT

Electromagnetic field computation as an aid to engineering design is an area of very active research. The more traditional techniques used for device analysis in practical design rely on the notion of magnetic circuit analogy, but the computer based numerical field analysis attracts increasing interest from existing and potential users of computational tools for a wide range of technical applications. Many companies, in diverse branches of industry, have already incorporated *Computer Aided Design in Magnetics* into their everyday design practice. Application areas include electrical machines, MRI magnets, electromechanical actuators, induction heating, power transmission, robotics, non-destructive testing, electromagnetic compatibility, recording and audio industry, transport systems, particle physics etc. The list is virtually endless. Advances in computational techniques for design purposes, as well as their implementations and applications, are regularly reported in learned society journals (IEEE Transactions, IEE Proceedings), professional journals (e.g. COMPEL [7.47; 7.71]), and specialist conferences such as COMPUMAG [7.12; 7.13; 7.14; 7.15], ISEF [7.36; 7.71], INTERMAG, ICEM and others. Finally, the reader is encouraged to consult other introductory and advanced books and publications on the subject (Binns, Lawrenson and Trowbridge [7.4], Biro and Richter [7.5], Hammond and Sykulski [7.30], Jin [7.37], Lowther and Silvester [7.40], Silvester and Ferrari [7.64], Silvester [7.67]). In this Chapter we shall review the more fundamental ideas and address some of the more advanced related issues, such as optimization and error estimation.

First, a designer has to build a computational model of his physical problem. This, of course, is a vital step, often underestimated, which may decide whether the whole process is a success or a failure. Clearly, the model must be adequate for the results to be meaningful. This emphasizes the significance of human input to the design and importance of an experiment and/or alternative models for solving the same problem, so that comparisons can be made. Once the model has been formulated the CAD system will facilitate finding the solution. A typical CAD system [7.30] is illustrated in Fig. 7.1.

There are three distinctive, although interlinked, stages in the magnetic CAD process:

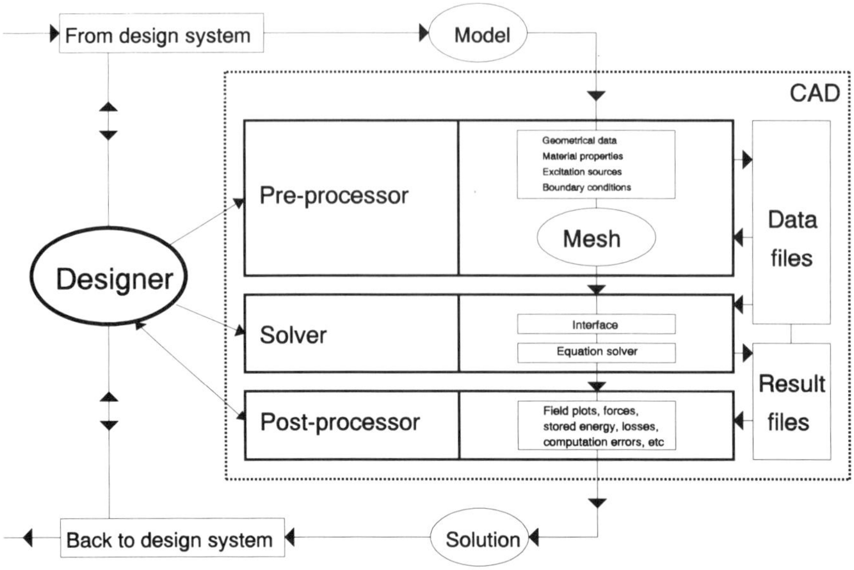

Fig. 7.1 *A typical CAD system for magnetics*

- *Pre-processor*: for defining geometry, input of material data and definition of sources and boundary conditions (including possible symmetries),
- *Solver*: for performing computation, e.g. solution of the system of algebraic equations in the finite-element or finite-difference formulation, or calculation of series/parallel connections of components in the 'tubes and slices' formulation (see Sections 1.10 and 1.11),
- *Post-processor*: for graphical display of field distributions, calculation of field components, and integrating along lines, over areas or inside volumes, to find integral parameters such as forces, stored energy, impedances etc.

In most contemporary CAD systems both pre- and post-processing are interactive, thus taking full advantage of graphics capabilities of modern computers. A combination of keyboard/mouse input is usually available (i.e. *command driven* versus *menu driven*), some systems offer direct transfer from drawing packages such as *AutoCad*. Results files are often transferred to mechanical stress or thermal analysis programs for further processing – such programs may be built-in or external. The rapid developments in graphics user interfaces (GUI) make the programs increasingly user-friendly and easier to use. The solution process, on the other hand, is very often 'hidden' from the user, i.e. run in the background, performed using batch processing, or sent, via a network, to a mainframe computer. (In the 'Tubes and Slices' method and program – see references [7.31], [7.72] or [7.76] – however, owing to

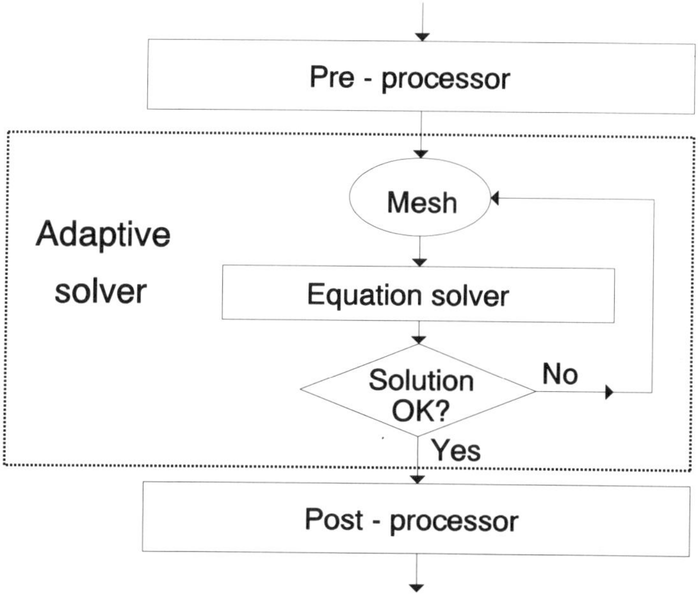

Fig. 7.2 *Adaptive meshing*

efficient formulation of the method, the solver is effectively amalgamated with the rest of the graphics interface, because solution times are extremely short.) All processors share a common database.

The engineering software used in such CAD systems is necessarily highly specialized, and its development is influenced, on the one hand, by advances in computer hardware and software (e.g. operating systems, window environments), and, on the other hand, by actual and perceived needs of potential users. The life-span of any software package is usually limited. A typical large-scale finite-element program contains hundreds of thousands of lines of code and is the result of many man–years of development. Its total life-cycle is estimated at about 10–12 years [7.40], which is quite a long time in a high-technology world.

Most commercially available software is based on a finite-element formulation (see the survey in Section 7.10 later in this Chapter). These programs offer various levels of automation, but of particular interest and importance is how they handle the mesh generation. We shall discuss various techniques in the next Section, but it would be appropriate to mention here that in an ideal design system the discretization should in fact form part of the solution stage, rather than pre-processing, as shown in Fig. 7.2. This is known as *adaptive meshing*, and some existing programs already offer this facility, although perhaps it may not be yet fully integrated into the design process. Adaptive meshing, although simple as a general concept, is rather difficult to implement in practice, especially for time-varying fields, and is an area of active research and development. Adaptive meshing is directly associated with error analysis. Both topics are discussed later in this Chapter (Section 7.6).

The difficulties hindering the wide-spread usage of electromagnetic software are that the underlying theory is conceptually complicated, mathematically demanding and computationally challenging. At the same time, and probably for the same reasons, electromagnetism has never been a favourite subject amongst students of engineering and physics, and a high proportion of designers in various sections of industry, who would otherwise benefit from using the available CAD systems, have insufficient electromagnetic background. More effort is invested into making the subject more 'user-friendly', and the high capabilities of modern computers in field visualization should make an important impact on improving the situation [7.65; 7.66]. A recent book by Hammond and Sykulski [7.30], and associated computer program TAS, both already mentioned in this book, make an attempt to bridge the existing gap by providing a simple and friendly computational tool, and at the same time emphasize the field geometry and structure, rather than its mathematical description.

At the same time researchers are looking at ways of simplifying the interaction between the software and the user. One such development has lead to the idea of *Design Environment Modules* (DEMs) [7.3]. A DEM facilitates the use of electromagnetic analysis software by providing an application-specific shell to guide a non-specialist through the geometric design and physical property specification of a class of device. After analysis it then provides values of the important design parameters. The DEMs are created by experts and contain a parametrized model of a device or group of devices with a set of decision making routines suggesting optimal representation of materials and boundary conditions, followed by automatic or adaptive meshing. The

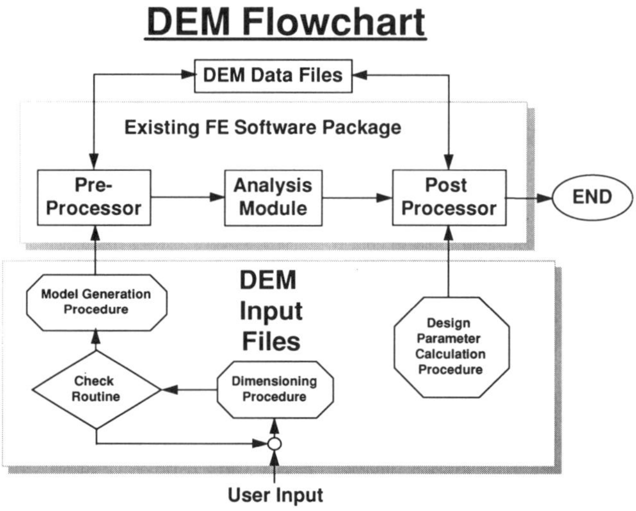

Fig. 7.3 *A general DEM and its relation to existing FE package*

post-processor offers top level commands for specific tasks such as calculation of device parameters. The command language of the software is used throughout. A general DEM is shown in Fig. 7.3, and more details about a particular implementation may be found in reference [7.3]. The design modules are currently under development for the package OPERA-2d [7.48].

The requirements imposed by a modern design environment on electromagnetic software are quite tough. A potential buyer or current user of a package expects speed of computation, reliable error estimation with guaranteed accuracy, practical and convenient ergonomics for data input and processing, accurate and meaningful field visualization and post-processing, high levels of automation, fully interactive communication and reasonable costs. Thus sophisticated and interchangeable database structures, interactive pre- and post-processing, automatic meshing, adaptive solvers, knowledge based systems, are all necessary components of contemporary systems for magnetic design.

7.2 PRE-PROCESSING AND MESH GENERATION

Most finite-element programs use three types of mesh generation: deterministic, semi-automatic and inter-active. The fourth possibility arises of a fully automatic procedure when combined with an adaptive solver (see Section 7.6). The choice of method often depends upon the type of geometric 'primitives' (i.e. basic geometrical shapes) which are used to construct the finite-element model. At the same time the designer may wish to have more control over the element distribution, for example to take proper account of the 'skin effect'. A 'manual' input may be easily implemented for a quadrilateral, where the definition of boundary subdivisions results in a predetermined internal meshing. Some simple examples in two dimensions are illustrated in Fig. 7.4. Clearly, there are more possibilities and the principle applies not only to a rectangle but to any quadrilateral, which may be achieved through iso-parametric mapping (see Fig. 7.5).

Interactive methods allow particular 'nodes' or 'construction lines' of the mesh to be repositioned directly on the screen of the terminal using a graphics interface, e.g. a mouse. This method is used to complement other methods, for example to apply 'finishing touches' to a manually or automatically generated mesh.

Semi-automatic methods usually apply to polygons of any shape, whilst the discretization of sides is either predetermined or automated using some simple rule (e.g. similar length of the 'edge' subdivisions). The objective is first to find the optimal distribution of internal 'nodes', and then the best way of connecting these points. The intuitive criterion is that all triangles should be as near to equilateral as possible, to avoid large obtuse angles, and thus to produce a mesh which is 'well proportioned'. An illustration of a 'bad' and a 'good' mesh for a given set of points is given in Fig. 7.6. Several algorithms have been developed and are well covered in specialist literature. Probably the most popular method is based on the so-called *Delaunay triangulation* [7.16].

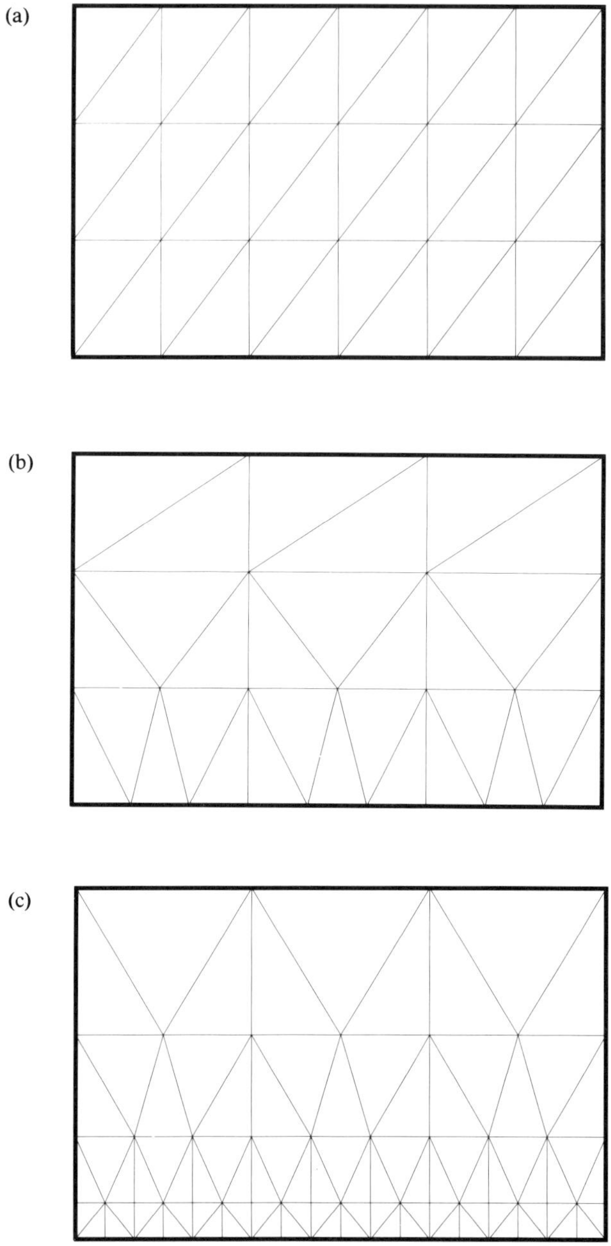

Fig. 7.4 *Simple meshes for rectangles [7.30]: (a) regular, (b) graded in one direction, (c) graded and non-uniform*

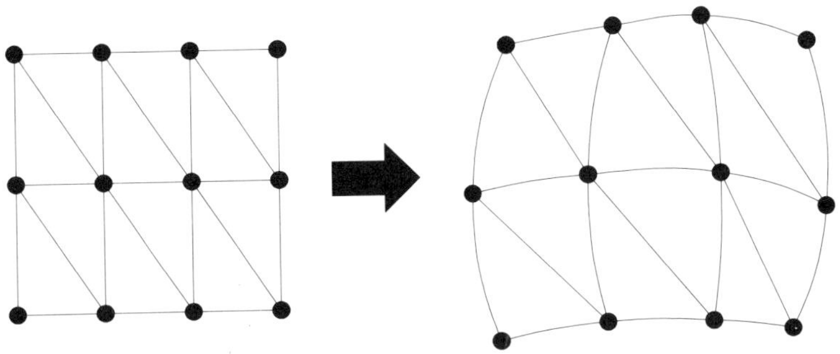

Fig. 7.5 *Isoparametric meshing*

The criterion is that the circumcircle of a Delaunay triangle may not contain another vertex in its interior. A well constructed mesh of triangles produced by the Delaunay triangulation maximizes the sum of the smallest angles in the grid and thus achieves as near an equilateral set of triangles as is possible using the given points. We shall illustrate the procedure using a particular scheme.

During the specification of the geometry of a problem, once a region in the form of a polygon has been defined, and boundary subdivisions entered (either manually or carried over from another region, so that continuity of discretization is preserved), the initial triangulation may commence. This may be done, for example, by selecting the first two boundary points as vertices of the first element, and then testing the suitability of all other points to become the third vertex, see Fig. 7.7b. Two tests are conducted:

- the lines connecting the first two vertices with the new point must not cross any existing internal or boundary lines, and
- no other boundary point is allowed on either of the two lines (as an example, point D in Fig. 7.7b would be rejected).

If the point has passed the two checks, the radius squared (r^2) of the circumscribing circle for the newly formed triangle is calculated and stored. The procedure is repeated for all remaining available boundary points. A point with the smallest value of r^2 is chosen as the third vertex.

When a new triangular element is constructed, either one or two internal lines are created. If one new internal line is formed (and thus the other one is an existing boundary line), it is taken as the base-line for the next element and the process is repeated, although there is now one point less available than previously. If two internal lines are created, one is used as the next base-line and the other is placed in a last-in first-out stack. If no suitable point can be found, a new base-line is taken from the stack. The process continues until no suitable point can be found and the stack is empty. An example of the initial triangulation is shown in Fig. 7.7a. The triangular elements constructed are of poor quality and further points need to be added inside the region.

(a)

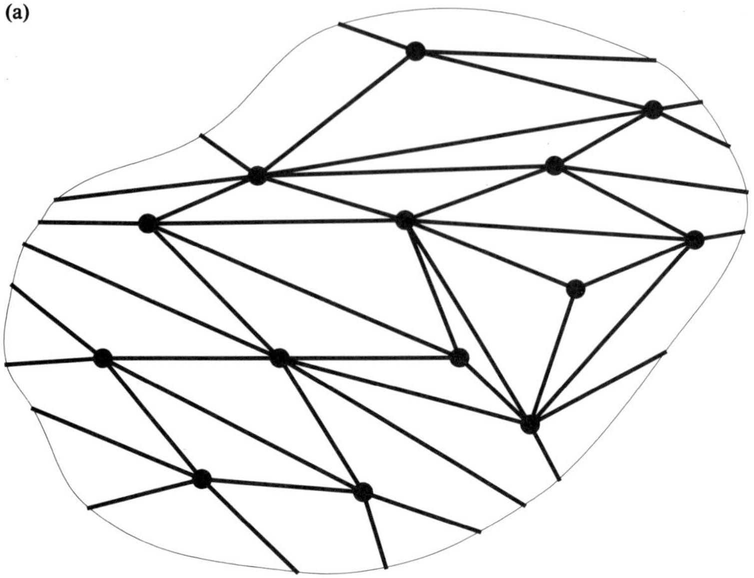

(b)

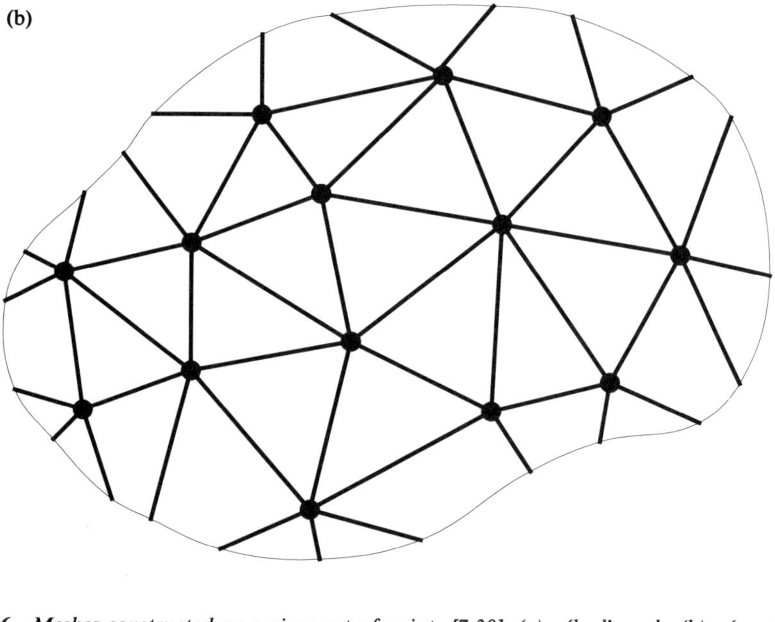

Fig. 7.6 *Meshes constructed on a given set of points [7.30]: (a) a 'bad' mesh, (b) a 'good' mesh*

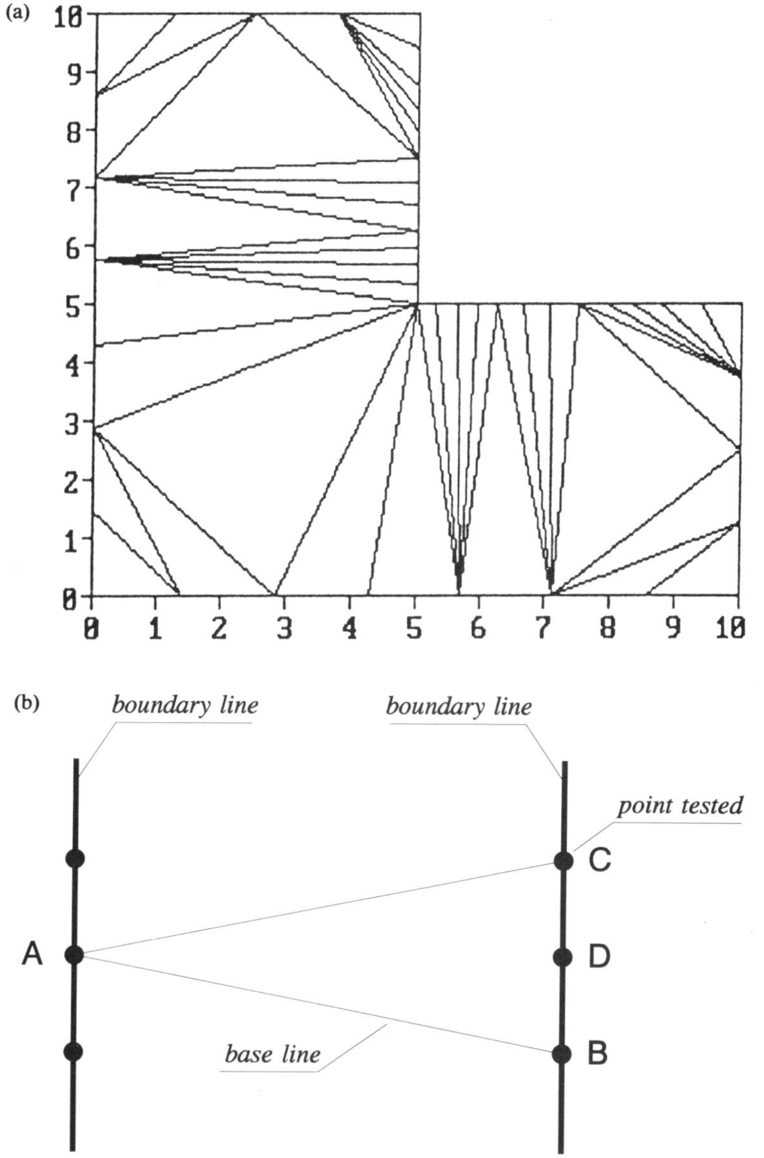

Fig. 7.7 *Initial triangulation: (a) final result, (b) test for a new point*

The procedure of adding interior points may be based on selecting the 'worst' triangular element, in terms of its shape, and inserting a new point somewhere inside or on the edge of this element. In order to determine the 'quality' of an element we shall follow a simple criterion suggested by Lindholm [7.38], where the quality of a triangle is measured as the ratio of the diameter of its inscribing circle to the radius of its circumscribing circle

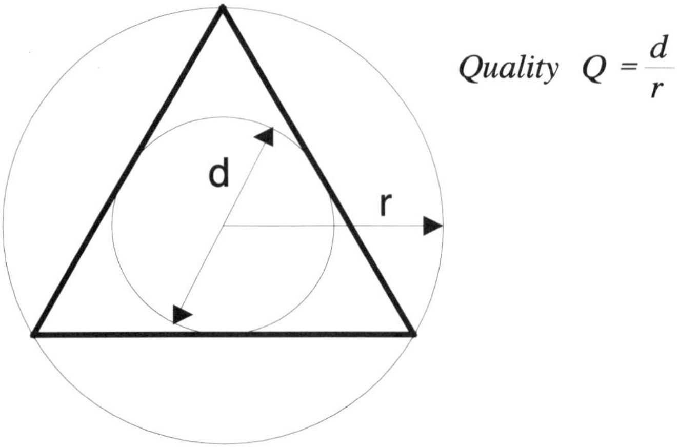

Fig. 7.8 *Measuring the quality of a triangle*

(see Fig. 7.8). Thus $Q = d/r$, and $0 \leqslant Q \leqslant 1$. Hence to find the 'worst' triangle we need to identify a triangle with the lowest value of the Q factor.

As an example, Fig. 7.9 shows a possible strategy, where first a triangle with the worst Q factor is found (element 5), and then a new point P is added at the midpoint of its longest side (Fig. 7.9a). The two elements which share this side are deleted, and four new triangles are created, as demonstrated in Fig. 7.9b. At this stage, the new mesh may not conform to the Delaunay criterion (for example point C in Fig. 7.9c is inside the circumscribing circle).

The 'swapping algorithm' tests all triangles having a vertex at P (by constructing the circumscribing circles), and if a case like that shown in Fig. 7.9c is found, i.e. if the fourth vertex of the quadrilateral falls inside the circumcircle of the other three, the existing diagonal of this quadrilateral (APBC in our case) is 'swapped'. The final position is shown in Fig. 7.9d. Any new triangles created by the swap are placed in a last-in first-out stack. A triangle is taken off the stack and the last step is repeated, swapping if necessary and adding triangles to the stack, until the stack is empty. The mesh now contains the new point P and conforms to the Delaunay criterion.

The process of adding internal points will continue until some quality condition has been achieved, e.g. the minimum acceptable value of the Q factor has been reached by all triangles (except, perhaps, very small triangles, i.e. some 'area restriction' condition could also be in operation). The triangulation may also be terminated manually or by setting an upper limit to the number of triangles in the mesh. At this stage, a good mesh will have been created that conforms to the Delaunay criterion. Further refinement is possible, for example by applying Laplacian smoothing (node relaxation) [7.25]. As shown in Fig. 7.10, internal nodes are 'relaxed', i.e. moved to the centroid of the polygon formed by its connected neighbours. This has the effect of maximizing the minimum angles of the elements forming the polygon.

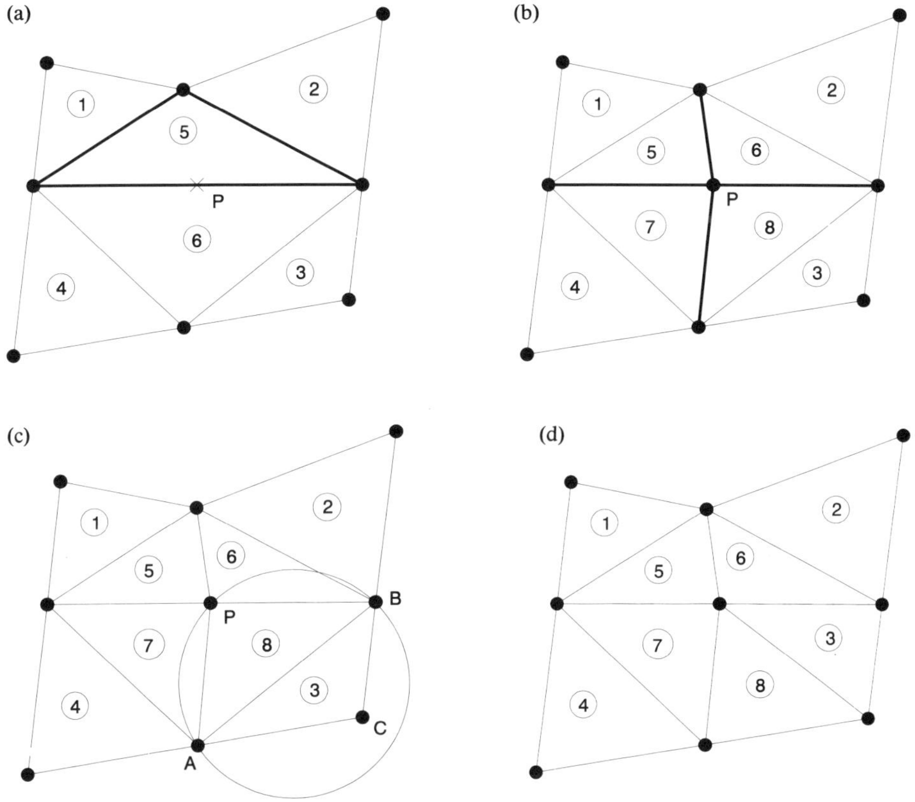

Fig. 7.9 *Addition of interior nodes in the FE mesh*

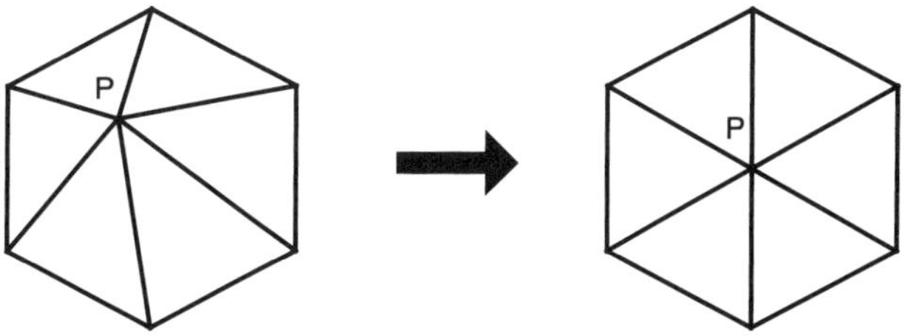

Fig. 7.10 *Laplacian smoothing*

The scheme described above is based on a procedure suggested by Sloan [7.70]. It leads to a triangulation of the type shown in Fig. 7.11a. Finally, in order to produce a denser mesh around re-entrant corners (where the field may vary most rapidly), a simple weighting could be applied, for example by setting a lower value of the 'area restriction' on elements close to such corners. A possible modified mesh is shown in Fig. 7.11b.

In three dimensions, the extension of the Delaunay procedure is possible

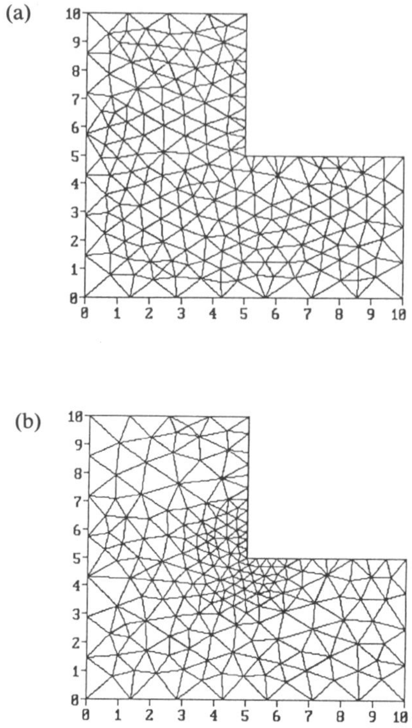

Fig. 7.11 *Examples of final meshes: (a) without weighting, (b) with weighting*

using tetrahedra and inscribing and circumscribing spheres. An appropriate algorithm may be found for example in reference [7.9].

In addition to the various techniques for meshing polygons, most commercially available programs offer useful extra features for pre-processing, such as copying, replication, rotation, mirror image, loop input, as well as

Fig. 7.12 *Finite element mesh for a Linear Variable Differential Transformer [7.75]*

libraries of shapes, material properties, magnetization curves etc. Moreover, the pre-processor usually performs several checks on the generated mesh to ensure continuity of element subdivision, completeness of material data, matching of interface boundaries, sufficiency of applied boundary conditions, consistency of geometrical representation, and many other. These checks are often highly automated. As a practical illustration Fig. 7.12 shows a typical mesh used in FE analysis of a Linear Variable Differential Transformer [7.75]. A combination of deterministic and automatic meshing is used.

The discussion of this section focused on geometrical criteria for mesh generation and refinement. An addition of a field criterion, i.e. a combination of geometric and field conditions, leads to adaptive meshing, discussed in Section 7.6.

7.3 CODE DEVELOPMENT AND LIMITATIONS

The most important, although perhaps least visible to the user, part of any CAD system is the *solution processor* or *solver*. Many different or alternative formulations are used for solution of particular types of equations. At the same time the processor must be equipped with all the necessary algorithms for discretization of the governing field equations, setting up of the appropriate matrix of coefficients, and finally the solution of the system of algebraic equations. This is an area of very active research and proceedings of specialist conferences, such as COMPUMAG [7.12; 7.13; 7.14; 7.15], should be consulted for up-to-date information. Another significant event in recent years has been the series of international workshops TEAM (Testing Electromagnetic Analysis Methods), sponsored by the COMPUMAG conferences, which started in 1985 as a means of aiding in development and validation of 3D eddy current codes, expanded later to include other aspects of electromagnetic field computation. A number of bench-mark problems have been defined and many of them solved using special or general methods. Currently the active TEAM problems include: coil above a crack, moving coil in a cylinder, plate over a coil, nonlinear steel channels, rectangular slot in a steel plate, magnetic damping in torsional mode, jumping ring experiment, waveguide loaded cavity and microwave field in a loaded cavity. More information may be found in the newsletter of the International Compumag Society [7.35].

Various possible field formulations have been discussed elsewhere in this book, so was the importance of choosing the appropriate gauge for solution potentials. Currently, the most popular formulations used for eddy-current calculations are as shown in Table 7.1 [7.4]. Some other possibilities are also explored in reference [7.5].

The historical development of codes for solution of two- and three-dimensional electromagnetic field problems is summarized in Table 7.2 [7.77].

Considerable research and development are still required in a number of areas, although significant advancements have already been made as reflected by Table 7.2. Work currently in progress includes strategies for dealing with

Table 7.1 Contemporary eddy-current field formulations [7.4]

Method	Defining equation	Gauge
$\mathbf{H}-\phi$	$\nabla\times\nabla\times\mathbf{H}=-\mu\sigma(\partial\mathbf{H}/\partial t)$	
$\mathbf{A}-V-\phi$	$\nabla\times(1/\mu)\nabla\times\mathbf{A}=-\sigma(\partial\mathbf{A}/\partial t+\nabla V)$	$\nabla\cdot\mathbf{A}=0$
	$\nabla\cdot\sigma(\partial\mathbf{A}/\partial t+\nabla V)=0$	$\nabla\cdot\mathbf{A}=-\mu\sigma V$
$\mathbf{A}^*-\phi$	$\nabla\times(1/\mu)\nabla\times\mathbf{A}^*=-\sigma(\partial\mathbf{A}/\partial t)$	$\nabla\cdot\sigma\mathbf{A}^*=0$
$\mathbf{T}^*-\phi$	$\nabla\times(1/\sigma)\nabla\times\mathbf{A}^*=-\mu(\partial/dt)(\mathbf{T}-\nabla\phi)$	$\mathbf{T}\cdot\mathbf{u}=0$
	$\nabla\cdot\mu(\mathbf{T}-\nabla\phi)=0$	

In free space in all cases ϕ satisfies $\nabla^2\phi=0$.

Table 7.2 Time evolution of code development [7.77]

	1960	1965	1970	1975	1980	1985	1990
Statics	D2Df I2D	D2D I3D	D2Dnp I3Dn	D2Da	D3Dna	D2Dse	D2Dv
Steady state a.c.			D2D I2D	D2Dn*	D3D†	D2De	D3D
Transient			I2D	D2D	D2Dn	D3D†	D3D
Full Maxwell						D2D	D3D
Motion					D2D‡	D3D‡	D3D
Coupled problems				D2Df	D2D		D3D

Key:

D2D	differential 2D	D3D	differential 3D
I2D	integral 2D	I3D	integral 3D
n	nonlinear	a	anisotropy
p	permanent magnet	f	finite difference
s	scalar hysteresis	v	vector hysteresis
e	error analysis	†	restricted formulation
*	approximate model	‡	uni-directional velocity

'far field', or 'open boundary', condition (e.g. infinite elements, recursive ballooning or mapping techniques), the analysis of moving boundaries, vector hysteresis, coupled problems, high performance computing, material handling, adaptive solving and optimization. The reader is encouraged to consult some of the references provided for this Chapter.

7.4 THE ART OF SPARSE MATRICES

A matrix is said to be *sparse* if many of its coefficients are zero and there is an advantage in exploiting the presence of these zeros. Sparse matrices arise in a wide range of physical and mathematical problems and are a natural product of methods dealt with in this book. Significant savings in computer memory requirements and computational times may be achieved as a result of using suitable sparse storage strategies.

Consider a simple regular finite-element mesh of Fig. 7.13, where nodes are numbered sequentially in a vertical direction and faces 1 and 2 have fixed potential values (Dirichlet condition). Within each quadrilateral either

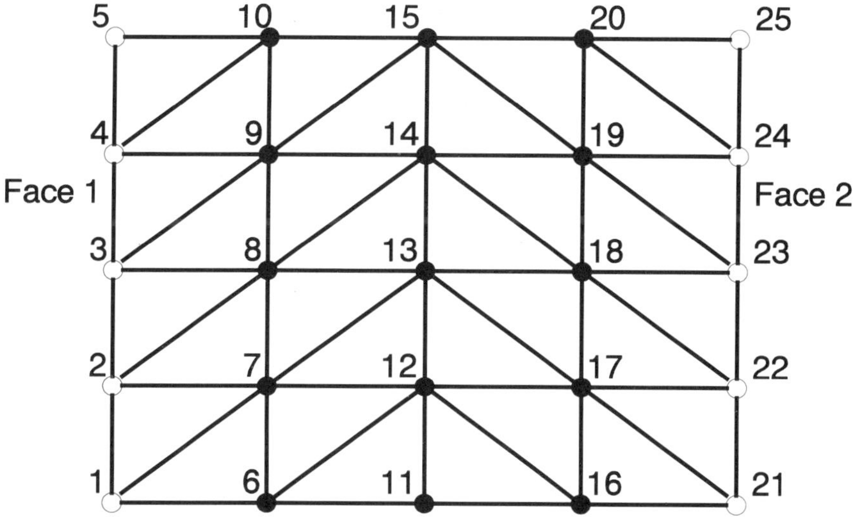

Fig. 7.13 *A regular finite-element mesh*

diagonal may be used and a particular choice is demonstrated. The mesh is regular in the topological, rather than geometrical, sense and could fit a more complicated shape. Potentials of nodes 6 to 20 are unknown.

In a finite-element formulation, using first-order triangular elements, this will result in a symmetric matrix for nodes 6 to 20 of the following form

$$\begin{bmatrix}
1 \\
16 & 2 \\
\cdot & 17 & 3 \\
\cdot & \cdot & 18 & 4 \\
\cdot & \cdot & \cdot & 19 & 5 \\
36 & 28 & \cdot & \cdot & \cdot & 6 \\
46 & 37 & 29 & \cdot & \cdot & 20 & 7 \\
\cdot & 47 & 38 & 30 & \cdot & \cdot & 21 & 8 \\
\cdot & \cdot & 48 & 39 & 31 & \cdot & \cdot & 22 & 9 \\
\cdot & \cdot & \cdot & 49 & 40 & \cdot & \cdot & \cdot & 23 & 10 \\
\cdot & \cdot & \cdot & \cdot & \cdot & 41 & 32 & \cdot & \cdot & \cdot & 11 \\
\cdot & \cdot & \cdot & \cdot & \cdot & 50 & 42 & 33 & \cdot & \cdot & 24 & 12 \\
\cdot & \cdot & \cdot & \cdot & \cdot & \cdot & 51 & 43 & 34 & \cdot & \cdot & 25 & 13 \\
\cdot & \cdot & \cdot & \cdot & \cdot & \cdot & \cdot & 52 & 44 & 35 & \cdot & \cdot & 26 & 14 \\
\cdot & \cdot & \cdot & \cdot & \cdot & \cdot & \cdot & \cdot & 53 & 45 & \cdot & \cdot & \cdot & 27 & 15
\end{bmatrix} \tag{7.1}$$

where the numbers refer to the non-zero entry positions, not their values. Only these non-zero values need to be retained, and this can be arranged easily in the form of two smaller arrays

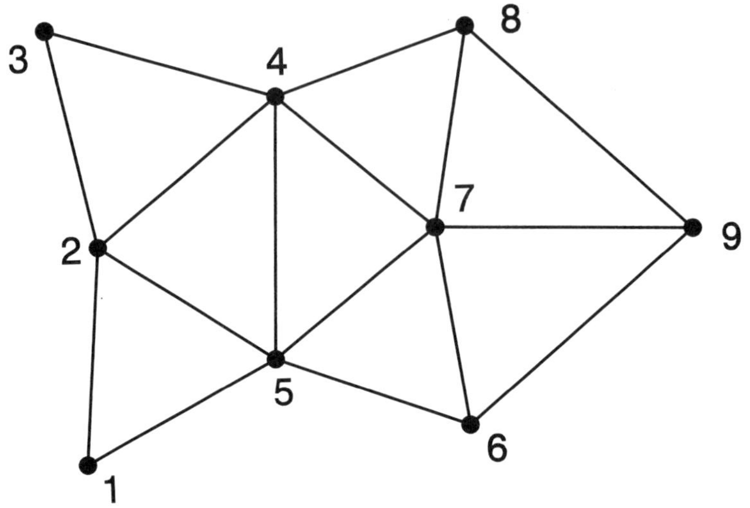

Fig. 7.14 *A finite-element mesh defined on nine nodes*

$$\begin{bmatrix} 1 & 2 & 3 & 4 & 5 & 6 & 7 & 8 & 9 & 10 & 11 & 12 & 13 & 14 & 15 \\ 16 & 17 & 18 & 19 & \cdot & 20 & 21 & 22 & 23 & \cdot & 24 & 25 & 26 & 27 & \cdot \end{bmatrix} \text{ and }$$

$$\begin{bmatrix} \cdot & 28 & 29 & 30 & 31 & \cdot & 32 & 33 & 34 & 35 \\ 36 & 37 & 38 & 39 & 40 & 41 & 42 & 43 & 44 & 45 \\ 46 & 47 & 48 & 49 & \cdot & 50 & 51 & 52 & 53 & \cdot \end{bmatrix} \tag{7.2}$$

The initial matrix contained $15 \times 15 = 225$ elements, which are now stored as $15 \times 2 + 10 \times 3 = 60$ numbers in two convenient tables. The scheme is very efficient; however, it is suitable only for regular meshes.

Consider now a more general case of Fig. 7.14, where nodes are connected and numbered in an arbitrary way, and first-order triangular elements are used. The resulting matrix will be of the following form

$$\begin{bmatrix} 11 \\ 21 & 22 \\ \cdot & 32 & 33 \\ \cdot & 42 & 43 & 44 \\ 51 & 52 & \cdot & 54 & 55 \\ \cdot & \cdot & \cdot & \cdot & 65 & 66 \\ \cdot & \cdot & \cdot & 74 & 75 & 76 & 77 \\ \cdot & \cdot & \cdot & 84 & \cdot & \cdot & 87 & 88 \\ \cdot & \cdot & \cdot & \cdot & \cdot & 96 & 97 & 98 & 99 \end{bmatrix} \tag{7.3}$$

11	21	51	22	32	42	52	33	43	44	54	74	84	55	65	75	66	76	96	77	87	97	88	98	99
1	2	5	2	3	4	5	3	4	4	5	7	8	5	6	7	6	7	9	7	8	9	8	9	9
1	1	1	2	2	2	2	3	3	4	4	4	4	5	5	5	6	6	6	7	7	7	8	8	9

Fig. 7.15 *Representation of the matrix of eqn (7.3) as three linear arrays*

In view of symmetry, only half of this 9×9 matrix needs to be stored, and there are 25 non-zero elements. A straightforward and popular storage method for general sparse matrices places the non-zero elements in a linear array and complements this array with two integer arrays which store the corresponding row and column numbers (see Fig. 7.15). In our example the storage is by columns; storage by rows is of course also possible.

The repetition of numbers in the second integer array may be exploited by storing only the information about where new columns begin. Thus the much more compact modified system becomes as shown in Fig. 7.16.

The second integer array now contains pointers identifying the beginning points of columns. If the matrix is very sparse the storage savings will be considerable.

Another group of desirable ordering strategies are band and variable-band (also called skyline, profile, and envelope) matrices. The matrix of eqn (7.3) may be shown in the form of Fig. 7.17.

Space is allocated for all elements of each row starting with the leftmost non-zero entry and continuing up to the diagonal, as shown by the shaded area in Fig. 7.17, called the matrix *profile*. Some zero elements are thus stored. However, the method is quite popular, in particular in conjunction with

11	21	51	22	32	42	52	33	43	44	54	74	84	55	65	75	66	76	96	77	87	97	88	98	99
1	2	5	2	3	4	5	3	4	4	5	7	8	5	6	7	6	7	9	7	8	9	8	9	9

1	4	8	10	14	17	20	23	25

Fig. 7.16 *A modified representation of the matrix of eqn (7.3)*

11	12			15				
21	22	23	24	25				
	32	33	34					
	42	43	44	45		47	48	
51	52		54	55	56	57		
				65	66	67		69
			74	75	76	77	78	79
			84			87	88	89
					96	97	98	99

Fig. 7.17 *Profile storage of sparse symmetric matrix of eqn (7.3)*

11	21	22	32	33	42	43	44	51	52	·	54	55	65	66	74	75	76	77	84	·	·	87	88	96	97	98	99

1	1	2	2	1	5	4	4	6

1	3	5	8	13	15	19	24	28

Fig. 7.18 *Profile storage using three linear arrays*

triangular decomposition algorithms. One of the reasons is that, although the lower triangular matrix will not normally have the same pattern of zeros as the original matrix, it will nevertheless have the same profile [7.67]. Thus the same storage and identical addressing functions can be used for both matrices.

A possible way of storing our matrix is demonstrated in Fig. 7.18. This time the elements (including some zeros) are stored in a linear array row by row. The two integer arrays point, respectively, to the column number of the beginning of each row and to the diagonal element (i.e. its position in the linear array) which forms the end of each row. Strictly speaking the second integer array is not necessary as the ending of a row is always on the main diagonal. However, the addressing functions become much more complicated and a system with both arrays may still be preferred.

Most practical finite-element programs exploit matrix sparsity and the profile storage, or its variation, is probably the most popular technique. The space efficiency of such storage depends crucially on numbering of matrix rows and columns, and thus on numbering of the nodes in the mesh. Moreover, the time required for triangular decomposition of a profile-stored matrix depends even more critically on the profile size. Hence, the initial random node numbering is usually followed by some systematic renumbering scheme in the search for an improved sparsity pattern. This process often starts by classifying nodes into sets called *levels* [7.64]. A starting node is assigned to level $L(0)$, and level $L(i)$ is made up from all available nodes connected to level $L(i-1)$. Each node can belong to one level only. For example, the levels for the mesh of Fig. 7.14 are

$$\begin{aligned} L(0) &= [1] \\ L(1) &= [2\ 5] \\ L(2) &= [3\ 4\ 6\ 7] \\ L(3) &= [8\ 9] \end{aligned} \tag{7.4}$$

The distance between nodes is thus given by the difference between their levels, e.g. the shortest path between node 1 and the most distant node requires traversing three links. The total number of levels, as well as the grouping of nodes within the levels, depends on the starting node. In our example, starting with node 5 yields

$$\begin{aligned} L(0) &= [5] \\ L(1) &= [1\ 2\ 4\ 6\ 7] \\ L(2) &= [3\ 8\ 9] \end{aligned} \tag{7.5}$$

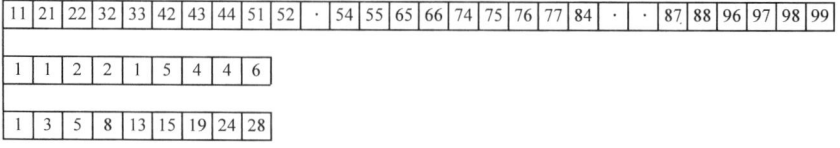

A matrix is said to have a *topological diameter* of d if the highest-numbered level reached for any starting node is $L(d)$ (in our case $d = 3$). Hence a possible strategy is to examine the matrix structure to find its topological diameter and choose a node at either end of this diameter as a starting node. Alternatively, a node with the fewest connections to other nodes could be selected. Subsequently, good numberings are usually obtained when widely different node numbers are widely separated. Such separation may be achieved by using the node levels in an orderly sequence, as indicated for example by eqn (7.4). Although the amount of work involved in the search for the 'best' sparsity pattern may appear considerable, it is usually amply recovered through decreased costs of the actual solution of the equations.

The art of sparse matrices and node numbering has its own specialist literature and a number of specific techniques exist. It is very much an area of continuing research. The reader will find more information in books [7.18], [7.64] and [7.67].

7.5 THE ICCG METHOD

If the coefficient matrix $[N]$, say, is symmetric, positive definite and sparse, as finite-element or finite-difference matrices typically are, the computational effort may be considerably reduced by application of the *incomplete Cholesky preconditioning*, followed by the *conjugate gradient* iterative solution. This combination is known as the ICCG [7.45; 7.4; 7.64] and has established itself as probably the most effective method of solving large systems of algebraic equations of the type encountered in this book.

In the standard Cholesky decomposition, the positive definite matrix $[N]$ can be factorized into the form

$$[N] = [L][L]^{\mathrm{T}} \tag{7.6}$$

where $[L]$ is a lower triangular matrix with positive diagonal elements. The elements of $[L]$ may be found by recursive application of the following relationships

$$L_{ik} = \frac{N_{ik} - \sum_{j=1}^{k-1} L_{ij} L_{kj}}{L_{kk}} \tag{7.7}$$

for every off-diagonal element in column k in each row i, and

$$L_{ii} = \sqrt{\left(N_{ii} - \sum_{j=1}^{i-1} L_{ij}^2\right)} \tag{7.8}$$

for the diagonal element in each row, where

$$L_{11} = \sqrt{N_{11}} \tag{7.9}$$

In the *conjugate gradient* (CG) method, the solution of the system of equations

$$[K]\,[x] = [b] \tag{7.10}$$

is achieved by computing successive solution estimates $[x]_k$, i.e. by putting

$$[x]_{k+1} = [x]_k + \alpha_k[p]_k \tag{7.11}$$

where the direction vectors $[p]_k$ are selected so as to make successive residuals orthogonal to each other

$$[r]_k^{\mathrm{T}}[r]_{k+1} = 0 \tag{7.12}$$

The residual vector $[r]_k$ is given by

$$[r]_k = [K]\,[x]_k - [b] \tag{7.13}$$

and the parameter α_k is chosen to make the best possible improvement at each step. The process starts with an initial estimate $[x]_0$ and chooses the initial search direction $[p]_0$ to coincide with $[r]_0$:

$$[r]_0 = [K]\,[x]_0 - [b] \tag{7.14}$$

$$[p]_0 = [r]_0 \tag{7.15}$$

There follows then a succession of residuals and search directions through the recursive step [7.64]

$$\alpha_k = -\frac{[p]_k^{\mathrm{T}}[r]_k}{[p]_k^{\mathrm{T}}[K]\,[p]_k} \tag{7.16}$$

$$[x]_{k+1} = [x]_k + \alpha_k[p]_k \tag{7.17}$$

$$[r]_{k+1} = [r]_k + \alpha_k[K]\,[p]_k \tag{7.18}$$

$$\beta_k = -\frac{[p]_k^{\mathrm{T}}[K]\,[r]_{k+1}}{[p]_k^{\mathrm{T}}[K]\,[p]_k} \tag{7.19}$$

$$[p]_{k+1} = [r]_{k+1} + \beta_k[p]_k \tag{7.20}$$

In the conjugate gradient method the scalar parameter β is chosen so that the search directions $[p]_k$ and $[p]_{k+1}$ are conjugate with respect to the matrix $[K]$, i.e.

$$[p]_{k+1}^{\mathrm{T}}[K]\,[p]_k = 0 \tag{7.21}$$

There are alternative choices for both α and β and they are discussed for example in reference [7.59].

The rate of convergence of the conjugate gradient method depends strongly on the eigenvalue spectrum of the matrix $[K]$. The process may be accelerated if an appropriate *preconditioning matrix* $[S]$ is chosen and the CG method applied to a matrix $[S]\,[K]\,[S]^{\mathrm{T}}$ instead of $[K]$. $[S]$ must be chosen so that $[S]\,[K]\,[S]^{\mathrm{T}}$ has a smaller eigenvalue spectrum than $[K]$ and thus more rapid convergence may be achieved. Equations (7.19) and (7.20) now take the form

$$\beta_k = -\frac{[p]_k^{\mathrm{T}}[K]\,[S]\,[r]_{k+1}}{[p]_k^{\mathrm{T}}[K]\,[p]_k} \tag{7.22}$$

$$[p]_{k+1} = [S]\,[r]_{k+1} + \beta_k[p]_k \tag{7.23}$$

and once the process has converged, the solution to the original problem may be easily recovered.

If the matrix $[K]$ is sparse, the *incomplete Cholesky decomposition* provides a very good choice for preconditioning. The lower triangular matrix $[L]$ is forced to have the same sparsity pattern (the same topology) as the lower triangular part of $[K]$, but otherwise the decomposition is carried out in the standard manner, except for the pre-assigned zeros. The preconditioning matrix is then chosen to be the inverse of $[L]$

$$[S] = [L]^{-1} \tag{7.24}$$

The Cholesky decomposition is no longer exact, because of the enforced sparsity pattern, and may be written as

$$[K] = [L]\,[L]^{\mathrm{T}} + [E] \tag{7.25}$$

where $[E]$ is an error matrix. The product $[S]\,[K]\,[S]^{\mathrm{T}}$ then becomes

$$[S]\,[K]\,[S]^{\mathrm{T}} = [I] + [L]^{-1}[E]\,[L]^{-\mathrm{T}} \tag{7.26}$$

and if $[E]$ is small the eigenvalue structure of the preconditioned matrix may be quite close to that of the identity matrix, and thus the eigenvalue spectrum has been narrowed.

Although the mechanism is not absolutely clear, it has been consistently found that the computational effort to solve the transformed system is considerably less than for the original system. This incomplete Cholesky preconditioning with conjugate gradients is known as ICCG and has been very successfully implemented in many finite-element codes, particularly for both 2D and 3D static fields.

Modifications to the method are required if the matrix is not positive definite. One such interesting adaptation of the Cholesky decomposition is described in reference [7.39] for coupled magnetic–electric equations.

7.6 ERROR ANALYSIS AND ADAPTIVE MESHING

Numerical solutions obtained using the finite-element method, or other methods discussed in this book, are, by their very nature, approximate. The difference between the numerical and the exact solution, locally in an element or subdomain, or globally in some average sense, is the actual error of the approximation. Such an error can be calculated only if the exact solution is known, e.g. in the form of an analytical expression, or accurate measurement. This is a routine phase in software development, testing and comparative assessment, where benchmark problems are used for this purpose (see for example TEAM activities described in reference [7.35]). Generally the exact solution will not be known and the question arises whether the numerical solution obtained is accurate enough. The level of acceptable error must be set by the designer and will depend on many factors, such as tolerances of the design, accuracy of material data and characteristics, approximations introduced to the model, sensitivity of resultant parameters etc. A perfect numerical solution will rarely be obtained, nor should it be required, as this would be uneconomical in view of the many other types of errors involved. Methods of reliable error estimation are thus required and it is currently an active area of research. The reader will find some suggestions for further reading in the references for this section, in particular the following publications: [7.4; 7.15; 7.17; 7.20; 7.21; 7.22; 7.23; 7.25; 7.26; 7.27; 7.28; 7.52; 7.53; 7.54; 7.57; 7.58].

The errors involved in approximate numerical solutions are of three principal kinds: truncation or discretization errors (although strictly these are slightly different as discussed in Chapter 2), round-off errors due to floating point arithmetics (i.e. limited precision), and approximations in the mathematical model of the real problem. Only discretization errors are considered in this Section. They are due to the incomplete satisfaction of the governing equations and boundary conditions, and are introduced by the limited accuracy of the trial function approximation. If the process of error estimation is completed *before* the solution of the problem is attempted, it is said to be *a priori*. If error is estimated *after* the numerical solution has been obtained, it is called *a posteriori*. An interesting example of the former type is a scheme based on a *neural network* used to predetermine the 'ideal' mesh density and thus to provide an initial mesh for an adaptive system [7.19]. The latter method is becoming more and more popular and is usually combined with adaptive meshing.

Intuitively, the accuracy of the solution should improve if discretization is increased or higher order elements are used. Indeed, these are the two fundamental strategies for adaptive meshing, i.e. mesh refinement or trial function improvement (e.g. hierarchical elements). In either case reliable and accurate error estimation is required to control the procedure and make sure that refinement is performed in such parts of the mesh where it really matters, so that the process is efficient. We shall now discuss briefly some possible methods of estimating the error in the finite element solution.

The first possibility is to exploit the discontinuity at element interfaces of selected field quantities. Consider two adjacent tetrahedral elements, say 1 and 2, with a common face as shown in Fig. 7.19.

We would expect the normal component of the magnetic flux density **B** to be continuous across any interface; in practice, depending on the actual formulation, vector **B** obtained from the finite-element calculation may present normal discontinuity. Similarly, the magnetic field strength **H** should have its tangential component continuous, except in the case of a surface current, but in the numerical solution this continuity may not necessarily be preserved. Hence this type of discontinuity may be used as an error estimator. Appropriate expressions could be

$$error = \int_S |(\mathbf{B}^{(1)} - \mathbf{B}^{(2)}) \cdot \boldsymbol{n}|^2 \, \mathrm{d}s \tag{7.27}$$

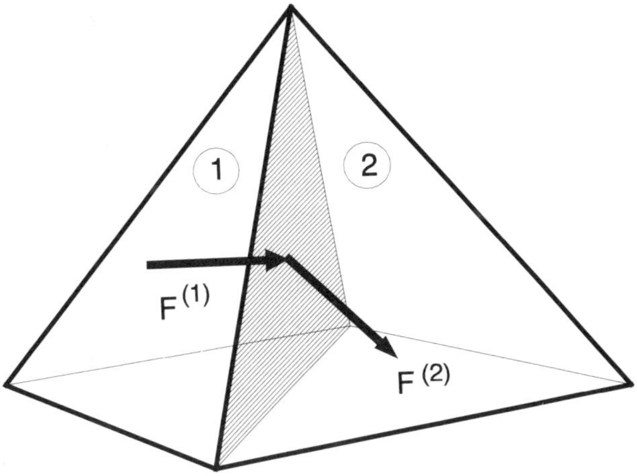

Fig. 7.19 *Discontinuity across element interface*

or

$$error = \int_S |\boldsymbol{n} \times (\mathbf{B}^{(1)} - \mathbf{B}^{(2)})|^2 \, \mathrm{d}s \tag{7.28}$$

where integration is carried over the face area. As each tetrahedra has four faces, the overall error for the element could be taken as the maximum of the component errors, or some sort of weighted average, or simply the sum of the four errors.

Another possibility arises if the software has the option of calculations with both first and second order elements. Clearly the second order approximation will be more accurate and the difference between the two solutions could be used as a measure of the error. A convenient parameter is the energy in each element [7.28], hence

$$error = |energy^{(1st\ order)} - energy^{(2nd\ order)}| \tag{7.29}$$

This approach could be easily extended to higher order elements if the code offers such elements.

Nodal perturbations between successive mesh refinements offer another straightforward alternative. An error indicator for each element could be chosen as [7.32]:

$$error = (|f^1_{new} - f^1_{old}| + |f^2_{new} - f^2_{old}| + |f^3_{new} - f^3_{old}| + |f^4_{new} - f^4_{old}|) \times volume \tag{7.30}$$

where the indices 1, 2, 3 and 4 refer to the four nodes of the tetrahedra and f is the solution function. Care has to be exercised if, during the process of mesh refinement, the position of nodes changes. One possibility is to calculate f by interpolation; this, however, leads to a computationally expensive scheme. An alternative is to solve the problem without node relaxation, then stabilize the mesh, calculate the error and finally refine the mesh [7.28].

A very elegant approach to error estimation exploits the dual and complementary variational formulations [7.52]. The duality is inherent in Maxwell's equations and was mentioned before in Section 1.8. It forms the basis of the method of 'Tubes and Slices' described briefly in Sections 1.9 and 1.10, and more fully in publications [7.31], [7.72] and [7.76], and a book [7.30]. Complementary and dual energy methods show that error bounded solutions are obtainable when complementary pairs of functionals are extremized. In magnetostatics, the suitable functionals have been suggested in [7.53] as

$$\Theta(\mathbf{A}) = \frac{1}{2}\int (\nabla \times \mathbf{A}) \cdot \frac{1}{\mu}(\nabla \times \mathbf{A})\, \mathrm{d}\Omega - \int \mathbf{J} \cdot \mathbf{A}\, \mathrm{d}\Omega + \int_{S_2} \mathbf{h} \cdot \mathbf{A}\, \mathrm{d}S_2 \qquad (7.31)$$

and

$$\Gamma(\Psi) = -\frac{1}{2}\int (\mathbf{H}_s - \nabla\Psi) \cdot \mu (\mathbf{H}_s - \nabla\Psi)\, \mathrm{d}\Omega - \int_{S_1} \mathbf{g} \cdot (n \times \mathbf{H})\mathrm{d}S_1 \qquad (7.32)$$

where $\mathbf{A}$ is the magnetic vector potential, $\mathbf{J}$ is current density, $\mathbf{H}_s$ is the reduced magnetic field intensity, Ψ is the reduced scalar potential, and $\mathbf{h}$ and $\mathbf{g}$ are specified on S_2 and S_1, respectively. The difference between functionals Γ and Θ indicates the degree of discretization error present in a problem. Thus, as recommended in reference [7.54], if two solutions are available, say *standard* and *complementary*, then the parameter chosen for evaluating error in each element, say $P(\mathbf{B})$, may be used as an error indicator by writing

$$error = P(\mathbf{B}_{comp}) - P(\mathbf{B}_{stan}) \qquad (7.33)$$

or

$$error = P(\mathbf{B}_{comp} - \mathbf{B}_{stan}) \qquad (7.34)$$

where the subscripts refer to the standard and complementary solutions. Equation (7.33) ignores errors in field direction, which may be important, therefore eqn (7.34) is more likely to provide better error indication. It is also possible, and may be beneficial, to weight the error indicators by element area, for example

$$error = \int_{\Omega} P(\mathbf{B}_{comp} - \mathbf{B}_{stan})\, \mathrm{d}\Omega \qquad (7.35)$$

The choice of the error parameter $P(\mathbf{B})$ is important. An energy density parameter, such as $\frac{1}{2}(\mathbf{B} \cdot \mathbf{H})$, provides one possible choice. This, however, may lead to relatively little mesh refinement in high-permeability regions, such as iron, because the energy density is much lower in iron than in air. A flux density parameter such as $(\mathbf{B} \cdot \mathbf{B})^{1/2}$, on the other hand, would have an opposite effect, as it enhances the importance of the flux in the iron in the final solution. Another approach is to weight the parameters by the element area. The main drawback of the complementary scheme is that it requires two finite-element solutions at each stage.

An interesting alternative derivation of complementary variational principles is put forward in reference [7.61]. Here the constitutive relationship $\mathbf{B} = \mu\mathbf{H}$ is used directly and is cast in error form. This assumes the existence of a constitution error, Λ, having the following properties

$$\Lambda \geq 0 \tag{7.36}$$

with

$$\Lambda = 0 \Leftrightarrow \begin{pmatrix} \mathbf{B} = \mu(\mathbf{H})\mathbf{H} + \mathbf{B}_r \\ \text{and} \\ \mathbf{H} = \nu(\mathbf{B})\mathbf{B} + \mathbf{H}_c \end{pmatrix} \text{ everywhere in } R \tag{7.37}$$

where $\mathbf{B}_r$ and $\mathbf{H}_c$ are known constants. Hence Λ is non-negative, and is zero, i.e. minimum, if and only if the fields $\mathbf{B}$ and $\mathbf{H}$ satisfy the constitutive relationship everywhere in the region. Such an error form provides a variational principle, so that we can solve

$$\delta\Lambda = 0 \tag{7.38}$$

to minimize Λ and thus obtain the fields. A suitable expression for the constitution error in linear media is

$$\Lambda = \frac{1}{2}\langle \mu\mathbf{H}, \mathbf{H}\rangle_R + \frac{1}{2}\langle \nu\mathbf{B}, \mathbf{B}\rangle_R - \langle \mathbf{H}, \mathbf{B}\rangle_R \tag{7.39}$$

where the brackets $\langle\ \rangle$ indicate integration through the region of interest. It has been shown in reference [7.61] that for well-posed problems Λ splits into complementary H- and B-system functionals; these provide complementary variational principles that may be solved independently of each other. The error, which is wholly attributed to the numerical constraints in an approximate solution, is always defined and computable locally and globally.

There exist less expensive methods based on a single solution and utilizing the polynomial interpolation theory. For example [7.4], using the first order finite elements with linear shape functions, an element average error can be evaluated by computing the integral over the element of the difference between the nodally interpolated fields B_s (using the first order element shape functions) and the element constant fields B (obtained directly from the potential solution). The local error thus becomes

$$error = \int_{\Omega} (B - B_s)\, d\Omega \tag{7.40}$$

This technique uses the information provided by neighbouring nodes to evaluate the magnitude of the higher order terms in the solution that have been neglected. In another paper [7.20], the error estimation is based on the evaluation of the term

$$g = |grad|grad\ U|| \tag{7.41}$$

where U is the computed finite-element solution. It is also suggested that the gradient of the solution is not calculated by differentiating the first order shape functions, but is obtained by assigning to the nodes field values calculated using the five-links generalized finite difference method suitable for irregular grids and linearly interpolating them on the elements. In the same paper other single solution methods are discussed, one based on approximate solution of a local boundary value problem with hierarchical shape functions, and two which utilize a residual evaluation.

The objective of an adaptive mesh generation is to produce an optimum mesh automatically, i.e. a mesh which guarantees the required level of accuracy achieved in the minimum time. This process is illustrated in Fig. 7.20.

A simple and practical approach has been suggested in reference [7.28]. Only elements with errors greater than a threshold value are refined. If the distribution of errors has a typical form like that shown in Fig. 7.21, then all elements on the right of the threshold line will be refined. A good threshold value has been suggested to be the mean value of all element error indicators possibly weighted by a constant. The procedure leads to an efficient number of elements being refined at each step.

To illustrate the implementation of the above algorithm, a problem of current flowing through the conductors junction, as used in metal welding, has been solved [7.78]. Figure 7.22 shows different levels of discretization

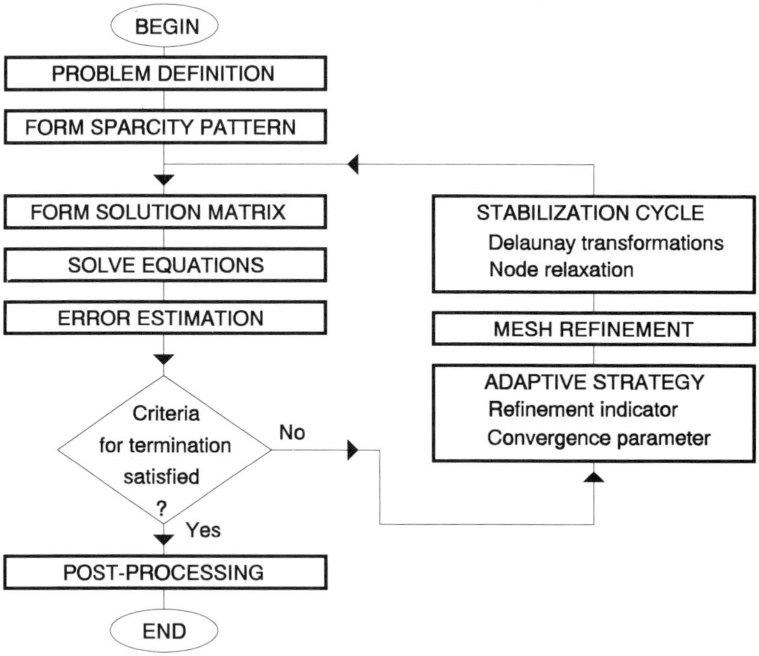

Fig. 7.20 *Self-adaptive refinement algorithm*

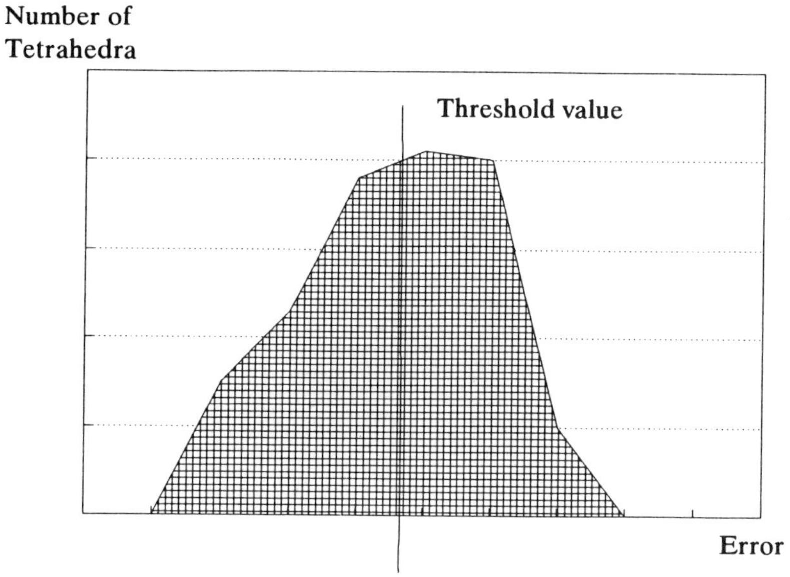

Fig. 7.21 *A typical distribution of element error*

achieved by the adaptive refinement procedure and Fig 7.23 the resultant distribution of equipotential lines.

A particular strategy favoured by a group of researchers from Genova in Italy [7.17; 7.21; 7.22; 7.23] uses a linear combination of dimensionless ratios, with suitable weighting factors, to introduce two key quantities: the refinement indicator and the convergence parameter. A standard quadratic norm is used, which for a generic scalar or vector quantity is defined as

$$\|u\| = \sqrt{\left(\int_{\Omega} |u|^2 \, d\Omega\right)} \tag{7.42}$$

The *refinement indicator* is then specified as

$$\eta_i = \sqrt{\left(\kappa \frac{\|\nabla e_i\|^2}{\sum_{j=1}^{N} \|\nabla \phi_j\|^2} + (1-\kappa) \frac{\|e_i\|^2}{\sum_{j=1}^{N} \|\phi_j\|^2}\right)} \tag{7.43}$$

where ϕ_j is the computed solution over the element j, $\nabla\phi_j$ is its gradient (or curl), e_i is the estimate of the error on the solution over the element i, ∇e_i is the estimate of the error of the gradient of the solution over the element i, N is the total number of elements of the domain and κ is the weighting factor. The *convergence parameter* is defined as

$$c = \sqrt{\left(\sum_{i+1}^{N} \eta_i^2\right)} \tag{7.44}$$

(a)

(b)

(c)

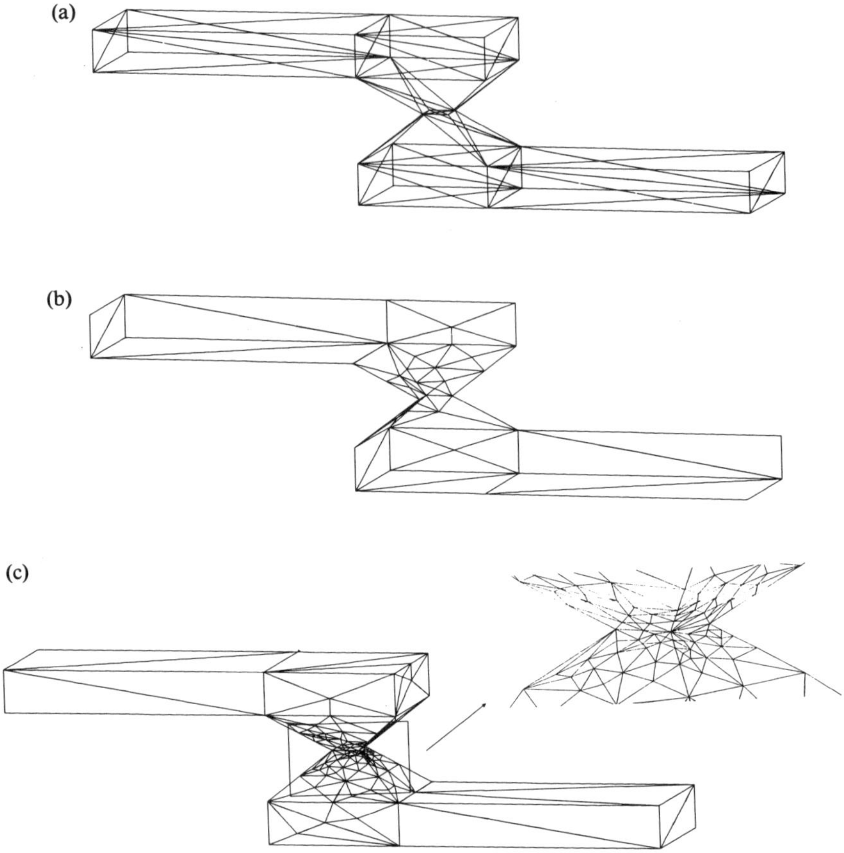

Fig. 7.22 *Discretization of the 'electric welding' geometry [7.78]: (a) 28 nodes, 36 tetrahedra; (b) 111 nodes, 420 tetrahedra; (c) 602 nodes; 3025 tetrahedra.*

and the adaptive process is stopped when c falls below the user defined value. For $\kappa = 1$, i.e. when only the estimate of the error on the gradient (or curl) is used, this reduces to

$$c = \sqrt{\left(\frac{\sum_{i=1}^{N} \|\nabla e_i\|^2}{\sum_{j=1}^{N} \|\nabla \phi_j\|^2} \right)} \tag{7.45}$$

The refinement criterion is that the element is subdivided if

$$\eta_i > \zeta \eta_{max} \tag{7.46}$$

where η_{max} is the maximum value of the refinement indicator η_i for all elements, and ζ is the threshold value, $0 < \zeta < 1$. In a modified version [7.23],

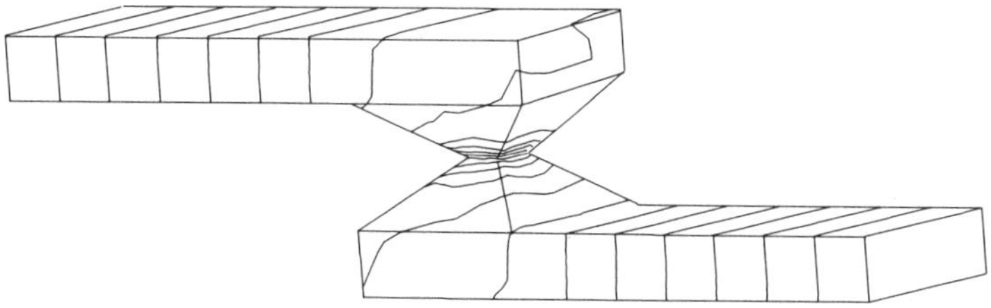

Fig. 7.23 *Distribution of equipotential lines [7.78]*

two threshold values are introduced, $0 < \zeta_1 < \zeta_2 < 1$, and the ith element is subdivided once if

$$\zeta_1 \eta_{max} < \eta_i < \zeta_2 \eta_{max} \tag{7.47}$$

and twice if

$$\eta_i > \zeta_2 \eta_{max} \tag{7.48}$$

The above procedure may be used with various error estimators mentioned before.

Methods relying on adding (subdividing) elements to the existing mesh are sometimes called the *h-type*; an alternative strategy involving increase of the order of element representation is known as the *p-type*. Combined algorithms, called the *h-p type*, are also emerging. The *h-p* adaption may be either decoupled (*h*- followed by *p*-adaption) or fully integrated. Some recent advances in all three methods are reported in reference [7.15].

7.7 OPTIMIZATION

Optimization of geometric boundaries and placement of conductors to achieve specific design goals is paramount in computational magnetics. The underlying theory and mathematical formulations are well developed and documented in specialist books, see for example reference [7.24]. Increasingly, investigators are exploiting optimization theory and apply it to magnetic designs in order to achieve automatically specified objectives. Many proven methods and techniques exist, such as direct-search methods, evolution algorithms, steepest-descent and conjugate gradient methods, Newton and quasi-Newton schemes, stochastic methods, simulated annealing techniques, and many other, not to mention variations and modifications. A selection of solved problems in electromagnetics may be found in references [7.62], [7.63], [7.68] and [7.69], or more generally in the proceedings of COMPUMAG conferences [7.12–7.15]. Simple optimization subroutines are often offered as part of a development software, such as MATLAB [7.44]; the main task is then to link such procedures with the magnetic field analysis software.

As an example, consider a typical solenoid actuator shown in Fig. 7.24. A suite of programs called EAMON (*E*lectromechanical *A*ctuator *M*odel

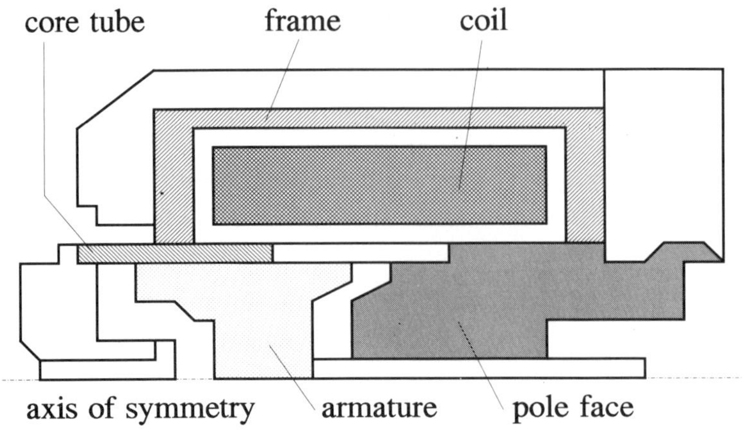

Fig. 7.24 *Radial cross-section of a solenoid actuator*

*O*ptimization [7.11]) has been developed to interface with a general purpose finite-element package OPERA-2d [7.48]. The optimizing algorithm employs a mixture of conjugate gradient and descent methods with Marquardt modifications [7.68] for minimization in a least squares sense of a set of objective functions.

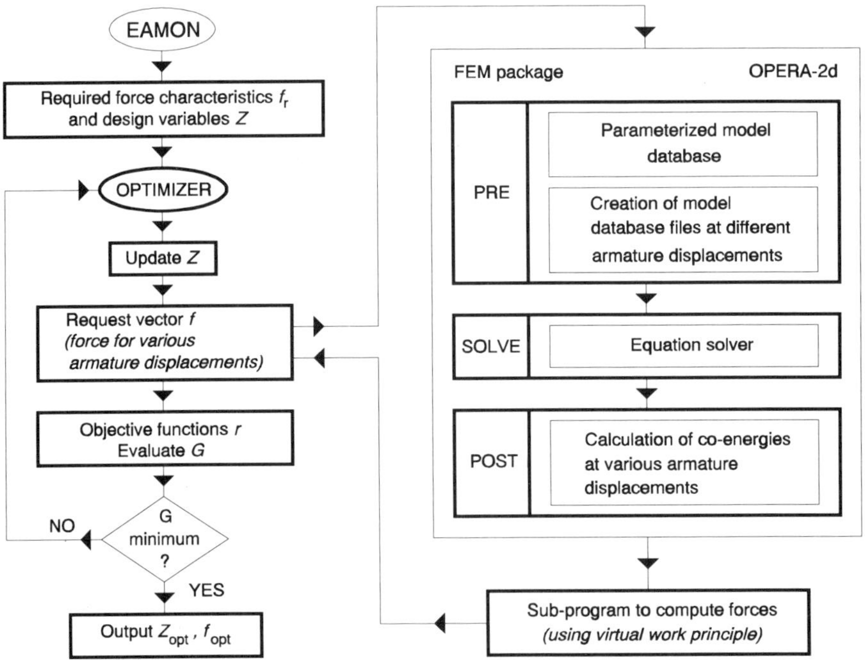

Fig. 7.25 *Flowchart of the optimization program EAMON [7.11]*

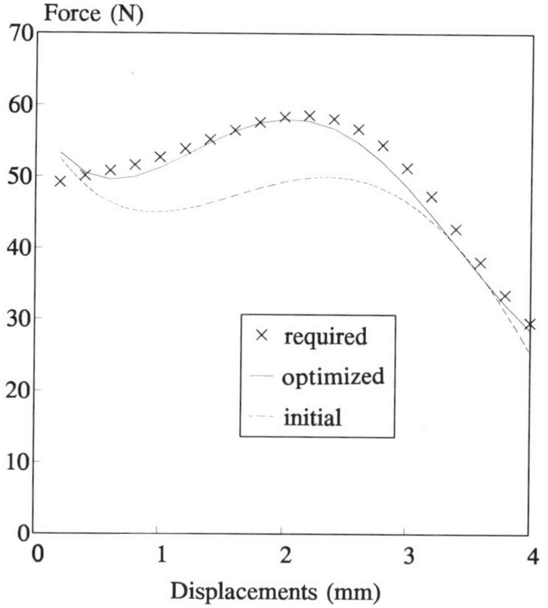

Fig. 7.26 *Optimization of an electromechanical actuator*

The field modelling of the actuator has been conducted using a customized shell developed for OPERA-2d [7.10]. The shell is designed for Sun Open-window (CusOP) and automates the entire process of magnetic design. Force calculations are based on the *Virtual Work Principle* (VWP) and calculation of co-energy, although the *Maxwell Stress Tensor* (MST) method has also been found helpful. The flowchart of the optimization program EAMON is shown in Fig. 7.25.

The optimizer operates on the objective functions which are selected as the difference between the required force curve, f_r, and the calculated force, f, using the VWP method, for a range of displacements, i.e.

$$r = f - f_r \tag{7.49}$$

The function G is then defined as

$$G = \sum_{i=1}^{N} (wr_i)^2 \tag{7.50}$$

where N is the number of armature displacements and w the weighting. The task is to find the variables set Z_{opt} for which $G(Z)$ is minimum. Because of the nonlinearity of $G(Z)$ an iterative approach is employed. As an illustration, Fig. 7.26 shows a force/displacement characteristics of an optimized actuator, where the optimization has been applied to the armature length.

7.8 POST-PROCESSING AND FIELD VISUALIZATION

Post-processing is the art of extracting useful and meaningful engineering information from the numerical results of a problem solution. The solution is usually provided in the form of a set of values of a potential (scalar or vector) calculated at a large number of nodes of a mesh, and thus it needs to be processed before it can be viewed or related in any way to design parameters of the device. Postprocessing techniques typically require that the problem is set up in a particular way. Thus pre- and post-processing are in a way inseparable, they usually share a common database and in contemporary commercial software they are often combined to form one large subprogram (see for example the Pre- and Post-Processor of OPERA-2d [7.48] or PC- OPERA [7.51]). For an engineer the post-processor is an invaluable interface for integrating the magnetic field computation into a design process.

One of the more familiar ways of displaying the solution to a magnetic field problem, in particular in two dimensions, is through contour plots. If vector potential **A** is used as a solution function, i.e. A_z in 2D plane or A_θ in 2D axisymmetric systems, then contours of constant A_z or rA_θ, respectively, become *tubes* of magnetic flux, and if they are equidistant in value then the same amount of flux is always contained between neighbouring lines. This type of field display is particularly helpful to designers of magnetic devices. As an example, consider a simple case of a magnetic hollow cylinder inserted into an applied uniform transverse field. This problem has an analytical solution, derived for example in reference [7.30]. A typical magnetic field distribution is shown in Fig. 7.27.

This type of a magnetic tube is often used to describe magnetic screening effects, and indeed we can see how the field has been effectively excluded from the interior of the tube. If magnetic nonlinearity comes into consideration, or if 'slits' or 'holes' are made in the tube, analytical solutions are no longer

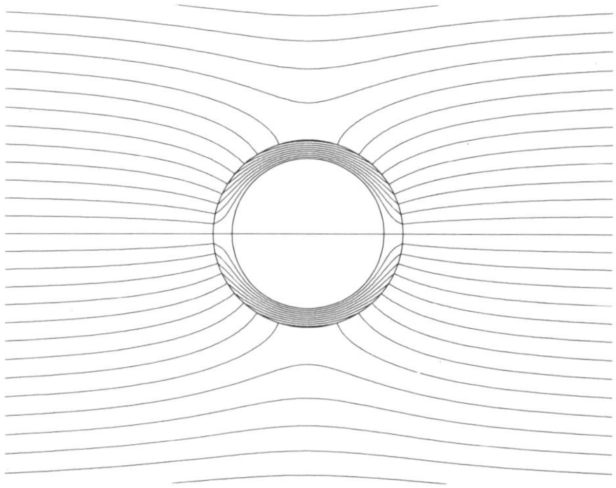

Fig. 7.27 *Screening effect of an unsaturated iron tube*

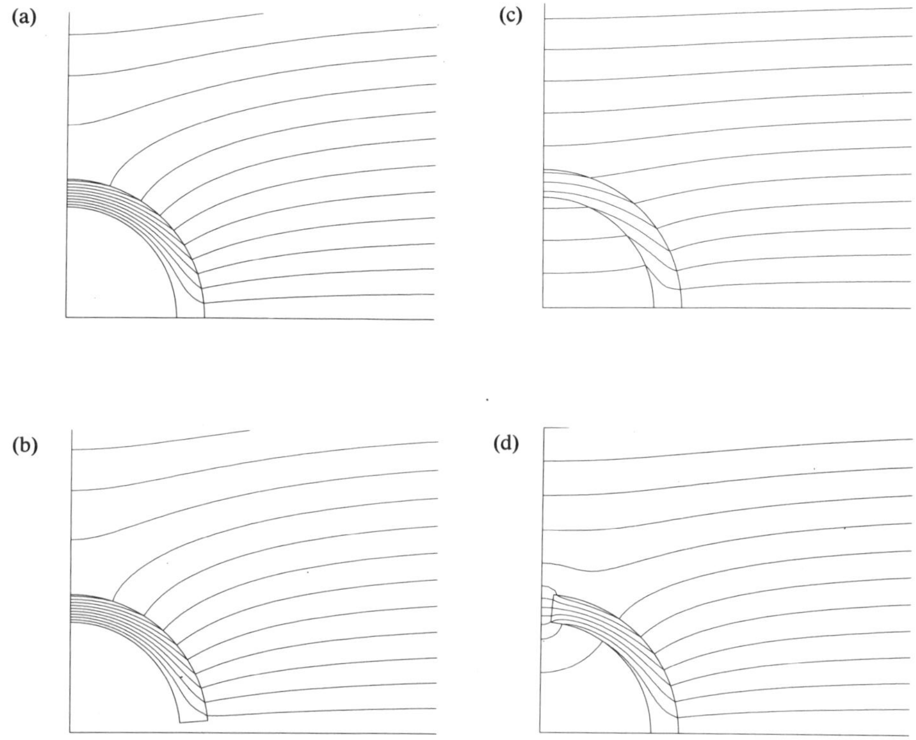

Fig. 7.28 *Field distribution in the magnetic tube: (a) unsaturated tube, (b) magnetically saturated tube, (c) tube with a slit at 0°, (d) tube with a slit at 90°*

available, but numerical analysis presents no difficulties. Figure 7.28 demonstrates some particular interesting cases; owing to symmetry only one quadrant of the solution has been shown in each case. These displays are useful as they demonstrate, in visual form, both qualitatively and quantitatively, important design principles for this type of screening. In this case they point to the dangers of impairing the screening performance by saturating the tube wall (case (b)) or by putting the slits in the wrong position (case (d)). Case (c), on the other hand, shows that if a more appropriate position of a slit is chosen it will hardly affect the screening. By counting the contour lines entering the interior of the tube it is even possible to estimate the degree of the worsened performance of the screen, although for this purpose a more accurate result may be obtained by other methods, for example through direct interrogation of field levels in different regions. In practice different ways of displaying results are often combined and cross-referenced.

A second example is for the Linear Variable Differential Transformer described previously in Fig. 7.12 and based on solutions presented in reference [7.75]. Field plots for two positions of the inner core are shown in Fig. 7.29.

Contour plots are frequently complemented by displays, usually in the form of coloured zones, showing distribution within regions of a specified

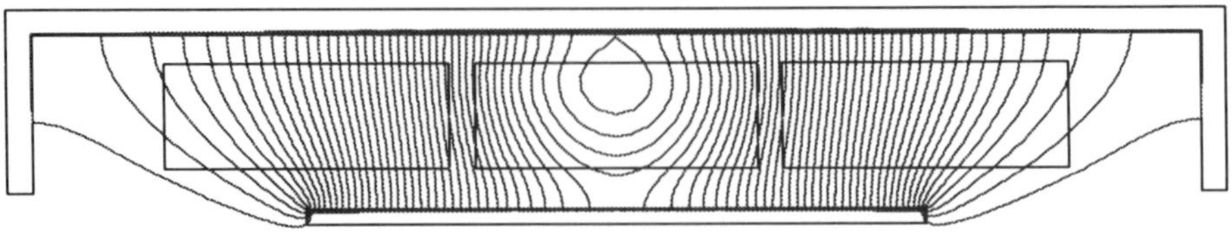

(a)

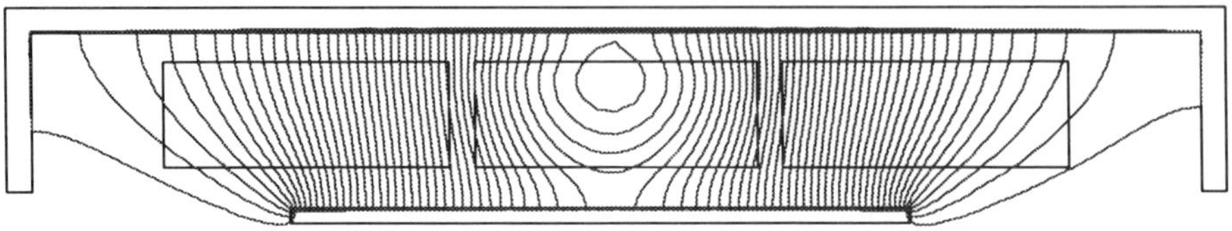

(b)

Fig. 7.29 *Magnetic field plots for a Linear Variable Differential Transformer: (a) displacement = 0, (b) displacement = 0.5 mm*

component. Values of interest are typically magnetic flux density, induced current density, permeability and error distribution (if error is calculated). Unfortunately, such coloured 'maps' do not reproduce well in black and white, and are not included here for illustration.

Another useful and informative way of displaying results is to show a distribution of a field component along specified lines (straight or curved). The component is calculated at a number of points along the line and shown as a piece-wise linear function, or first interpolated or smoothed and then displayed.

In three dimensions, field visualization is much more complicated. Two-dimensional techniques are used to some extent, i.e. fields are displayed on identified or specified surfaces or planes. Again, colour is usually very helpful. Line graphs are also used. Fully three-dimensional displays are much harder to achieve. Researchers are trying various methods and one of such interesting attempts is described in reference [7.80], where a generic 'cube' is defined on part of a solution and then various field displays are applied to this cube. These include coloured contour maps on a series of planes defined on the cube, splitting the cube into sub-cubes with colours determined by field values at the particular location, and transparent displays on planes parallel to cube walls. Probably the most popular method of showing the fully three-dimensional field patterns is based on the idea of a three-dimensional arrow shown in perspective (see Fig. 7.30). The size of the arrow is proportional to the magnitude of the field quantity. Thus both the value and the direction of a vector quantity are indicated on the display.

Examples of images of a three-dimensional arrow are shown in Fig. 7.31. Cases (a) and (b) refer to some general 'positive' and 'negative' directions,

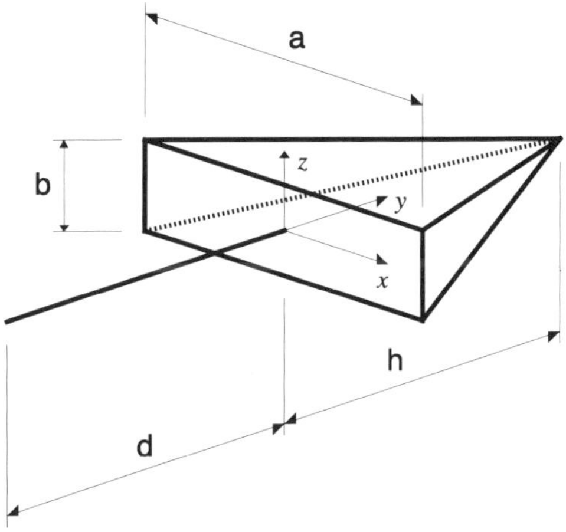

Fig. 7.30 *A three-dimensional arrow representing a vector quantity*

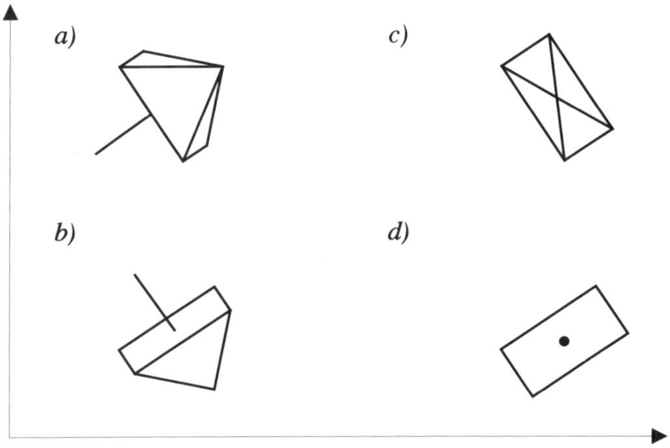

Fig. 7.31 *Examples of positive and negative images of the three-dimensional arrow: (a) positive image, (b) negative image, (c) positive degenerate image, (d) negative degenerate image*

respectively, whereas cases (c) and (d) show particular positions of the arrow, when it is perpendicular to the plane of display and pointing outwards or inwards.

A popular variation of the three-dimensional arrow uses a circular rather than rectangular base for the arrowhead. Figure 7.32 shows how such arrows may be used to describe the eddy-current distribution in a conducting cube immersed in a homogeneous harmonic magnetic field [7.78]. This particular solution has been obtained using an adaptive procedure applied to the skin-effect problem, which led to a mesh consisting of 3100 tetrahedra and 1008 nodes.

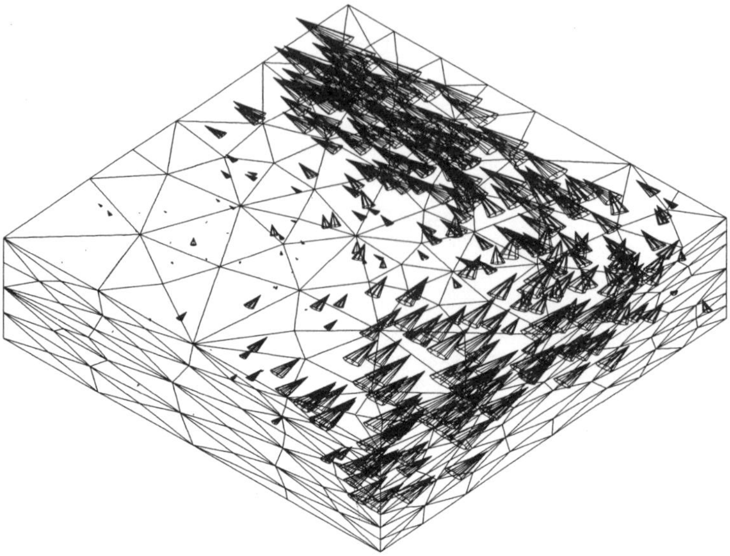

Fig. 7.32 *Eddy-current distribution in a conducting cube*

Modern post-processors offer much more than post-viewing and are capable of manipulating the field solutions. Hence global parameters such as stored energy (and thus inductance or capacitance) or ohmic power loss (resistance) may be calculated via volume integration of appropriate field components. Surface integration, on the other hand, is required for finding stresses, and thus for calculation of forces and torques. In order to improve the accuracy of quantities whose computation relies on differentiation of the solution potential, smoothing techniques may be applied to refine the quality of the output data. Finally, specific values or their distributions may need to be prepared in a particular way if transfer of output data is required to another program or output from magnetic design is going to be used as input to mechanical stress or thermal analysis programs. For example, forces of electromagnetic origin will constitute sources of a mechanical stress field, or power loss in conducting regions will give rise to a thermal field. These are not yet fully coupled fields as there is no feed-back loop; nevertheless in many practical problems a single sequential solution is quite sufficient as the change of magnetic or electric properties due to mechanical stress or change in temperature may be negligible. Some commercial programs (e.g. OPERA-2d [7.48]) have such stress or thermal analysis solvers built-in as extensions to the electromagnetic solving procedures.

Error analysis and adaptive meshing add to the variety of tasks which may be required during the post-processing stage. Errors have to be calculated first, using perhaps one of the techniques discussed in Section 7.6, and may be interrogated using available display options. If a mesh refinement procedure is then activated and information and decisions fed back to the pre-processor, then an adaptive meshing loop has been completed.

Finally, optimization techniques come into consideration. The post-processor then provides a feed-back link where design parameters are retrieved from the magnetic solution and passed back to the optimization algorithm. The pre- and post-processor become even more fully integrated, as demonstrated before in Fig. 7.3 in relation to the Design Environment Module or Fig. 7.25 for the optimization program EAMON.

7.9 THE IMPACT OF PARALLEL COMPUTING

Until the early 1970s most computers were strictly serial, i.e. one arithmetic operation was completed before the next commenced. The execution time of a program implementing a numerical algorithm could be approximated by the formula

$$time = (average\ time\ per\ operation) \times (number\ of\ operations) \quad (7.51)$$

In recent years, however, many new advanced computer architectures have been developed which employ one or more degrees of parallelism. For example, arithmetic operations may be segmented into several distinct phases and functional units designed so that different operands can be in different segments of the same operation at the same time. This technique, called pipelining, is employed by machines commonly called *vector processors*, since pipelining is particularly useful when performing calculations with vectors. The model for vector computation can be written as [7.18]

$$time = (n + s)c \quad seconds \quad (7.52)$$

where n is the length of a vector, s the number of segments, and c the clock cycle time in seconds. The time sc before any result is produced is usually called the *start-up time*, and s is sometimes termed the $n_{1/2}$ value. This is because the rate of computation attains half of its peak performance of $1/c$ (sometimes denoted r_∞) at a vector length of s. With the $n_{1/2}$ and r_∞ notation, formula (7.52) becomes

$$time = (n + n_{1/2})/r_\infty \quad (7.53)$$

The present generation of parallel computers appear to be based on three main architectural characteristics [7.8; 7.77]: memory structure, grain size and control. Memory can either be distributed, with each processor controlling its own memory, or it can be shared among several processors. *Grain size* refers to the computational power of each node (each processor and its associated memory). Control can either be SIMD (Single Instruction, Multiple Data) in which the processors execute identical instructions in *lock-step*, or MIMD (Multiple Instruction, Multiple Data) in which the processors run different codes asynchronously. Examples are:

- distributed memory, small grain size, SIMD machines like the Digital Array Processors (DAPs),
- shared memory, small grain size SIMD machines like the family of vector processors,

- shared memory, large grained MIMD machines like the CRAY-2,
- distributed memory, large grained MIMD machines like the INMOS transputer.

The implementation of finite-element codes on parallel machines is a natural development that has been accelerated by availability of a variety of parallel architectures. Much of the effort goes into parallelizing of the solution of the system of equations, although element and matrix assembly have also been implemented. An interesting comparison is presented in reference [7.33]. First, results are reported obtained using the Massively Parallel Processor (MMP) which is a very large SIMD processor. The speedup in solution time was obtained by reducing the number of sequential operations and exploiting the intrinsic parallelism of the Gauss elimination and the Gauss–Jordan algorithm. While the number of operations on sequential computers is of the order of n^3, where n is the number of equations, the number of operations on the MPP is of the order of n^3/N^2, where N^2 is the number of processors. For the MPP, this translates to an ideal speedup of 16 384. In practice, the figure is much lower and depends on the communication required, as well as on the size of the problem. Improvements of up to three orders of magnitude have nevertheless been reported.

In the second comparison, solutions were obtained using a MIMD computer, where each computer works independently, and thus efficiency of processors is very important. For this experiment the following definitions were used

$$speedup = \frac{\text{CPU } (one\ processor)}{\text{CPU } (N\ processors)} \tag{7.54}$$

while efficiency was defined by the ratio

$$efficiency = \frac{speedup}{N} \times 100\% \tag{7.55}$$

The results have shown that the speedup is almost linear with the number of processors. At the same time efficiencies ranging from 70% to 98% were achieved, depending on the problem. For larger size problems, an almost constant efficiency ranging between 80% and 90% with up to nine processors has been obtained. These results are very encouraging and clearly demonstrate possible benefits of parallel computation.

Parallel computing is a rapidly developing branch of computer science and its potential impact on CAD in electromagnetics cannot be overestimated. As a demonstration of a very recent initiative, PAFEC – one of the leading developers of general purpose finite element analysis software packages – and the Parallel Applications Centre at Southampton [7.50], are collaborating to produce a parallel version of the PAFEC-FE finite element package. A generic, scalable prototype Parallel PAFEC-FE made its first public appearance at Supercomputing'93 conference in November 1993. The prototype allows parallel computation to improve performance for both static and dynamic structural analyses. The parallelized code is fully integrated into the original sequential package, with parallel efficiency over 90%. Substantial improvements in turn-around time for complete analyses have been obtained for 16 processor runs for models with as few as 15 000 degrees of freedom. Larger models and greater number of processors will no doubt yield even better performance gains.

7.10 A SHORT SURVEY OF CAD SYSTEMS FOR MAGNETICS

This section is not intended to provide a full catalogue of all available software in electromagnetics. Nevertheless, it seems worthwhile to include some very brief information about programs which perhaps have made the greatest impact on the progress in the field of computational magnetics. There now exist quite a few commercially available systems offering integrated tools for CAD in magnetics.

A typical commercial package will have most of the following components:

- *pre- and post-processor*: fully interactive with colour graphics, windows environment, sophisticated post-viewing facilities, comprehensive range of supported output devices, automatic and adaptive meshing;
- *statics*: magnetostatics and electrostatics analysis with nonlinear materials, including permanent magnets, special versions for laminated materials using a 'packing factor';
- *steady-state eddy currents*: steady-state a.c. eddy-current analysis, including complex permeabilities, approximate nonlinear solutions (fundamental harmonic field), background d.c. fields, external circuits for voltage-driven problems;
- *transient eddy currents*: full transient eddy-current analysis, with non-linear materials, multiple drives and background d.c. fields;
- *motional eddy currents*: uniform motion induced eddy-current analysis (with constant or varying topology);
- *stress and thermal*: mechanical stress using forces, or thermal analysis using ohmic heating, calculated from electromagnetic solutions;
- 2D, 2D axisymmetric and 3D formulations.
- implementations on a variety of hardware platforms, including 386/486/Pentium DOS machines, HP, Sun SPARC, Apollo, DEC and IBM workstations, and Cray supercomputers.

Two particular commercial systems will be described here in some detail, but several other will also be mentioned. Finally, examples will be given of in-house systems, in particular those used in academic and research institutes. Such systems are often very advanced and use state of the art formulations and new developments. However, they usually lack the robustness and reliability of commercial systems and the graphic interfaces are usually much less sophisticated.

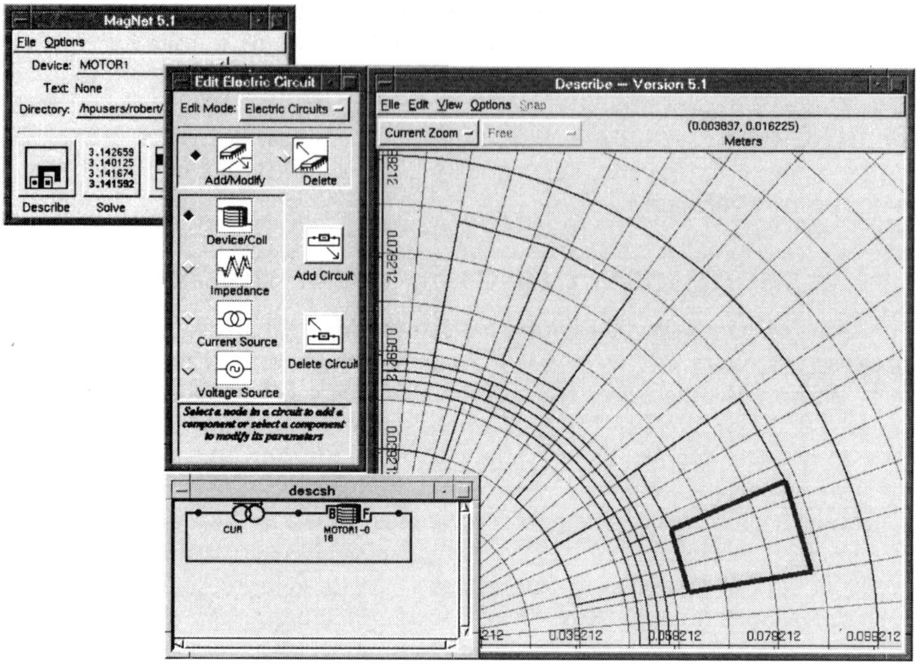

Fig. 7.33 *User interface in MagNet 5*

One of the oldest electromagnetic analysis software systems on the market, available commercially since 1978 from Infolytica Corporation in Canada and Infolytica Ltd in the UK, is a suite of programs called MagNet. The latest release is MagNet 5 [7.41; 7.42; 7.43]. The package employs a modular approach and a common operating environment on all hardware platforms. Hence, the user sees a similar interface, and has all software facilities available, on hardware ranging from a PC386 to a SuperComputer. Two special modes of operation are available, known as Fast-Track and Tool-Box.

Fast-Track allows new users to gain quick access to the package and offers experienced users a quick start on more advanced problems. It is entirely windows-based (see Fig. 7.33), and input is via mouse-button selection, together with data entry in dialogue boxes, where required. An object-oriented approach to model building is employed, whereby both simple and compound closed regions with associated material properties, known as devices, may be stored for future use. The ability to create new devices from assemblies of old ones, together with a scaling facility, gives a rapid and powerful model-building capability. An electric-circuit facility allows devices to be interconnected, or to be connected to external sources and components. Fast-Track is applicable to 2D or axisymmetric problems, employing vector or scalar potentials, and solves static and linear time-harmonic problems. Variable element order is available, up to fourth. To further simplify the process adaptive meshing is an option, whereby repeated solutions are performed,

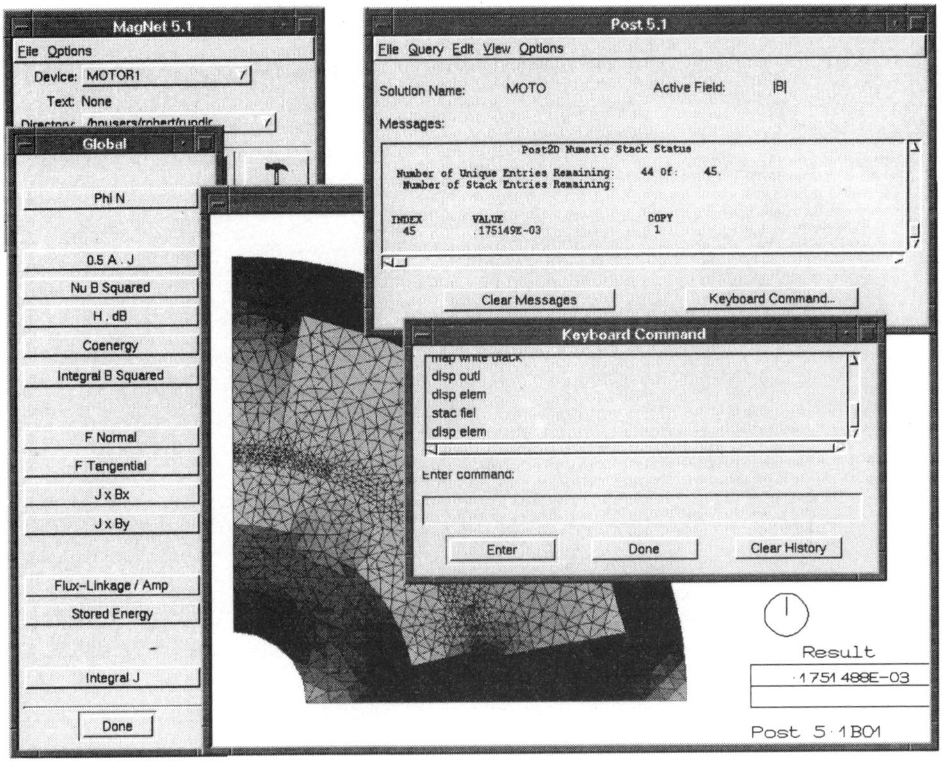

Fig. 7.34 *Postprocessing in MagNet 5*

with mesh refinement being undertaken in regions where fields are seen to be varying most. The Fast-Track Post Processor detects the problem type, and selects the appropriate menus and display characteristics. It then gives access to all standard line, arrow, and density plots. Point values of all variables are available, as well as displays of quantities along contour lines, or enlarged displays over local areas (zones). Global quantities, such as energy and its related quantities, such as force and torque, are also available at the click of a button.

Tool-Box makes available script facilities to run multiple problems in batch mode. A full set of macros is also provided, together with the facility for the user to develop new ones for specific requirements. All the modules are available directly in Tool-Box, covering mesh generating, problem defining, post processing (Fig. 7.34), as well as file managing. Four 3D solvers are also available in the Tool-Box. PM3D covers 3D magneto- and electrostatics, and TH3D is the corresponding linear time harmonic solver. TR3D is a 3D electromagnetic transient solver allowing voltage or current input of arbitrary waveshape, and interconnection of resistors or inductors. Both linear anisotropic, and nonlinear isotropic materials may be modelled. HF3D is a high frequency 3D solver, designed to calculate the scattering parameters of passive microwave components of arbitrary shape. Lossy dielectrics may be included, and

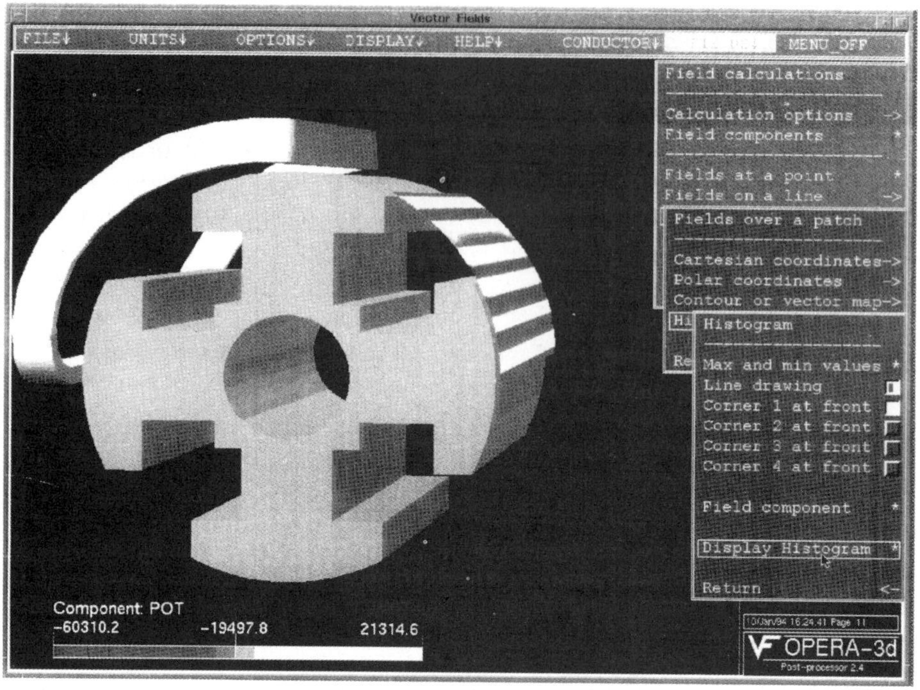

Fig. 7.35 *User interface in OPERA-3d*

boundary conditions available are perfect electric and magnetic conductors. Typical quantities calculable are energy density and the Poynting Vector.

MagNet 5 employs the robust solver technology developed over the years by Infolytica. In particular, the 3D solvers largely employ accurate and very fast scalar-potential edge-elements, as well as hierarchical techniques which vary element order, giving the finest detail in the most critical regions. Increasingly, MagNet 5 appears as an invisible inner core of a much larger package, for which the input comes directly from the designer's standard data sheet. This is referred to as the shell mode of operation, and this variety of modes of usage is believed to be unique to the MagNet system.

Another famous name in software for electromagnetics worth mentioning in this Section, is associated with the company Vector Fields Ltd, Oxford, UK, and Vector Fields Inc, Aurora, USA, and a suite of programs called OPERA [7.48; 7.51; 7.79]. This includes several programs: OPERA-2d, and its version for personal computers called PC-OPERA, with integrated pre- and post-processing, for solution of two-dimensional static, eddy-current and transient fields, including axisymmetric problems; OPERA-3d pre- and post-processor; TOSCA for solution of static three-dimensional fields; ELEKTRA for three-dimensional eddy-current problems. Vector Fields is well established in the international market and the software is highly regarded for its extensive functionality, versatility and good customer support. The software is command

and/or menu driven and a typical user interface is shown in Fig. 7.35.

The most recent release of the OPERA software [7.79] contains many new features, mainly improvements and enhancements to user interface and communications. For example, a new interface to the Autocad CAD package enables geometric data to be transferred to the OPERA-2d pre-processor. Both 2D and 3D solvers can now cope with high speed motion using a numerical technique known as 'upwinding'. A new analysis program for modelling beams of charged particles, including the effects produced by space charge, has also been introduced. Simplified construction of models has been achieved in a 2D pre-processor through the introduction of a 'background' region (for example filling in the 'empty' space with air) and automatic region matching. Finally, an adaptive solver has been introduced for 2D static fields.

This new adaptive process uses a 'three step' refinement procedure. First, the adaption is implemented, where the refinement of an element is based on the local error criterion. An improvement process is then applied, where two new criteria (based on area and length) are introduced to avoid deformed elements. Finally a regularization process is activated, where for each internal node a new optimal position is found at the centre of gravity of the polygon formed by the neighbouring nodes. A new solution is found for the refined mesh, and the process is repeated until the error in each element is less than that specified by the user, or the number of element exceeds a maximum value, or the maximum number of iterations have been performed. In addition, a limit is placed on the smallest element allowed. To reduce the number of refinement iterations the value of the potential for each new node is interpolated from the surrounding nodal values. Thus the process always starts from a previous non-zero solution. As an example, Fig. 7.36a shows the initial mesh for a motor problem, and Fig. 7.36b demonstrates the final mesh (and flux lines) after the adaptive refinement process has been completed.

The following list of other popular commercial systems is not exhaustive, but is provided here for reference.

- *ANSYS* [7.1] – developed and maintained by Swanson Analysis Systems, Inc. (SASI) of Houston, Pennsylvania. One of the oldest commercial finite-element systems, ANSYS began in 1970 as a structural and thermal analysis package. Magnetic solvers were first introduced in 1982. Of particular interest are the coupled magneto-structural-thermal field analysis options and solid modelling capabilities of the software. A PC-based subset of the ANSYS program, the ANSYS-PC/MAGNETIC, was designed specifically for electromagnetic analysis and has been available since 1991.
- *COSMOS/M* – by the Structural Research and Analysis Corporation, specialists in desktop finite-element analysis software. The package includes electro-thermal and magneto-structural couplings.
- *Maxwell* – by Ansoft Corporation, Pittsburgh, USA. The Maxwell 2D Simulator can run under Microsoft Windows 3.1 graphics interface.

(a)

(b)

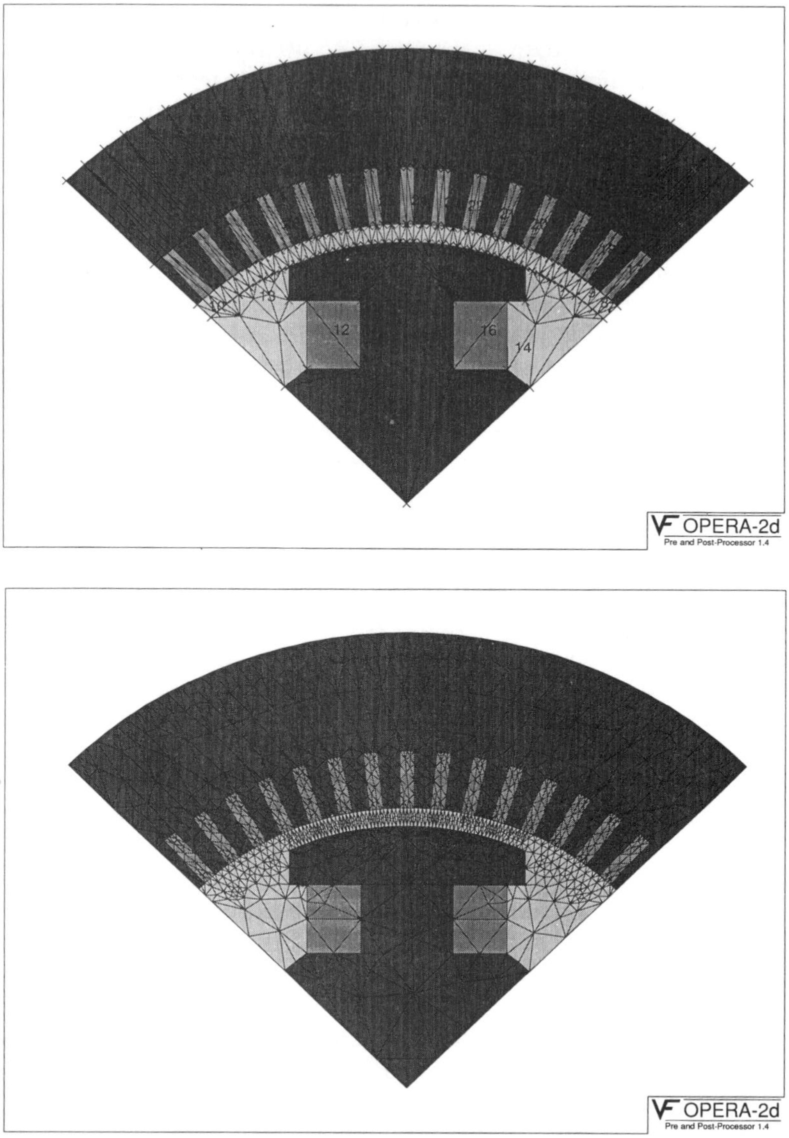

Fig. 7.36 *Adaptive meshing using OPERA-2d [7.79]: (a) initial mesh for a motor problem, (b) final mesh after refinement process*

- *MSC/EMAS* – by the MacNeal-Schwendler Corporation, who also developed MSC/NASTRAN. This 2D and 3D solver is particularly useful for solving problems in electromagnetic imaging and compatibility, radio frequency shielding, microwave heating, magnetic signature, antennas and other radiating structures and is available on most popular workstations and larger computers.

- *BEASY* [7.2] – The Wessex Institute of Technology, Southampton, UK, is recognized as the world's leader for boundary element technology, and its software development arm, the BEASY Computational Mechanics Company, offers very advanced software for a variety of applications. The range of analysis capabilities includes stress, thermal, mechanical fracture, acoustics, potential flow and electro-magnetic analysis. One of the advantages of using boundary elements is simplified modelling as only surfaces have to be defined. Advances in the boundary element method are regularly reported in professional conferences (see for example references [7.65] and [7.66]) and are the subject of specialist books (e.g. [7.6; 7.7]).
- *Integrated Engineering Software* [7.34] – with a suite of programs ELECTRO (2D), OERSTED (2D), MAGNETO (2D), COULOMB (3D) and AMPERES (3D), all based on a boundary element method. Each program is fully integrated around a convenient user-interface and includes geometric modellers and appropriate field solvers. Geometry files may be exchanged with other CAD packages.
- *MEGA* – by Bath University Applied Electromagnetics Centre, UK. MEGA is a very comprehensive suite of 2D and 3D electromagnetic programs, with many unique formulations, especially for transient 3D eddy currents and moving media.
- *SLIM* – an integrated collection of software modules developed by GEC Alsthom Engineering Research Centre, Stafford, UK,
- plus a number of other packages.

From the many in-house systems developed in academic and research institutions, two will be mentioned here.

- *SONMAP* [7.29] – This is an advanced suite of programs, written at the Technical University of Szczecin, Poland, incorporating many state-of-the-art developments in finite-element techniques, such as adaptive meshing, infinite elements, surface impedance boundary condition etc. The package is available commercially and may be supplied in a source code version.
- *ARISTOTLE-EUKLID* [7.78] – ARISTOTLE is a research package developed at the Aristotelian University of Thessaloniki. Simplex elements are employed with node and edge shape function expansions for solution of static and time varying fields at both low and high frequency. EUKLID implements a three-dimensional adaptive refinement on tetrahedral meshes. Examples obtained from these programs have been shown in this Chapter as Figs. 7.22, 7.23 and 7.32.

There is a continuously growing market for electromagnetic software as more and more design offices in many sections of industry invest in Computer Aided Design of electromechanical devices. Thus further improvements to the existing finite-element and other codes, as well as new CAD systems, are to be expected.

8
Experimental Methods

Kazimierz Zakrzewski

8.1 INTRODUCTION

Experiments are the prime source of information about physical processes and are conducted to assess material properties, identify equivalent parameters or formulate empirical relationships. Experimental investigations are indispensable as a means of verifying various computational methods and models. Operational characteristics of devices are usually established through tests and measurements and they provide an ultimate answer to whether the design objectives have been fulfilled.

In this book we have discussed various approximation methods used to obtain field distributions via the solution of differential or integral equations. Such solutions may also be obtained by mathematical or physical analogy using analogue computers or special analysers.

The final verification of the design is performed by investigating prototypes. These are usually objects with a high degree of complexity both in terms of geometrical configuration and in view of the electromagnetic, mechanical or thermal processes.

8.2 PHYSICAL MODELLING

Experimental investigations on real engineering objects, especially high power electrical devices, may be very expensive. Moreover, from the technical point of view, it is not always possible to prepare a versatile programme of investigations which is applicable to all cases. Therefore, in addition to mathematical modelling, physical modelling continues to play an important role. Measurements are taken on a model of reduced dimensions but similar physical properties and configuration. The relevant theory enables the results obtained on a small model to be transposed to the full-scale original device, without the need to experiment on a large prototype [8.10; 8.22].

8.2.1. Scaling criteria

A set of dimensionless numbers resulting from the application of fundamental equations of electromagnetics to both the original system and its model, provide the scaling criteria for physical modelling [8.10; 8.22].

Maxwell's equations applied to the original system, taking into account the velocity term, yield

$$curl\,\mathbf{H} = \sigma\mathbf{E} - \frac{\partial \mathbf{D}}{\partial t} + \sigma(\mathbf{v} \times \mathbf{B}) + \mathbf{J}_i + \mathbf{v}\,div\mathbf{D} + curl(\mathbf{D} \times \mathbf{v})$$

$$curl\,\mathbf{E} = -\frac{\partial \mathbf{B}}{\partial t} - curl(\mathbf{B} \times \mathbf{v}) \tag{8.1}$$

$$\mathbf{B} = \mu\mathbf{H}, \qquad \mathbf{D} = \varepsilon\mathbf{E}$$

For the reduced-size model

$$\frac{m_H}{m_l}\,curl\,\mathbf{H} = m_\sigma m_E \sigma\mathbf{E} + \frac{m_\varepsilon m_E}{m_t}\frac{\partial \mathbf{D}}{\partial t} + m_\sigma m_v m_\mu m_H \sigma(\mathbf{v} \times \mathbf{B}) +$$

$$m_{ji}\mathbf{J}_i + m_v m_\varepsilon m_E \mathbf{v}\,div\,\mathbf{D}$$

$$\frac{m_E}{m_l}\,curl\,\mathbf{E} = -\frac{m_\mu m_H}{m_t}\frac{\partial \mathbf{B}}{\partial t} - \frac{m_\mu m_H m_v}{m_l}\,curl(\mathbf{B} \times \mathbf{v}) \tag{8.2}$$

$$m_B\mathbf{B} = m_\mu m_H \mathbf{H}, \qquad m_D\mathbf{D} = m_\varepsilon m_E \mathbf{E}$$

The quotient of a given physical quantity in the model to its value in the full-scale original is called the scaling factor and is designated by the letter 'm' with an appropriate index. For example m_l denotes a scaling factor of a linear dimension. If we introduce two further self-explanatory relationships

$$m_f = \frac{1}{m_t} \qquad \text{and} \qquad m_v = m_l m_f$$

we can clearly see that equivalence of Maxwell equations for the model and the original necessitates the fulfilment of the following conditions

$$m_\sigma m_E m_l / m_H = 1 \tag{8.3}$$

$$m_\varepsilon m_E m_f m_l / m_H = 1 \tag{8.4}$$

$$m_\sigma m_\mu m_f m_l^2 = 1 \tag{8.5}$$

$$m_{ji} m_l / m_H = 1 \tag{8.6}$$

$$m_\mu m_H m_f m_l / m_E = 1 \tag{8.7}$$

Rearranging terms leads to a set of scaling criteria

$$\pi_1 = m_\mu m_\sigma m_f m_l^2 = 1 \tag{8.8}$$

$$\pi_2 = m_\varepsilon m_\mu m_f^2 m_l^2 = 1 \tag{8.9}$$

$$\pi_3 = m_{ji} m_l / m_H = 1 \tag{8.6a}$$

It will be noticed that in practice it may be difficult or virtually impossible to meet criteria (8.8) and (8.9) simultaneously. In particular, assuming the decrease of dimensions coupled with increase of frequency, and in addition keeping $m_\varepsilon = m_\mu = 1$, we obtain a condition $m_\sigma = m_f$, which in general is an impossible requirement, because a material of the postulated electric conductivity is unlikely to be available. Hence, for eddy-current problems, we often consider the criterion (8.8) only, whereas the condition (8.9) will usually be applied to electric rather than magnetic effects. The third criterion, equation (8.6a), resulting from the comparison of the imposed current density, defines the scale for the magnetic field strength m_H on the basis of current excitation. Following the above procedure it is also possible to define scaling criteria of processes and physical effects described by other types of equations.

8.2.2 Modelling fields and losses in ferromagnetics

In the majority of cases, various iron components of electrical devices, when subjected to a harmonic magnetic field, maintain the properties of solid iron. Moreover, it is often possible to neglect the effect of eddy-currents on the excitation flux. From a practical point of view, it is usually sufficient to apply the plane wave theory in order to calculate the local power loss [8.24; 8.29]. As shown in Fig. 8.1 the penetration of flux into solid iron is governed by the normal component of flux density.

The leakage flux per unit length along the z-axis in the cross-section of iron at position y equals

$$\phi_m(y) = \int_{y_0}^{y} |\mathbf{B}_n(y)| \, dy \tag{8.10}$$

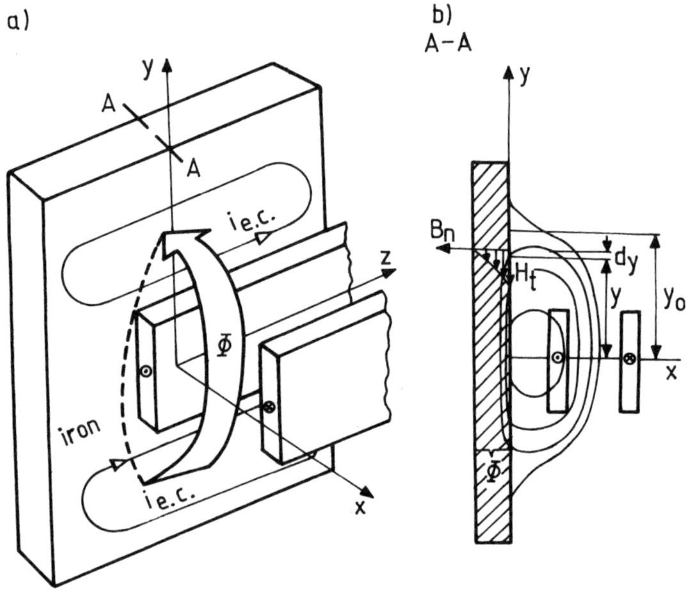

Fig. 8.1 *Penetration of magnetic field into solid iron [8.29]: (a) iron plate near current leads, (b) cross-section*

where y_0 is the coordinate at which $H_{mt}(y)=0$. The same flux from the solid iron side may be expressed as

$$\phi_m(y) = \psi\sqrt{\left(\frac{\mu_m}{\omega\sigma}\right)}H_{mt} \tag{8.11}$$

where:

μ_m static magnetic permeability $\mu_m(H_{mt})$;

ψ a semi-empirical coefficient describing an increase of leakage flux due to the magnetic non-linearity of iron (see reference [8.25]).

Comparison of eqns (8.10) and (8.11) leads to the following relationships:

$$m_\phi = m_{Bn}m_l \tag{8.12}$$

$$m_\phi = m_\psi\sqrt{\left(\frac{m_\mu}{m_f m_\sigma}\right)}m_{Ht} \tag{8.13}$$

We adopt, after Neyman [8.16], the approximation:

$$H = C_1 B^n \tag{8.14}$$

where C_1 is a constant dependent on the actual B/H curve. Following the author's investigations [8.27; 8.28] we write:

$$m_\mu = m_B^{(1-n)} = m_H^{(1-n)/n} \tag{8.15}$$

Taking into account eqns (8.7) and (8.15) we find a condition for frequency:

$$m_f = m_H^{(n-1)/n} m_\sigma^{-1} m_l^{-2} \tag{8.16}$$

Finally, comparison of eqns (8.12) and (8.13), together with the identity $m_\psi = 1$, yields:

$$m_{Ht} = m_{Hn} = (m_{Bn})^n \tag{8.17}$$

which implies consistency of modelling of both components of magnetic field at frequency defined by eqn (8.16).

The flux density is modelled according to the formula:

$$m_{Bn} = m_j m_l m_n \tag{8.18}$$

where:

m_j scaling factor for current density;

m_n scaling factor for a number of turns.

Thus the frequency is modelled as

$$m_f = (m_j m_n)^{n-1} m_\sigma^{-1} m_l^{n-3} \tag{8.19}$$

The local power loss, in W m^{-2}, will satisfy

$$m_{ps} = \frac{(m_j m_n)^{2n}}{m_\sigma}(m_l)^{2n-1} \tag{8.20}$$

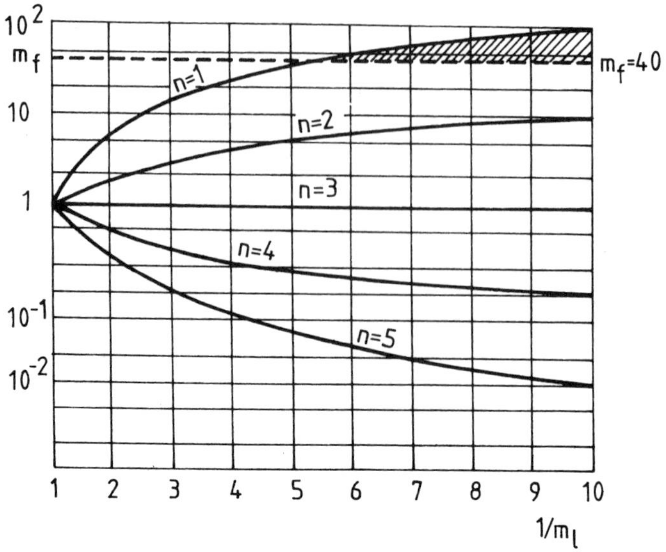

Fig. 8.2 *Frequency scale m_f as a function of linear dimensions [8.29] ($m_j = m_n = m_\sigma = 1$)*

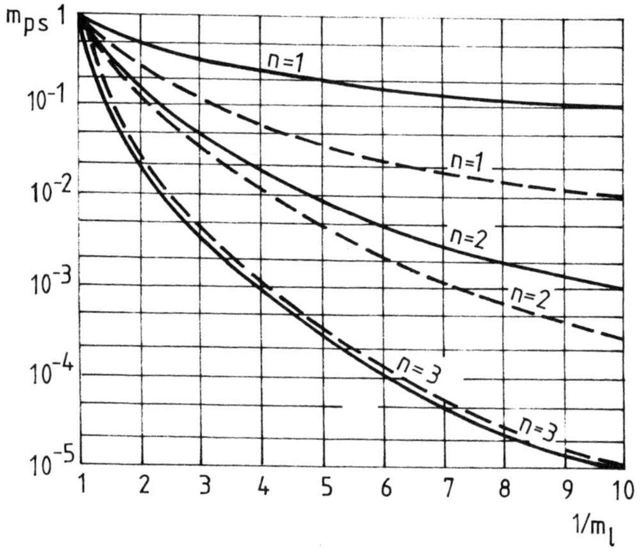

Fig. 8.3 *Local power loss scale m_{ps} as a function of linear dimensions at frequency given by eqn (8.19) – continuous line, and at the frequency scale $m_f = 1$ – dashed line [8.29] ($m_j = m_n = m_\sigma = 1$)*

Coefficients m_f and m_{ps} are plotted in Figs 8.2 and 8.3 respectively, and can be used to design reduced-scale models.

A typical electrical power device contains elements of different levels of magnetic saturation. Therefore, it will not be possible to fulfil the frequency requirement for the entire object. Modelling can be carried out as a function of frequency; however, in a given experiment only one frequency of the external field may be used at a time.

The general formula for the power loss scale m_{ps} is as follows:

$$m_{ps} = \sqrt{\left(\frac{m_f}{m_\sigma}\right)} m_{Bn}^{(3n+1)/2}$$

For a particular case of $m_f = 1$, i.e. the same frequency in the model and the original system, the variation of the coefficient m_{ps} is demonstrated with a dashed line in Fig. 8.3.

To determine the maximum allowed modelling frequency at which the reaction of eddy-current may still be neglected, numerical calculations were made using a simplified model shown in Fig. 8.4 with the additional assumption that $m_\mu = m_\sigma = 1$ [8.31].

The magnetic field strength should obey $m_H = m_l$, because $m_j = m_n = m_\mu = 1$. The calculated scaling factor for tangential and normal components are m_{Ht} and m_{Hn} respectively. As long as the eddy-current reaction on the excitation field is not observed, the coefficients $w_t = m_{Ht}/m_H$ and $w_n = m_{Hn}/m_H$ have values close to unity.

It follows from Fig. 8.5, that for a non-screened steel, the maximum admissible frequency is equal about 2000 Hz.

8.2.3 Linear motors

A significant contribution in the area of electromagnetic field modelling in rotating machines was made by Ivanov-Smolensky [8.10]. In the field of linear

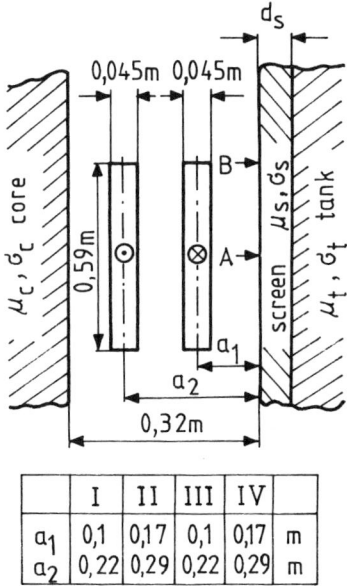

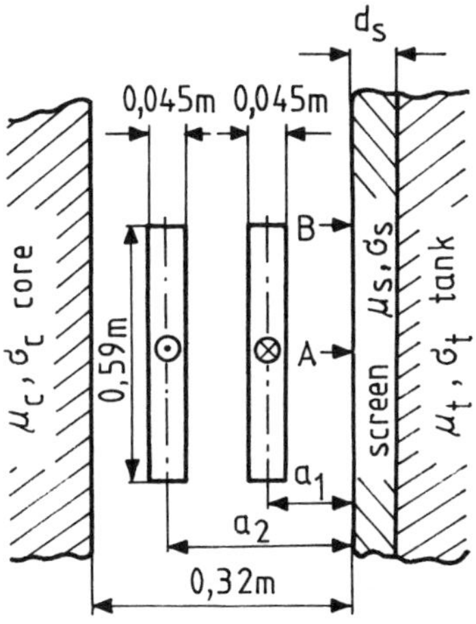

	I	II	III	IV	
a_1	0,1	0,17	0,1	0,17	m
a_2	0,22	0,29	0,22	0,29	m

Fig. 8.4 *Dimensions of the transformer system used to calculate maximum modelling frequency [8.31]*

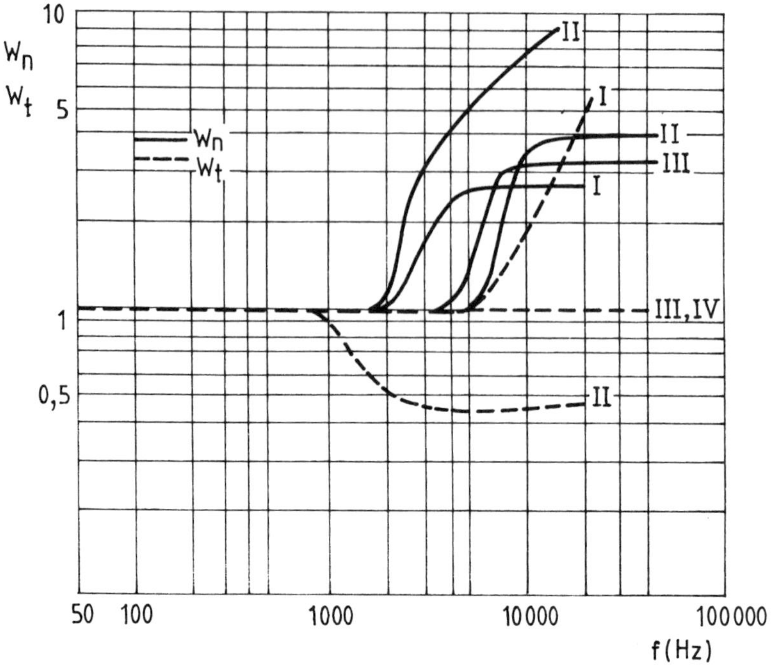

Fig. 8.5 *Coefficients w_t and w_n for the system from Fig. 8.4 as functions of frequency [8.31]*

motors some interesting papers were published by Bland, Lowther, Freeman and Laithwaite [8.1, 8.5]. Further studies which take account of magnetic non-linearity of solid iron are reported in [8.1, 8.5]. In particular the scaling factor for the magnetomotive force is

$$m_{mf} = \frac{m_j m_l^2 m_{k\mu} m_r m_n m_k}{m_g}, \tag{8.21}$$

where:

- $m_{k\mu}$ scale for the magnetizing current;
- m_r scale for the coefficient of demagnetizing armature reaction;
- m_n scale for the number of winding turns;
- m_k scale for the winding coefficient;
- m_g scale for the number of parallel paths.

The scaling factors for a magnetic field strength in laminated core, iron armature and yoke are:

$$m_{Hc} = m_{Ha} = m_{Hy} = \frac{m_{mf}}{m_l} \tag{8.22}$$

In the air-gap, where for technological reasons m_δ may be different from m_l, we have

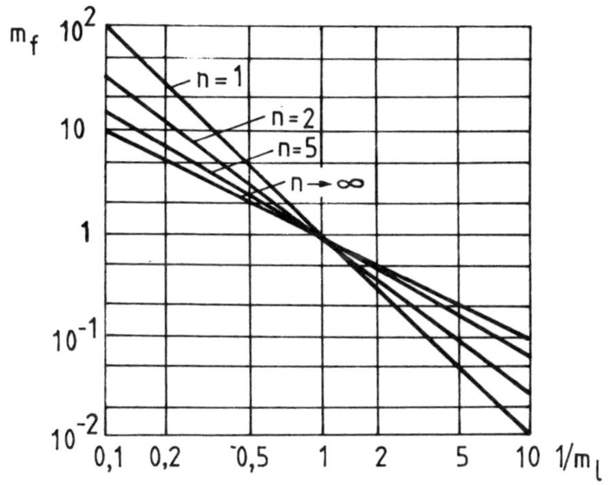

Fig. 8.6 *Frequency scale m_f as a function of linear dimensions [8.28] ($m_\sigma = m_j = m_{k\mu} = m_r = m_n = m_g = m_k = 1$)*

$$m_{H\delta} = \frac{m_{mf}}{m_\delta} \tag{8.23}$$

Substituting (8.21) and (8.22) into eqn (8.16) gives the frequency condition

$$m_f = \frac{(m_j m_{k\mu} m_r m_n m_k)^{(n-1)/n}}{(m_g)^{(n-1)/n}(m_l)^{(n+1)/n}(m_\sigma)} \tag{8.24}$$

The scaling factor for a linear speed may be found from eqn (8.5) as

$$m_v = \frac{(m_j m_{k\mu} m_r m_n m_k)^{(n-1)/n}}{(m_g)^{(n-1)/n}(m_l)^{1/n}(m_\sigma)} \tag{8.25}$$

The plots of m_f and m_v are presented in Figs 8.6 and 8.7.

The thrust force of the motor may be calculated from the power of the travelling field penetrating into the solid armature, and the synchronous linear speed. This power is calculated by integration of the complex Poynting vector over one or both sides of the armature surface, depending on whether it is a single-sided or double-sided motor. Thus

$$P = \mathrm{Re}\left\{\frac{1}{2}\int \mathbf{E}_{ma}\mathbf{H}^*_{ma}\,\mathrm{d}S\right\} \tag{8.26}$$

in the model

$$m_p P = m_{Ea} m_{Ha} m_l^2 P \tag{8.27}$$

Hence the scaling factor for a motor thrust force is

$$m_F = \frac{m_p}{m_v} \tag{8.28}$$

which after substitution and rearrangement gives

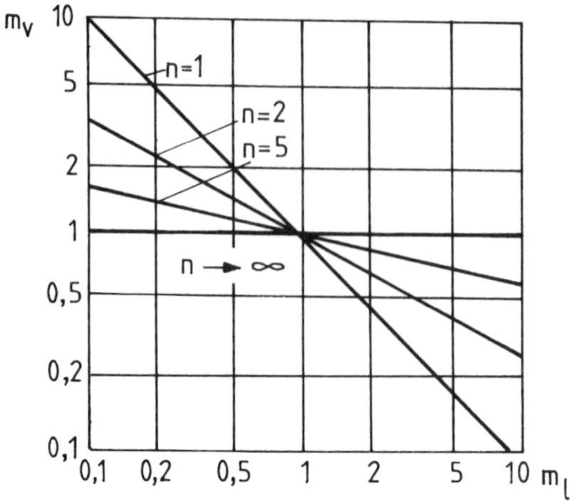

Fig. 8.7 *Linear speed scale m_v as a function of linear dimension [8.28] ($m_\sigma = m_j = m_{k_\mu} = m_r = m_n = m_g = m_k = 1$)*

$$m_F = m_l^{(3n+1)/n} \left(m_j m_{k\mu} m_r \frac{m_n}{m_g} m_k \right)^{(n+1)/n} \tag{8.29}$$

The plots of the scaling factor for the motor thrust m_F for non-magnetic or highly-saturated ferromagnetic armature are presented in Fig. 8.8. Following a similar procedure it is possible to define modelling scales for power loss in the inductor winding, iron loss, leakage reactance, supply voltage and apparent power of the motor.

8.3 EXPERIMENTAL TECHNIQUES

8.3.1 Experimental models

Experiments are often conducted on special models for the purpose of verifying computational methods and techniques. Such models are of a special

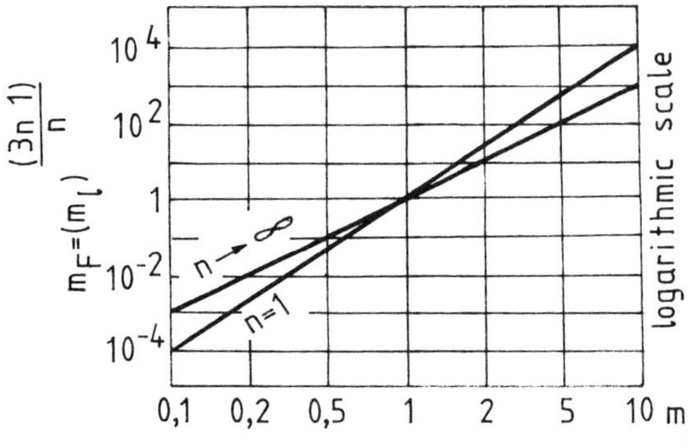

Fig. 8.8 *Motor thrust scale m_F as a function of linear dimensions ($m_\sigma = m_j = m_{k\mu} = m_r = m_n = m_g = m_k = 1$)*

type because from the point of view of physical modelling theory they are actually originals. In this case, for distinction, the name experimental models will be used.

The level of simplification in experimental models varies. The majority of these models resemble practical devices such as electrical machines, some will simply be industrial products as such. They are used to assess the adequacy and accuracy of mathematical models.

Physical fields are described by both space distribution and integral parameters. Thus the experimental methods must be able to provide measurements of distributed as well as integral values.

8.3.2 Basic measurements

Basic investigations include experiments on ring or tube samples to define the relation between the electric and magnetic field strength in metals, especially in iron. In such experiments the boundary and initial conditions are precisely defined.

The ring sample is excited by a voltage supply (primary winding 1 of n_1 turns), whereas the tube sample by a current flow (see Fig. 8.9). The shape of a sample enables the calculation of magnetic field strength from simple analytical formulae. The time wave-forms of currents can be displayed on an oscilloscope. Electric field strength in the ring sample is found from an electromotive force induced in the secondary winding 2 of n_2 turns or in the case of a tube, by point measurement of electric potential drop along the current flow. The application of compensation techniques is necessary to measure accurately such small potentials [8.22; 8.26].

The magnetic field strength, averaged for both surfaces of the ring sample, is calculated from the formula:

$$H = \frac{in_1}{\pi D} \tag{8.30}$$

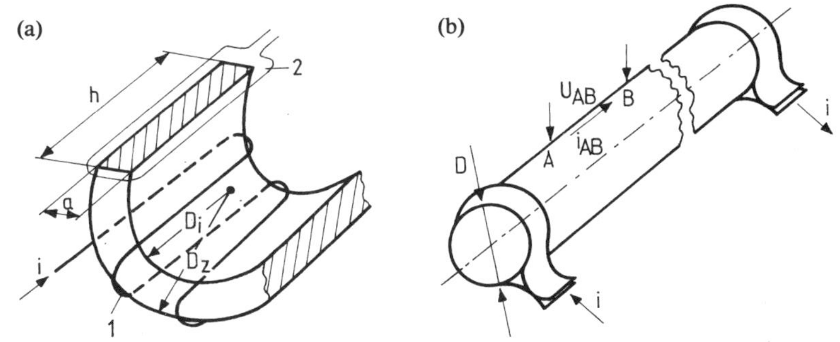

Fig. 8.9 *Samples used for basic measurements: (a) section through a ring sample (1 – exciting winding, 2 – voltage winding); (b) tube sample (AB – points of potential measurements)*

in which

$$D = \frac{\sqrt{2}D_0 D_i}{\sqrt{(D_0^2 + D_i^2)}} \tag{8.31}$$

The electric field strength equals

$$E = \frac{U_2}{2(h + a)n_2} \tag{8.32}$$

where a and h are shown in Fig. 8.9.

For the tube sample

$$H = \frac{i}{\pi D} \tag{8.33}$$

and

$$E = \frac{U_{AB}}{I_{AB}} \tag{8.34}$$

The power loss resulting from the Poynting vector may be measured using a wattmeter or from the oscilloscope by multiplying E by H and integrating over the surface.

For sinusoidal time variation:

$$P = \mathrm{Re}\left\{ \oint_S \frac{1}{2} (\mathbf{E}_m \cdot \mathbf{H}_m^*)\, \mathrm{d}S \right\}. \tag{8.35}$$

The loss in a sector of the ring sample of length l, according to the field penetration model, may be written as

$$P = \left(\frac{1}{\sqrt{2}} \int_0^{2(h+a)} E_m \,\mathrm{d}y \right) \left(\frac{1}{\sqrt{2}} \int_0^{l} H_m \,\mathrm{d}z \right) \cos \varphi \tag{8.36}$$

The first term in eqn (8.36) gives the rms voltage per turn e_t, whereas the second yields the mmf drop (magnetic potential difference) U_μ. Therefore the formula for power loss in the sample reads

$$P = e_t U_\mu \cos \varphi \tag{8.37}$$

where φ is the phase angle between the magnetic potential difference and the induced voltage. If the oscilloscope is used then the power loss is proportional to the surface of Lissajou figure. In practice, the total power loss is usually measured using a wattmeter and then divided by the surface of the sample, to obtain power loss per unit area, which gives a correct result providing the sample is homogeneous and isotropic. In the ring sample, the excitation current and induced voltage u_2 are used to measure the power. In the case of the tube, the conduction current and the tube terminal voltage are used [8.22; 8.26].

8.3.3 Flux and power loss in magnetic cores

Ferromagnetic cores of transformers, rotating machines, reactors and other electrical devices are made from laminated electrical steel. The *cold-rolled grain-oriented* sheet and strip materials have magnetic properties and power loss in the rolling direction superior to those on any other axis. Such sheets have been used in transformers for over 30 years and have also been recently introduced to rotating machines.

The Epstein's apparatus is the basic instrument enabling measurements of average unit loss in samples for comparison of different types of magnetic sheets. However, the measurements tend not to be very accurate and are not particularly helpful in explaining physical phenomena.

A modified experiment uses single sheets as shown in Fig. 8.10. The advantage is that we can define exactly the conditions of field excitation. This overcomes the problem inherent to the standard Epstein's instrument of non-uniformity of flux distribution in the packet of laminations and the anisotropy effect of the corners [8.26].

In Fig. 8.10 the so called *Rogowski's coil* is presented. The measuring coil is wound on an elastic non-magnetic core. The flux linking the coil is given by

$$\psi = \mu_0 s n' U_{\mu_{AB}} \tag{8.38}$$

where:

n' number of turns per coil unit length;
s cross-section area of the coil;
$U_{\mu_{AB}}$ magnetic potential difference (mmf).

The flux linkage in an alternating field is obtained from the electromotive force e induced in the coil:

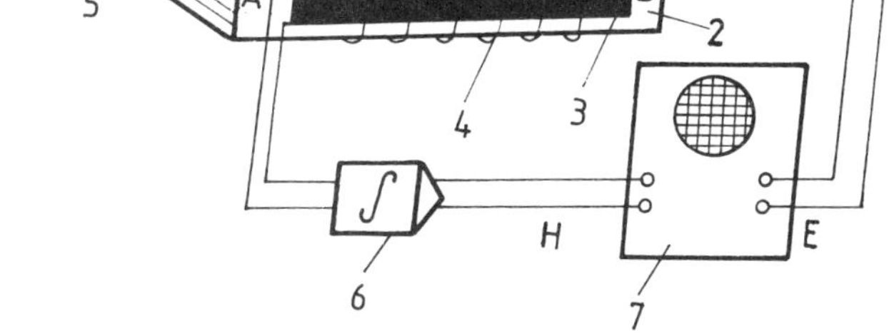

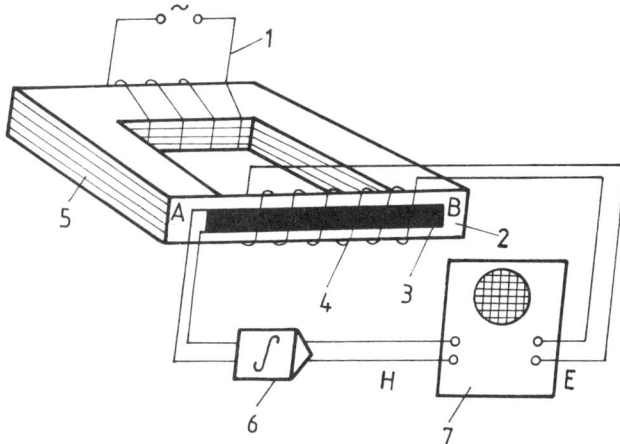

Fig. 8.10 *Modified Epstein's apparatus for measurements on single sheets: 1 – exciting winding, 2 – sample (single sheet), 3 – Rogowski's coil, 4 – voltage winding, 5 – core, 6 – integrator, 7 – oscilloscope*

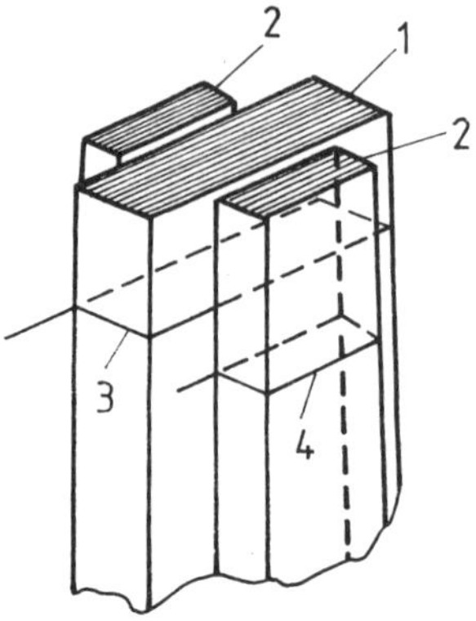

Fig. 8.11 *Coils for measuring flux in packets of laminations: 1, 2 – packets of lamination, 3, 4 – multiwire voltage probes*

$$\psi = \frac{e}{4\sigma_k f} \tag{8.39}$$

where σ_k is the form coefficient of the electromotive force.

In the magnetostatic field

$$\psi = c_\psi \alpha \tag{8.40}$$

where:

c_ψ constant of the ballistic galvanometer;
α galvanometer indication.

The electric field strength is measured using the winding number 4 shown in Fig. 8.10. Using a two-beam oscilloscope we can record simultaneously both the magnetic field strength **H** and the electric field strength **E**.

The packets of laminations forming a core may have different reluctances. Thus it is necessary to measure the total flux and flux distribution between individual packets. Tests show that even if the total flux in the core is sinusoidal, the component fluxes may be distorted [8.13]. Multiwire voltage pick-up coils enclosing or stitched through the lamination are often used for such measurements (Fig. 8.11). The application of a vector-meter or an oscilloscope with an integrator enables the time recording of the average flux density in the core.

The *Hall probe* may also be used for measuring power loss distribution in a transformer core. The magnetic field strength **H** must be perpendicular to the probe plane. The terminals must be connected to the measuring coil enclosing a part of the core adjacent to the probe. Current is then proportional to electric field strength on the core surface [8.6; 8.22].

8.3.4 Measurements in air-gaps of electrical machines

Mechanical characteristics of rotating machines and linear motors depend on the magnetic field distribution in their air-gap. Slots in both stator and rotor

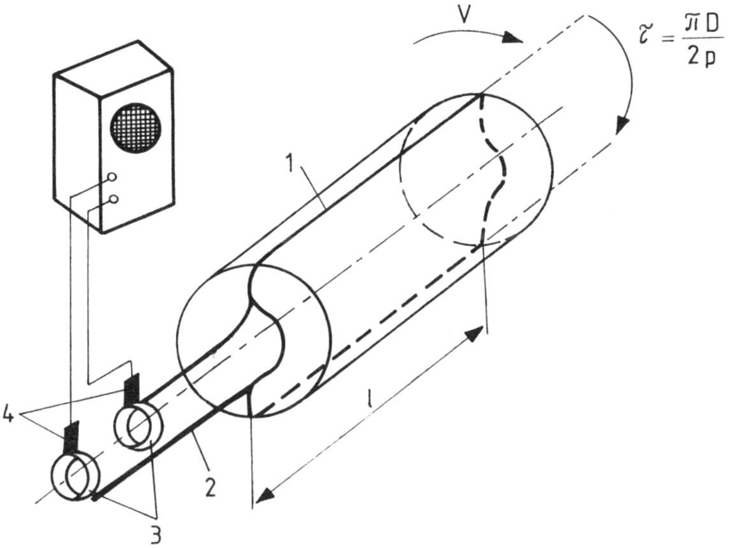

Fig. 8.12 *Fully-pitched coil for measurements of magnetic flux density in the air-gap of rotating machines 1 – multiwire coil, 2 – terminals, 3 – slip-rings, 4 – brushes*

generate additional harmonics of flux density. The distribution of the normal component of flux density may be registered using fixed or movable Hall's probes in the air-gap region. Other ways of investigating flux distribution are search coils fixed to the stator surface, or coils wound on the tooth heads to measure flux entering the iron core.

One of the oldest methods is the application of a fully-pitched coil located on the rotor and connected to the slip-rings (Fig. 8.12). The electromotive force of rotation, induced in the coil, represents the air-gap distribution of flux density in one pole pitch:

$$e = Blvn \tag{8.41}$$

where:

- B magnetic flux density;
- l axial length of the coil;
- v speed of rotation of the coil;
- n number of turns.

8.3.5 Power loss in transformer tanks

Using the eddy-current flow analogy it is possible to build a simplified model for investigating eddy-current in a single-phase transformer tank. As shown in Fig. 8.13, the model is built from a thick steel plate placed between excitation coils. The eddy-currents on both surfaces of the middle plate of the model have similar distribution to eddy-currents on the inner surface of a real transformer tank. The transformer core is modelled using laminations 5. For

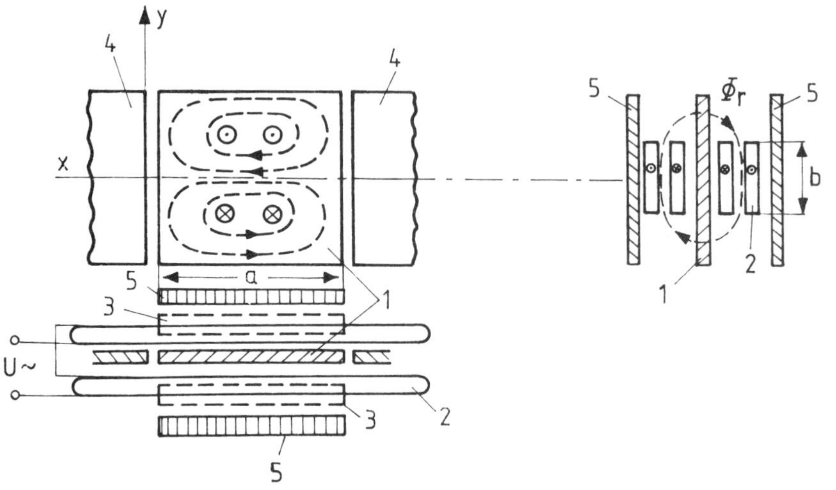

Fig. 8.13 *Model of a single-phase transformer tank for power loss investigations: 1 – steel plate tank, 2 – exciting winding, 3 – voltage windings, 4 – additional steel plates for balancing the field distribution, 5 – laminations (transformer core)*

power loss measurements the idea of Epstein's apparatus is used. The voltage winding 3, which has the same number of turns as the excitation winding 2 and the same length as the steel plate, is connected to the voltage coil of the wattmeter. The current coil of the wattmeter is supplied by the excitation current directly or via a current transformer. It is possible, for example, to investigate the effect of eddy-current screening of the tank by covering the steel plate with aluminium or copper sheets of different thickness.

Similarly, measurements can be made with magnetic shunts which partially or fully cover the steel plate, to assess the magnetic screening efficiency.

Measurements of normal and tangential components of magnetic field strength are carried out by means of simple search coils, placed near the core, shunts, windings etc. Coils stitched through the steel plate enable measurements of electric field strength on the surface, as well as flux distribution along the x-axis inside the tank.

8.3.6 Leakage fields

The three-dimensional character of leakage fields of most practical devices makes experiments laborious. The space configuration of leakage fields can be investigated by Hall probes moved in space relative to a fixed coordinate system. An alternative system consists of three search coils fixed to a stand so that the position of a 'measurement point' can be easily defined. Such investigations are carried out near the transformer windings, inside and outside the 'window', and in the coil end region in alternating current machines (Fig. 8.14).

In the case of large power transformers successful measurements have been reported for leakage fields in tanks, covers, yokes and shields [8.18; 8.2].

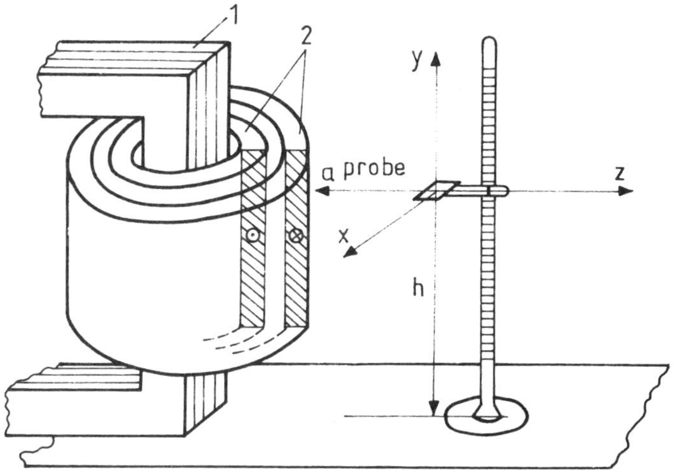

Fig. 8.14 *Measurements of the transformer leakage field with probes fixed to a stand: 1 – transformer core, 2 – windings*

Flux investigation in the closed volume of a transformer tank, or rotating machine casing, requires a large number of fixed search coils. However these fields are usually quite strong, well above the background noise level, which, in addition, is usually well screened by the metallic cover.

The situation is very different with external leakage fluxes which penetrate through the cover and are thus present in the neighbourhood of electrical devices. These fields are very weak and are often disturbed by background fields of, for example, geomagnetic origin. Measurements are made using special equipment with geomagnetic field compensation (e.g. Forster's probes). Usually there is automatic control and microcomputer processing of results [8.3]. The results are often presented graphically as three-dimensional field plots [8.9].

8.4 EXPERIMENTS FOR COUPLED FIELDS

8.4.1 Forces and stresses

Electromagnetic force is one of the integral parameters of an electromagnetic field. As distributed systems, these are fields of space force density, surface stress or tension. Both the distributed and the integral forces or stresses are measured with tensometers fixed directly or indirectly to appropriate elements of the device. The tensometer technique requires sensitive bridge methods and electronic amplifiers for registration of transients [8.20; 8.21].

Each force-measuring system has an oscillating mechanical part (elastic mass causing the deformation of the element with attached tensometers), and an electrical part (bridge, amplifier, oscilloscope), see Fig. 8.15.

Because of negligible non-linearity and omission of mechanical hysteresis,

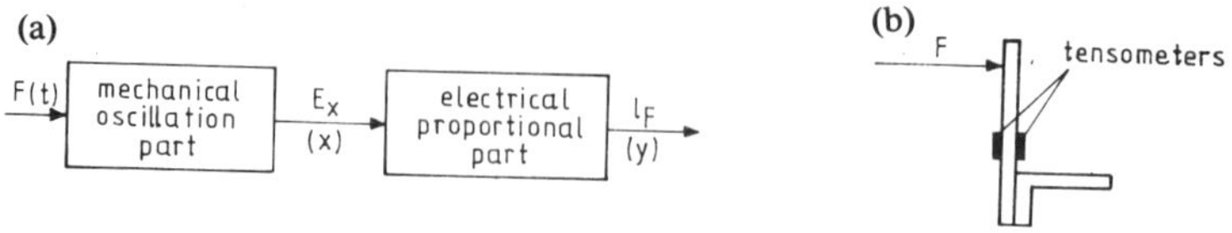

Fig. 8.15 *(a) Measuring system for the linear-motor thrust force, and (b) the measuring beam*

the static sensitivity is often identified with the amplification coefficient K, defined as a ratio of the output Y_u to the value F_u measured in the steady state [8.8; 8.19]:

$$K = \frac{Y_u}{F_u} \tag{8.42}$$

The following equation describes the system

$$\ddot{y} - 2\beta\omega_0\dot{y} + \omega_0^2 y = \omega_0^2 KF(t) \tag{8.43}$$

where:

$\beta = d/2\sqrt{(K_s m_s)}$	relative damping;	(8.44)
$\omega_0 = \sqrt{(K_s/m_s)}$	angular frequency of oscillations;	(8.45)
m_s	reaction mass;	
d	damping coefficient;	
K_s	equivalent rigidity.	

The time response to the force step impulse is given by

$$F(t) = F_u\left[1 - \frac{1}{\sqrt{(1-\beta^2)}}\,e^{-\beta\omega_0 t}\sin(\omega_T t + 2\varphi)\right]1(t) \tag{8.46}$$

where:

$$\varphi = \arctan\frac{\omega_T}{\beta\omega_0} \tag{8.47}$$

and ω_T is the angular frequency of damped oscillations.

By recording the system's response to the force step input it is thus possible to evaluate the frequency ω_0 of free oscillations, the relative damping, and the frequency ω_T of damped oscillations.

Better dynamic performance can be achieved by increasing the free vibration frequency and increasing the damping. On the other hand, however, such increases are limited by the sensitivity of the measuring and registering equipment.

The dynamic error

$$F = \left|\frac{Y_u/K - F(t)}{F_u}\right|\cdot 100\% \tag{8.48}$$

should be investigated carefully in every case for a particular measurement.

Vibration measurements are made using capacitive sensors or piezoelectric transducers [8.21].

The vibroacoustics problems resulting from electromagnetic fields are a separate area of scientific research [8.14].

Occasionally, observation of winding vibrations in electrical machines and transformers involves using high speed cameras (thousands of frames per second).

8.4.2 Temperature fields

The temperature field, resulting from and coupled with the electromagnetic field in electrical machines and devices, is usually investigated using contact methods (thermometers, thermoelements). The connections (welding or other coupling) of copper–nickel or iron–constantan wires, have temperature dependent electromotive force characteristics [8.21]. Thermocouples with very small dimensions favour these sensors in applications not only on open surfaces but also inside the core or windings. Each thermocouple must be individually calibrated by measuring the voltage–temperature characteristic. Full automatic control and numerical processing of results is possible and quite common.

The measurements of high temperature fields are made indirectly by means of photo-optical pyrometers, such as those used in metallurgy and electrothermics. The accuracy at this compared with the contact methods is considerably smaller.

Recently an extremely sensitive indirect method based on thermal radiation has become more popular. Thermovision cameras enable measurements and presentation of thermal field charts of the external surfaces of electrical devices.

8.4.3 Thermometric method

The field of power loss may be investigated experimentally by a thermometric method developed by Turowski [8.12; 8.17; 8.22]. The measurement can be conducted in a short time without heating the entire object.

The thermal balance equation for an arbitrary point of the metal sheet may be written as

$$P_1 = c\rho_m \frac{\partial \theta}{\partial t} + \lambda \Delta \theta - \alpha' (\theta - \theta_0) \tag{8.49}$$

where:

- P_1 unknown power loss;
- c specific heat;
- ρ_m specific mass;
- λ specific heat conductivity;
- α' representative coefficient of the heat exchange with the environment by convection and radiation;
- θ body temperature;
- θ_0 ambient temperature;
- t time.

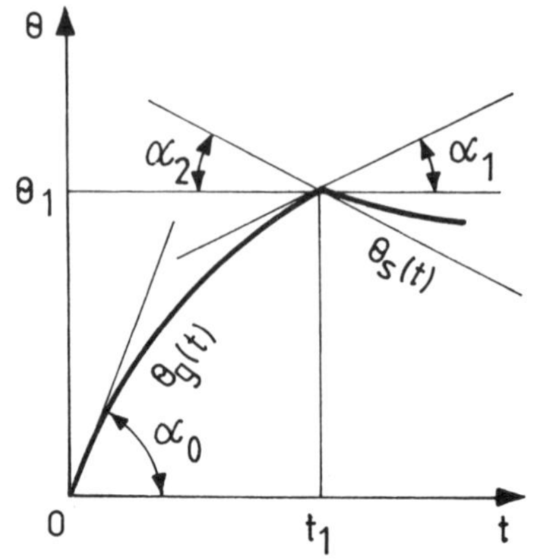

Fig. 8.16 *Heating and cooling curves of a homogenous object*

Assuming that before switching on, the temperature at each point was the same, then for the time $t = 0$ eqn (4.49) simplifies to:

$$P_1 = c\rho_m \left(\frac{\partial \theta}{\partial t} \right)_{t=0} \tag{8.50}$$

or

$$P_q = c\rho_m \tan \alpha_0, \tag{8.51}$$

where $\tan \alpha_0$ is the initial slope of the temperature curve as shown in Fig. 8.16.

The power loss is then calculated using the initial temperature increase. Before the measurement can be repeated it is necessary to cool the object. The power loss can also be determined from the rate of temperature changes caused by switching on and off. This method is known as the two-tangent method, where

$$\tan \alpha_0 = \tan \alpha_1 + \tan \alpha_2 \tag{8.52}$$

It is found from experience that the two-tangent method is less accurate than the initial temperature increase method.

Errors associated with the use of eqn (8.50) increase with high values of λ and α', as well as being due to the non-linearity of power density distribution on the surface of the body. Other errors depend on the sensitivity and accuracy of the measuring and data processing instruments.

8.5 ANALOGUE MODELLING

In the early days of magnetostatic field measurements, iron filings were used to visualize fields of salient poles or multiwire lines. This method however gave no information about the values of flux density. Thus additional measurements were taken with a search coil and a ballistic galvanometer. Govorkov

[8.7] describes a method of calculating field lines and equipotential lines under certain conditions, because the two systems of lines are orthogonal and the integral $\int \mathbf{B}\, dl$ along any equipotential line between two adjacent field lines has the same value everywhere.

Analogue modelling uses the mathematical similarity of different physical phenomena. The reluctance network method, presented earlier in this book, is an example of an analogy between the differential form of Poisson's equation and the Kirchoff equation in electrical circuits.

The analogue field modelling as an experimental procedure is carried out on continuous or discrete field analysers. The oldest analysers, which used the flow of electric current, were an electrolytic tank and a sheet of semi-conducting paper. These were intended primarily for investigating electrostatic and magnetostatic fields in two dimensions. In the case of an electrolytic tank with a sloping bottom the volume field of axial symmetry could also be modelled.

From the first Maxwell's equation we know that in the current-free regions ($curl\, \mathbf{H} = 0$) the magnetic field can be expressed in terms of a magnetic scalar potential as V_m:

$$\mathbf{H} = -grad\, V_m \tag{8.53}$$

In a homogeneous isotropic medium the magnetic scalar potential obeys the Laplace's equation:

$$\nabla^2 V_m = 0 \tag{8.54}$$

The steady current flow field may be described by the equation $I = GU$,

where:
- I current;
- G electric conductance;
- U potential difference.

In the electrostatic field we have the equation $Q = CU$,

where:
- C capacitance;
- Q charge.

In the magnetic field, by analogy, we find:

$$\phi = \left(\frac{1}{R_\mu}\right) U_\mu$$

where:
- R_μ reluctance;
- U_μ magnetic potential difference (mmf);
- ϕ magnetic flux.

There is an analogy between current I, charge Q and flux ϕ. There is also an analogy between the electric voltage U and the magnetic potential difference U_μ. An example of the modelling of a part of a synchronous

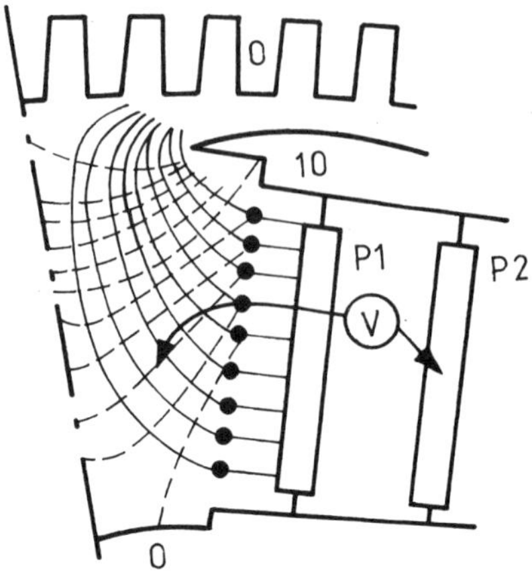

Fig. 8.17 *Modelling of magnetic field of a synchronous generator with the use of an electrolytic tank [8.7]*

generator with non-equipotential boundary conditions is presented in Fig. 8.17.

The discrete analysers built as networks of resistors or capacitors are used to provide two- or three-dimensional solutions of the boundary value problems. The networks are plane or three-dimensional respectively.

The idea of the method of analogy for the case of an eddy-current problem will be presented using the diffusion equation.

For a section of a transmission line shown in Fig. 8.18 the following four equations may be written:

(1) for the nonlinear resistance $u_r(i) = r(i)i$:

$$\frac{\partial^2 i}{\partial x^2} = c\frac{\partial u_r(i)}{\partial t} + g u_r(i) + gl\frac{\partial i}{\partial t} + cl\frac{\partial^2 i}{\partial t^2} \tag{8.55}$$

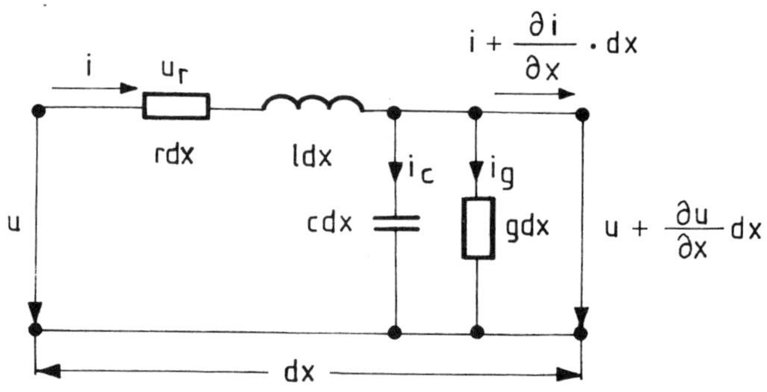

Fig. 8.18 *Elementary section of a transmission line [8.32]*

(2) for the nonlinear inductance $\psi(i) = l(i)i$:

$$\frac{\partial^2 i}{\partial x^2} = g\frac{\partial \psi(i)}{\partial t} + gri + cr\frac{\partial i}{\partial t} + c\frac{\partial^2 \psi(i)}{\partial t^2} \tag{8.56}$$

(3) for the nonlinear capacitance $q(u) = c(u)u$:

$$\frac{\partial^2 u}{\partial x^2} = r\frac{\partial q(u)}{\partial t} + rgu + lg\frac{\partial u}{\partial t} + lc\frac{\partial^2 u}{\partial t^2}, \tag{8.57}$$

(4) for the nonlinear conductance $i_g(u) = g(u)u$:

$$\frac{\partial^2 u}{\partial x^2} = l\frac{\partial i_g(u)}{\partial t} + ri_g(u) + rc\frac{\partial u}{\partial t} + lc\frac{\partial^2 u}{\partial t^2} \tag{8.58}$$

The modelling of a diffusion equation in one dimension, at some depth x for a thin layer of thickness $\mathrm{d}x$, may be based on the following four elementary circuits:

(1) quadrupole $r(i)$, c; $l = 0$, $g = 0$:

$$\frac{\partial^2 i}{\partial x^2} = c\frac{\partial u_r(i)}{\partial t} \tag{8.59}$$

(2) quadrupole $l(i)$, g; $c = 0$, $r = 0$:

$$\frac{\partial^2 i}{\partial x^2} = g\frac{\partial \psi(i)}{\partial t} \tag{8.60}$$

(3) quadrupole r, $c(u)$; $l = 0$, $g = 0$:

$$\frac{\partial^2 u}{\partial x^2} = r\frac{\partial q(u)}{\partial t} \tag{8.61}$$

(4) quadrupole l, $g(u)$; $c = 0$, $r = 0$:

$$\frac{\partial^2 u}{\partial x^2} = l\frac{\partial i_g(u)}{\partial t} \tag{8.62}$$

The first circuit is described in reference [8.32] and is shown in Fig. 8.19, where a variable resistor (varistor) is used as the nonlinear element to model the magnetic characteristic of iron (see Fig. 8.20).

From the similarity of the circuit equation:

$$i_{k,k+1} - 2i_{k-1,k} + i_{k-2,k-1} = c\frac{\mathrm{d}U_{k-1,k}}{\mathrm{d}i_{k-1,k}}\frac{\mathrm{d}i_{k-1,k}}{\mathrm{d}t} \tag{8.63}$$

and the differential-difference field equation:

$$\frac{H(x+\Delta x, t) - 2H(x,t) + H(x-\Delta x, t)}{\Delta x^2} = \sigma\frac{\mathrm{d}B(x,t)}{\mathrm{d}H(x,t)}\frac{\mathrm{d}H(x,t)}{\mathrm{d}t} \tag{8.64}$$

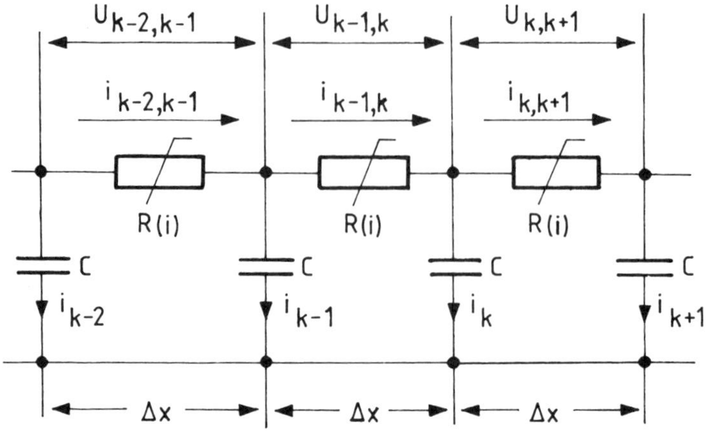

Fig. 8.19 *Simple sections forming an analyser for the diffusion equation*

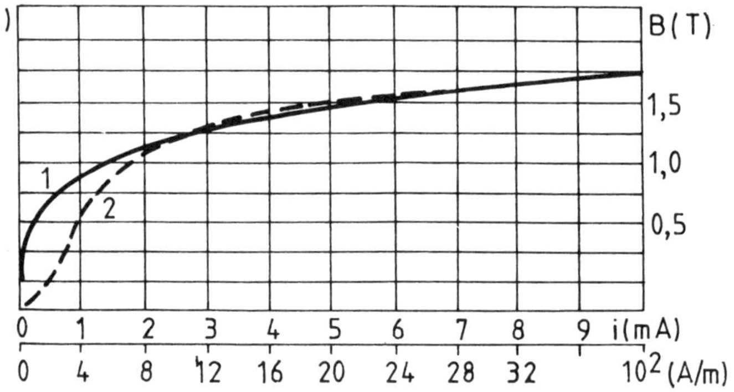

Fig. 8.20 *The characteristic of a variable resistor (varistor) – curve 1, and a typical B-H curve – curve 2 [8.32]*

the condition of analogy results in

$$c = (\Delta x)^2 \sigma \frac{\mathrm{d}B(x, t)}{\mathrm{d}H(x, t)} \bigg/ \frac{\mathrm{d}u_{k-1,k}}{\mathrm{d}i_{k-1,k}} \tag{8.65}$$

The magnetic hysteresis effect may also be included by connecting a capacitor (selected experimentally) in parallel with the resistor in each section of the analyser. For two- and three-dimensional fields more elaborate networks are required but they will follow the same principle and similar equations.

8.6 EXPERIMENTAL ERRORS

Suppose the *objective value* of a given parameter can be defined in some way. Then the discrepancy between the result obtained from a measurement and this objective value of the quantity is called the measurement error. Ideally, the objective value should be chosen as the actual value. This value, however, is not known, and alternative definitions are sought. Very often the objective value is established computationally, especially if calculation is considered to be more accurate than other methods. Sometimes the average of several measurements is taken. It is important to realize that it is not possible to eliminate experimental errors, we can only try to reduce them or minimize their effect.

In reference [8.21] the following classification of measurement errors is suggested.

(1) In view of the relationship between the measured result W and the objective value W_0:

- *absolute errors* (expressed in units of the measured quantity):

$$\Delta W = W - W_0 \tag{8.66}$$

- *relative errors* (expressed as percentage):

$$\delta_W = \frac{W - W_0}{W_0} \cdot 100\% \tag{8.67}$$

(2) Depending on the type of variation of the measured quantity and properties of the measurement system:

- *static errors* occurring during steady-state after transients have died away,
- *dynamic* errors under transient conditions.

(3) Depending on the regularity of the occurrence of errors during the multiple repetition of the measurement under the same conditions:

- *systematic errors* of constant sign and value, changing according to a known function with varying measurement conditions,
- *stochastic errors*, which are non-recurrent even under exactly the same conditions.

(4) Depending on the physical conditions at the time of experiment (reference conditions):

- *basic errors* existing in reference conditions,
- *supplementary errors* resulting from disturbances in reference conditions.

References

Chapter 1

1.1 Baldomir D.: Differential forms and electromagnetism in 3-dimensional Euclidean space R^3, *IEE Proc. A*, 1986, **133**, (3), pp. 139–143.
1.2 Baldomir D. and Hammond P.: Geometrical formulation for variational electromagnetics, *IEE Proc. A*, 1990, **137**, (6), pp. 321–330.
1.3 Baldomir D. and Hammond P.: Geometrical approach to eddy-current systems, *IEE Proc. A*, 1993, **140**, (2), pp. 166–172.
1.4 Baldomir D. and Hammond P.: Global geometry of electromagnetic systems, *IEE Proc. A*, 1990, **137**, (6), pp. 321–330.
1.5 Binns K.J., Lawrenson P.J. and Trowbridge C.W.: *The Analytical and Numerical Solution of Electric and Magnetic Fields*, John Wiley & Sons, Chichester, 1992.
1.6 Bozorth R.M.: *Ferromagnetism*, Van Nostrand, Princeton, USA, 1951, p. 476.
1.7 Carpenter C.J.: Comparison of alternative formulations of 3-dimensional magnetic field and eddy-current problems at power frequencies, *Proc. IEE*, 1977, **124**, (11), pp. 1026–1034.
1.8 Carpenter C.J.: Electromagnetic energy changes due to charges moving through constant, or zero, magnetic field, *IEE Proc. A*, 1991, **138**, (1), pp. 55–70.
1.9 Carpenter C.J.: Electromagnetic theory without electric flux, *IEE Proc. A*, 1992, **139**, (4), pp. 189–209.
1.10 Carter R.G.: *Electromagnetism for Electronic Engineers*, Chapman & Hall, London, 1992.
1.11 Carter G.W.: *The Electromagnetic Field in its Engineering Aspects*, Longmans, Green and Co, London, 1957.
1.12 Chikazumi S.: *Physics of Magnetism*, Wiley, New York, 1964, p. 245.
1.13 Christopoulos C.: *An Introduction to Applied Electromagnetism*, John Wiley & Sons, Chichester, 1990.
1.14 Dobbs E.R.: *Electricity and Magnetism*, Routledge & Kegan Paul, London, 1984.
1.15 Ferrari R.L.: Complementary variational formulation for eddy-current problems using the field variables E and H directly, *IEE Proc. A*, 1985, **132**, (4), pp. 157–164.
1.16 Hammond P. and Sykulski J.K.: *Engineering Electromagnetism: Physical Processes and Computation*, Oxford University Press, Oxford, 1994.
1.17 Hammond P. and Sykulski J.K.: Tubes and Slices: a new way of teaching the principles of electric and magnetic fields, *IEEE Trans. Educ.*, 1992, **35**, (4), pp. 300–306.
1.18 Hammond P.: *Electromagnetism for Engineers*, Pergamon Press, Oxford, 1978.
1.19 Hammond P.: *Applied Electromagnetism*, Pergamon Press, Oxford, 1988.
1.20 Hammond P.: *Energy Methods in Electromagnetism*, Clarendon Press, Oxford, 1981.
1.21 Hammond P.: Use of potentials in calculation of electromagnetic fields, *IEE Proc. A*, 1982, **129**, (2), pp. 106–112.
1.22 Hammond P. and Baldomir D.: Dual energy methods in electromagnetism using tubes and slices, *IEE Proc. A*, 1988, **135**, (3), pp. 167–172.
1.23 Harrington R.F.: *Introduction to Electromagnetic Engineering*, McGraw-Hill Book Company, New York, 1958.
1.24 Hayt W.H.: *Engineering Electromagnetics*, McGraw-Hill Book Company, 1974.

1.25 Krawczyk A. and Tegopoulos J.A.: *Numerical Modelling of Eddy Currents,* Clarendon Press, Oxford, 1993.
1.26 Lammeraner J. and Stafl M.: *Eddy Currents*, Iliffe Books Ltd, London, 1966.
1.27 Lancaster G.: *Introduction to Fields and Circuits*, Oxford University Press, New York, 1992.
1.28 Liao S.Y.: *Engineering Applications of Electromagnetic Theory*, West Publishing Company, St Paul, 1988.
1.29 Lorrain P., Corson D.P. and Lorrain F.: *Electromagnetic Fields and Waves*, W.H. Freeman and Company, New York, 1988.
1.30 Maxwell J.C.: *A Treatise on Electricity and Magnetism*, Clarendon Press, Oxford, 1892.
1.31 Moon P. and Spencer D.E.: *Field Theory for Engineers*, D. Van Nostrand Company, New York, 1961.
1.32 Penman J. and Fraser J.R.: Unified approach to problems in electromagnetism, *IEE Proc. A*, 1984, **131,** (1), pp. 55–61.
1.33 Penman J. and Fraser J.R.: Dual and complementary methods in electromagnetism, *IEEE Trans.*, 1983, **MAG-19**, pp. 2311–2316.
1.34 Popovic B.D.: *Introductory Engineering Electromagnetics*, Addison-Wesley, 1971, pp. 416–418.
1.35 Rikabi J., Bryant C.F. and Freeman E.M.: Error-based derivation of complementary formulations for the eddy-current problem, *IEE Proc. A*, 1988, **135**, (4), pp. 208–216.
1.36 Smythe W.R.: *Static and Dynamic Electricity*, 2nd edn, McGraw-Hill, New York, 1950.
1.37 Solymar L. and Walsh D.: *Lectures on the Electrical Properties of Matters*, 4th edn, Oxford University Press, 1988.
1.38 Spiegel M.R.: *Theory and Problems of Vector Analysis*, Schaum Publishing Company, New York, 1959, p. 121.
1.39 Stoll R.L.: *The Analysis of Eddy Currents*, Clarendon Press, Oxford, 1974.
1.40 Stoll R.L., Bodner B. and Al-Khoury A.H.N.: Method of magnetic sources applied to the basic configuration of a rotating electrical machine, *IEE Proc. B*, 1993, **140**, (3), pp. 217–231.
1.41 Stratton J.A.: *Electromagnetic Theory*, McGraw-Hill, New York, 1941.
1.42 Sykulski J.K. and Turowski J.: *ISEF'91 Proceedings*, COMPEL, Vol. 11, No. 1, James & James Science Publishers, London, March 1992.
1.43 Sykulski J.K.: Computer package for calculating electric and magnetic fields exploiting dual energy bounds, *IEE Proc. A*, 1988, **135**, (3), pp. 145–150.
1.44 Sykulski J.K.: Tubes and Slices: a software package for dual field calculations, in *Software Applications in Electrical Engineering*, Computational Mechanics Publications, Southampton, 1993, pp. 247–254.

Chapter 2

2.1 Abrams M.D. and Gillott D.H.: Numerical analysis of hysteresis and eddy current losses in solid cylindrical rods of No 1010 steel, *IEEE Trans.*, 1967, **PAS-86**, (9), pp. 1077–1082.
2.2 Ames W.F.: *Numerical Methods for Partial Differential Equations*, Nelson, London, 1969.
2.3 Binns K.J. and Lawrenson P.J.: *Analysis and Computation of Electric and Magnetic Field Problems*, Pergamon Press, Oxford, 1963.
2.4 Binns K.J., Lawrenson P.J. and Towbridge C.W.: *Analytical and Numerical Solution of Electric and Magnetic Field Problems*, John Wiley, Chichester, 1992.
2.5 Burais N. and Grellat G.: Numerical modelling of iron losses in ferromagnetic steel plate, *IEEE Trans.*, 1982, **MAG-18**, pp. 558–562.
2.6 Buzbee B., Dorr F., George J. and Golub G.: The discrete solution of Poisson equation on irregular regions, *SIAM J. Numer. Anal.*, 1971, **8**(4), pp. 722–736.
2.7 Carpenter C.J.: Comparison of alternative formulations of 3-dimensional magnetic-field and eddy-current problems at power frequencies, *Proc. IEE*, 1977, **124**, (11), pp. 1026–1034.

2.8 Carpenter C.J. and Wyatt E.A.: *Efficiency of Numerical Techniques for Computing Eddy currents in Two and Three dimensions*, COMPUMAG, Oxford, 1976, pp. 242–250.

2.9 Carré B.A.: The determination of the optimum accelerating factor for successive over-relaxation, *Computer Journal*, 1961, **4**, pp. 73–78.

2.10 Coulson M.A., Slater R.D. and Simpson R.R.S.: Representation of magnetic characteristic, including hysteresis, using Priesach's theory, *Proc. IEE*, 1977, **124**, (10), pp. 895–898.

2.11 Conte S.D. and De Boor C.: *Elementary Numerical Analysis*, McGraw-Hill Kogakusha, Tokyo, 1972.

2.12 Davey K.R. and King E.I.: A 3-dimensional scalar potential field solution and its application to the turbine generator end region, *IEEE Trans.*, 1981, **PAS-100**, (5), pp. 2302–2310.

2.13 De la Vallée Poussin F. and Lion A.: Calcul itératif de l'induction magnétique dans les machines électriques, *Rev. Générale de l'Electricité*, April 1967, pp. 731–739.

2.14 Del Vecchio R.M.: An efficient procedure for modelling complex hysteresis processes in ferromagnetic materials, *IEEE Trans.*, 1980, **MAG-16**, (5), pp. 809–811.

2.15 DuFort E.C. and Frankel S.P.: Stability conditions in the numerical treatment of parabolic differential equations, *Mathl. Tabl. Natn. Res. Coun.*, Washington, 1953, **7**, pp. 135–152.

2.16 Erdélyi E.A. and Fuchs E.F.: Nonlinear magnetic field analysis of DC machines. Part I: Theoretical fundamentals, *IEEE Trans.*, 1970, **PAS-89**, (7), pp. 1546–1554.

2.17 Fischer J. and Moser H.: The representation of the magnetization curve by simple algebraic and transcendental functions, *Arch. Elektrotech*, 1956, **42**, pp. 286–299.

2.18 Fuchs E.F. and Erdélyi E.A.: Nonlinear magnetic field analysis of DC machines. Part II: Application of the improved treatment, *IEEE Trans.*, 1970, **PAS-89**, (7), pp. 1555–1564.

2.19 Garadedian P.R.: Estimation of the relaxation factor for small mesh size, *Math. Tabl. Natn. Res. Coun.*, Washington, 1956, **10**, pp. 183–185.

2.20 Gourlay A.R.: Hopscotch: A fast second-order partial differential equation solver, *J. Inst. Math. Appl.*, 1970, **6**, pp. 375–390.

2.21 Hockney R.W.: A fast direct solution of Poisson's equation using Fourier analysis, *J. Assoc. Comput. Mach.*, 1965, **12**, (1), pp. 95–113.

2.22 Jack A.G. and Stoll R.L.: Negative-sequence currents and losses in the solid rotor of a turbogenerator, *IEE Proc. C*, 1980, **127**, (2), pp. 53–64.

2.23 Kozakoff D.J. and Simons F.O.: Three-dimensional nonlinear magnetic field boundary value problem and its numerical solution, *IEEE Trans.*, 1970, **MAG-6**, (4), pp. 828–833.

2.24 Kraus J.D.: *Electromagnetics*, McGraw-Hill, New York, 1991.

2.25 Krawczyk A. and Tegopoulos J.A.: *Numerical Modelling of Eddy Currents*, Clarendon Press, Oxford, 1993.

2.26 Kronsjö L: *Algorithms*; *their Complexity and Efficiency*, John Wiley, Chichester, 1979.

2.27 Leurs L. and Stoll R.L.: Three-dimensional quasi-static magnetic field in superconducting-rotor synchronous generators with a magnetic steel rotor screen, *IEE Proc. C*, 1986, **133**, (2) pp. 69–80.

2.28 Lim K.K. and Hammond, P.: Universal loss chart for the calculation of eddy-current losses in thick steel plates, *Proc. IEE*, 1970, **117**, (4), pp. 857–864.

2.29 Lim K.K. and Hammond, P.: Numerical method for determining the electromagnetic field in saturated steel plates, *Proc. IEE*, 1972, **119**, (11), pp. 1667–1674.

2.30 Lukaniszyn M.: Analiza pola elektromagnetycznego przekladnikow pradowych o rdzeniach ramkowych, Politechnika Lodzka, Praca doktorska, Lodz 1985.

2.31 Peaceman D.W. and Rachford H.H.: The numerical solution of parabolic and elliptic differential equations, *J. Soc. Ind. Appl. Math.*, 1955, **3**, pp. 28–41.

2.32 Poritsky H. and Butler J.M.: A.C. flux penetration into magnetic materials with saturation, *IEEE Trans.*, 1964, **COM-83**, pp. 99–111.

2.33 Proskurowski W. and Widlund O.: On the numerical solution of Helmholtz's equation by the capacitance matrix method, *Math. Comput.*, 1976, **30**, (135), pp. 433–468.

2.34 Rhyner J.: Magnetic properties and AC losses of superconductors with power law current-voltage characteristics, *Physica C*, 1993, **212**, pp. 292–300.

2.35 Richtmyer R.D. and Morton K.W.: *Difference Methods for Initial Value Problems*, John Wiley-Interscience, New York, 1967.

2.36 Rivas J.: Simple approximations for magnetization curves and hysteresis loops, *IEEE Trans.*, 1981, **MAG-17**, (4), pp. 1498–1502.
2.37 Sarma M.S.: Potential functions in electromagnetic field problems, *IEEE Trans.*, 1970, **MAG-6**, (3), pp. 513–518.
2.38 Saulyev V.K.: *Integration of Equations of Parabolic Type by the Method of Nets*, Pergamon Press, Oxford, 1964.
2.39 Schumann U. (ed.): Computers fast elliptic solvers and applications, *Proc. of the GAMM-Workshop on Fast Solution Methods for the Discretized Poisson Equation*, Karlsruhe, 1977, Advance Publ., London 1978.
2.40 Schumann U. and Sweet R.: A direct method for solution of Poisson's equation with Neumann boundary conditions on a staggered grid of arbitrary size, *J. Comput. Phys.*, 1976, **20**, (2), pp. 171–182.
2.41 Smith G.D.: *Numerical Solution of Partial Differential Equations: Finite Difference Methods*, Clarendon Press, Oxford, 1978.
2.42 Stoll R.L.: Solution of linear steady-state eddy-current problems by complex successive overrelaxation, *Proc. IEE*, 1970, **117**, (7), pp. 1317–1323.
2.43 Stoll R.L.: *The Analysis of Eddy Currents*, Clarendon Press, Oxford, 1974.
2.44 Stoll R.L.: Modelling very small airgaps between magnetic surfaces in finite-difference field solutions, *COMPEL*, 1992, **11**, (1), pp. 77–80.
2.45 Stoll R.L. and Hindmarsh R.: Reduction of local eddy-current losses in turbogenerator stator windings using composite subconductors, *Proc. IEE*, 1976, **123**, (11), pp. 1217–1222.
2.46 Stone H.L.: Iterative solution of implicit approximations of multidimensional partial differential equations, *SIAM J. on Numer. Anal.*, 1968, **5**, (3), pp. 530–558.
2.47 Tegopoulos J.A. and Kriezis E.E.: *Eddy Currents in Linear Conducting Media*, Elsevier, Amsterdam, 1985.
2.48 Temperton C.: Direct methods for the solution of the discrete Poisson equation: some comparisons, *J. Comput. Phys.*, 1979, **31**, (1), pp. 1–20.
2.49 Trutt F.C., Erdeli E.A. and Hopkins R.E.: Representation of the magnetic characteristics of dc machines for computer use, *IEEE Trans.*, 1968, **PAS-87**, pp. 665–669.
2.50 Varga R.S.: *Matrix Iterative Analysis*, Prentice-Hall, New Jersey, 1962.
2.51 Von Rosenberg D.U.: *Methods for the Numerical Solution of Partial Differential Equations*, Elsevier, New York, 1969.
2.52 Wachspress, E.L.: *Iterative Solution of Elliptic Systems and Applications to the Neutron Diffusion Equations of Reactor Physics*, Prentice-Hall, New Jersey, 1966.
2.53 Wiak S.: Method of calculation of the transient electromagnetic field in ferromagnetic materials for the known vector of induction B on the surface, *IEE Proc. A*, 1984, **131**, (5), pp. 293–300.
2.54 Wiak S.: A method of calculation of transient electromagnetic field in non-linear ferromagnetic medium, *Arch. Elektrotech.*, 1985, **68**, pp. 113–120.
2.55 Wiak S.: Stability, convergence and accuracy of the DuFort and Frankel difference diagram approximating a class of non-linear parabolic field equations, *Int. J. Numer. Methods Eng.*, 1987, **24**, pp. 1421–1437.
2.56 Wiak S.: Analysis of difference diagrams stability approximating a class of one- and two-dimensional non-linear parabolic field equations, *Int. J. Numer. Methods Eng.*, 1989, **28**, pp. 1995–2020.
2.57 Wiak S. and Pelikant A.: Difference method in integral technique in field analysis of doubly salient switched reluctance motor, *Arch. Elektrotech.*, 1989, **72**, pp. 381–387.
2.58 Widger G.F.T.: Representation of magnetization curves over extensive range by rational-fraction approximations, *Proc. IEE*, 1969, **116**, (1), pp. 156–160.
2.59 Wood, W.L.: *Practical Time-Stepping Scheme*, Clarendon Press, Oxford, 1990.
2.60 Zakrzewski K. and Lukaniszyn M.: Three-dimensional model of three-phase transformer for leakage field calculation, *Arch. Elektrotech.*, 1990, **73**, pp. 319–324.
2.61 Zakrzewski K. and Lukaniszyn M.: Three-dimension model of one- and three-phase transformer for leakage field calculation, *IEEE Trans.*, 1992, **MAG-28**, (2), pp. 1344–1347.

2.62 Zakrzewski K. and Pietras F.: Method of calculating the electromagnetic field and power losses in ferromagnetic materials, taking into account magnetic hysteresis, *Proc. IEE*, 1971, **118**, (11), pp. 1679–1685.

Chapter 3

3.1 Alarcon E., Reverter A. and Molina J.: Hierarchical boundary elements, *Computer and structures*, 1985, **20**, (1-3), pp. 151–156.
3.2 Babuska T.: Trends in finite elements, *IEEE Trans. on Mag.*, 1989, **25** (4).
3.3 Bettess P.: Infinite elements, *Int. J. Num. Methods Eng.*, 1977, **11**, pp. 53–64.
3.4 Binns K.J., Lawrenson P.J. and Trowbridge C.W.: *The Analytical and Numerical Solution of Electric and Magnetic Fields*, John Wiley & Sons, Chichester, 1993.
3.5 Brebbia C.A.: *The Boundary Element Method for Engineers*, Pentech. Press, London, 1980.
3.6 Brebbia C.A.: *Boundary Element Methods in Engineering*, Springer-Verlag, Berlin, 1982.
3.7 Chari M.V.K.: Non linear magnetic field analysis of d.c. machines, *IEEE Trans.*, 1970, **PAS-89**, (7), pp. 85–89.
3.8 Collatz L.: *Numerischel Behandlung von Differentialgleichungen*, Springer-Verlag, Berlin, 1955.
3.9 Collins R.J.: Bandwidth reduction by automatic renumbering, *Int. J. Num. Methods Eng.*, 1973, **6**, pp. 345–356.
3.10 Coulson M.A., Preston T.W. and Reece A.B.J.: 3-dimensional finite-element solvers for the design of electrical equipment, *Compumag*, 1985, Fort Collins, USA. (GEC Power Transformers, Stafford, UK).
3.11 Dabrowski M.: *Electrical Machines Magnetic Fields and Circuits*, PWN, Warsaw, 1971 (in Polish).
3.12 Douglas J.Jr. and Dupont Z.: Galerkin methods for parabolic equation, *SIAM J. Numer. Anal.*, 1970, **7**, (4), pp. 575–626.
3.13 Enderle G., Kansy K. and Pfaff G.: *Computer Graphic Programming*, Springer Verlag, Berlin, 1984.
3.14 Feirweather G.: *Finite Element Galerkin for Differential Equations*, Marcel Dekker, NY, 1978.
3.15 Finlayson B.A.: *The Method of Weighted Residuals and Karietional Principles*, Academic Press, NY, 1972.
3.16 Foley J. and van Dam A.: *Fundamentals of Interactive Computer Graphics*, Addison Wesley, London, 1982.
3.17 Gramz M. and Ziótkowski M.: *Calculation of Three-Dimensional Electromagnetic Fields*, TU of Szczecin, 1991 (in Polish).
3.18 Gratkowski S.: More on a simple infinite element, *Int. J. for Comp. Matut. in El. and Elec. Eng., COMPEL*, 1986, **5**, (191–194).
3.19 Gratkowski S. and Ziókowski M.: A three-dimensional infinite element for modelling open boundary field problems, *IEEE Trans. Magn.*, 1992, **28**, pp. 1675–1678.
3.20 Janke, Emde, Losch.: *Tafeln Hoherer Funktionen*, B.G. Teubner Verlagsgesellsehaft, Stuttgart, 1960.
3.21 Jennings A.: *Matrix Computation for Engineers and Scientists*, J. Wiley, London, 1977.
3.22 Kasper M. and Kost A.: Finite element field analysis in ferromagnetic materials under consideration of anisotropy and saturation; *Electrosoft*, 1990, **1**, (3).
3.23 Komeza K. and Krusz G.: Hermitian finite elements in numerical analysis of transformer leakage magnetic field, *Proc. Internat. Symposium on Electromagn. Fields in Electrical Eng., ISEF'89*, 1989, pp. 109–112.
3.24 Mackerle J. and Brebbia C.A.: *The Boundary Element Reference Book*, Springer-Verlag, Berlin, 1988.
3.25 Nakata T.: 3-D electromagnetic field analysis, *COMPEL*, **9**, Sup. A James and James, 1990.
3.26 Norris D.H. and de Vries G.: *An Introduction to Finite Element*, Academic Press, NY, 1978.
3.27 Pissanetzky S.: A simple infinite element, *COMPEL*, 1984, **3**, pp. 107–114.

3.28 Ratnajeevan S. and Hoole S.: *Computer-Aided Analysis and Design of Electromagnetic Devices*, Elsevier, New York, 1989.

3.29 Sikora R.T.: Electromagnetic field theory, WNT, Warsaw, 1985 (in Polish).

3.30 Silvester P.P. and Chari M.V.K.: Finite element solution of saturable for potential problems, *IEEE Trans.*, 1970, **PAS-89**, (7), pp. 1642–1951.

3.31 Silvester P. and Ferrari R.: *Finite Elements for Electrical Engineers*, Cambridge University Press, London, 1983.

3.32 Strany G. and Fix G.J.: *An Analysis of the Finite Element Method*, Prentice-Hall, 1973.

3.33 Tandon S.C., Armor A.F. and Chari M.L.K.: Nonlinear transient finite element field computation for electrical machines and devices, *IEEE Trans.*, 1983, **PAS-102**, (5), pp. 1089–1096.

3.34 Turowski J., Pawluk K., Sikora R. and Zakrzewski K.: *Elektromagnetic Field Analysis and Synthesis*, Polish Academy of Science, 'Ossolineum' Wrocaw 1990 (in Polish).

3.35 Walker C. and Brebbia C.A.: *Boundary Element Technique in Engineering*, Butterworth, London, 1980.

3.36 Yuan J.S. and Fitzsimons C.J.: *A Mesh Generator for Tetrahedral Elements using Delaunay Triangulation*, Rutheford Appleton Laboratory, Sept. 1992.

3.37 Young D.M.: *Iterative Solution of Large Linear Systems*, Academic Press, NY, 1971.

3.38 Zienkiewicz O.C.: *The Finite Element Method*, McGraw-Hill, London, 1977.

3.39 Zienkiewicz O.C., Kelly D.W., Gago J. and Babuska I.: Hierarchical finite element approaches, error estimates and adaptive refinement, *Proc. 3rd Congress on Finite Element Methods*, Los Angeles, 1981.

3.40 Akin J.E.: *Application and Implementation of Finite Element Methods*, Academic Press, London, 1982.

3.41 Hannakam L. and Albach M.: Lösung des Randwertproblems bei Anwesenheit leitender Körper hoher Permeabilität im erregenden magenetischen Wechselseld, *Archiv für Elektrotechnik*, 65, 1982, pp. 117–122 (in German).

3.42 Albach M. and Hannakam L.: Bestimmung induzierter Wirbelströme in Körpern hoher Permeabilität nach dem Differenzenverfahren, *Archiv für Elektrotechnik*, 66, 1983, pp. 67–74 (in German).

3.43 Forghani B., Freeman E.M., Lowther D.A. and Silvester P.P.: Interactive modelling of magnetization curves, *IEEE Trans.* Mag-18, 1982, pp. 1070–1072.

3.44 Nakata T. and Kawas Y.: Numerical analysis of nonlinear transient magnetic field using the finite element method, *Electrical Engineering in Japan*, 1984, **104**, (4), pp. 81–87.

3.45 Ziółkowski M.: Electromagnetic field analysis in non linear conducting media, *PhD thesis*, 1984, TU of Szczecin (in Polish).

3.46 Aldefeld B.: Electromagnetic field diffusion in ferromagnetic materials, *Proc. IEE*, 1978, **125**, (4), pp. 278–282.

3.47 Aldefeld B.: Numerical calculation of electromagnetic actuators, *Archiv für Electrotechnik*, 64, 1979, pp. 347–352.

3.48 Krawczyk, A. and Tegopoulos J.A.: *Numerical Modelling of Eddy Currents*, Oxford Science Publications, 1993.

3.49 Lim K.K. and Hammond P.: Numerical method for determining the electromagnetic field in saturated steel plates, *Proc. IEE*, 1972, **119**, (11), pp. 1667–1674.

3.50 Lavers J.D.: Finite element solution of nonlinear two dimensional TE-mode eddy current problems, *Trans. IEEE*, 1983, **Mag-19**, (5), pp. 2201–2203.

3.51 Luetke-Daldrup B.: Comparison of exact and approximate finite-element solution of the two-dimensional nonlinear eddy-current problem with measurements, *INTERMAG 1984*, Hamburg.

3.52 Chari M.V.K.: Finite element solution of eddy-current problem in magnetic structures, *IEEE Trans*, 1974, **PAS-93**, (1), pp. 62–72.

3.53 Foggia A., Sabonnadiere J.C. and Slivester P.: Finite element solution of saturated travelling magnetic field problems, *IEEE Trans*, 1975, **PAS-94**, (3), pp. 866–871.

3.54 Yamada S., Kanamaru Y. and Bessho K.: The transient megnetization process and operation in the plunger type electromagnet, *IEEE Trans*, 1976, **Mag-12**, (6), pp. 1056–1058.

Chapter 4

4.1 Allan D.J. and Harrison T.H.: Design for reliability of high-voltage power transformers and reactors, *GEC Rev.*, 1985, **1**, (1), pp. 44–50.
4.2 Anuszczyk J.: Rotational magnetization in magnetic circuits of electrical machines, *Proc. ISEF'89*, (eds: J. Turowski and K. Zakrzewski), James & James Science Publishers, London, 1990, pp. 147–150.
4.3 Carpenter C.J.: Magnetic equivalent circuits, *Proc. IEE*, 1968, **115**, (10), pp. 1503–1512.
4.4 Carpenter C.J.: Finite-element network models and their application to eddy-current problems, *Proc. IEE*, 1975, **122**, (4), pp. 455–462.
4.5 Chua L.O. and Lin P.M.: *Computer-Aided Analysis of Electronic Circuits: Algorithms and Computational Techniques*, Prentice-Hall, Englewood Cliffs, 1975.
4.6 Csendes Z.J., Freeman E.M., Lowther D.A. and Silvester P.P.: Interactive computer graphics in magnetic field analysis and electric machine design, *IEEE Winter Power Conference*, 1980.
4.7 Davey K.R. and King E.J.: A three dimensional scalar potential field solution and its application to turbine generator end region, *IEEE Trans.*, 1981, **PAS-100**, (5), pp. 2302–2310.
4.8 Demenko A.: Equivalent RC networks with mutual capacitances for electromagnetic field simulation of electrical machine transients, *IEEE Trans. Magn.*, 1992, **28**, (2), pp. 1406–1409.
4.9 Djuroviz M. and Carpenter C.J.: 3-dimensional computation of transformer leakage fields and associated losses, *IEEE Trans. Magn.*, 1975, **MAG-11**, (5), pp. 1535–1537.
4.10 Djuroviz M. and Monson J.E.: Stray losses in the step of a transformer yoke with a horizontal magnetic shunt, *IEEE Trans.*, 1982, **PAS-101**, (8), pp. 2995–3000.
4.11 Eisenstat S., Gursky M., Schultz M. and Sherman A.: *Yale Sparse Matrix Package*, Dept of Comp. Science, Yale Univ., New Haven, CT, Techn. Report 114, 1977.
4.12 Hammond P.: *Applied Electromagnetism*, Pergamon Press, Oxford, 1979.
4.13 Hammond P. and Sykulski J.K.: Tubes and slices: a new way of teaching the principles of electric and magnetic fields, *IEEE Trans. Educ.*, 1992, **35**, (4), pp. 300–306.
4.14 Ho C.W., Ruehli A.E. and Brennan P.A.: The modified nodal approach to network analysis, *IEEE Trans. Circuits Syst.*, 1975, **CAS-22**, (6), pp. 504–509.
4.15 King E.I.: Equivalent circuits for two-dimensional magnetic fields. I. The static field. II. The sinusoidally time-varying field, *IEEE Trans.*, 1966, **PAS-85**, (9), pp. 927–945.
4.16 Komeza K., Krusz G. and Turowski J.: Comparison of network and finite element approach to the solution of stray problems, *Proc. ICEM'84*, Part I, Lausanne, 1984, pp. 17–19.
4.17 Kubusch N.: Berechnung der Verlustdichtverteilung und der Temperaturfelder inaktiver Konstruktionsteile von Transformatoren, *Proc. Int. Symp. Electrodynamic, Forces and Losses in Transformers*, Lodz, 1979, pp. 84–95.
4.18 Oberretl K.: Evaluation of magnetic fields, eddy-currents and forces in complicated cases by simulation on grid models (in German), *Archiv fur Elektrotechnik*, 1963, **XLVIII**, (5), pp. 297–313.
4.19 Laithwaite E.R.: Magnetic equivalent circuits for electrical machines, *Proc. IEE*, 1967, **114**, (11), pp. 1805–1809.
4.20 Rais V.R., Turowski J. and Turowski M.: Reluctance network analysis of coupled fields in reversible electromagnetic pump, *Proc. ISEF'87*, Plenum Press, New York, 1988, pp. 279–283.
4.21 Tinney W.F., Brandwajn V. and Chan S.M.: Sparse vector methods, *IEEE Trans.*, 1985, **PAS-104**, (2), pp. 295–301.
4.22 Turowski J.: *Technical Electrodynamics*, WNT, Warsaw, 1993 (in Polish).
4.23 Turowski J.: *Electromagnetic Calculation of Components of Electrical Machines and Devices*, WNT, Warsaw, 1982 (in Polish).
4.24 Turowski J. and Kopec M.: 3-D hybrid, analytically-numerical computation of tank losses in 3-phase power transformers, *Proc. ISEF'93*, Warsaw, 1993, pp. 131–134.

4.25 Turowski J. and Turowski M.: A network approach to the solution of stray field problem in large transformers, *Rozprawy Elektrotechniczne*, Polish Academy of Science, No. 2, 1985, pp. 405–422

4.26 Turowski J., Turowski M. and Kopec M.: Method of three-dimensional network solution of leakage field in three-phase transformers, *IEEE Trans. Magn.*, 1990, **26**, (5), pp. 2911–2919.

4.27 Turowski J., Zakrzewski K., Sikora R. and Pawluk K.: *Analysis and Synthesis of Electromagnetic Fields*, (ed. J. Turowski), Polish Academy of Science, Ossolineum, Wroclaw, 1990 (in Polish).

4.28 Wiak S., Pelikant A. and Turowski J.: Solution of TEAM Workshop Problem 7 by Reluctance Network Method, *Proc. ISEF'93*, Warsaw, 1993, pp. 37–40.

4.29 Worotynski J., Turowski M. and Mendrela E.A.: The accuracy of calculation of magnetic fields, inductance and forces in electromagnetic devices using the reluctance network method (RNM), *Proc. ISEF'93*, Warsaw, 1993, pp. 159–162.

Chapter 5

5.1 Adamiak K.: Application of coils with current to the magnetic field synthesis, *Sci. Electr.*, 1981, **27**, pp. 15–30.

5.2 Adamiak K.: Method of the magnetic field synthesis on the axis of cylinder solenoid, *J. Appl. Phys.*, 1978, **16**, pp. 417–423.

5.3 Adamiak K.: Synthesis of homogeneous magnetic field in internal region of cylindrical solenoid, *Arch. Elektrotech.*, 1980, **62**, pp. 75–79.

5.4 Armstrong A., Fan M.W., Simkin J. and Trowbridge C.W.: Automated optimization of magnet design using the boundary integral method, *IEEE Trans. Magn.*, 1982, **MAG-18**, (2), pp. 620–623.

5.5 Arumugam G., Neittaanmäki P. and Salmenjoki K.: Optimal shape design of an eletromagnet, *ZAMM*, 1989, **69**, (4), pp. 234–237.

5.6 Bellina F., Campostrini P., Chitarin G., Stella A. and Trevisan F.: Automated optimal design techniques for inverse electromagnetic problems, *IEEE Trans. Magn.*, 1992, **MAG-28**, (2), pp. 1549–1552.

5.7 Borghi C.A., Reggiani U. and Zama G.: A method for the solution of an axisymmetric magnetic field synthesis problem, *IEEE Trans. Magn.*, 1991, **MAG-27**, (5), pp. 4093–4096.

5.8 Di Barba P., Navarra P., Savini A. and Sikora R.: Optimum design of iron-core electromagnets, *IEEE Trans. Magn.*, 1990, **MAG-26**, (2), pp. 646–649.

5.9 Grago G., Manella A., Nervi M., Repetto M. and Secondo G.: A combined strategy for optimization in non linear magnetic problems using simulated annealing and search techniques, *IEEE Trans. Magn.*, 1992, **MAG-28**, (2), pp. 1541–1544.

5.10 Gitosusastro S., Coulomb J.L. and Sabonadière J.C.: Performance derivative calculations and optimization procedures, *IEEE Trans. Magn.*, 1989, **MAG-25**, (4), pp. 2834–2389.

5.11 Gottvald A.: Optimal magnet design for NMR, *IEEE Trans. Magn.*, 1991, **MAG-26**, (2), pp. 399–401.

5.12 Gottvald A., Preis K., Magele C., Biro O. and Savini A.: Global optimization methods for computational electromagnetics, *IEEE Trans. Magn.*, 1992, **MAG-28**, (2), pp. 1537–1540.

5.13 Guarnieri M., Stella A. and Trevisan F.: A methodological analysis of different formulations for solving inverse electromagnetic problems, *IEEE Trans. Magn.*, 1990, **MAG-26**, (2), pp. 622–625.

5.14 Gyimesi M.: Inverse field calculation for structure positioning, *IEEE Trans. Magn.*, 1991, **MAG-27**, (5), pp. 4170–4172.

5.15 Hadamard J.: Sur les problèmes aux dérivées partielles et leur signification physique, *Bull. Univ. Princeton*, 1902, p. 13.

5.16 Henneberger G., Meunier G., Sabonnadière J.C., Settler P. and Shen D.: Sensitivity analysis of the nodal position in the adaptive refinement of finite element meshes, *IEEE Trans. Magn.*, 1990, **MAG-26**, (2), pp. 787–790.

5.17 Il-han Park, Beom-taek Lee and Song-yop Hahn: Design sensitivity analysis for nonlinear magnetostatic problems using finite element method, *IEEE Trans. Magn.*, 1992, **MAG-28**, (2), pp. 1533–1536.
5.18 Iwamura Y. and Miya K.: Numerical approach to inverse problem of crack shape recognition based on the electrical potential method, *IEEE Trans. Magn.*, 1990, **MAG-26**, (2), pp. 618–621.
5.19 Kahler G.R. and Della Torre E.: Minimizing the deformation of a static magnetic field by the presence of a ferromagnetic body, *IEEE Trans. Magn.*, 1991, **MAG-27**, (6), pp. 5025–5027.
5.20 Kasper M.: Shape optimization by evolution strategy, *IEEE Trans. Magn.*, 1992, **MAG-28**, (2), pp. 1556–1559.
5.21 Korn G.A. and Korn T.M.: *Mathematical Handbook*, McGraw Hill, New York, Toronto, London, 1961.
5.22 Krawczyk A.: Application of the boundary element method to transient electromagnetic problems, *Pr. Inst. Elektrotech.*, 1984, **132**, pp. 85–95.
5.23 Lattes R. and Lions J.: *Méthode de Quas-Réversibilité et Applications,* Dunod, Paris, 1967.
5.24 Lord W., Nath S., Shin Y.K. and You Z.: Electromagnetic methods of defect detection, *IEEE Trans. Magn.*, 1990, **MAG-26**, (5), pp. 2070–2075.
5.25 Marrocco A. and Pironneau O.: Optimum design with Lagrangian finite elements: design of an electromagnet, *Comput. Methods Appl. Mech. Eng.*, 1978, **15**, 277–308.
5.26 Michalski A., Sikora J. and Wincenciak S.: Optimal shape design of electromagnetic flow gauge, *IEEE Trans. Magn.*, 1988, **MAG-24**, (1), pp. 565–568.
5.27 Mittra M.: *Computer Techniques for Electromagnetics*, Pergamon Press, Oxford, 1973.
5.28 Miyata K., Tawara Y., Matsui Y. and Ohashi K.: Optimum design for magnetic resonance imaging circuit using permanent magnets, *COMPEL*, 1990, **9**, Suppl. A, pp. 115–118.
5.29 Mustarelli P., Rudnicki M., Savini A., Savoldi F. and Villa M.: Synthesis of magnetic gradients for NMR tomography, *Magn. Reson. Imaging*, 1990, **8**, pp. 101–105.
5.30 Nakata T. and Takahashi N.: New design method of permanent magnets by using the finite element method, *IEEE Trans. Magn.*, 1983, **MAG-19**, (6), pp. 2494–2497.
5.31 Nakata T., Takahashi N., Fujiwara K., Kawashima T. and Morii A.: Optimal design of injection mold for plastic bonded magnet, *IEEE Trans. Magn.*, 1991, **MAG-27**, (6), pp. 4992–4994.
5.32 Osama M. Mohammed: Optimal design method of magnetic circuit by finite elements and dynamic search procedure, *COMPEL*, 1990, **9**, Suppl. A, pp. 107–110.
5.33 Palka R.: Application of finite element technique to continuation problems of stationary fields, *IEEE Trans. Magn.*, 1983, **MAG-19**, (6), pp. 2356–2359.
5.34 Palka R.: Extrapolation of static fields measurements by means of the finite element method, *Arch. Elektrotech.*, 1984, **67**, pp. 247–251.
5.35 Palka R.: Synthesis of electrical field by optimization of the shape of region boundaries, *Proc. IEE A*, 1985, **132**, pp. 28–32.
5.36 Palka R.: Synthesis of magnetic field by optimization of the shape of current areas, *Arch. Elektrotech.*, 1989, **72**, pp. 293–300.
5.37 Palka R.: Synthesis of magnetic field due to the direct current, *ETZ Archiv*, 1985, **7**, pp. 299–302.
5.38 Palka R. and Sikora R.: Reverse problem of penetration of magnetic field into conducting region, *IEEE Trans. on Magn.*, 1978, **MAG-14**, pp. 566–568.
5.39 Pawluk K.: Synthesis of electromagnetic fields, Analiza i synteza pol elektromagnetycznych, (ed. J. Turowski), Ossolineum, Wroclaw, 1990 pp. 201–276 (in Polish).
5.40 Pawluk K.: 3-D magnetic field of coils with an open metallic core in boundary-integral approach, *Boundary Element Technology VIII*, (eds: H. Sina and C. Brebbia), Computational Mechanics Publications, Southampton, Boston, 1993, pp. 147–156.
5.41 Pawluk K. and Rudnicki M.: The state of art in the synthesis of electromagnetic fields, *Electromagnetic Fields in Electrical Engineering*, (eds: J. Turowski and A. Savini), Plenum Press, New York, 1988, pp. 287–292.

5.42 Phillips D.L.: A technique for the numerical solution of certain integral equation of the first kind, *J. Assoc. Comp. Mech.*, 1962, **9-1**, pp. 84–97.

5.43 Preis K., Biro O., Friedrich M., Gottvald A. and Magele C.: Comparison of different optimization strategies in the design of electromagnetic devices, *IEEE Trans. Magn.*, 1991, **MAG-27**, (5), pp. 4154–4157.

5.44 Preis K., Magele C. and Biro O.: FEM and evolution strategies in the optimal design of electromagnetic devices, *IEEE Trans. Magn.*, 1990, **MAG-26**, (5), pp. 2181–2183.

5.45 Preis K. and Ziegler A.: Optimal design of electromagnetic devices with evolution strategies, *COMPEL*, 1990, **9**, Suppl. A, pp. 119–122.

5.46 Ratnajeevan S. and Hoole H.: Optimal design inverse problems and parallel computers, *IEEE Trans. Magn.*, 1991, **MAG-27**, (5), pp. 4146–4149.

5.47 Ratnajeevan S., Hoole H. and Sirikumaran S.: Reflections off aircraft and the shape optimization of a ridged waveguide, *IEEE Trans. Magn.*, 1991, **MAG-27**, (5), pp. 4150–4153.

5.48 Ratnajeevan S., Hoole H. and Subramaniam S.: Higher finite element derivatives for the quick synthesis of electromagnetic devices, *IEEE Trans. Magn.*, 1992, **MAG-28**, (2), pp. 1565–1568.

5.49 Ratnajeevan S., Hoole H. and Subramaniam S.: Inverse problems with boundary elements: synthetising a capacitor, *IEEE Trans. Magn.*, 1992, **MAG-28**, (2), pp. 1529–1532.

5.50 Rudnicki M.: Synthesis of electromagnetic field using generalized discrepancy, *Pr. Inst. Elektrotech.*, 1985, **137**, pp. 19–33.

5.51 Rudnicki M., Savini A., Di Barba P. and Pawluk K.: Optimal synthesis of electric devices using the boundary element method, *Computational Modelling of Free and Moving Boundary Problems - 2*, (eds: L.C. Wrobel and C.A. Brebbia), Walter de Gruyter, Berlin, New York, 1988, pp. 269–279.

5.52 Russenschuck S.: Application of Lagrange multiplier estimation to the optimization of permanent magnet synchronous machines, *IEEE Trans. Magn.*, 1992, **MAG-28**, (2), pp. 1525–1528.

5.53 Saldanha R., Coulomb J.L. and Sabonadière J.C.: An ellipsoidal algorithm for the optimum design of magnetostatic problems, *IEEE Trans. Magn.*, 1992, **MAG-28**, (2), pp. 1573–1576.

5.54 Saldanha R., Coulomb J.L. and Sabonadière J.C.: A dual method for constrained optimization design in magnetostatic problems, *IEEE Trans. Magn.*, 1991, **MAG-27**, (5), pp. 4136–4141.

5.55 Salon S.J. and Istfan B.: Inverse non-linear finite element problems, *IEEE Trans. Magn.*, 1986, **MAG-22**, (5), pp. 817–818.

5.56 Savini A., Di Barba P. and Rudnicki M.: On the optimal design of air-cored solenoid inductors of rectangular cross-section, *COMPEL*, 1992, **11**, 1, pp. 205–208.

5.57 Se-yun Kim, Hyun-chul Choi, Jung-woong Ra and Soo-young Lee: Electromagnetic imaging of 2-D inhomogeneous dielectric objects by an improved spectral inverse technique, *IEEE Trans. Magn.*, 1990, **MAG-26**, (2), pp. 634–637.

5.58 Siebold H., Hübner H., Söldner L. and Reichert T.: Performance and results of a computer program for optimizing magnets with iron, *IEEE Trans. Magn.*, 1988, **MAG-24**, (1), pp. 419–422.

5.59 Sikora J.: Minimax approach to the optimal shape design of electromagnetic devices, *COMPEL*, 1990, **9**, Suppl. A, pp. 67–81.

5.60 Sikora J.: Sensitivity approach to the optimal shape design of a magnetic pole contour, *Arch. Elektrotech.*, 1989, **72**, pp. 27–32.

5.61 Sikora J. Skoczylas J., Sroka J. and Wincenciak S.: The use of the singular value decomposition method in the synthesis of the Neumann's boundary problem, *COMPEL*, 1985, **4**, pp. 19–27.

5.62 Sikora R. and Palka R.: Reverse problems for diffusion equation, *Arch. Elektrotech.*, 1980, **62**, pp. 177–180.

5.63 Sikora R. and Palka R.: Synthesis of one- and two-dimensional field, *Arch. Elektrotech.*, 1981, **64**, pp. 105–108.

5.64 Sikora R., Purczynski J. and Adamiak K.: The magnetic field synthesis on a cylinder solenoid axis by means of Tichonov's regularization method, *Arch. Elektrotech.*, 1978, **60**, pp. 83–86.

5.65 Simkin J. and Trowbridge C.W.: Optimizing electromagnetic devices combining direct search methods with simulated annealing, *IEEE Trans. Magn.*, 1992, **MAG-28**, (2), pp. 1545–1548.

5.66 Simkin J. and Trowbridge C.W.: Optimization problems in electromagnetics, *IEEE Trans. Magn.*, 1991, **MAG-27**, (5), pp. 4016–4019.

5.67 Takahashi N., Nakata T. and Uchiyama N.: Optimal design method of 3-D nonlinear magnetic circuit by using magnetization integral equation method, *IEEE Trans. Magn.*, 1989, **MAG-25**, (5), pp. 4144–4146.

5.68 Tichonov A.N. and Arsenin W.J.: *Solution Methods of Non-Correct Problems*, Nauka, Moskva, 1974 (in Russian).

5.69 Twomey S.: On the numerical solution of Fredholm integral equations of the first kind by the inversion of the linear system produced by quadrature, *J. Assoc. Comp. Mech.*, 1963, **10-1**, pp. 97–101.

5.70 Twomey S.: The application of numerical filtering to the solution of integral equations encountered in indirect sensing measurements, *J. Franklin Inst.*, 1965, **2792**, pp. 95–101.

5.71 Yin Liu, Lian Gong, Ping Liu and Keqian Zhang: An algorithm and simulation of the inverse problem of low-frequency current field, *IEEE Trans. Magn.*, 1990, **MAG-26**, (2), pp. 630–633.

Chapter 6

6.1 Brebbia C.A.: *Progress in Boundary Element Method, Vol. 1*, Pentech Press, London, 1981.

6.2 Ceolini F. and Lupi S.: The mutually-coupled circuits method for calculations regarding inductors for induction heating with special configurations, *IEEE IAS Conf. Proc.*, Toronto, Canada, 1–5 October, 1978, pp. 1151–1157.

6.3 Dobrogowski J. *et al.*: *Stamping of Metals using Impulse Magnetic Field*, PWN Warsaw-Poznan, 1979 (in Polish).

6.4 *Essays on the Formal Aspects of Electromagnetic Theory*, (ed: A. Lakhtakia), World Scientific. Singapore, London, 1993.

6.5 Gieras J.: *Linear Induction Drives*, Oxford University Press, 1993.

6.6 Gierczak E., Fleszar J. and Turowski J.: Current asymmetry in the three-phase linear motor, *Proc. ICEM'82*, Budapest. 1982, Paper LM1/6, pp. 961–963.

6.7 Hammond P.: *Energy Methods in Electromagnetism*, Clarendon Press, Oxford, 1981.

6.8 Jackson J.D. *Classical Electrodynamics*, John Wiley & Sons, 1975.

6.9 Kazmierski M., Kozlowski M., Lasocinski J., Pinkiewicz I. and Turowski J.: Hot spot identification and overheating hazard preventing when designing a large transformer, *CIGRE 1984 Session*, Paris, Rep. 12–12, pp. 1–6.

6.10 Kacki E.: *Thermokinetics*, WNT, Warsaw 1967, (in Polish).

6.11 Knoepfel H.: *Pulsed High Magnetic Fields*, North-Holland, Amsterdam 1970.

6.12 Komęza K., Turowski J. and Wiak S.: The experimental prediction of power losses distribution in metal elements of electrical machines, *Proc. Int. Conf. on Reliability and Lifetime of Electrical Machines*, Budapest, 1984, pp. 72–76.

6.13 Krajewski W.: The method of computation of coupled electromechanical fields in metal tubes by a magnetic impulse, *Pr. Inst. Elektrotech.*, 1984, **132**, pp. 71–84.

6.14 Lavers J.: *Numerical Solution Methods for Electroheat Problem*, COMPUMAG, Santa Margherita Ligure, 1983.

6.15 Lawrynowicz J.: Variation Calculus with Introduction to Mathematical Programming, WNT, Warsaw 1977, (in Polish).

6.16 Mankin E.A. and Morozov D.N.: Stray losses in large power transformer cores, *CIGRE 1964 Session*, Paris, Rep. 125.

6.17 Mendrela E.A. and Gierczak E.: Influence of primary twist on linear induction motor performance, *IEEE Trans. Magn.*, 1985, **MAG-21**, (6), pp. 2664–2671.

6.18 Napieralska-Juszczak E.: The chosen problem of modeling of the electromagnetic and thermal fields, *Proc. of ISEF'85*, Warsaw, 1985.

6.19 Nenadovic V. and Riches E.E.: Maglev at Birmingham Airport, *GEC Revue*, No. 1, 1985, pp. 3–17.

6.20 Niewierowicz N. and Turowski J.: New thermometric method of measuring power losses in solid metal elements, *Proc. IEE*, 1972, **119**, (5), pp. 629–636.

6.21 Nowacki W.: *Electromagnetic Effects in Deformed Solid Bodies*, PWN, Warsaw, 1983 (in Polish).

6.22 Ooi B.T. and White D.C.: Traction and normal forces in the linear induction motor, *IEEE Trans.*, No. 4, 1970.

6.23 Panofsky K. and Phillips: *Classical Electricity and Magnetism*, Addison-Wesley, Cambridge, Mass, 1942.

6.24 Parcus H.: *Magnetothermoelasticity*, CISM Unide, Springer-Verlag, Wien, 1972.

6.25 Rawa H.: Limitations following from application of Maxwell's equations in their classical form to the description of electromagnetic fields in non-uniform media, *Arch. Elektrotech.*, 1985, **34**, (314), pp. 765–775 (in Polish).

6.26 Reichert K.: Numerical procedure at the calculation of devices for induction heating, *Elektrowaerme Int.*, 1968, **26**, pp. 113–123 (in German).

6.27 Reichert K.: *Sci. Electr.*, 1970, **16**, pp. 126–146.

6.28 Sajdak C.: Electrodynamic forces in liquid metal with inductive mixing in the processes of semi-continuous or continuous casting of cylindrical ingots, *Rozpr. Elektrotech.*, 1985, **31**, (2), pp. 423–436 (in Polish).

6.29 Sajdak C.: Influence of non-uniform distribution of speed of liquid metal on electrodynamic forces in mixers and induction pumps, *Rozpr. Elektrotech.*, 1986, **32**, (2), pp. 455–470 (in Polish).

6.30 Turowski J.: *Technical Electrodynamics*, WNT, Warsaw, 1993, (in Polish).

6.31 Turowski J.: Generalized equations for electromechanical and electrodynamic energy conversion in transformers, *Rozpr. Elektrotech.*, 1980, **26**, (4), pp. 911–927.

6.32 Turowski J.: *Electromagnetic Computation of Elements of Electrical Machines and Devices*, WNT, Warsaw, 1982, (in Polish).

6.33 Turowski J., Zakrzewski K., Sikora R. and Pawluk K.: *Analysis and Synthesis of Electromagnetic Fields*, Polish Academy of Science. Ossolineum, Wroclaw -Warszawa -Krakow -Gdansk -Lodz, 1990, (in Polish).

6.34 White D.C. and Woodson H.H.: *Electromechanical Energy Conversion*, J. Wiley, NY, 1959.

6.35 Voldek A.I.: Induction Magnetohydrodynamic Machines with Liquid Metallic Secondary, Energia, Leningrad, 1970, (in Russian).

6.36 Zakrzewski K. and Pietras G.: Method of calculating the electromagnetic field and power losses in ferromagnetic materials, taking into account magnetic hysteresis, *Proc. IEE*, 1971, (11), pp. 1679–1685.

Chapter 7

7.1 *ANSYS News*, Fourth Issue, Swanson Analysis Systems, Inc., Houston, 1991.

7.2 *BEASY Engineering Analysis*, BEASY Computational Mechanics, Ashurst Lodge, Southampton, 1993.

7.3 Biddlecombe C.S., Sykulski J.K. and Taylor S.C.: Design Environment Modules for non-specialist users of EM software, *Record of the 9th Compumag Conference on the Computation of Electromagnetic Fields*, Miami, Florida October 31–November 4, 1993, pp. 222–223.

7.4 Binns K.J., Lawrenson P.J. and Trowbridge C.W.: *The Analytical and Numerical Solution of Electric and Magnetic Fields*, John Wiley & Sons, Chichester, 1992.

7.5 Biro O. and Richter R.: CAD in Electromagnetism, *Adv. Electron. Electron Phys.*, 1991, **82**, Academic Press.

7.6 Brebbia C.A. and Dominguez J.: *Boundary Elements*, Computational Mechanics Publications, Southampton, 1992.

7.7 Brebbia C.A. and Aliabadi M.H. (eds): *Adaptive Finite and Boundary Element Methods*, Computational Mechanics Publication, Southampton, 1992.
7.8 Bryant C.F., Emson C.R., Diserens N.J., Molinari G. and Trowbridge C.W.: *Parallel Processing and the Integration of Analysis and Design in Electromagnetics Computation*, COMPEL, Volume 9, Supplement A, James & James Science Publishers, London, 1990, pp. 171–176.
7.9 Cavendish J.C. *et al.*: An approach to automatic 3d fe mesh generation, *IJNME*, 1985, **21**, p. 329.
7.10 Cheng Y.B. and Sykulski J.K.: A design shell for force calculations of solenoid actuators, in *Software Applications in Electrical Engineering*, Computational Mechanics Publications, Southampton, 1993, pp. 39–46.
7.11 Cheng Y.B., Sykulski J.K. and Stoll R.L.: Force optimization in dc actuators, *Proc. Int. Symp. Electromagnetic Fields in Electrical Engineering*, Warsaw, September 16–18, 1993, pp. 231–234.
7.12 COMPUMAG'87-Graz, Proceedings of the Conference on the Computation of Electromagnetic Fields, *IEEE Trans. Magn.*, 1988, **24**, (1).
7.13 COMPUMAG'89-Tokyo, Proceedings of the Conference on the Computation of Electromagnetic Fields, *IEEE Trans. Magn.*, 1990, **26**, (2).
7.14 COMPUMAG'91-Sorrento, Proceedings of the Conference on the Computation of Electromagnetic Fields, *IEEE Trans. Magn.*, 1992, **28**, (2).
7.15 COMPUMAG'93-Miami, *Record of the 9th Compumag Conference on the Computation of Electromagnetic Fields*, Miami, Florida October 31–November 4, 1993.
7.16 Delaunay B.: *Sur la Sphere Vide*, Izvestiya Akademii Nauk, USSR, Math. and Nat. Sci. Div., No. 6, 1934, p. 793.
7.17 Drago, Molfino P., Nervi M. and Repetto M.: A 'local field error problem' approach for error estimation infinite element analysis, *IEEE Trans. Magn.*, 1992, **28**, (2), pp. 1743–1746.
7.18 Duff I.S., Erisman A.M. and Reid J.K.: *Direct Methods for Sparse Matrices*, Clarendon Press, Oxford, 1990.
7.19 Dyck D.N., Lowther D.A. and McFee S.: Determining an approximate finite element mesh density using neural network techniques, *IEEE Trans. Magn.*, 1992, **28**, (2), pp. 1767–1770.
7.20 Fernandes P., Girdinio P., Molfino P. and Repetto M.: Local error estimates for adaptive mesh refinement, *IEEE Trans. Magn.*, 1988, **24**, (1), pp. 299–302.
7.21 Fernandes P., Girdinio P., Molfino P., Molinari G. and Repetto M.: A comparison of adaptive strategies for mesh refinement based on 'a posteriori' local error estimation procedures, *IEEE Trans. Magn.*, 1990, **26**, (2), pp. 795–798.
7.22 Fernandes P., Girdinio P., Molinari G. and Repetto M.: Local error estimation procedures as refinement indicators in adaptive meshing, *IEEE Trans. Magn.*, 1991, **27**, (5), pp. 4189–4192.
7.23 Fernandes P., Girdinio P., Repetto M. and Secondo G.: Refinement strategies in adaptive meshing, *IEEE Trans. Magn.*, 1992, **28**, (2), pp. 1739–1742.
7.24 Fletcher R.: *Practical Methods of Optimization*, 2nd edn, John Wiley & Sons, 1987.
7.25 Golias N.A. and Tsiboukis T.D.: Adaptive refinement in 2-D finite element applications, *Int. J. Numer. Model. Electron. Netw. Devices Fields*, 1991, **4**, pp. 81–95.
7.26 Golias N.A. and Tsiboukis T.D.: *A-Posteriori Adaptive Mesh Refinement In the Finite Element Eddy Current Computation*, COMPEL, Volume 11, No. 1, James & James Science Publishers, London, 1992, pp. 249–252.
7.27 Golias N.A. and Tsiboukis T.D.: Three-dimensional automatic adaptive mesh generation, *IEEE Trans. Magn.*, 1992, **28**, (2), pp. 1700–1703.
7.28 Golias N.A. and Tsiboukis T.D.: Adaptive refinement strategies in three dimensions, *IEEE Trans. Magn.*, 1993, **29**, (2), pp. 1886–1889.
7.29 Gramz M. and Ziółkowski M.: *SONMAP V.2.0, User Manual*, Szczecin, 1988 (in Polish).
7.30 Hammond P. and Sykulski J.K.: *Engineering Electromagnetism: Physical Processes and Computation*, Oxford University Press, Oxford, 1994.

7.31 Hammond P. and Sykulski J.K.: Tubes and slices: a new way of teaching the principles of electric and magnetic fields, *IEEE Trans. Educ.*, 1992, **35**, (4), pp. 300–306.

7.32 Hoole S.R.H.: Nodal perturbations in adaptive expert finite element mesh generation, *IEEE Trans. Magn.*, 1987, **23**, pp. 2635–2637.

7.33 Ida N.: *Finite Element Computation on SIMD and MIMD Parallel Computers*, COMPEL, Volume 9, Supplement A, James & James Science Publishers, London, 1990, pp. 177–180.

7.34 *Integrated Engineering Software*, Winnipeg, Canada, 1993.

7.35 *International Compumag Society*, Newsletter, October 1993, **1**, (1).

7.36 *ISEF'93, Proc. Int. Symp. Electromagnetic Fields in Electrical Engineering*, Warsaw, September 16–18, 1993.

7.37 Jin J.M.: *The Finite Element Method in Electromagnetics*, John Wiley & Sons, New York, 1993.

7.38 Lindholm D.A.: Automatic mesh generation on surfaces of polyhedra, *IEEE Trans. Magn.*, 1983, **19**, (6), pp. 2539–2542.

7.39 Long W., Piriou F. and Razek A.: *An Adapted Cholesky Decomposition Method for the Solution of Coupled Magnetic Electric Equations*, COMPEL, Volume 8, No. 4, James & James Science Publishers, London, 1989, pp. 203–208.

7.40 Lowther D. and Silvester P.: *Computer-Aided Design in Magnetics*, Springer-Verlag, New York, 1986.

7.41 *MagNet 5.0, User Manual*, Infolytica Limited, Swindon, 1993.

7.42 *Magnetics Update*, Newsletter, 1993, **5**, (3), Infolytica Corporation, Montreal, Canada.

7.43 *Magnetics Update (Europe)*, 1993, **1**, (1), Newsletter, Infolytica Ltd, London, UK.

7.44 *MATLAB and Optimization Toolbox, User Manual*, The MathWorks, Inc., California, 1994.

7.45 Meijerink J.A. and der Vorst V.: An iterative solution method for systems of which the coefficient matrix is a symmetric in matrix, *Math Comp*, 1977, **31**, p. 148.

7.46 Marchand C. and Razek A.: Optimal torque operation of digitally controlled permanent magnet synchronous motor drives, *IEE Proc. A*, 1993, **140**, (3), pp. 232–240.

7.47 Nakata T. (ed.): *3-D Electromagnetic Field Analysis*, COMPEL, Volume 9, Supplement A, James & James Science Publishers, London, 1990.

7.48 *OPERA*, *User Manual*, Vector Fields Ltd, Oxford, 1993.

7.49 Pan Q., Kladas A. and Razek A.: Magneto-thermal coupled nonlinear finite element analysis in squirrel cage induction motors, *Math. Engng Ind.*, 1989, **2**, (3), pp. 203–217.

7.50 Parallel Applications Centre, University of Southampton, 1994.

7.51 *PC-OPERA*, *User Manual*, Vector Fields Ltd, Oxford, 1993.

7.52 Penman J. and Fraser J.R.: Dual and complementary energy methods in electromagnetism, *IEEE Trans. Magn.*, 1983, **19**, (6), pp. 2311–2316.

7.53 Penman J. and Grieve M.D.: An approach to self adaptive mesh generation, *IEEE Trans. Magn.*, 1985, **21**, (6), pp. 2567–2570.

7.54 Penman J. and Grieve M.D.: Self-adaptive mesh generation technique for the finite element method, *IEE Proc. A*, 1987, **134**, (8), pp. 634–650.

7.55 Pichon L. and Razek A.: Electromagnetic field computations in a three-dimensional cavity with a waveguide junction of a frequency standard, *IEE Proc. A*, 1992, **139**, (4), pp. 343–346.

7.56 Pichon L. and Razek A.: Hybrid finite-element method and boundary-element method using time-stepping for eddy-current calculation in axisymmetric problems, *IEE Proc. A*, 1989, **136**, (4), pp. 217–222.

7.57 Pinchuk A.M. and Silvester P.: Error estimation for automatic adaptive finite element mesh generation, *IEEE Trans. Magn.*, 1985, **21**, (6), pp. 2551–2554.

7.58 Raizer A., Meunier G. and Coulomb J.L.: An approach for automatic adaptive mesh refinement in the finite element computation of magnetic fields, *IEEE Trans. Magn.*, 1989, **25**, (4), pp. 2965–2967.

7.59 Reid J.K.: *On the Method of Conjugate Gradients for the Solution of Large Sparse Systems of Equations*, Academic Press, New York, 1971.

7.60 Rencis J.J. and Brebbia C.A. (eds): *Boundary Elements XV*, Computational Mechanics Publications, Southampton, 1993.

7.61 Rikabi J., Bryant C.F. and Freeman E.M.: An error-based approach to complementary formulations of static field solutions, *Int. J. Numer. Methods Eng.*, 1988, **26**, pp. 1963–1987.

7.62 Russenschuck: Mathematical optimization techniques for the design of permanent magnet synchronous machines based on numerical field calculation, *IEEE Trans. Magn.*, 1990, **26**, (2), pp. 638–641.

7.63 Russenschuck: Application of Lagrange multiplier estimation to the design optimization of permanent magnet synchronous machines, *IEEE Trans. Magn.*, 1992, **28**, (2), pp. 1525–1528.

7.64 Silvester P. and Ferrari R.: *Finite Elements for Electrical Engineers*, Cambridge University Press, Cambridge, 1990.

7.65 Silvester P. (ed.): *Advances in Electrical Engineering Software*, Computational Mechanics Publications, Southampton, 1990.

7.66 Silvester P. (ed.): *Software Applications in Electrical Engineering*, Computational Mechanics Publications, Southampton, 1993.

7.67 Silvester P.: *Data Structures for Engineering Software*, Computational Mechanics Publications, Southampton, 1993.

7.68 Simkin J. and Trowbridge C.W.: Optimization problems in electromagnetics, *IEEE Trans. Magn.*, 1991, **27**, (5), pp. 4016–4019.

7.69 Simkin J. and Trowbridge C.W.: Optimizing electromagnetic devices combining direct search methods with simulated annealing, *IEEE Trans. Magn.*, 1992, **28**, (2), pp. 1545–1548.

7.70 Sloan S.W.: A fast algorithm for constructing Delaunay triangulations in the plane, *Advanced Engineering Software*, Ch. 9, 1987, pp. 34–55.

7.71 Sykulski J.K. and Turowski J. (eds): *ISEF'91 Proceedings*, COMPEL, Volume 11, No. 1, James & James Science Publishers, London, March 1992.

7.72 Sykulski J.K.: Computer package for calculating electric and magnetic fields exploiting dual energy bounds, *IEE Proc. A*, 1988, **135**, (3), pp. 145–150.

7.73 Sykulski J.K. and Stoll R.L.: Magnetic field modelling and calculation of reflected impedance of inductive sensors, *IEEE Trans. Magn.*, 1992, **28**, (2), pp. 1426–1429.

7.74 Sykulski J.K. and Stoll R.L.: *Finite Element Modelling of Inductive Sensors*, COMPEL, Volume 11, No. 1, James & James Science Publishers, London, 1992, pp. 69–72.

7.75 Sykulski J.K., Sykulska E. and Hughes S.T.: *Application of Finite Element Modelling in LVDT Design*, COMPEL, Volume 11, No. 1, James & James Science Publishers, London, 1992, pp. 73–76.

7.76 Sykulski J.K.: Tubes and Slices: a software package for dual field calculations, in *Software Applications in Electrical Engineering*, Computational Mechanics Publications, Southampton, 1993, pp. 247–254.

7.77 Trowbridge C.W.: *The Work in Progress: Main Development Trends in Electromagnetic Analysis*, Technical Course: Field Computation for Low Frequency Devices, CEMM (IEE) course, Southampton, June 1993.

7.78 Tsiboukis T.D.: Automatic finite element analysis, Private communication, 1993.

7.79 *Vector*, 1993, **9**, (2), Electromagnetics Newsletter, Vector Fields Ltd, Oxford, UK.

7.80 Ziolkowski M.: 3D scalar or vector magnetic field – how one can see it, *ISEF'93, Proc. Int. Symp. Electromagnetic Fields in Electrical Engineering*, Warsaw, September 16–18, 1993, pp. 273–276.

Chapter 8

8.1 Bland T.G. and Freeman E.M.: Scale-model linear induction motors with solid iron secondaries, *Proc. IEE*, 1978, **125**, (11), pp. 1223–1226.

8.2 Boriu N.W.: Modelling of leakage power loss of high power transformers, *Elektrichestvo*, 1960, (9), pp. 38–42 (in Russian).

8.3 Dobkowski J.: External leakage fluxes of DC machines, Technical University of Lodz, Ph.D. thesis, 1977 (in Polish).

8.4 Filtshakov P.F. and Pantshishin B.N.: *EGDA integrators. Potential field modelling on Semi-Conducting Paper*, Kiev, 1961 (in Russian).

8.5 Freeman E.M., Lowther D.A. and Laithwaite E.R.: Scale-model linear induction motors, *Proc. IEE*, 1975, **122**, (7), pp. 721–726.

8.6 Gilbert A.J.: A method of measuring loss distribution in electrical machines, *Proc. IEE*, 1961, **108A**, pp. 239–244.

8.7 Govorkov W.A.: *Electric and Magnetic Fields*, WNT, Warsaw, 1984 (in Polish).

8.8 Hagel R. and Zakrzewski J.: *Dynamic Measurements*, WNT, Warsaw, 1962 (in Polish).

8.9 Hippner M.: *Graphical Presentation of Magnetic Field Calculation Results*, Z.N. ATR, Bydgoszcz (in Polish).

8.10 Ivanow-Smolensky A.W.: *Electromagnetic Fields and Phenomena in Elecrical Machines, and their Modelling*, Energia, Moscow, 1969 (in Russian).

8.11 Jablonski M.: *Transformer Tests in Industry and Exploitation*, WNT, Warsaw, 1969 (in Polish).

8.12 Komeza K., Turowski J. and Wiak S.: The experimental prediction of power loss distribution in metal elements of electrical machines, *Proc. of ICEM'84*, Budapest, 1984, pp. 72–76.

8.13 Kozlowska A.: Measurements and analysis of flux density distribution in 3-phase transformer core made of grain oriented steel, *Rozpr. Elektrotech.*, (2), 1967, p. 347 (in Polish).

8.14 Laczkowski R.: *Vibroacoustic of Machines and Devices*, WNT, Warsaw, 1983 (in Polish).

8.15 Nalecz M. and Jaworski J.: *Magnetic Measurements*, PWN, Warsaw-Lodz, 1965 (in Polish).

8.16 Neyman L.R.: *Skin-Effect in Ferromagnetic Materials*, GEI, Leningrad-Moscow, 1949 (in Russian).

8.17 Niewierowicz N. and Turowski J.: New thermometric method of measuring power losses in solid metal elements, *Proc. IEE*, 1972, **119**, (5), pp. 629–636.

8.18 Nowaczynski J.: Local overheating of construction parts of transformers during overloading and overexcitation, Technical University of Lodz, PhD thesis, 1977 (in Polish).

8.19 Pietrzak S. and Zakrzewski K.: Measurement system of static characteristics of double-sided linear motor, *Z.N. Elektryka*, 1983, (74), pp. 337–345 (in Polish).

8.20 Styburski W.: *Tensometric transducers, construction, design, exploitation*, WNT, Warsaw, 1976 (in Polish).

8.21 Szumilewicz B., Slomski B. and Styburski W.: *Electronic Measurements in Engineering*, WNT, Warsaw, 1982 (in Polish).

8.22 Turowski J.: *Technical Electrodynamics*, WNT, Warsaw, 1968 (in Polish).

8.23 Turowski J., Kazmierski M. and Ketner A.: Thermometric method for power loss measurements in construction parts of transformers, *Prz. Elektrotech.*, (10), 1964, pp. 439–448 (in Polish).

8.24 Venikov W.A. and Ivanov-Smolensky A.W.: *Physical Modelling of Electrical Systems*, GEI, Moscow, 1956 (in Russian).

8.25 Zakrzewski K.: Practical method of calculating active and reactive power in solid iron, *Rozpr. Elektrotech.*, 1975, **21**, (1), pp. 215–233 (in Polish).

8.26 Zakrzewski K.: Electromagnetic field in ferromagnetic materials, *Z.N. Elekryka*, (38), 1972 (in Polish).

8.27 Zakrzewski K. Physical modelling of field and loading loss in transformers, *Rozpr. Elektrotech.*, 1979, **25**, (2), pp. 401–418. (in Polish).

8.28 Zakrzewski K. Scale modelling of linear induction motors taking into account magnetic nonlinearity of iron secondaries, *Rozpr. Elektrotech.*, 1983, **29**, (2), pp. 469–489.

8.29 Zakrzewski K.: Physical modelling of leakage field and stray losses in steel constructional parts of electrotechnical devices, *Arch. Elektrotech.*, 1986, (69), pp. 129–135.

8.30 Zakrzewski K. and Pietras F.: Method of calculating the electromagetic field and power losses in ferromagnetic materials taking into account magnetic hysteresis, *Proc. IEE*, 1971, **118**, (11), pp. 1679–1785.

8.31 Zakrzewski K. and Sykulski J.: The effect of eddy currents on physical modelling of magnetic field strength at increased frequencies, *Rozpr. Elektrotech.* 1981, **72**, (1), pp. 53–63.

8.32 Zakrzewski K.: Mathematical modelling of electromagnetic field in solid iron, *Rozpr. Elektrotech.*, 1970, **16**, (1), pp. 27–43 (in Polish).

Index